Recent Advances in Polyphenol Research

Recent Advances in Polyphenol Research

A series for researchers and graduate students whose work is related to plant phenolics and polyphenols, as well as for individuals representing governments and industries with interest in this field. Each volume in this biennial series focuses on several important research topics in plant phenols and polyphenols, including chemistry, biosynthesis, metabolic engineering, ecology, physiology, food, nutrition, and health.

Volume 7 Editors:
Jess Dreher Reed (University of Madison–Wisconsin, USA), Victor Armando Pereira de Freitas (University of Porto, Portugal), and Stéphane Quideau (University of Bordeaux, France)

Recent Advances in Polyphenol Research

Volume 7

Edited by

Jess Dreher Reed
Professor, Nutrition and Phytochemistry
Department of Animal Sciences,
College of Agricultural and Life Sciences
University of Wisconsin–Madison, Madison, USA
Complete Phytochemical Solutions LLC,
Cambridge, Wisconsin, USA

Victor Armando Pereira de Freitas
Professor, Food Chemistry
Chemistry and Biochemistry Department, Faculty of Sciences
University of Porto, Porto, Portugal

Stéphane Quideau
Professor, Organic and Bio-organic Chemistry
Institut des Sciences Moléculaires, CNRS-UMR 5255
University of Bordeaux, Talence, France
& Institut Universitaire de France, Paris, France

WILEY Blackwell

This edition first published 2021

Registered Office(s)
John Wiley & Sons, Inc., 111 River Street, Hoboken, NJ 07030, USA
John Wiley & Sons Ltd, The Atrium, Southern Gate, Chichester, West Sussex, PO19 8SQ, UK

Editorial Office
The Atrium, Southern Gate, Chichester, West Sussex, PO19 8SQ, UK

For details of our global editorial offices, customer services, and more information about Wiley products visit us at www.wiley.com.

Wiley also publishes its books in a variety of electronic formats and by print-on-demand. Some content that appears in standard print versions of this book may not be available in other formats.

Library of Congress Cataloging-in-Publication data applied for

ISBN: 9781119545927
ISSN: 2474-7696

Cover Design: Wiley
Cover Images: © Reed Research Group

Set in 9.5/12.5pt STIXTwoText by SPi Global, Pondicherry, India
Printed and bound by CPI Group (UK) Ltd, Croydon, CR0 4YY

C9781119545927_310321

Contents

Contributors

Nick W. Albert
Plant & Food Research, Palmerston North, New Zealand

C. Peter Constabel
Centre for Forest Biology and Biology Department, University of Victoria, Victoria, Canada

Kevin M. Davies
Plant & Food Research, Palmerston North, New Zealand

Kriya Dunlap
Department of Biochemistry, University of Alaska Fairbanks, Fairbanks, USA

Daniel Esquivel-Alvarado
Department of Animal Sciences, University of Wisconsin–Madison, Madison, USA

Ana Gutiérrez
Instituto de Recursos Naturales y Agrobiología de Sevilla (IRNAS), CSIC, Seville, Spain

Rubina Jibran
Plant & Food Research, Palmerston North, New Zealand

James A. Kennedy
Functional Phenolics, LLC, Corvallis, USA

Christian G. Krueger
Complete Phytochemical Solutions LLC, Cambridge, Wisconsin, USA
Department of Animal Sciences, University of Wisconsin–Madison, Madison, USA

Hoon Kim
Department of Energy Great Lakes Bioenergy Research Center, Wisconsin Energy Institute
University of Wisconsin–Madison, Madison, USA

Wu Lan
Department of Biological System Engineering University of Wisconsin–Madison, Madison, USA & Department of Energy Great Lakes Bioenergy Research Center, Wisconsin Energy Institute
University of Wisconsin–Madison, Madison, USA

Mary Ann Lila
Department of Food Bioprocessing and Nutrition Sciences, Plants for Human Health Institute, North Carolina State University, Kannapolis, USA

Richard L. Lindroth
Department of Entomology, University of Wisconsin–Madison, Madison, USA

Dawei Ma
Centre for Forest Biology and Biology Department, University of Victoria, Victoria, Canada

Johan Mendoza
Department of Chemistry, Nova School of Science and Technology, Caparica Portugal

Irene Mueller-Harvey
School of Agriculture, Policy and Development, University of Reading, Reading, UK

Ken Ohmori
Department of Chemistry, Tokyo Institute of Technology, Tokyo, Japan

Marisa S. Otegui
Department of Botany, University of Wisconsin–Madison, Madison, USA

Fernando Pina
Department of Chemistry, Nova School of Science and Technology, Caparica, Portugal

John Ralph
Department of Energy Great Lakes Bioenergy Research Center, Wisconsin Energy Institute, University of Wisconsin–Madison, Madison, USA & Department of Biochemistry, University of Wisconsin–Madison, Madison, USA

Jess Dreher Reed
Department of Animal Sciences, University of Wisconsin–Madison, Madison, USA
Complete Phytochemical Solutions LLC, Cambridge, Wisconsin, USA

Jorge Rencoret
Instituto de Recursos Naturales y Agrobiología de Sevilla (IRNAS), CSIC, Seville, Spain

José C. del Río
Instituto de Recursos Naturales y Agrobiología de Sevilla (IRNAS), CSIC, Seville, Spain

Kennedy F. Rubert-Nason
Division of Natural and Behavioral Sciences University of Maine–Fort Kent, Fort Kent, USA

Kathy E. Schwinn
Plant & Food Research, Palmerston North, New Zealand

Keisuke Suzuki
Department of Chemistry, Tokyo Institute of Technology, Tokyo, Japan

Wayne E. Zeller
ARS-USDA, U.S. Dairy Forage Research Center, Madison, USA

Yanfei Zhou
Plant & Food Research, Palmerston North, New Zealand

Preface

Every two years, Groupe Polyphénols (GP) hosts the International Conference on Polyphenols (ICP). The XXIX ICP was the first one to be held in the United States in Madison, Wisconsin, on the campus of the University of Wisconsin–Madison (UW–Madison), from July 16 to 20, 2018. Groupe Polyphénols also hosted the 9th Tannin Conference (TC) concurrently with the XXIX ICP. Groupe Polyphénols was founded in 1972 and is the world's premier society of scientists in the fields of polyphenol chemistry, synthesis, bioactivity, nutrition, industrial applications, and ecology.

Madison is Wisconsin's state capital (the capitol building is shown on the front cover) and one of the nicest cities in the great lakes region. UW–Madison is a top ranked University (25th worldwide and 19th in the USA) and has a lovely campus with miles of lakefront and beautiful scenery adjacent to the state capitol. This venue for the XXIX ICP and 9th TC was fitting because Wisconsin's cranberry industry provides 60 percent of the world's supply of cranberries and is the state's largest fruit industry. The cranberry industry is also strongly dependent on the polyphenolic composition of the fruit. Cranberries are harvested in the fall after they turn from yellow-green to bright red, as shown on the front cover. The fruits are harvested by flooding the marsh (also called cranberry bogs). After removing the fruits from the vine, they float to the surface and are corralled with a floating boom and conveyed into trucks (as depicted on the front cover). The fruits are either transferred to a packaging facility for the fresh fruit market or to a frozen storage facility for subsequent processing into juice or sweetened dried cranberries (SDC). In both cases the bright red color of the fruit is a critical component of processing because the fruit is sorted based on color before packaging as fresh fruit or processing for juice and SDC (a processing line after sorting is also shown on the front cover). The color is a function of six anthocyanins, cyanidin 3-O-galactoside, cyanidin 3-O-glucoside, cyanidin 3-O-arabinoside, peonidin 3-O-galactoside, peonidin 3-O-glucoside, and peonidin 3-O-arabinoside. In addition to the anthocyanins, cranberries contain a large diversity of other monomeric polyphenols, especially flavonol glycosides, and contain simple phenols such as hydroxycinnamic acids and hydroxybenzoic acids. Cranberries also contain proanthocyanidins, which are just as important to the economic value of the fruit as the anthocyanins. The importance of proanthocyanidins to the cranberry market is a result of pioneering research from the late 1990s in which "A-type" interflavan bonds were discovered to be the structural feature of cranberry proanthocyanidins that is associated with the prevention of adherence of P-fimbriated *E. coli* to uroepithelial cells, the putative mechanism in the prevention of urinary tract infections. Proanthocyanidin content is now used to market cranberry products (including juice, sweetened dried cranberries, and dietary supplements) and consumers widely recognize cranberries as healthy. Therefore, all of the subjects that were discussed at the XXIX ICP and 9th TC and the chapters of this volume of *Recent Advances in Polyphenol Research* are of direct importance to Wisconsin's cranberry industry. The role of polyphenols in this industry is an excellent example of the importance of polyphenol research in general.

The XXIX ICP and 9th TC were attended by 189 registrants from 23 countries, with 62 invited and contributed presentations and 104 posters. This seventh edition of *Recent Advances in Polyphenol Research* presents 11 chapters that represent the work of the invited speakers at the XXIX ICP and 9th TC and reflect the depth of science in this important field of natural product chemistry. The conference included sessions on the chemistry and physical chemistry of polyphenols; synthesis, genetics and metabolic engineering of polyphenols; the effects of polyphenols on the nutrition and health of humans and animals; the role of polyphenols in plants and ecosystems; applied research on polyphenols; and a special session devoted to the 9th Tannin Conference.

We owe a special thanks to Hannah Scott and Laura Richards from the Campus Events Services, UW–Madison, for their professional and excellent organization of the conference. Finally, we thank all of the participants, some who traveled a great distance to come to Madison, for making the conference a very enjoyable event and a wonderful learning experience.

Jess Dreher Reed
Victor Armando Pereira de Freitas
Stéphane Quideau

Acknowledgements

The editors wish to thank all of the members of the Groupe Polyphénols Board Committee (2016–2018 & 2018–2020) for their guidance and assistance throughout this project.

Groupe Polyphénols Board 2016–2018 & 2018–2020

Dr. Denis Barron
Dr. Luc Bidel
Dr. Catherine Chèze
Dr. Peter Constabel
Prof. Olivier Dangles
Dr. Kevin Davies
Prof. M Teresa Escribano
Prof. Victor Armando Pereira de Freitas
Prof. Kazuhiko Fukushima
Dr. David Gang
Dr. Sylvain Guyot
Prof. Ann E. Hagerman
Prof. Heidi Halbwirth
Prof. Amy Howell
Dr. Stefan Martens
Dr. Fulvio Mattivi
Dr. Irene Mueller-Harvey
Prof. Stéphane Quideau
Prof. Jess Dreher Reed
Dr. Erika Salas
Prof. Juha-Pekka Salminen
Prof. Kathy Schwinn
Dr. David Vauzour
Prof. Kristiina Wähälä

1

Achieving Complexity at the Bottom Through the Flavylium Cation-Based Multistate

A Comprehensive Kinetic and Thermodynamic Study

Johan Mendoza and Fernando Pina

Department of Chemistry, Nova School of Science and Technology, Caparica, Portugal

1.1 Introduction

Complexity is ubiquitous in biological systems. The main strategy to study complexity has been carried out using a top-down approach. Though the top-down approach the simpler components of the complex systems are identified, and whenever possible, up to the molecular level. In contrast, supramolecular chemistry, a concept well established and recognized after the 1987 Nobel Prize awarded to Donald J. Cram, Jean-Marie Lehn, and Charles J. Pedersen, is a bottom-up approach (Figure 1.1). Supramolecular chemistry studies how molecules interact to form higher-dimension entities and tends to fill the gap between "classical chemistry" and biology (Lehn, 1995).

A beautiful example of supramolecular chemistry is the structure of the metalloanthocyanin that gives color to *Commelina communis* (Kondo et al. 1992; Yoshida et al. 2009). An anthocyanin, a flavone, and a metal ion in a ratio 6:6:2 are organized into two parallel plans, each one containing three anthocyanins, three flavones, and one metal ion that organizes the space Figure 1.1.

There is an alternative to achieve complexity that we coin *metamorphosis* (Petrov et al. 2012). When a molecule (generator) is able to be transformed into other molecules by means of successive conversions and as a response to external stimuli, new molecules are formed. The complexity results from the number of the species and everything takes place at the bottom.

The pH-dependent multistate of species of anthocyanins and related compounds is a paradigm of the metamorphosis concept; see Scheme 1.1.

1.2 Flavylium Cation as a Metamorphosis Generator

The flavylium cation, $\mathbf{AH^+}$, is the most stable species at very low pH values, in anthocyanins generally for pH<1. The system is conveniently studied by *direct pH jumps* when base is added to the flavylium cation, and *reverse pH jumps*, defined as addition of acid to

Recent Advances in Polyphenol Research, Volume 7, First Edition. Edited by Jess Dreher Reed, Victor Armando Pereira de Freitas, and Stéphane Quideau.

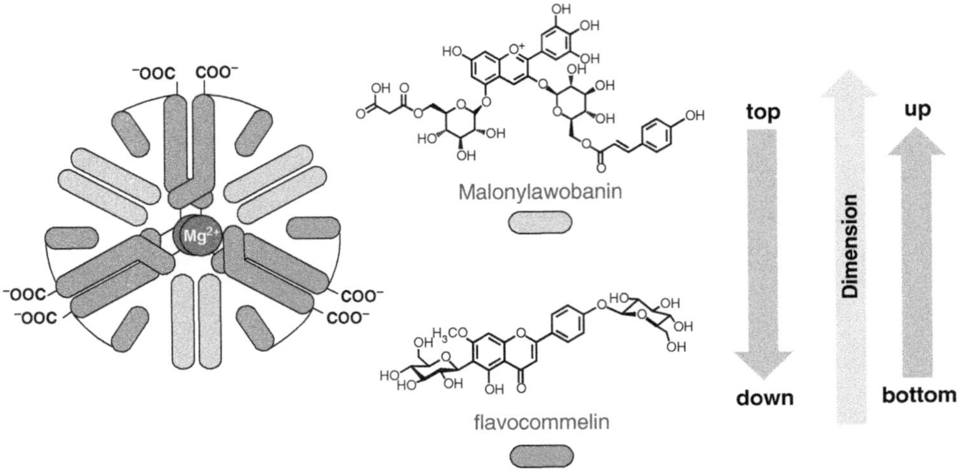

Figure 1.1 Sketch of the metalloanthocyanin responsible for the color in *Cummelina communis*. The building blocks self-associate to create the supramolecule in a bottom-up approach. *Source:* Courtesy of Prof. Kumi Yoshida. (*See insert for color representation of the figure.*)

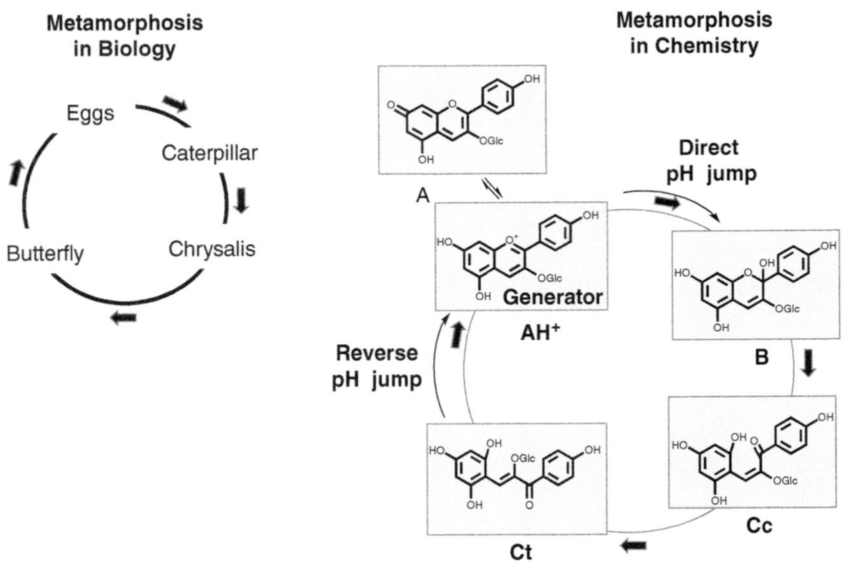

Scheme 1.1 The metamorphosis concept in biology and in chemistry applied to anthocyanins and related compounds in acidic medium. *Source:* Reproduced from Mendoza et al. (2018), with permission.

equilibrated solutions at higher pH values. After a direct pH jump to moderately acidic pHs, the flavylium cation equilibrates in microseconds with quinoidal base, **A** eq. (1). The next step is the formation of the hemiketal, **B**, through the hydration of $\mathbf{AH^+}$ (min) eq. (2), followed by the ring opening to form *cis*-chalcone, **Cc**, (ms) eq. (3). The fact that the quinoidal base does not open in acidic medium is a breakthrough discovery (Brouillard and Dubois 1977) crucial for the comprehension of anthocyanins and related compounds systems. The **Cc** isomerization to *trans*-chalcone, **Ct**, in anthocyanins takes place in several hours eq. (4). When the system is equilibrated in moderately acidic pH values, a reverse pH jumps restores the flavylium cation. The following set of equilibrium reactions accounts for the system:

$$\mathbf{AH^+} + H_2O \underset{k_{-a}}{\overset{k_a}{\rightleftarrows}} \mathbf{A} + H_3O^+ \qquad K_a \quad \text{proton transfer} \tag{1}$$

$$\mathbf{AH^+} + 2H_2O \underset{k_{-h}}{\overset{k_h}{\rightleftarrows}} \mathbf{B} + H_3O^+ \qquad K_h \quad \text{hydration} \tag{2}$$

$$\mathbf{B} \underset{k_{-t}}{\overset{k_t}{\rightleftarrows}} \mathbf{Cc} \qquad K_t \quad \text{tautomerization} \tag{3}$$

$$\mathbf{Cc} \underset{k_{-i}}{\overset{k_i}{\rightleftarrows}} \mathbf{Ct} \qquad K_i \quad \text{isomerization} \tag{4}$$

A few years ago we introduced an energy level diagram that accounts for the thermodynamic of the anthocyanin system in acidic medium (Pina et al. 1997; Pina 2014a). This diagram can be straightforwardly constructed provided that the equilibrium constants, eq. (1) to eq. (4), of the system have been determined, see Scheme 1.2.

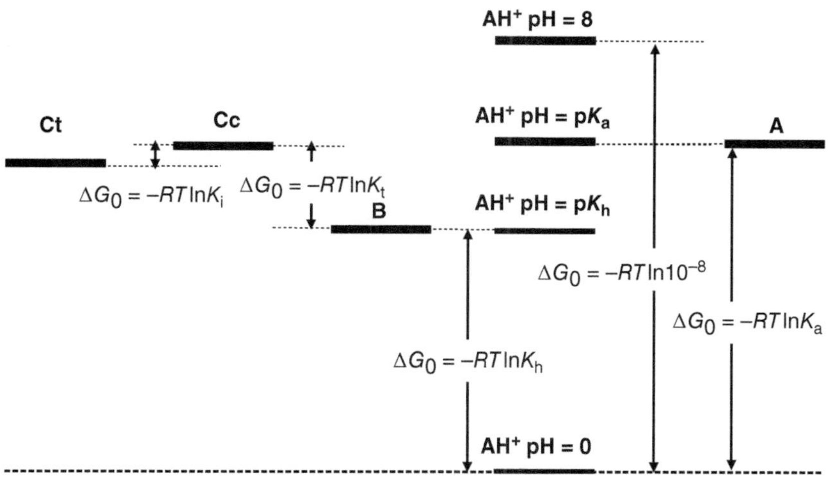

Scheme 1.2 Energy level diagram for anthocyanins and related compounds in acidic medium. *Source:* Adapted from Pina 2014a. © 2014 John Wiley & Sons.

1.3 Extending the Multistate of Anthocyanins and Related Compounds to the Basic Region

In many flavylium derivatives from natural or synthetic origin, including anthocyanins, it is indispensable to extend the multistate study to basic medium.

In order to account for these new species, eight equilibrium equations should be added to eq. (1) through eq. (4).

For the formation of the mono-anionic species[1]

$$\mathbf{A} + H_2O \rightleftharpoons \mathbf{A}^- + H_3O^+ \qquad K_{A/A^-} \qquad \text{proton transfer} \tag{5}$$

$$\mathbf{B} + H_2O \rightleftharpoons \mathbf{B}^- + H_3O^+ \qquad K_{B/B^-} \qquad \text{proton transfer} \tag{6}$$

$$\mathbf{Cc} + H_2O \rightleftharpoons \mathbf{Cc}^- + H_3O^+ \qquad K_{Cc/Cc^-} \qquad \text{proton transfer} \tag{7}$$

$$\mathbf{Ct} + H_2O \rightleftharpoons \mathbf{Ct}^- + H_3O^+ \qquad K_{Ct/Ct^-} \qquad \text{proton transfer} \tag{8}$$

And for the formation of the di-anionic species

$$\mathbf{A}^- + H_2O \rightleftharpoons \mathbf{A}^{2-} + H_3O^+ \qquad K_{A^-/A^{2-}} \qquad \text{proton transfer} \tag{9}$$

$$\mathbf{B}^- + H_2O \rightleftharpoons \mathbf{B}^{2-} + H_3O^+ \qquad K_{B^-/B^{2-}} \qquad \text{proton transfer} \tag{10}$$

$$\mathbf{Cc}^- + H_2O \rightleftharpoons \mathbf{Cc}^{2-} + H_3O^+ \qquad K_{Cc^-/Cc^{2-}} \qquad \text{proton transfer} \tag{11}$$

$$\mathbf{Ct}^- + H_2O \rightleftharpoons \mathbf{Ct}^{2-} + H_3O^+ \qquad K_{Ct^-/Ct^{2-}} \qquad \text{proton transfer} \tag{12}$$

The system can be generalized for higher charged anionic species.

In spite of the complexity of this system, the set of eqs. 1 through 12 can be simplified considering a triprotic acid, eq. (13) through eq. (15), with constants K'_a, eq. (19) K''_a, eq. (20), and K'''_a, eq. (21). The complete mathematical development of the system above was previously reported (supplementary information, Mendoza et al. 2019) and is straightforwardly obtained from a mass balance and representation of all species as a function of $\mathbf{AH}^+$.

$$\mathbf{AH}^+ + H_2O \rightleftharpoons \mathbf{CB} + H_3O^+ \qquad K'_a \tag{13}$$

$$\mathbf{CB} + H_2O \rightleftharpoons \mathbf{CB}^- + H_3O^+ \qquad K''_a \tag{14}$$

$$\mathbf{CB}^- + H_2O \rightleftharpoons \mathbf{CB}^{2-} + H_3O^+ \qquad K'''_a \tag{15}$$

Where

$$[\mathbf{CB}] = [\mathbf{A}] + [\mathbf{B}] + [\mathbf{Cc}] + [\mathbf{Ct}] \tag{16}$$

$$[\mathbf{CB}^-] = [\mathbf{A}^-] + [\mathbf{B}^-] + [\mathbf{Cc}^-] + [\mathbf{Ct}^-] \tag{17}$$

$$[\mathbf{CB}^{2-}] = [\mathbf{A}^{2-}] + [\mathbf{B}^{2-}] + [\mathbf{Cc}^{2-}] + [\mathbf{Ct}^{2-}] \tag{18}$$

and

$$K'_a = K_a + K_h + K_h K_t + K_h K_t K_i \tag{19}$$

$$K''_a = \frac{K_{A/A-} K_a + K_{B/B-} K_h + K_{Cc/Cc-} K_h K_t + K_{Ct/Ct-} K_h K_t K_i}{K'_a} \tag{20}$$

$$K'''_a = \frac{K_{A-/A2-}K_{A/A-}K_a + K_{B-/B2-}K_{B/B-}K_h + K_{Cc-/Cc2-}K_{Cc/Cc-}K_hK_t + K_{Ct-/Ct2-}K_{Ct/Ct-}K_hK_tK_i}{K'_aK''_a} \tag{21}$$

The mole fraction distribution X_R of all species can be expressed in terms of the 12 linearly independent constants reported in Scheme 1.3. Since the flavylium cation and the quinoidal bases are in very fast equilibrium (microseconds scale), it is convenient to consider them altogether. The same is valid for the other species related through the proton transfer reaction.

$$X_{AH+} + X_A + X_{A-} + X_{A2-} = \frac{\left[H^+\right]^3 + K_a\left[H^+\right]^2 + K_{A/A-}K_a\left[H^+\right] + K_{A-/A2-}K_{A/A-}K_a}{D} \tag{22}$$

where

$$D = \left[H^+\right]^3 + K'_a\left[H^+\right]^2 + K'_aK''_a\left[H^+\right] + K'_aK''_aK'''_a \tag{23}$$

$$X_B + X_{B-} + X_{B2-} = \frac{K_h\left[H^+\right]^2 + K_{B/B-}K_h\left[H^+\right] + K_{B-/B2-}K_{B/B-}K_h}{D} \tag{24}$$

$$X_{Cc} + X_{Cc-} + X_{Cc2-} = \frac{K_hK_t\left[H^+\right]^2 + K_{Cc/Cc-}K_hK_t\left[H^+\right] + K_{Cc-/Cc2-}K_{Cc/Cc-}K_hK_t}{D} \tag{25}$$

$$X_{Ct} + X_{Ct-} + X_{Ct2-} = \frac{K_hK_tK_i\left[H^+\right]^2 + K_{Ct/Ct-}K_hK_tK_i\left[H^+\right] + K_{Ct-/Ct2-}K_{Ct/Ct-}K_hK_tK_i}{D} \tag{26}$$

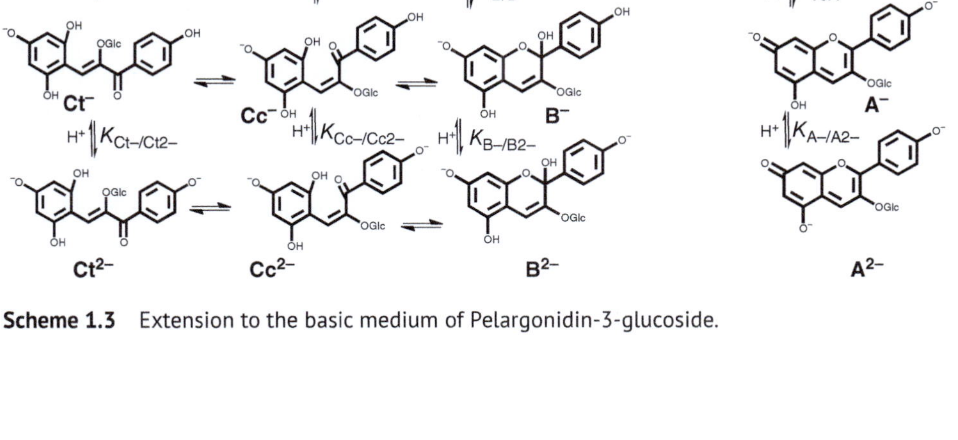

Scheme 1.3 Extension to the basic medium of Pelargonidin-3-glucoside.

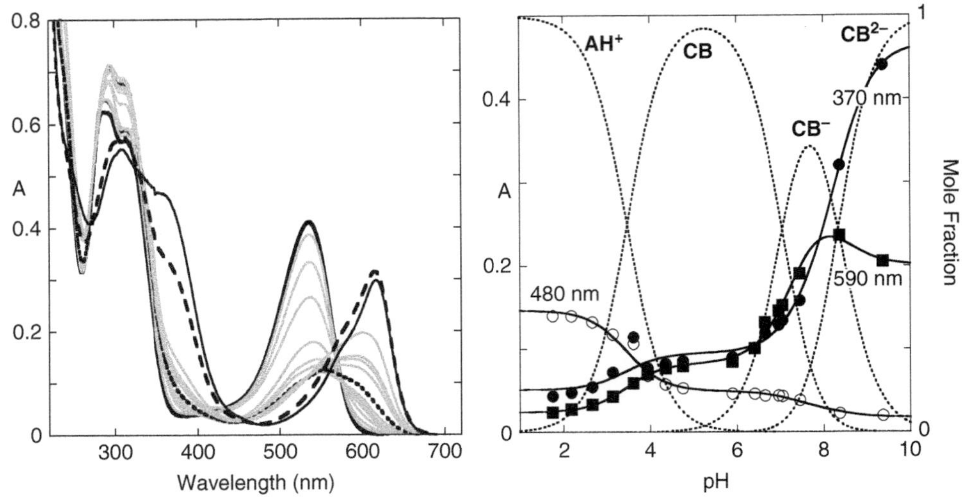

Figure 1.2 Absorption spectrum of heavenly blue anthocyanin, a peonidin derivative, black full line, flavylium cation; black pointed line, quinoidal base; black traced line, ionized quinoidal base. pK'_a=3.47; pK''_a=7.05; pK'''_a=8.30. *Source:* Mendoza et al. 2018.

Since the complex system shown in Scheme 1.3 behaves as a simple triprotic acid, the respective apparent equilibrium constants K'_a, K''_a, and K'''_a are experimentally obtained from the inflection points of the absorbance representation as a function of the pH. Consequently, the term ***D*** is a parameter obtained experimentally. In Figure 1.2 the example of the heavenly blue anthocyanin is shown (Mendoza et al. 2018).

The question now is to define the experimental strategy to calculate the equilibrium constants of the system.

1.3.1 Reverse pH Jumps from Pseudo-equilibrium Followed by Stopped Flow UV-visible Spectroscopy

Recently we have reported a new experimental procedure that allows the experimental determination of all equilibrium constants (as shown in Scheme 1.3) of the flavylium-based multistates including anthocyanins (Mendoza et al. 2019; Mendoza et al. 2018; Slavcheva et al. 2018). It is based on the reverse pH jumps defined above, followed by stopped flow. In Figure 1.3 the stopped flow traces of the model compound 4'-hydroxyflavylium are shown. The initial solutions should be equilibrated or pseudo-equilibrated. The reverse pH jumps consist of the addition of acid to make the solutions with pH=1, where flavylium cation is the sole species. In both cases of Figure 1.3 the initial absorbance is due to the quinoidal bases (independently on their protonation state) that give flavylium cation (absorption at 450 nm) during the mixing time of the stopped flow together with some flavylium cation present at the initial equilibrium (at lower pH values) prior to the jump; see also Scheme 1.3. This is the reason why the mole fraction distribution of the flavylium cation and quinoidal bases are represented together in eq. (22). At the final very low pH jump (pH=1) the hydration reaction becomes faster than the tautomerization because it is directly proportional to the proton concentration (Pina 2014b). Therefore, the faster trace is due to the conversion of **B** into $\mathbf{AH^+}$. The slower trace is the formation of more flavylium cation from **Cc** via **B** (Scheme 1.4) (Mendoza et al. 2019).

In anthocyanins and most flavylium derivatives the *cis-trans* isomerization is much slower than the other kinetic processes. It is possible thus to define a transient state (pseudo-equilibrium) where the mole fraction of the *trans*-chalcones is very small. Consequently, it is more convenient to carry out the reverse pH jumps from

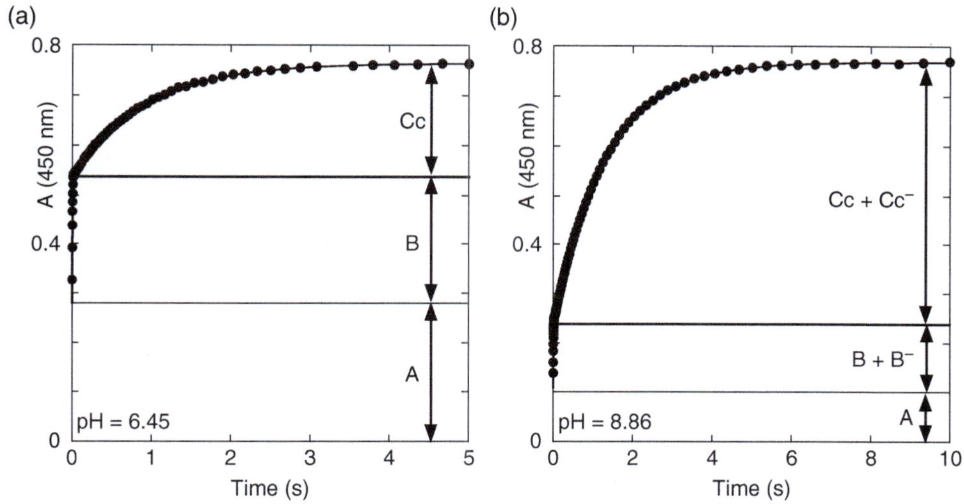

Figure 1.3 Stopped flow traces 4'-hydroxyflavylium (at pseudo-equilibrium, where no significant amounts of **Ct** were formed) after a reverse pH jump from pH=6.45 (a) and pH=8.9 (b) to the final pH=1.0.

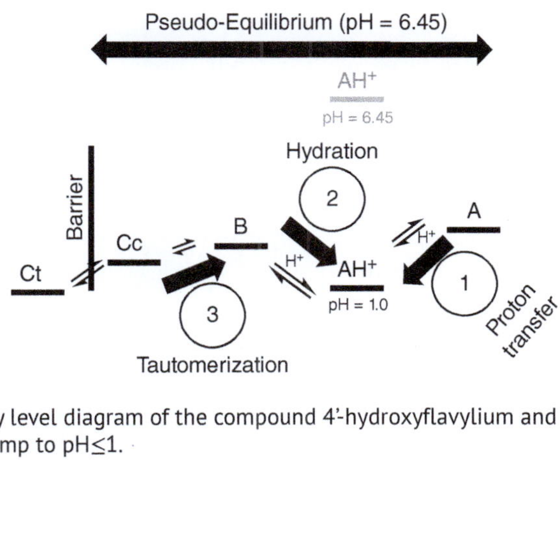

Scheme 1.4 Energy level diagram of the compound 4'-hydroxyflavylium and the kinetic processes after a reverse pH jump to pH≤1.

pseudo-equilibrium. Scheme 1.4 illustrates the question in acidic medium, but it is generalized to higher pH values. Even if some **Ct** is formed, the only drawback is the loss of sensitivity, because the kinetics of the **Ct** transformation in flavylium cation is much slower and is not detected in the stopped flow experiments. In conclusion, the data reported in Figure 1.3, extended to other pH values, allows the calculation of the mole fraction distribution of the species **A**, **B**, and **Cc** as well as the respective anionic forms.

The mole fraction distribution of these species can be represented as a function of the initial pH of the reverse pH jump (Figure 1.4).

The fitting of Figure 1.4 was carried out by considering for $\mathbf{AH^+}$, $\mathbf{CB^{\wedge}}$, $\mathbf{CB^{\wedge -}}$, and $\mathbf{CB^{\wedge 2-}}$ the contributions of the respective forms of quinoidal bases, hemiketals, and *cis*-chalcones. For example, the mole fraction distribution of $\mathbf{CB^{-\wedge}}$ is given by eq. (27) (Mendoza et al. 2019 supplementary information).

$$X_{CB-^{\wedge}} = X_{A-} + X_{B-} + X_{Cc-} = a_1 \frac{K_a^{\wedge} K_a^{\wedge\wedge}\left[H^+\right]}{D^{\wedge}} + b_1 \frac{K_a^{\wedge} K_a^{\wedge\wedge}\left[H^+\right]}{D^{\wedge}} + c_1 \frac{K_a^{\wedge} K_a^{\wedge\wedge}\left[H^+\right]}{D^{\wedge}} \tag{27}$$

with

$$D^{\wedge} = \left[H^+\right]^3 + K_a^{\wedge}\left[H^+\right]^2 + K_a^{\wedge} K_a^{\wedge\wedge}\left[H^+\right] + K_a^{\wedge} K_a^{\wedge\wedge} K_a^{\wedge\wedge\wedge} \tag{28}$$

and

$$a_1 + b_1 + c_1 = 1 \tag{29}$$

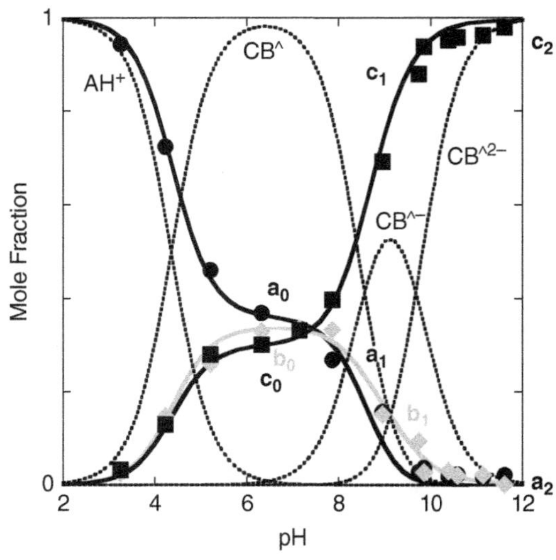

Figure 1.4 Representation of the mole fraction distribution of the compound 4'-hydroxyflavylium on the basis that the reverse pH jumps at pseudo-equilibrium. The symbol ^ is used to differentiate pseudo-equilibrium from equilibrium (').

The mole fractions of the more colored forms, eq. (30), as well of those of hemiketals, eq. (31) and *cis*-chalcones, eq. (31) are thus obtained.

$$X_{AH+}+X_{A}+X_{A-}+X_{A2-}=\frac{\left[H^{+}\right]^{3}+a_{0}K_{a}^{\wedge}\left[H^{+}\right]^{2}+a_{1}K_{a}^{\wedge}K_{a}^{\wedge\wedge}\left[H^{+}\right]+a_{2}K_{a}^{\wedge}K_{a}^{\wedge\wedge}K_{a}^{\wedge\wedge\wedge}}{D^{\wedge}} \tag{30}$$

$$X_{B}+X_{B-}+X_{B2-}=\frac{b_{0}K_{a}^{\wedge}\left[H^{+}\right]^{2}+b_{1}K_{a}^{\wedge}K_{a}^{\wedge\wedge}\left[H^{+}\right]+b_{2}K_{a}^{\wedge}K_{a}^{\wedge\wedge}K_{a}^{\wedge\wedge\wedge}}{D^{\wedge}} \tag{31}$$

$$X_{Cc}+X_{Cc-}+X_{Cc2-}=\frac{c_{0}K_{a}^{\wedge}\left[H^{+}\right]^{2}+c_{1}K_{a}^{\wedge}K_{a}^{\wedge\wedge}\left[H^{+}\right]+c_{2}K_{a}^{\wedge}K_{a}^{\wedge\wedge}K_{a}^{\wedge\wedge\wedge}}{D^{\wedge}} \tag{32}$$

Considering that the apparent equilibrium constants are experimentally obtained from the inflection points of the absorption spectra as a function of pH, the fitting of eq. (30) to eq. (32) permits us to obtain the constants a_n, b_n and c_n (n = 0, 1 and 2).

On the other hand, eq. (22) to eq. (25) can be re-written for the pseudo-equilibrium:

$$X_{AH+}+X_{A}+X_{A-}+X_{A2-}=\frac{\left[H^{+}\right]^{3}+K_{a}\left[H^{+}\right]^{2}+K_{A/A-}K_{a}\left[H^{+}\right]+K_{A-/A2-}K_{A/A-}K_{a}}{D^{\wedge}} \tag{33}$$

$$X_{B}+X_{B-}+X_{B2-}=\frac{K_{h}\left[H^{+}\right]^{2}+K_{B/B-}K_{h}\left[H^{+}\right]+K_{B-/B2-}K_{B/B-}K_{h}}{D^{\wedge}} \tag{34}$$

$$X_{Cc}+X_{Cc-}+X_{Cc2-}=\frac{K_{h}K_{t}\left[H^{+}\right]^{2}+K_{Cc/Cc-}K_{h}K_{t}\left[H^{+}\right]+K_{Cc-/Cc2-}K_{Cc/Cc-}K_{h}K_{t}}{D^{\wedge}} \tag{35}$$

Comparing eq. (30) with eq. (33), eq. (31) with eq. (34), and eq. (32) with eq. (35) the following relations are obtained:

$$K_{a}=a_{0}K_{a}^{\wedge};\ K_{h}=b_{0}K_{a}^{\wedge};\ K_{t}=\frac{c_{0}K_{a}^{\wedge}}{K_{h}} \tag{36}$$

$$K_{A/A-}=\frac{a_{1}K_{a}^{\wedge}K_{a}^{\wedge\wedge}}{K_{a}};\ K_{B/B-}=\frac{b_{1}K_{a}^{\wedge}K_{a}^{\wedge\wedge}}{K_{h}};\ K_{Cc/Cc-}=\frac{c_{1}K_{a}^{\wedge}K_{a}^{\wedge\wedge}}{K_{h}K_{t}} \tag{37}$$

$$K_{A-/A2-}=\frac{a_{2}K_{a}^{\wedge}K_{a}^{\wedge\wedge}K_{a}^{\wedge\wedge\wedge}}{K_{A/A-}K_{a}};\ K_{B-/B2-}=\frac{b_{2}K_{a}^{\wedge}K_{a}^{\wedge\wedge}K_{a}^{\wedge\wedge\wedge}}{K_{B/B-}K_{h}};\ K_{Cc-/Cc2-}=\frac{c_{2}K_{a}^{\wedge}K_{a}^{\wedge\wedge}K_{a}^{\wedge\wedge\wedge}}{K_{Cc/Cc-}K_{h}K_{t}} \tag{38}$$

1.3.2 Reverse pH Jumps from Equilibrium

From eq. (36) to eq. (38) all equilibrium constants except those regarding the *trans*-chalcones can be obtained. Moreover, the *cis-trans* isomerization constants can be calculated from a reverse pH jump from the equilibrated solutions. Considering that formation of the flavylium cation from the *trans*-chalcones is very slow, this kinetics should be followed by a standard spectrophotometer. The quinoidal bases, hemiketals, and *cis*-chalcones are transformed to flavylium cation much faster than the *trans*-chalcones and appear as an initial absorption.[2] From this point all the equilibrium constants have been calculated. The mole fraction of *trans*-chalcone is thus obtained from the ratio of the absorbance of the trace amplitude/total absorbance. The mole fractions of the other species at equilibrium are obtained from those at pseudo-equilibrium, calculating the respective proportion.[3] For example, if at pseudo-equilibrium **A**=0.3, **B**=0.2, and **Cc**=0.5 and the mole fraction of **Ct** at equilibrium is 0.5, the mole fractions of **A**, **B**, and **Cc** at equilibrium are the following: **A**=0.15, **B**=0.1, and **Cc**=0.25.

1.4 The Kinetic Processes

Scheme 1.5 represents the four kinetic processes of anthocyanins and related compounds in acidic medium. It is worth noting that, like in the case of the formation of the quinoidal base from flavylium cation, all the other anionic species are formed as in Scheme 1.3, from proton transfer. This reaction represents step 1 in the kinetic process and takes place in microseconds during the mixing time of the stopped flow. Only using special techniques

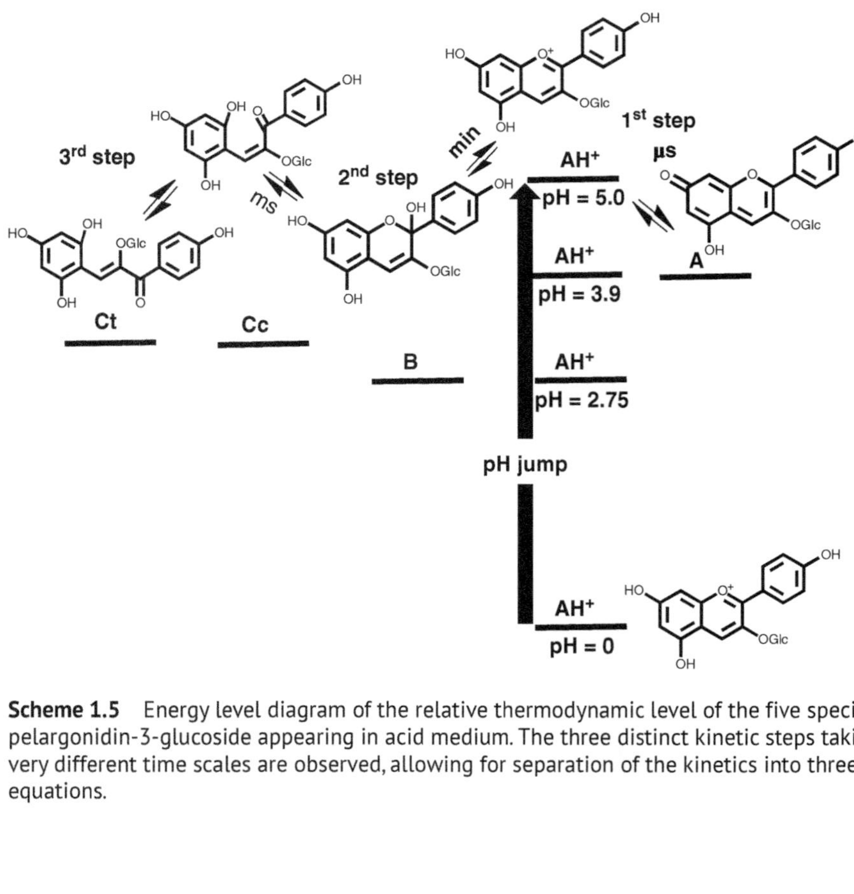

Scheme 1.5 Energy level diagram of the relative thermodynamic level of the five species of pelargonidin-3-glucoside appearing in acid medium. The three distinct kinetic steps taking place in very different time scales are observed, allowing for separation of the kinetics into three kinetic equations.

such as temperature jumps (Brouillard and Dubois 1977), and in some favourable cases flash photolysis, are these constants obtained.[4] This fact makes the kinetics reported in Scheme 1.5 the only relevant ones upon direct pH jumps, since the formation of the anionic species is immediate when compared with hydration, tautomerization, and isomerization. Moreover, the first process after a direct pH jump (from flavylium cation) is the formation of the quinoidal base, which equilibrates with the flavylium cation. In the subsequent kinetic steps these two species behave as a single one.

The following kinetic step is the hydration followed by tautomerization (Scheme 1.5). Except in very acidic solutions (not accessed by direct pH jumps), the tautomerization reaction is faster than hydration and by consequence this last one is the rate-determining step of this kinetic process. This kinetic step can thus be considered as in eq. (39).

During the hydration both $\mathbf{AH^+/A}$ and $\mathbf{B/Cc}$ can be considered as a single species.

$$k_2 = X_{AH+}k_h + X_{Cc}k_{-h}\left[H^+\right] = \frac{\left[H^+\right]}{\left[H^+\right]+K_a} + \frac{1}{1+K_t}k_{-h}\left[H^+\right] \tag{39}$$

where $X_{\mathrm{AH+}}$ is the mole fraction of $\mathbf{AH^+}$ in its equilibrium with **A**, and X_{B} is the mole fraction of **A** in its equilibrium with **Cc**.

In eq. (39) the forward reaction takes place only from the reaction of $\mathbf{AH^+}$ to form **B**, because, as mentioned above, the quinoidal base **A** does not hydrate in acidic medium (Brouillard and Dubois 1977).

For anthocyanins and many of the flavylium derivatives, the last step is controlled by the isomerization of chalcones, which is by far the slowest process of the kinetics. A similar reasoning used for step 2 can be made for step 3. In this case all species except **Ct** can be considered equilibrated.

$$k_3 = X_{Cc}k_i + k_{-i} = \frac{K_h K_t}{\left[H^+\right]+\hat{K}_a}k_i + k_{-i} \tag{40}$$

1.4.1 Heavenly Blue Anthocyanin

The experimental procedure above reported was used to rationalize the multistate of heavenly blue anthocyanin and two derivatives (Scheme 1.6).

Heavenly blue anthocyanin, HBA1, has attracted the attention of the scientific community due to its peculiar properties, specifically the fact that the same anthocyanin is used by the plant to confer purplish color to the buds and blue color to the petals (Yoshida et al. 1995; Goto and Kondo 1991; Kondo et al. 1992). Moreover, *in vitro* the blue color is persistent in neutral and moderately basic solutions (Kondo et al. 1992; Yoshida et al. 2009). Structural information regarding HBA1 fully supports the intramolecular stacking shown in Scheme 1.7.

The system was studied up to the mono-anionic forms because at higher pH values a slow decomposition takes place and the data does not have sufficient accuracy. In spite of equilibrium being reached in one to two weeks, the neutral and mono-anionic species are relatively stable. Table 1.1 summarizes the data.

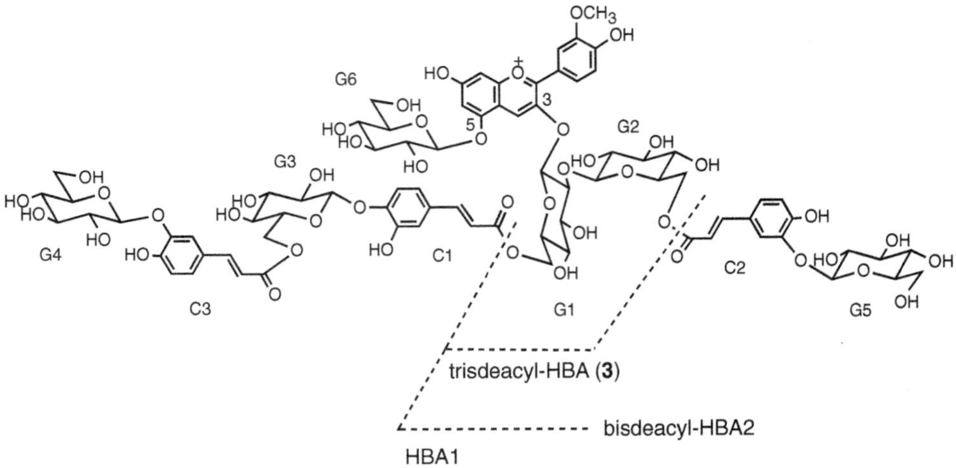

Scheme 1.6 Heavenly blue anthocyanin HBA1 and their derivatives bis-deacyl-HBA2 and tris-deacyl-HBA3. *Source:* Mendoza et al. 2018.

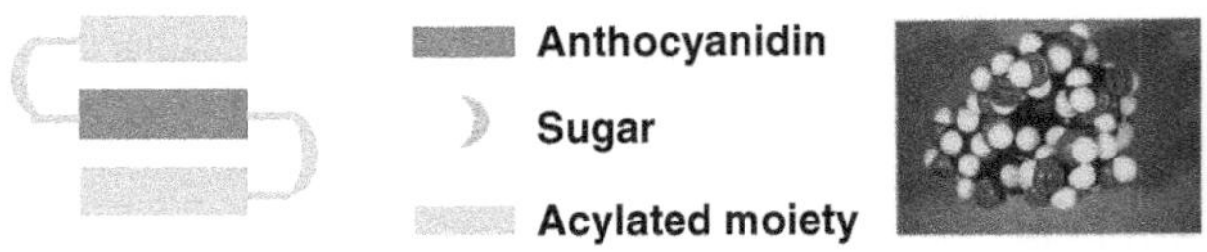

Scheme 1.7 Sketch representing the intramolecular copigmentation in polyacylated anthocyanins; CPK models of heavenly blue anthocyanin. *Source:* Mendoza et al. 2018. (*See insert for color representation of the figure.*)

Table 1.1 Equilibrium constants of heavenly blue anthocyanin and their derivatives.

	pK'_a	pK''_a	$pK^{\wedge}_a$	pK_a	pK_h	K_t
HBA1	3.5	7.3	3.6	3.8	4.6	1.1
HBA2	—	—	2.92	4.23	3.1	0.35
HBA3	—	—	1.95	4.19	2.1	0.37
	K_i	$pK^{\wedge\wedge}_a$	$pK_{A/A-}$	$pK_{B/B-}$	$pK_{Cc/Cc-}$	$pK_{Ct/Ct-}$
HBA1	4.0	7.35	7.35	7.5	7.25	7.36

Estimated error 10%.
Source: Mendoza et al. 2018

In Table 1.1 the equilibrium constants of the non-acylated, di-acylated and tri-acylated derivatives of heavenly blue anthocyanin are also reported (Scheme 1.8). HBA2 and HBA3 behave as common anthocyanins, being relatively stable only in acidic medium, preventing the calculation of the data regarding the anionic species at equilibrium.

The mole fraction distribution for HBA1 of the several species is represented in Figure 1.5. This distribution is in line with the previous observation (Yoshida et al. 1995) that the buds of heavenly blue anthocyanin are purple while the petals are blue. In fact the pH of the vacuoles in buds is around 6.6, while in petals pH=7.7 (Yoshida et al. 1995). In that pH

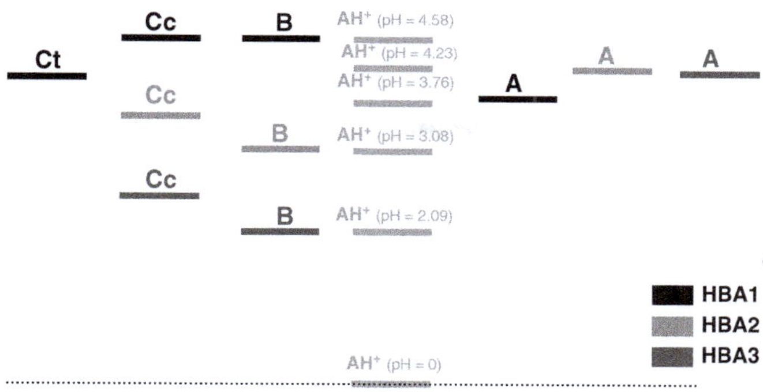

Scheme 1.8 Energy level diagrams of HBA**1** (black), HBA**2** (blue), and HBA**3** (red). In HBA**1** there is an inversion of the relative stability between the quinoidal base and hemiketal. The energy levels of **Ct** cannot be measured with the necessary accuracy due to some decomposition in HBA**2** and HBA**3** for longer reaction times. *Source:* Mendoza et al. 2018. (*See insert for color representation of the figure.*)

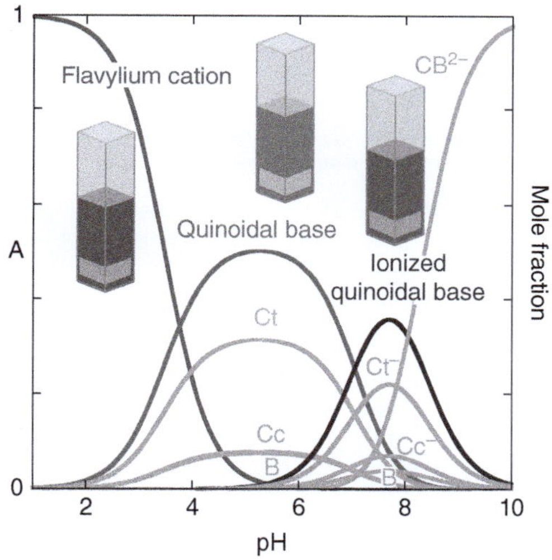

Figure 1.5 Mole fraction distribution of heavenly blue anthocyanin. *Source:* Mendoza et al. 2018. (*See insert for color representation of the figure.*)

region a small pH change gives different contributions of the quinoidal base and anionic quinoidal base and significant color changes can be observed.

The rate constants of the kinetic steps of HBA1 and derivatives were also determined and are reported in Table 1.2, together with the data for HBA2 and HBA3. The isomerization is very slow and some decomposition was observed, preventing the determination of the isomerization constants with accuracy.

Inspection of Table 1.2 shows that the hydration constant of HBA1 decreases 35-fold when compared with HBA3, while the dehydration increases c. nine-fold. This kinetic effect is compatible with the previous explanation considering a π-π stacking effect of the caffeoyl residues that protect positions C2 and C4 from the water attack. This kinetic effect also has thermodynamic consequences, because the equilibrium constant K_h increases 306-fold. In addition, the equilibrium constant of the quinoidal base does not change

Table 1.2 Rate constants between **AH+** and **CB** (estimated error 10%). Reproduced from Mendoza et al. (2018), with permission.

	k_h / s^{-1}	k_{-h} / M^{-1} s^{-1}	k_t / s^{-1}	k_{-t} / s^{-1}	k_i / s^{-1}	k_{-i} / s^{-1}
HBA1	0.01	377	0.09	0.086	2x10^{-6}	5x10^{-7}
HBA2	0.12	145	0.08	0.23	—	—
HBA3	0.35	43	0.12	0.33	—	—

Source: Mendoza et al. 2018.

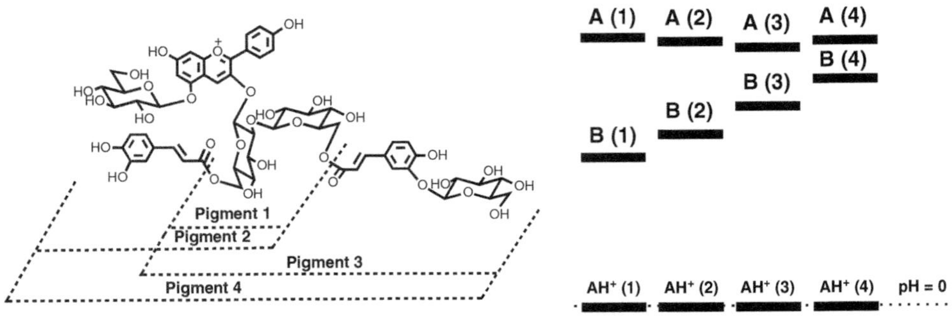

Scheme 1.9 (Left) The polyacylated anthocyanins (Dangles et al. 1993); (Right) representation of the relative energy levels at equilibrium using the hydration and acidity constants reported by the authors. The construction of this type of energy level diagram is shown in Scheme 1.2. It is worth noting that two acylated units are not enough to invert the energy levels between the hemiketal (B) and the quinoidal base (A). *Source:* Dangles et al. 1993.

significantly. The most interesting feature in HBA1 is the inversion of the energy levels of the quinoidal base relative to the hemiketal (Scheme 1.8).

The protection effect of the acylated sugars for the hydration reaction was studied previously (Fernandes et al. 2015; Dangles et al. 1993), but according to our knowledge no other study of the physical chemistry of three acylated sugars has been reported. The most complete study previously published, regarding this type of molecules, was reported by Dangles et al. (1993) (Scheme 1.9). In this work, the proton transfer as well as the hydration constants were determined, allowing the construction of an energy level diagram as in Scheme 1.8.

The same type of effect reported for HBA1 was observed: raising of the energy level of the hemiketal and a more or less constant energy level of the quinoidal base as the acylated units increase.[5] However, no inversion of the energy levels of the hemiketal compared with quinoidal base was observed, suggesting that three acylated units are necessary.[6]

The energy level diagram of the HBA1 was extended to the basic region; Scheme 1.10. Details can be found elsewhere (Mendoza et al. 2019).

After a pH jump to pH=5.5, for example, the system equilibrates between **A**, **B**, **Cc**, and **Ct**, but the most stable species is the quinoidal base **A**. In the case of a pH jump to the

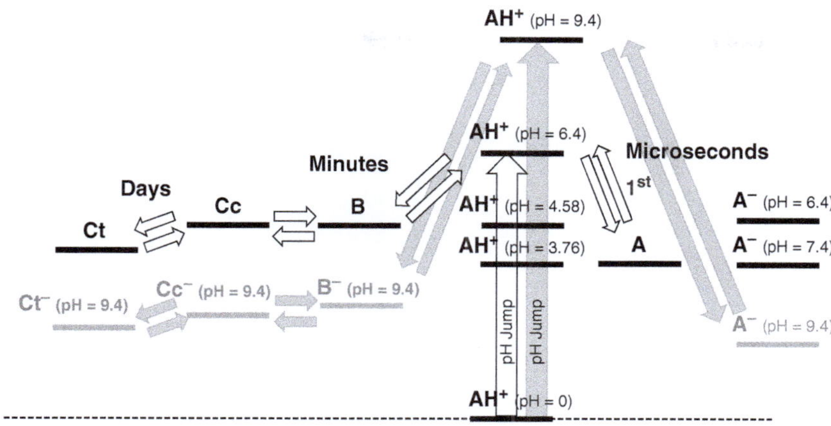

Scheme 1.10 Energy level diagram of the compound HBA**1** extended to the mono-anionic species. In the case of moderately acidic pHs the purple color of the quinoidal base is observed; at higher pH values it is the blue anionic quinoidal base that becomes more stable. *Source:* Mendoza et al. 2018.

region of the mono-anionic species, pH=8.5, the anionic quinoidal base becomes the most stable species.

1.4.2 2-Hydroxyflavylium Derivatives and Flavanones

The 2'-hydroxyflavylium derivatives give rise to the formation of flavanones as shown in Scheme 1.11 and Figure 1.6 (Petrov et al. 2008; Slavcheva et al. 2018).

Figure 1.6 shows the spectral variations after a direct pH jump to pH=8.1 (Slavcheva, et al. 2018). A few seconds after the pH jump the anionic hemiketal and *cis*-chalcone are formed (Figure 1.6a). The system evolves and gives the anionic *trans*-chalcone after c. 40 minutes. However, equilibrium is reached four hours later, only after formation of the flavanone (FLV; Figure 1.6b), indicating that the flavanone is formed from the anionic *trans*-chalcone. Acidification of the flavanone does not give any spectral variation, but basification to pH=12.0 very quickly gives $\mathbf{Ct}^{2-}$ ($k_{obs} = 3.3\ s^{-1}$). When the titration of $\mathbf{Ct}^{2-}$ by reverse pH jump is carried out, the absorption spectra of $\mathbf{Ct}^{-}$ and **Ct** are observed according to the final pH. The spectrum of **Ct** is stable and the one of $\mathbf{Ct}^{-}$ gives the flavanone as in Figure 1.6b.

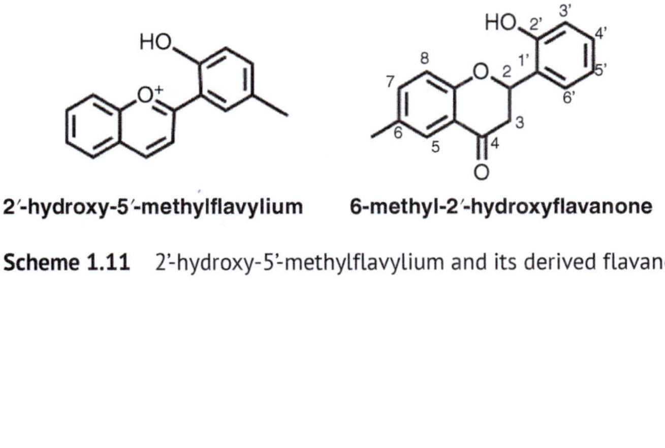

Scheme 1.11 2'-hydroxy-5'-methylflavylium and its derived flavanone form.

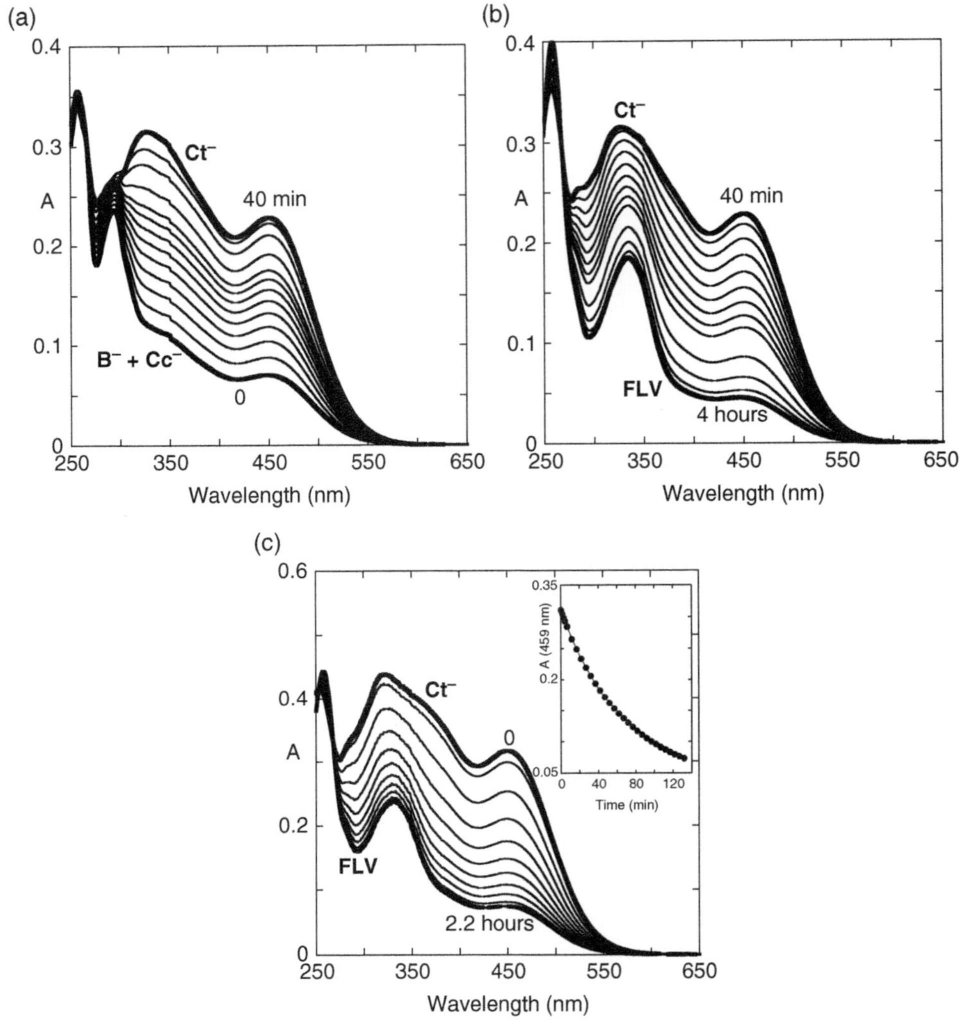

Figure 1.6 (a) Spectral variations after a direct pH jump from pH=1 to pH=8.1 of the compound 2'-hydroxy-5-methylflavylium up to 40 minutes; (b) the same after 40 minutes up to 4 hours; (c) reverse pH jump from pH=12 to pH=8.4. *Source:* Slavcheva, et al. 2018. © 2018 Elsevier.

This peculiar behavior permits to illustrate the concept of a timer at the molecular level with reset capacity (Scheme 1.12).

The initial species is **Ct** at pH=4.0. The activation of the timer is made by changing the pH to pH=8.0, which allows the flavanone formation to start. To stop the timer the pH is reversed back to pH=4.0. The unreacted **Ct**$^-$ immediately gives the stable **Ct** and the flavanone does not react. The flavanone/*trans*-chalcone molar ratio gives the extension of the reaction and the elapsed time.[7] To reset the timer two steps are needed: (i) the unlock step should be achieved by changing the pH to 12; the unreacted **Ct** and the flavanone (more

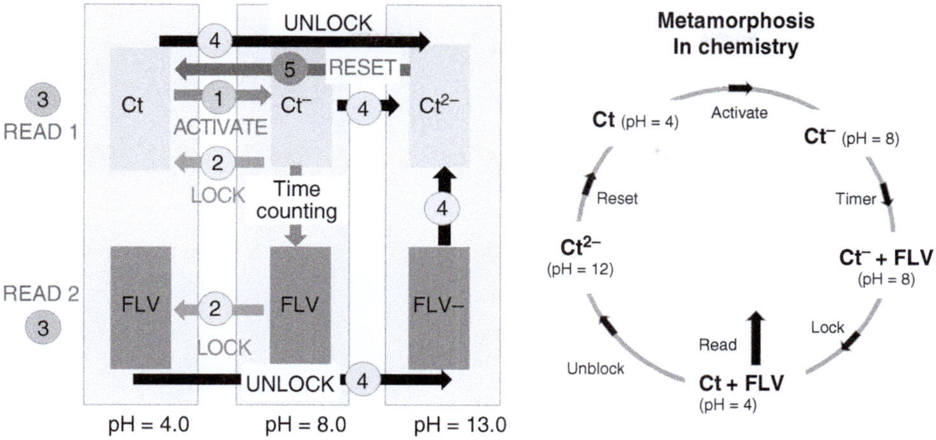

Scheme 1.12 Illustrating the concept of a timer with reset capacity through a metamorphosis cycle. *Source:* Slavcheva, et al. 2018. © 2018 Elsevier. (*See insert for color representation of the figure.*)

precisely the anionic flavanone) give **Ct**$^{2-}$, and (ii) after the end of the first step, the pH is shifted to 4.0 and the reaction goes back to **Ct**.

1.4.3 6,8 Rearrangements

In Scheme 1.13, the 6,8 rearrangement is shown. Two necessary requirements are needed to observe this reaction in flavylium-derived systems: (i) a hydroxyl group in position C5 and (ii) the lack of symmetry through the binary axis identified in Scheme 1.13 by the traced line. For example, with the natural symmetrical 3-deoxyanthocyanidins such as luteolinidin and apigeninidin, the 6,8 rearrangement is not observed. The 6,8 rearrangement in the flavylium cation was first reported (Jurd 1963) for 4',5,7,8-tetrahydroxyflavylium and for 6/8-*C*-glycosyl-3-deoxyanthocyanidins (Bjorøy et al. 2009). More recently, a complete study extended to the other species of the multistate present in moderately acidic medium was reported (Scheme 1.13), R=Br (Cruz et al. 2016, 2017).

In very acidic medium, equilibrium is established between c. 50% of each flavylium cation. The two flavylium cation isomers were separated by HPLC and lyophilized. The usual pH jumps followed by stopped flow permitted us to obtain the absorption spectra of the two neutral quinoidal base isomers and study the respective kinetic processes. At pH=3.7 (Figure 1.7), the equilibrium absorption spectrum can be fitted with the contributions of $0.45\mathbf{A_6} + 0.45\mathbf{A_8} + 0.1\mathbf{Ct}$.

In the case of the 6,8 rearrangements for R=CH_3 or phenyl (Basílio et al. 2017) the 8 derivative is the only one observed and equilibrium in moderately acidic medium is established between $\mathbf{A_8}$ and **Ct** (c. 20%). The pH-dependent equilibrium between flavylium cation and a mixture of quinoidal base (major species) and *trans*-chalcone was previously reported for 5-deoxyanthocyanidins (Sousa et al. 2014; Brouillard et al. 1982). At moderately acidic pH values there is some spectral evidence that the two quinoidal base isomers are in equilibrium, but the spectral variations are relatively small. In order to unveil the two isomers other strategies should be used; see below.

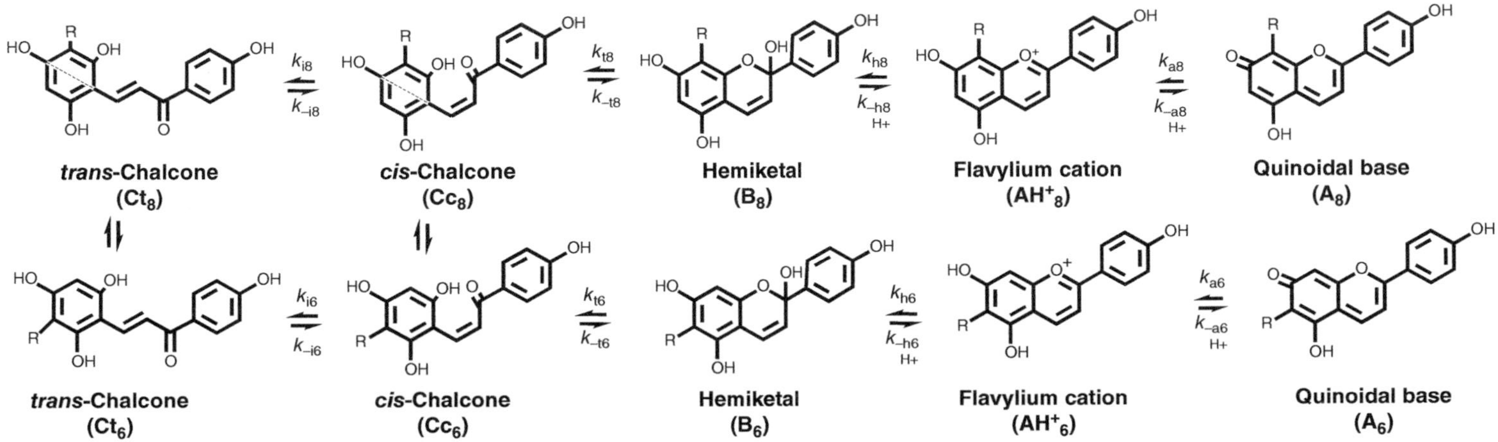

Scheme 1.13 General scheme of the 6,8 rearrangement. R=Br or R=CH_3 or R=Phenyl. *Source:* Adapted from Cruz et al. 2016. © 2016 John Wiley & Sons.

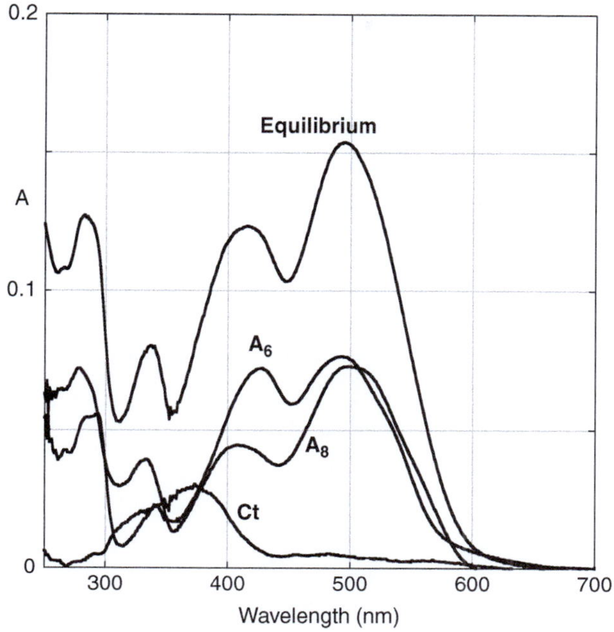

Figure 1.7 The absorption spectra of equilibrated solutions of the compound 8-bromo,4′,5,7 trihydroxyflavylium R=Br (3.3×10^{-5} M) at pH=3.7 can be fitted with a contribution of 0.45$\mathbf{A_6}$+0.45$\mathbf{A_8}$ +0.1**Ct**. *Source:* Adapted from Cruz et al. 2016. © 2016 John Wiley & Sons.

1.4.3.1 Unveiling the 6,8 Rearrangement Through Host-Guest Complexation

A step further to the study of the 6,8 rearrangement was achieved by host-guest complexation. It is known that while cucurbiturils complex preferentially the flavylium cation (Basílio and Pina 2014), cyclodextrins prefer the chalcones (Petrov 2013b). The sulfobutylether derivative of β-cyclodextrin (captisol) was used due to its better solubility in water, when compared with β-cyclodextrin. The complexation with captisol was used to unveil the 6-phenyl isomer from the 8-phenyl-5,7-dihydroxyflavylium (Basílio et al. 2017). The NMR and UV-vis spectrophotometry indicates that the fraction of **Ct** increases very significantly in the presence of captisol. Irradiation of this solution at pH=5.0 leads to a photostationary state, constituted by quinoidal bases A_6 and A_8 (Scheme 1.14). In fact, immediate acidification of the photostationary state to pH=1.0 shows the presence of the two respective flavylium cation isomers, with the 6-phenyl isomer reverting slowly to the more stable 8-phenyl isomer.

The host-guest strategy was used to form the 6-phenyl isomer from the 8-phenyl-5,7-dihydroxyflavylium profiting from the capacity of cucurbit[7]uril to complex preferentially the former (Basílio et. al 2017) as well as for 6,8-bromo-apigeninidin (Basílio et al. 2016).

From bottom to top in Scheme 1.15, the ^{1}H NMR spectra of the two isomers are shown. After addition of the host only the 6-bromo isomer is observed. Addition of adamantine, which possesses a very high association constant with this host, permits to release and isolate the 6-bromo isomer (Scheme 1.16). The system slowly reaches equilibrium by formation of the 8-bromo isomer.

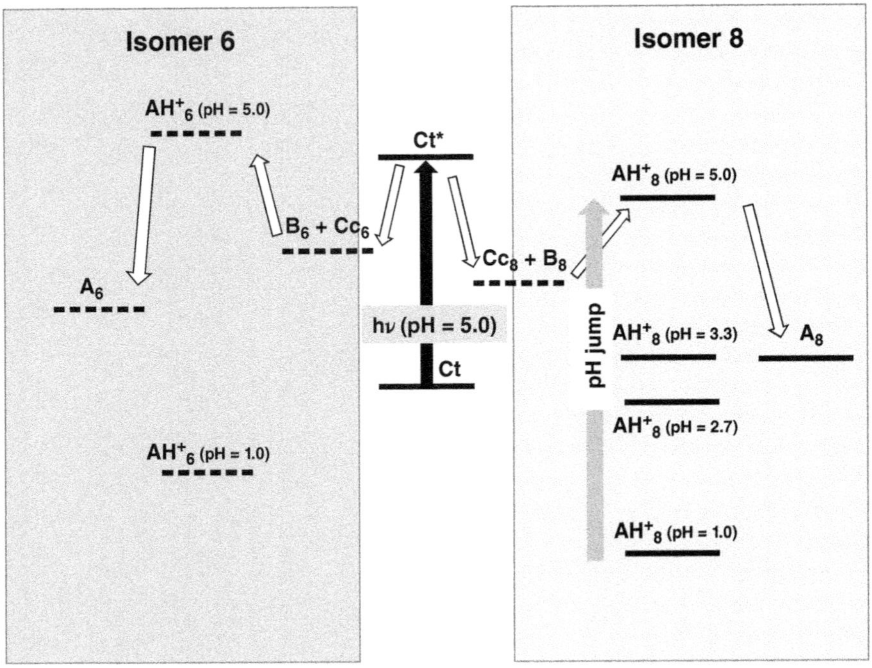

Scheme 1.14 Qualitative energy level diagram to account for the pH jumps and the photochemistry of the 6,8 rearrangement of 8-phenyl-5,7-dihydroxyflavylium in the presence of the captisol host. Traced energy levels are tentatively positioned. The arrow of the excitation energy is not in scale.

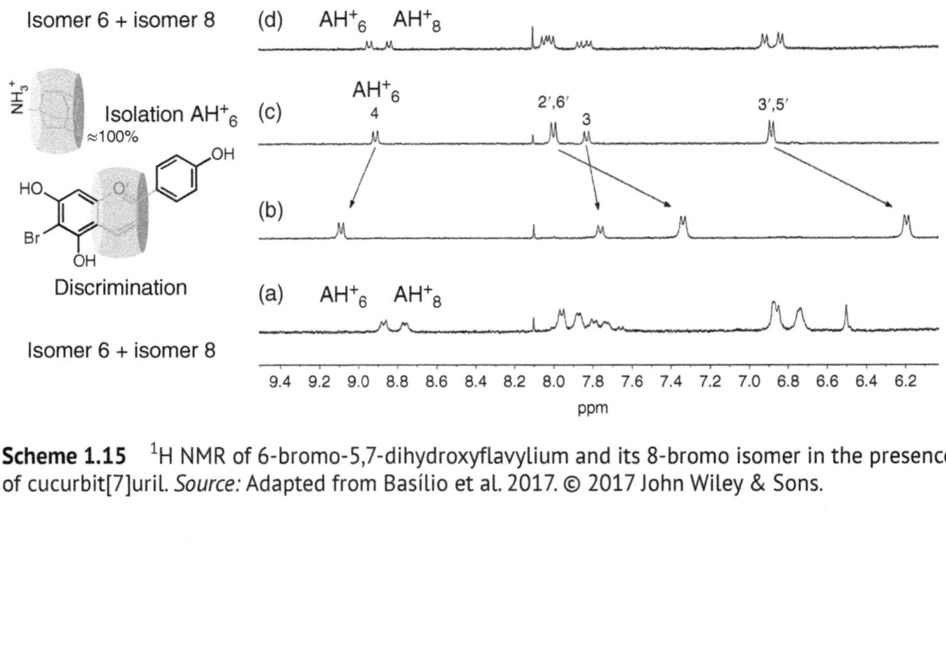

Scheme 1.15 ^{1}H NMR of 6-bromo-5,7-dihydroxyflavylium and its 8-bromo isomer in the presence of cucurbit[7]uril. *Source:* Adapted from Basílio et al. 2017. © 2017 John Wiley & Sons.

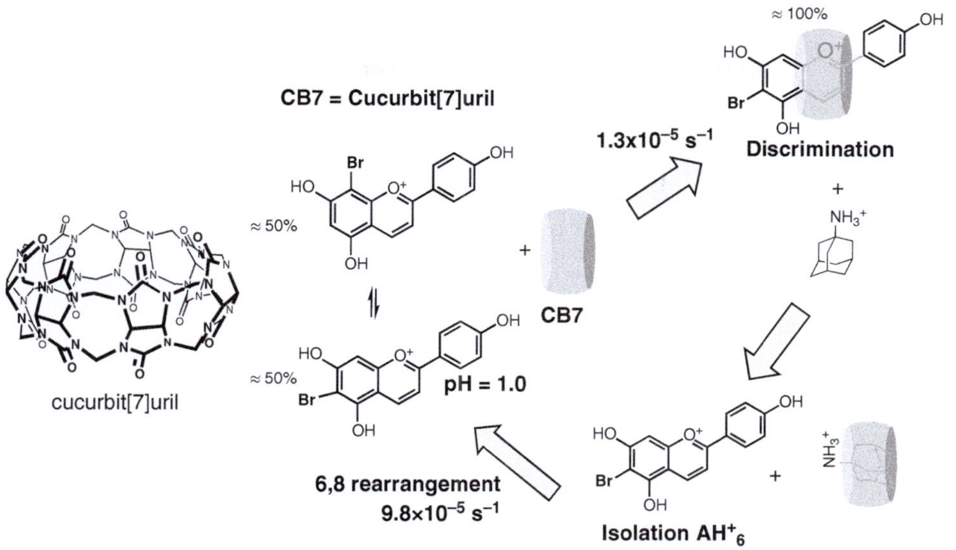

Scheme 1.16 Discrimination, isolation, and 6,8 rearrangement of 6-bromo-5,7 dihydroxyflavylium and its 8-bromo isomer in the presence of cucurbit[7]uril. Reproduced from Basílio et al. (2017), with permission. *Source:* Adapted from Basílio et al. 2017. © 2017 John Wiley & Sons.

1.5 Conclusions and Perspectives

The multistate of chemical species followed by anthocyanins is not restricted to this family of compounds. Besides those of natural origin such as anthocyanidins and deoxyanthocyanidins, anthocyanins have been a source of bio-inspiration for the synthesis of new derivatives that permit us to achieve complexity at the bottom, through this multistate of chemical species. The multistate of chemical species can be operated by pH changes, light, temperature, and host-guest chemistry. The light response is particularly challenging, since there are two photochromic systems: (i) the ring opening of the hemiketal to produce *cis*-chalcone, and (ii) *cis*-*trans* isomerization of the chalcones. The multistate response can be modulated by host-guest chemistry, increasing the potentialities of the system as photochromic materials, or in the domain of the health sciences.

Endnotes

1 In previous work we have used the term *mono ionized*, in the sense of formation of ions. This was the term used for ions resulting from the acid-base reactions and the respective constants were defined as ionization constants.

2 In any case, both processes are well separated in time and the respective amplitude can be obtained.

3 In the case of 4'-hydroxyflavylium the *trans*-chalcones are almost 100% of **CB** and thus equilibrium is achieved with **AH**$^+$ and the *trans*-chalcones.

4 Flash photolysis is applied when there is excited state flavylium/quinoidal base proton transfer.
5 In the case of two derivatives bearing one acylated unit, the compound with the three glucoside units has the highest effect.
6 It is possible to observe the blue color at pseudo-equilibrium in other systems. The peculiar behavior of HBA1 is that the purplish and blue colors are observed at equilibrium.
7 In principle, other analytical techniques, such as nuclear magnetic resonance (NMR), can be used to access the ratio **FLV/Ct**. The timer can be used only for a certain time interval (a few hours), because the spectral variations at the end of the reaction become very small.

References

Basílio, N., Cruz, L., Freitas, V., et al. (2016). A multistate molecular switch based on the 6,8 rearrangement in bromo-apigeninidin operated with pH and host–guest inputs. *Journal of Physical Chemistry*, 120 (29): 7053–7061.

Basílio, N., Lima, J.C., Cruz, L., et al. (2017). Unveiling the 6,8 rearrangement in 8-phenyl-5,7-dihydroxyflavylium and 8-methyl-5,7-dihydroxyflavylium through host-guest complexation. *European Journal of Organic Chemistry*, 37: 5617–5626.

Basílio, N., and Pina, F. (2014). Flavylium network of chemical reactions in confined media: modulation of 3',4',7-trihydroxyflavilium reactions by host–guest interactions with cucurbit[7]uril. *ChemPhysChem*, 15 (11): 2295–2302.

Bjorøy, Ø., Rayyan, S., Fossen, T., et al. (2009). Structural properties of anthocyanins: rearrangement of c-glycosyl-3 deoxyanthocyanidins in acidic aqueous solutions. *Journal of Agricultural and Food Chemistry*, 57 (15): 6668–6677.

Brouillard, R., and Dubois, J.E. (1977). Mechanism of structural transformations of anthocyanins in acidic media. *Journal of the American Chemical Society*, 99 (5): 1359–1364.

Brouillard, R., Iacobucci, G.A., and Sweeny, J.G. (1982). Chemistry of anthocyanin pigments. 9. UV-visible spectrophotometric determination of the acidity constants of apigeninidin and three related 3-deoxyflavylium salts. *Journal of the American Chemical Society*, 104 (26): 7585–7590.

Cruz, L., Basílio, N., Freitas, V., et al. (2016). Extending the study of the 6,8 rearrangment in flavylium compounds to higher pH values. Inter-conversion between 6-bromo and 8-bromo-apigeninidinin. *Chemistry Open*, 5 (3): 236–246.

Sousa, J.L., Gomes, V., Mateus, N., et al. (2017). Synthesis and equilibrium multistate of new pyrano-3-deoxyanthocyanin-type pigments in aqueous solutions. *Tetrahedron*, 73 (41): 6021–6030.

Cruz, L., Sousa, J.L.C., Gomes, V., et al. (2017). Chemical equilibrium characterization of pyrano-3-deoxyanthocyanin derivatives. *Tetrahedron* 73: 6012–6030.

Dangles, O., Saito, N., and Brouillard, R. (1993). Kinetic and thermodynamic control of flavylium hydration in the pelargonidin-cinnamic acid complexation. Origin of the extraordinary flower color diversity of Pharbitis nil. *Journal of the American Chemical Society*, 115 (8): 3125–3132.

Fernandes, A., Brás, N., Mateus, N., et al. (2015). A study of anthocyanin self-association by NMR spectroscopy. *New Journal of Chemistry*, 39: 2602–2611.

Goto, T., and Kondo, T. (1991). Structure and molecular stacking of anthocyanins – Flower color variation. *Angewandte Chemie International Edition*, 30 (1): 17–33.

Jurd, L., (1963). Anthocyanins and related compounds. I. Structural transformations of flavylium salts in acidic solutions. *The Journal of Organic Chemistry*, 28 (4): 987–991.

Hondo, T., Yoshida, K., Nakagawa, A., et al. (1992). Structural basis of blue-colour development in flower petals from Commelina communis. *Nature*, 358 (6386): 515–518.

Lehn, J.M. (1995). From molecular to supramolecular chemistry. In: *Supramolecular Photochemistry: Concepts and Perspectives*. Weinheim: Wiley-VCH.

Mendoza, J., Basílio, N., Pina,F., et al. (2018). Rationalizing the color in heavenly blue anthocyanin. A complete kinetic and thermodynamic study. *Journal of Physical Chemistry .B*, 122 (19): 4982–4992.

Mendoza, J., Basílio, N., de Freitas, V., et al. (2019). New procedure to calculate all equilibrium constants in flavylium compounds: application to the copigmentation of anthocyanins. *ACS omega*, 4 (7): 12058–12070.

Petrov, V., Diniz, A.M., Cunha-Silva, L., et al. (2013). Kinetic and thermodynamic study of 2'-hydroxy-8-methoxyflavylium. Reaction network interconverting flavylium cation and flavanone. *RSC Advances*, 3 (27): 10786–10794.

Petrov, V., Gomes, R., Parola, A.J., et al. (2008). 2'-hydroxyflavylium: introducing flavanones into the flavylium network of chemical reactions. *Tetrahedron*, 64 (4): 714–720.

Petrov, V., Parola, A.J., and Pina, F. (2012). Isomerization between 2-(2,4-Dihydroxystyryl)-1-benzopyrylium and 7-Hydroxy-2-(4-hydroxystyryl)-1-benzopyrylium. *Journal of Physical Chemistry*, 116 (31): 8107–8118.

Petrov, V., Stanimirov, S., Petrov, I.K., et al. (2013). Emptying the β-cyclodextrin cavity by light. Photochemical removal of the *trans*-chalcone of 4',7-dihydroxyflavylium. *Journal of Physical Chemistry*, 117 (41): 10692–10701.

Pina, F. (2014a). Thermodynamic and kinetic processes of anthocyanins and related compounds and their bio-inspired applications. In: *Recent Advances in Polyphenols Research. Vol. 4, Chapter 11*, (eds. Romani, A., Lattanzio, V., and Quideau, S.), 341–370. West Sussex: Wiley Blackwell.

Pina, F. (2014b). Anthocyanins and related compounds. Detecting the change of regime between rate control by hydration or tautomerization. *Dyes and Pigments*, 102: 308–314.

Pina, F., Melo, M.J., Maestri, M., et al. (1997). Photochromism of 4'-methoxyflavylium perchlorate. A "write-lock-read-unlock-erase" molecular switching device. *Journal of the American Chemical Society*, 119 (24): 5556–5561.

Slavcheva, S., Mendoza, J., Stanimirov, S., et al. (2018). On the multistate of 2'-hydroxyflavylium-flavanone system. illustrating the concept of a timer with reset at the molecular level. *Dyes and Pigments*, 158: 465–473.

Sousa, A., Cabrita, L., Araújo, P., et al. (2014). Color stability and spectroscopic properties of deoxyvitisins in aqueous solution. *New Journal of Chemistry*, 38 (2): 539–544.

Yoshida, K., Kondo, T., Okazaki, Y., et al. (1995). Cause of blue petal colour. *Nature*, 373 (6512): 291.

Yoshida, K., Mori, M., and Kondo, T. (2009). Blue flower color development by anthocyanins: from chemical structure to cell physiology. *Natural Product Reports*, 26 (7): 884–915.

2

Proanthocyanidin Oligomers with Doubly Linked (A-Type) Interflavan Connectivity

Structure and Synthesis

Ken Ohmori and Keisuke Suzuki

Department of Chemistry, Tokyo Institute of Technology, Japan

2.1 Introduction

The oligomeric proanthocyanidins (PAs), i.e. the condensed tannins, constitute a large class of natural polyphenols with an extreme structure diversity. Although some of these compounds proved to show significant biological activities, such as antiviral, antibacterial, and antitumor effects, detailed studies on their biological effects at the molecular level have remained premature due to the scarce availability of pure samples. The natural products are generally obtained as a complex mixture of closely related compounds, which are hardly separable even with the aid of modern separation methods. In view of the potentially interesting bioactivities in individual compounds, this is an unfortunate situation, but may be solved by organic synthesis if general, effective synthetic routes to these compounds were developed. Given a sufficient supply of homogeneous samples for the biological study, one may expect discovery of enhanced biological activities and have access to synthetic congeners and probes for further research on the mode of action.

The tremendous structural diversity of oligomeric PAs has its origin in the combination of several elements:

1) Difference in the constituting monomers: Figure 2.1 shows the representative flavan-3-ols contained as the monomers in the naturally occurring oligomeric PAs; that is, afzelechin (AZ), epiafzelechin (EZ), catechin (CA), epicatechin (EC), gallocatechin (GC), epigallocatechin (EGC) and their C(5)-deoxy congeners, guibourtinidol, fisetinidol, and robinetinidol. They differ in the hydroxylation level of the A-ring (one or two hydroxy groups) and the B-ring (one, two, or three hydroxy groups), and in the relationship of the C-ring substituents, *cis* or *trans*.
2) Single connectivity (B-type): Figure 2.2 shows the linear oligomers characterized by a single C–C interflavan linkage at the C(4)–C(8) position. This connectivity is schematically represented in section **I** of the figure. By a historic convention from the early days of catechin research, these compounds with single connectivity are called the B-type

Recent Advances in Polyphenol Research, Volume 7, First Edition. Edited by Jess Dreher Reed, Victor Armando Pereira de Freitas, and Stéphane Quideau.

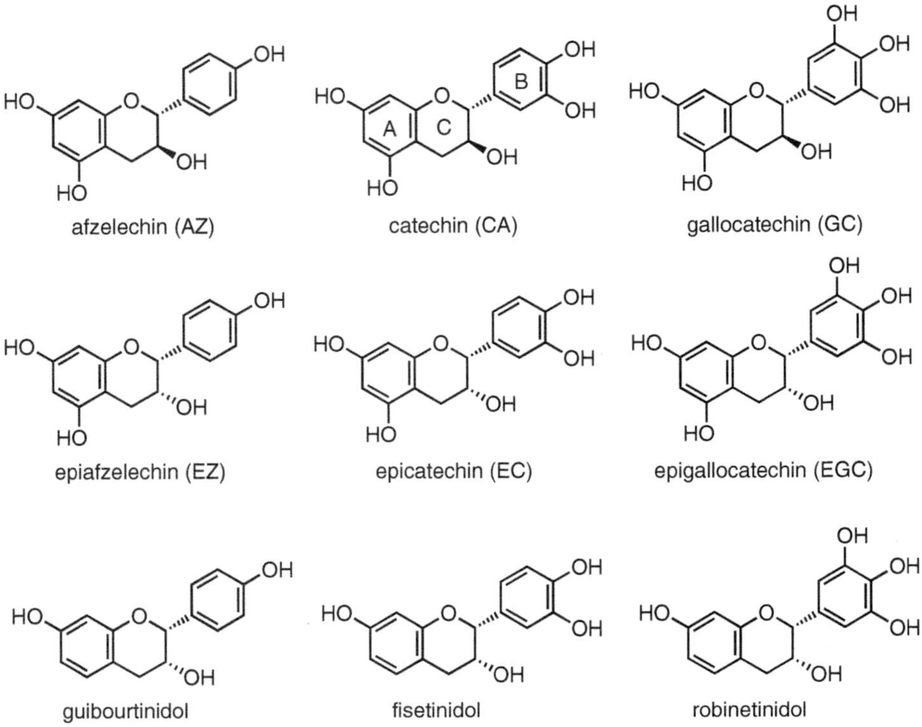

Figure 2.1 Origin of the structural diversity in oligomeric PAs: structural diversity of monomers.

(*vide infra*). The linear oligomers formed by an array of the single (B-type) interflavan linkages constitute a huge variety of molecular entities, differing in the degree of oligomerization and in the constituent monomers. A broad range of oligomers—small to large—is seen, ranging from dimers to higher oligomers, exceeding dodecamers. Various monomers, such as AZ, EZ, CA, EC, GC, EGC, are uniformly included to form homo-oligomers or heterogeneously included to form hetero-oligomers. In addition, as schematically expressed by section **II** of Figure 2.2, the oligomers with the C(4)–C(6) interflavan bonds are also found as the minor members with type-B linkages.

3) Dual connectivity (A-type): Figure 2.3 shows a rarer class of oligomeric PAs, in which all or part of the monomer units are doubly connected through a C–C bond and a C–O bond, thereby forming a characteristic dioxabicyclo[3.3.1]nonane skeleton. Such a dual interflavan connectivity is called an A-type. These compounds constitute a small but potentially interesting molecular class. As represented by section **III** of Figure 2.3, most congeners of this class have a C–C bond [C(4)–C(8)] and a C–O bond {C(2)–O[C(7)]}. An increasing number of the natural PAs with the A-type linkages has been identified. Taking the trimers as examples in Figure 2.3, cinnamtannin B1 has one double linkage and one single linkage, while aesculitannin C has two double linkages. For some other higher-order structures, see Figure 2.6 (*vide infra*). In addition, there are minor congeners having a different dual connectivity; i.e. a C–C bond [C(4)–C(6)] and a C–O bond {C(2)–O[C(7)]} as represented by section **IV** of Figure 2.3.

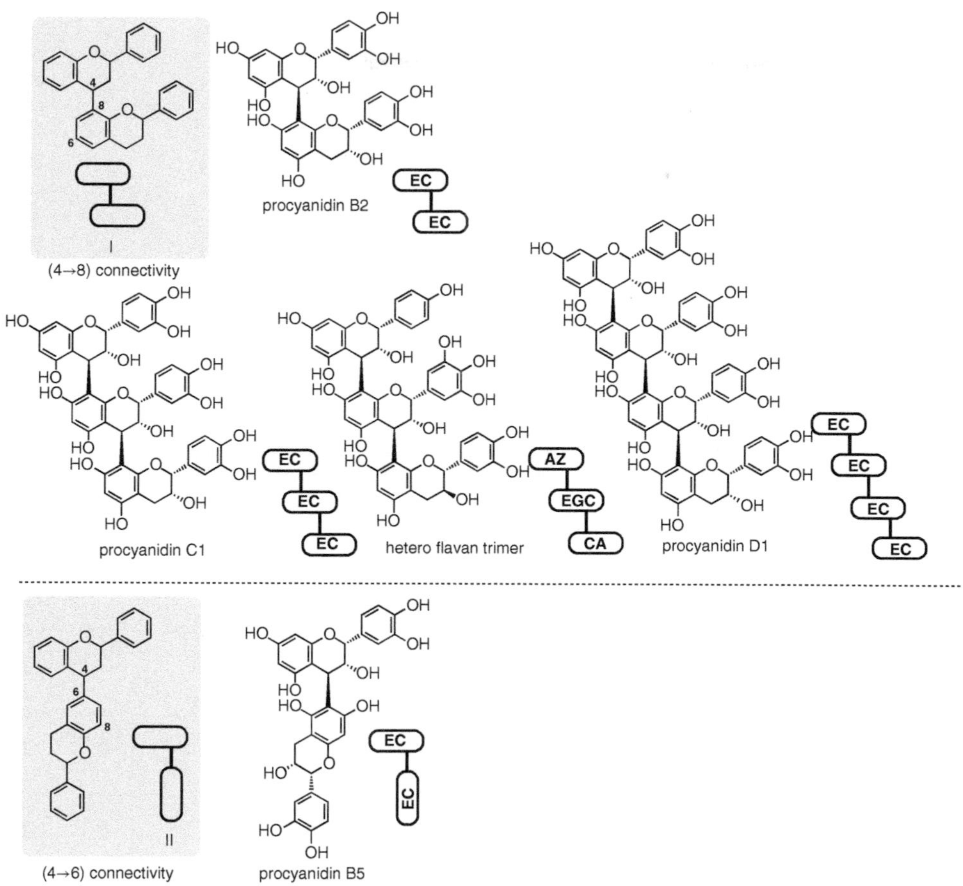

Figure 2.2 Origin of the structural diversity in oligomeric PAs: B-type connectivity.

This chapter will focus on the oligomeric PAs with an A-type structure. Section 2.2 explores the history of the structure elucidation, and section 2.3 summarizes the postulated biosyntheses and the model studies for construction of the dioxabicyclo[3.3.1]nonane skeleton, with several examples of the completed total syntheses.

2.2 Structure

This section describes a brief history of the oligomeric PAs with A-type linkages. In 1966, Mayer and coworkers isolated a crystalline phenolic compound from the seed shells of a horse chestnut (*Aesculus hippocastanum* L.). Using mass spectroscopy, the chemical composition was determined to be $C_{30}H_{24}O_{12}$ (Mayer et al. 1966). The compound was hydrolyzed with conc. hydrochloric acid, and paper chromatography identified two hydrolysates, epicatechin (**1**) and cyanidin (**2**) (Figure 2.4). At this stage, the compound was named procyanidino-(−)-epicatechin without specifying the parent dimer structure. Three possible dimeric structures, **I–III**, were postulated, sharing a common feature that two epicatechin

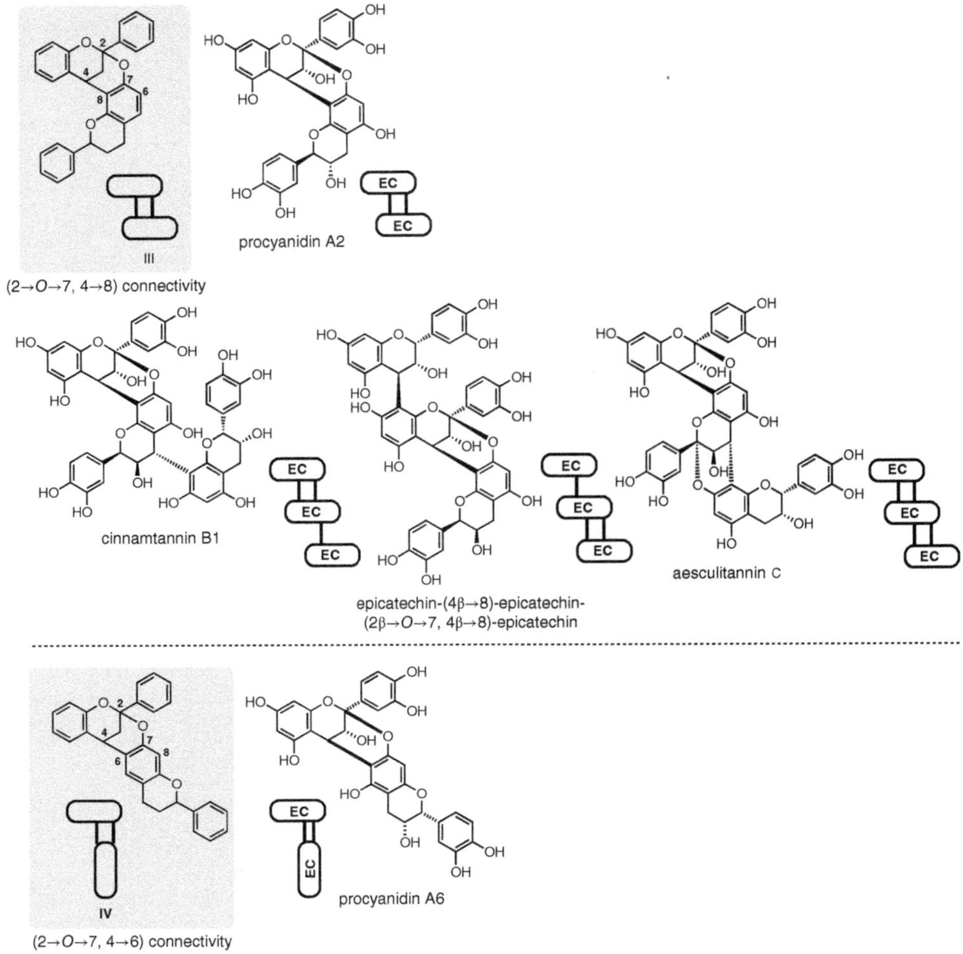

Figure 2.3 Origin of the structural diversity in oligomeric PAs: compounds having the A-type connectivity.

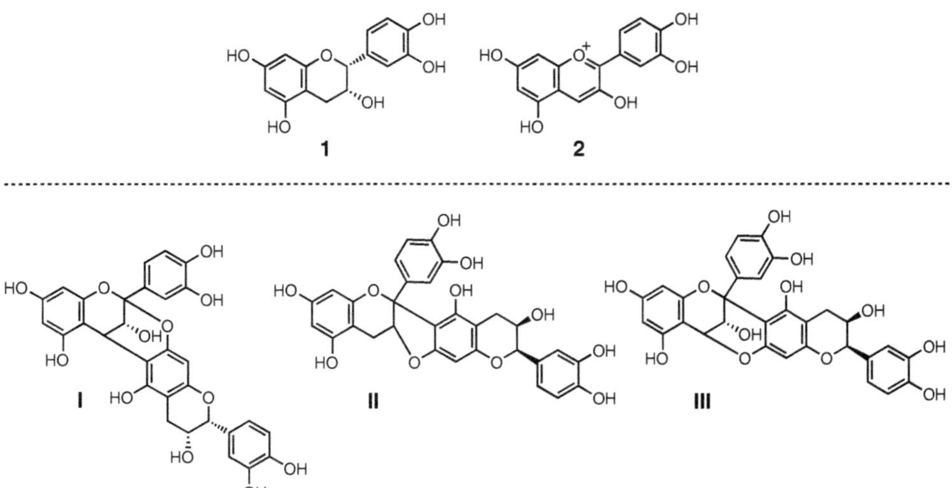

Figure 2.4 Acid hydrolysates of the dimeric proanthocyanidin isolated from *Aesculus hippocastanum*, and three possible dimeric structures proposed by Mayer.

units are linked through a double linkage. They stated that, among these, the structure **III** would be most likely.

Later, Weinges and coworkers extracted several related compounds from cranberries (*Vaccinium vitis-idaea*), which included the "Mayer's dimer" mentioned above (Weinges et al. 1968). They arbitrarily classified those compounds into two categories by the number of hydrogen atoms (Figure 2.5): the dimeric proanthocyanidin compounds with molecular formula $C_{30}H_{24}O_{12}$ as the A-type, and those with $C_{30}H_{26}O_{12}$ as the B-type. Thus, the above-stated Mayer's compound, i.e. procyanidino-(−)-epicatechin, was renamed procyanidin A2.

Haslam and coworkers later proposed the currently accepted structure **3** (Figure 2.5a) with a characteristic bicyclic skeleton formed by two interflavan bonds {C(4)–C(8) and C(2)–O[C(7)]} based on the ^{1}H- and ^{13}C-NMR data and some chemical evidences (Jacques et al. 1973, 1974). A decade later, the structure was unequivocally verified by X-ray crystallography (Van Rooyen and Redelinghuys 1983).

With advances in the separation and analytical methods, the more complicated oligomeric PAs having A-type linkages have been identified from various plant sources. Figure 2.6 shows some of the tetramers with A-type structure, which are homo- or hetero-oligomers (Nonaka et al. 1983; Morimoto et al. 1985; Morimoto et al. 1987; Balde et al. 1995; Nam et al. 2017). It should be noted that the molecular diversity arises not only from the elements stated before, but also from the inclusion of the enantiomeric flavan-3-ols (e.g. *ent*-AZ, *ent*-EC, *ent*-CA in pavetannin C5) as constituent monomers. The diversity will increase in the future, and some of these compounds may show potentially significant biological activities.

2.3 Synthetic Studies

2.3.1 Hypothetical Biosynthetic Routes

On the biogenesis of the characteristic double linkages (A-type), two putative pathways have been proposed (Paths I and II, Figure 2.7) (Selenski and Pettus 2006). Different consequences would be expected on the reactivity and the stereochemistry in the formation of

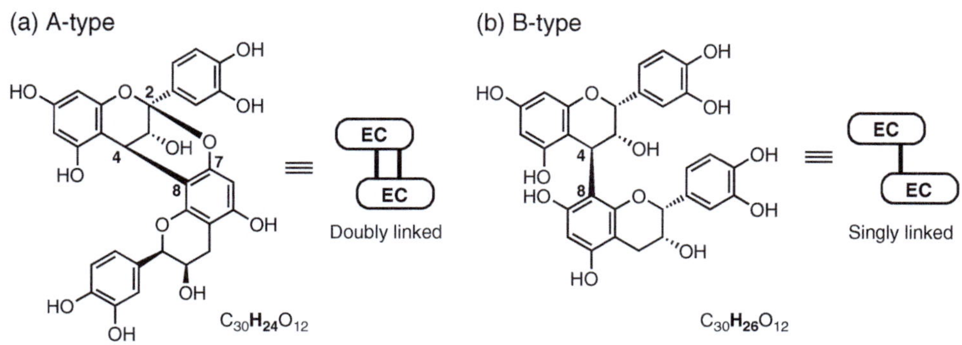

procyanidin A2 (**3**)
[epicatechin-(2β→O→7, 4β→8)-epicatechin]

procyanidin B2 (**4**)
[epicatechin-(4β→8)-epicatechin]

Figure 2.5 Mayer's PA (procyanidin A2): the terminological origin of A/B-type structures.

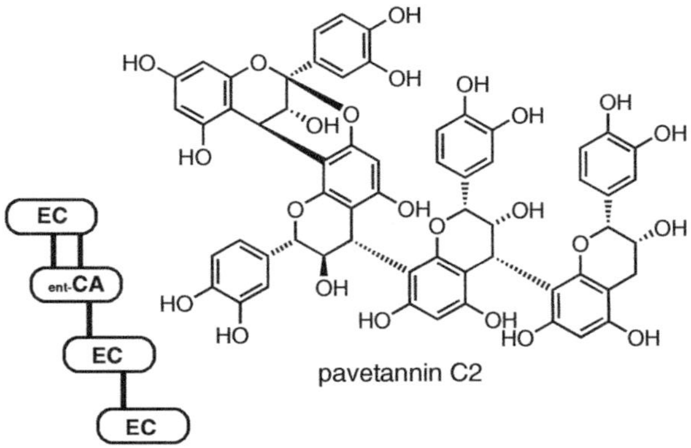

pavetannin C2

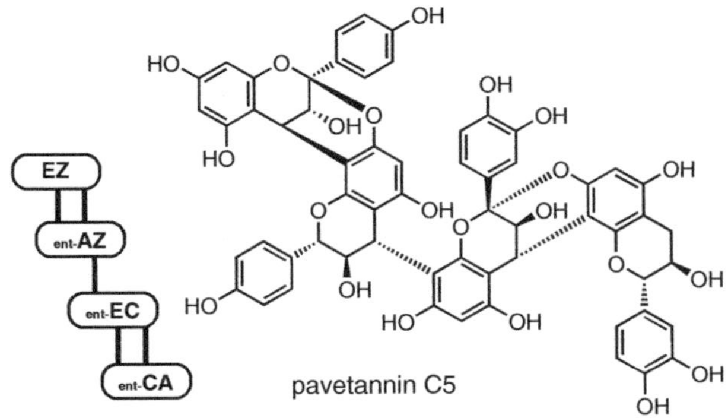

pavetannin C5

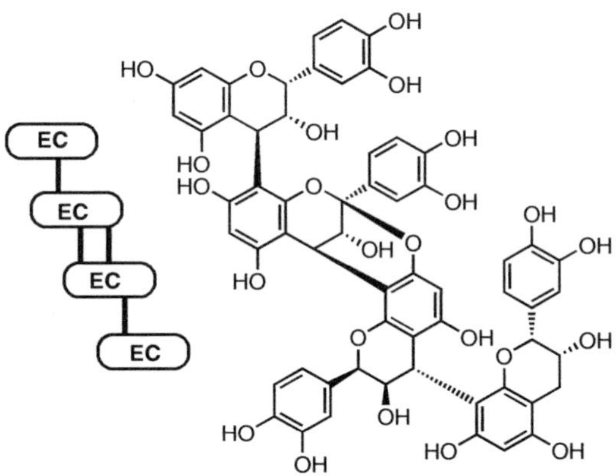

cinnamtannin B2

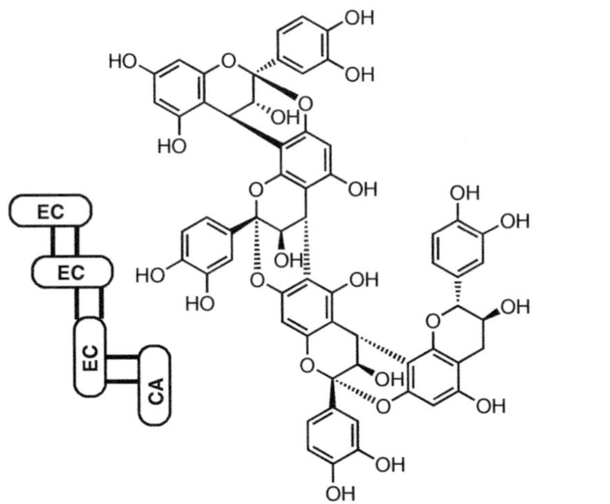

epicatechin-(2β→*O*→7, 4β→8)-epicatechin-(2β→*O*→7, 4β→6)-epicatechin-(2α→*O*→7, 4α→8)-catechin

Figure 2.6 Structures of the tetramers with A-type linkages.

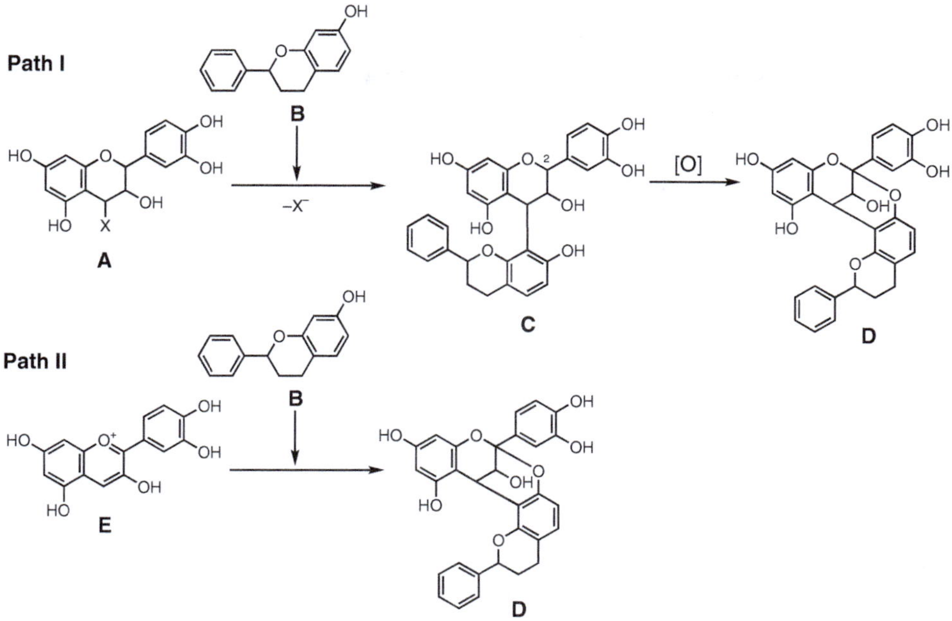

Figure 2.7 Two plausible biosynthetic pathways forming the A-type structure.

the [3.3.1]bicyclo skeleton **D**. Path I entails the addition of the flavan nucleophile **B** to the electrophilic partner **A**, producing the singly linked dimer **C** (the B-type structure). Oxidation at the C(2) position of the upper flavan unit in **C** allows the formation of a doubly linked derivative **D** (the A-type structure). From the stereochemical standpoint, if the initial C–C bond is formed in a stereoselective manner, the stereochemistry generated by the subsequent C–O bond formation would be settled spontaneously, due to the steric constraints in the bicyclic skeleton. On the other hand, Path II is based on a formal [3+3]-cycloaddition of the flavylium **E** with the flavan unit **B**. Since the flavylium **E** lacking any stereogenic centers is achiral (prochiral), the [3+3]-cycloaddition reaction needs to proceed with enantiofacial selectivity, which may be regulated by enzymes in the biogenesis. These putative biosynthetic pathways would give hints to chemical synthesis.

2.3.2 Retrosynthesis

Figure 2.8 illustrates a synthetic analysis en route to the A-type structure, focusing on the key dioxabicyclic skeleton. Three potential pathways are shown.

Route I is relevant to the Path I biosynthesis discussed in Section 2.3.1 (see Figure 2.7), disconnecting the C–O bond i in **A** to **B** with a single connection and the C(2) cation center, which could be traced back to a B-type structure **B'** as a precursor. In executing the synthesis, this approach has an advantage, that the corresponding B-type structures are synthetically well accessible (Ohmori et al. 2004, 2011; Oyama et al. 2008; Kozikowski and Tückmantel 2009; Saito et al. 2009; Yano et al. 2012; Makabe 2013). However, a concern is that the site-specific oxidation at the C(2) benzylic center on the upper flavan unit may be challenging.

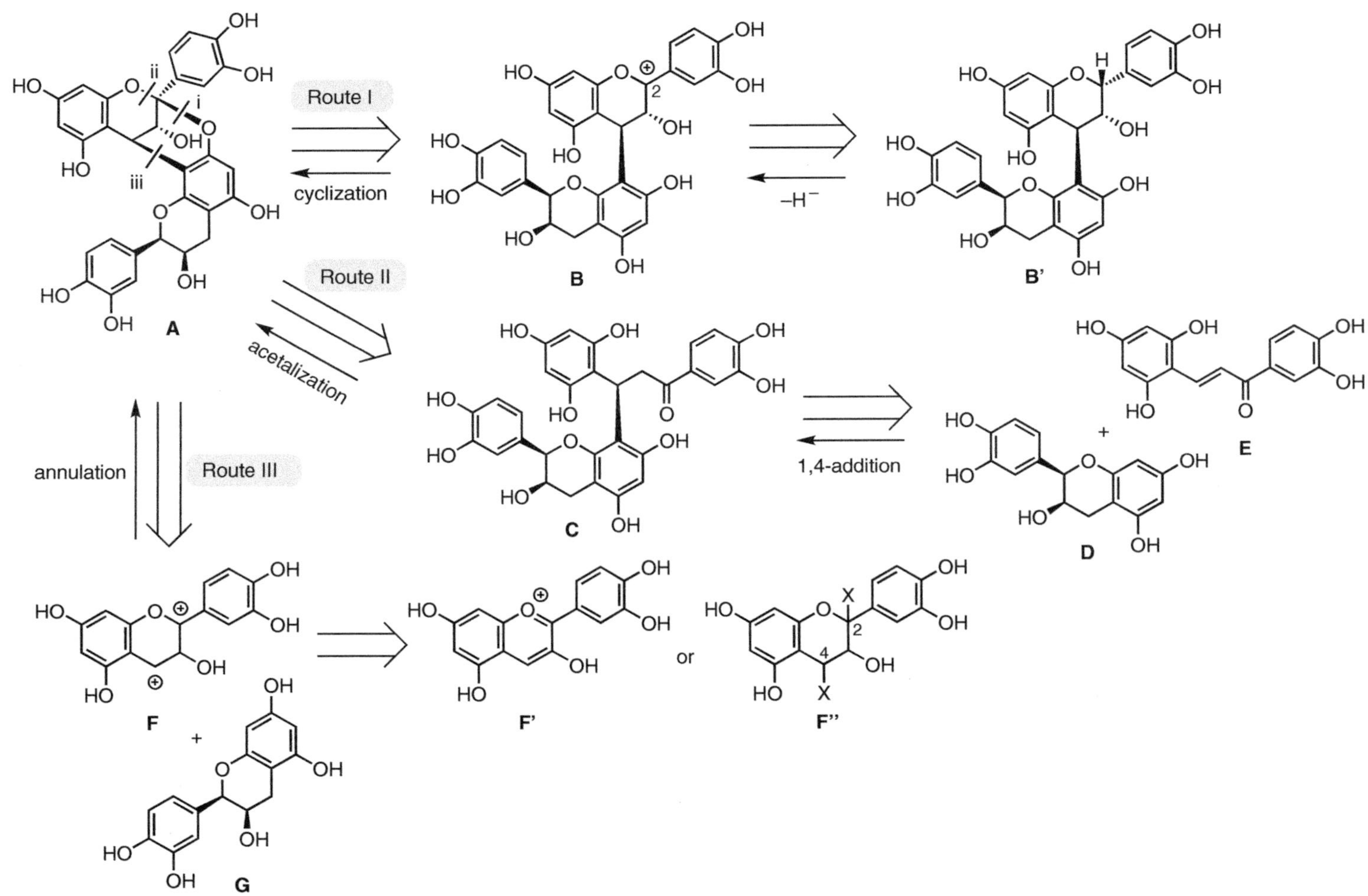

Figure 2.8 Retrosynthetic analyses of the A-type structure.

Route II relies on the retrosynthetic hydrolysis of the acetal moiety in **A**. Dissection of the bonds i and ii suggests ketone **C** as the precursor. Assuming the Michael addition, ketone **C** is accessible by combining an electrophilic chalcone unit **E** and a nucleophilic flavan unit **D**. In this approach, rigorous stereocontrol at the Michael addition stage is necessary.

Route III corresponds to another biomimetic pathway (Path II, Figure 2.7), based on the two-bond disconnection at bonds i and iii in **A**, assuming a formal [3+3]-cycloaddition of a dicationic species **F** and a nucleophilic partner **G**. As the possible synthetic equivalents to the key dicationic species **F**, one could conceive flavylium salt **F′** or flavan unit **F″** with two leaving groups at the C(2) and C(4) positions. This approach would realize direct conversion to the key bicyclic skeleton. If flavylium salt **F′** were used, the enantiocontrol would inevitably pose a serious problem. In contrast, use of the flavan unit **F″** could achieve a stereoselective reaction, as will be discussed later (see Section 2.3.5).

2.3.3 Oxidative Conversion from B-type PAs (Route I)

This section describes the reported reactions based on Route I in Figure 2.8. In early attempts, Nonaka et al. (1987) used the oxidative conversion of the B-type to the A-type structure (Figure 2.9). Upon treatment of procyanidin B1 (**5**) with hydrogen peroxide under basic conditions, an oxidative conversion proceeded to give procyanidin A1 (**6**) in 13% yield. This protocol was further applied to other B-type structures. For example, procyanidin B5 (**7**) and aesculitannin A (**10**) were converted to the corresponding compounds having the A-type structure, i.e. procyanidin A7 (**9**) and aesculitannin C (**11**), albeit in low yields.

As other means of converting the B-type structure into the A-type structure, a radical-mediated reaction was exploited (Figure 2.10) (Kondo et al. 2000). Upon treatment of procyanidin B1 (**5**) with 2,2-diphenyl-1-picrylhydrazyl (DPPH), the C(2) hydrogen atom in the upper epicatechin unit was abstracted, inducing an oxidative cyclization to give procyanidin A1 (**6**), though the chemical yield was not reported.

2.3.4 Approaches via an Acyclic Precursor (Route II)

Weinges and Theobald (1971) reported a stepwise construction of the dioxabicyclo[3.3.1]nonane skeleton (A-type) of PAs (Figure 2.11). The Michael addition of the *o*-benzyloxyphenylmagnesium bromide to chalcone **12** gave ketone **13**. After hydrogenolytic removal of the benzyl protecting groups, the resulting bisphenol was exposed to dehydrating conditions, giving bicycle **14** in 50% yield.

Xia et al. (2014) reported a cascade reaction of 2-hydroxychalcones with phloroglucinol derivatives by using a catalytic amount of ethylenediammonium diacetate (EDDA, 10 mol%), constructing the dioxabicyclo[3.3.1]nonane skeleton (Figure 2.12). The reaction of **15** with **16** proceeded in refluxing toluene via the Michael reaction followed by an internal acetal formation, giving the condensation product **17** in 87% yield.

Kraus and Geraskin (2017) reported a facile one-pot formation of the A-type structure (Figure 2.13). Under acidic conditions, twofold nucleophilic reactions of phloroglucinol to acetylenic aldehyde **18** generated bis-arylated **19** as an intermediate, which underwent acid-catalyzed acetal formation to give bicycle **20**. After acetylation, tetraacetate **21** was obtained in high yield.

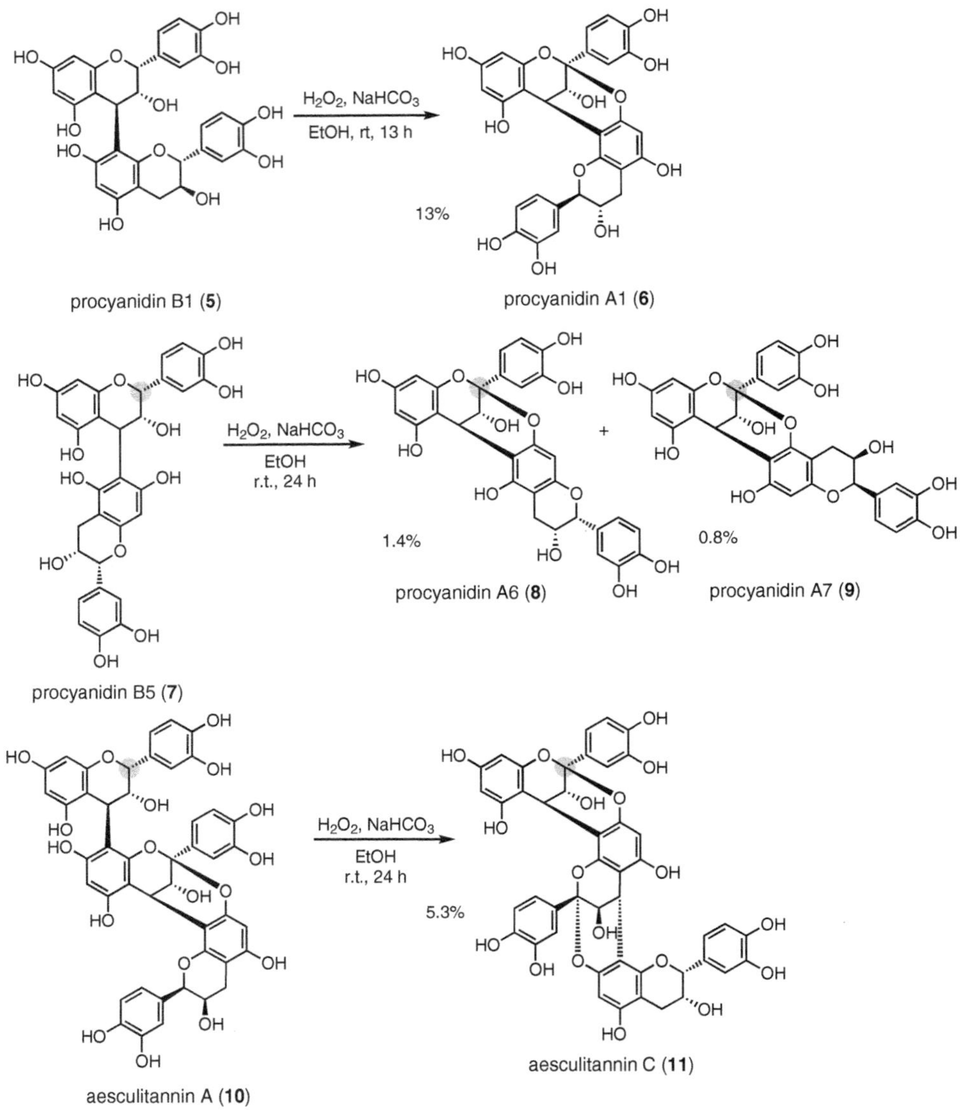

Figure 2.9 Oxidative conversion of the B-type to the A-type structure.

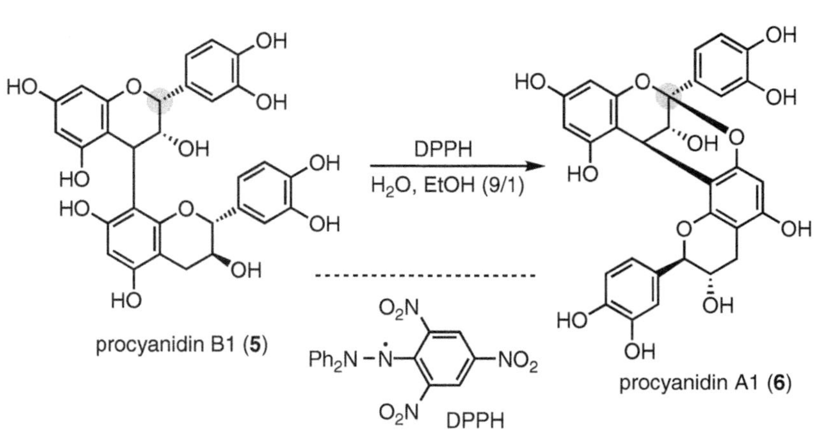

Figure 2.10 Radical-mediated oxidative conversion.

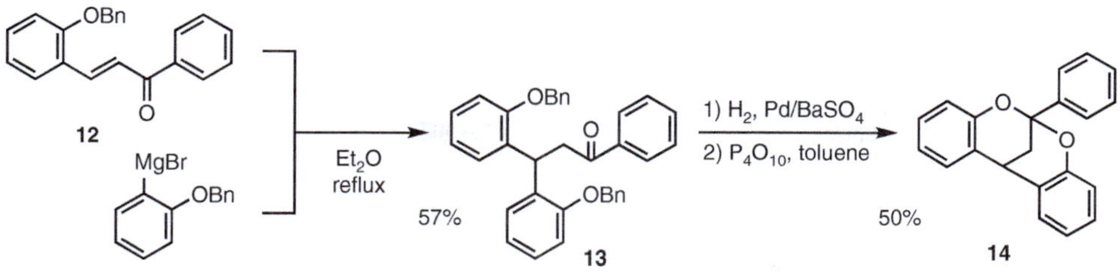

Figure 2.11 Stepwise construction of the A-type structure.

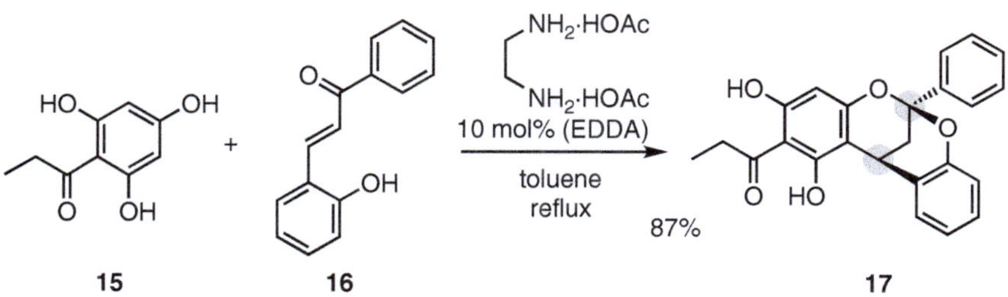

Figure 2.12 Cascade reaction of 2-hydroxychalcones with a phloroglucinol derivative

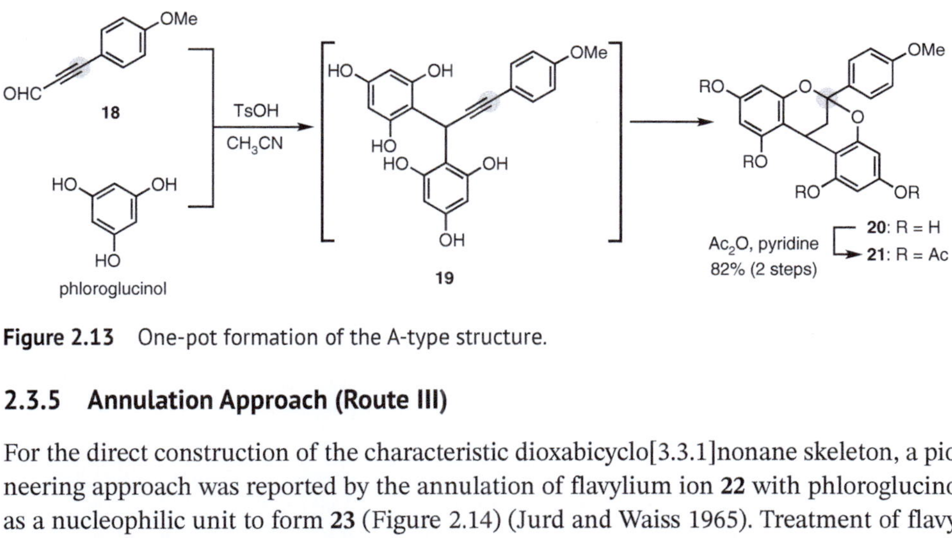

Figure 2.13 One-pot formation of the A-type structure.

2.3.5 Annulation Approach (Route III)

For the direct construction of the characteristic dioxabicyclo[3.3.1]nonane skeleton, a pioneering approach was reported by the annulation of flavylium ion **22** with phloroglucinol as a nucleophilic unit to form **23** (Figure 2.14) (Jurd and Waiss 1965). Treatment of flavylium salt **22** with phloroglucinol in aqueous MeOH under weakly acidic conditions (pH 5.8, 60 °C, 15 minutes) gave, after acetylation, the annulation product **23** as colorless prisms (23% yield). The stereochemistry was not clarified.

A similar reaction was reported by Pomilio et al. (1977) (Figure 2.15), carrying out the annulation of flavylium **24** and (+)-catechin under mild acidic conditions (pH 5.8). The reaction was sluggish, and isolation of the product after thorough protection of hydroxy groups resulted in a poor yield of annulation product **25**.

Kraus et al. (2009) improved this reaction (Figure 2.16). Treatment of flavylium salt **24** with phloroglucinol followed by treatment with silica gel gave the annulation product **26** in good yield. Moreover, use of (+)-catechin as a nucleophile gave a diastereomer mixture of

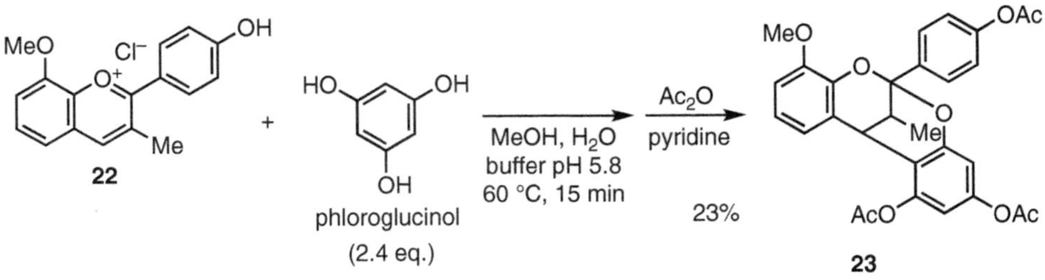

Figure 2.14 Direct annulation approach to the A-type structure.

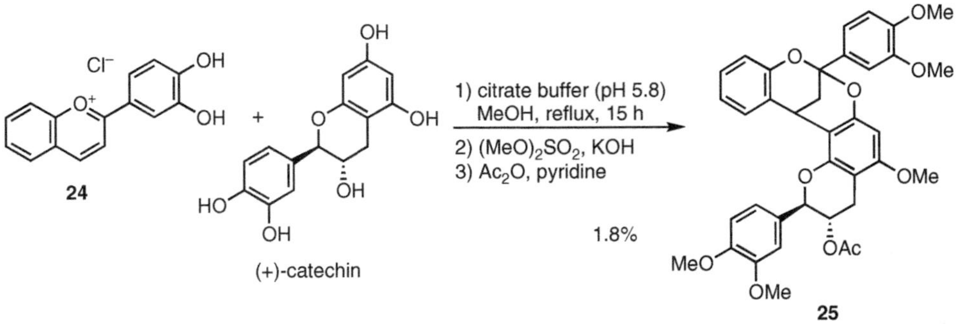

Figure 2.15 Early studies on annulation reaction of flavylium **24** with (+)-catechin.

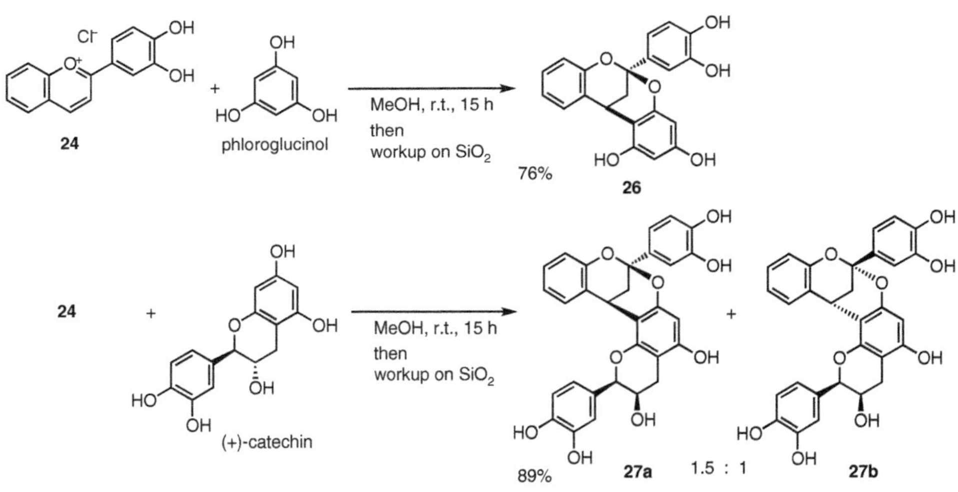

Figure 2.16 Annulation reaction by Kraus.

two annulation products **27a** and **27b** in 89% combined yield. Recently, another research group reported a similar protocol (Alejo-Armijo et al. 2018).

An asymmetric version of this approach appeared recently by using a salt of binaphthol-derived chiral phosphoric acid as a chiral phase transfer catalyst (Figure 2.17) (Yang et al. 2016). Reaction of flavylium salt **28** with phloroglucinol derivative **29** in the presence of **A**

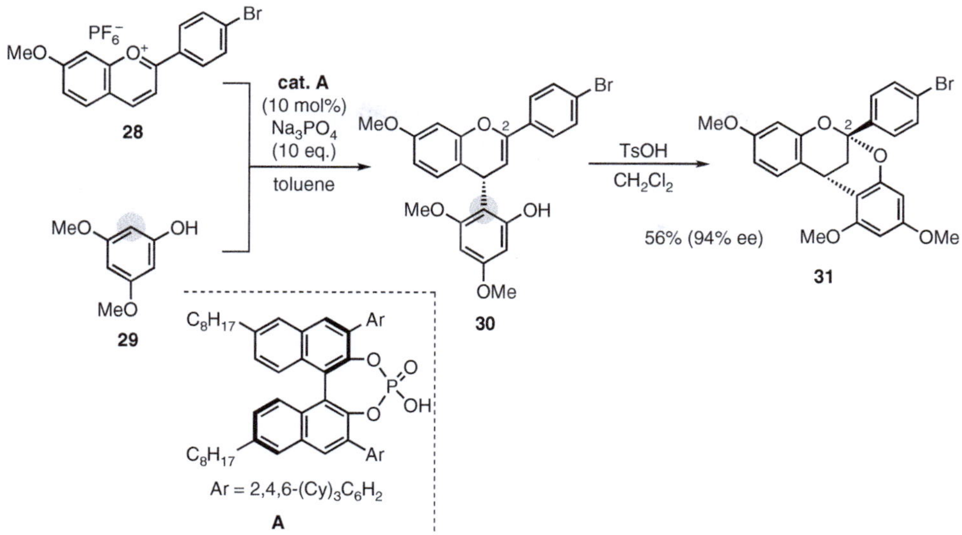

Figure 2.17 Asymmetric annulation approach.

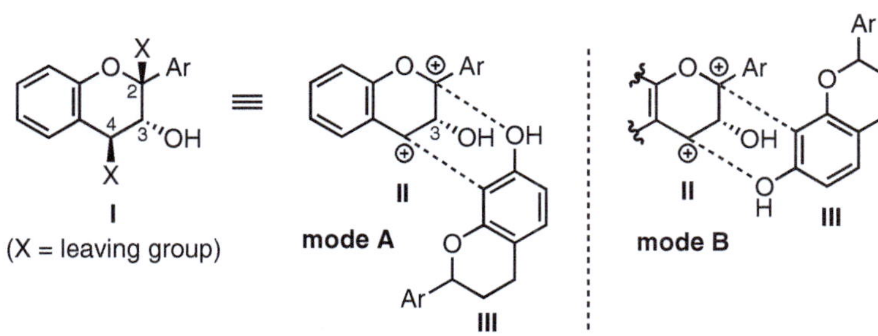

Figure 2.18 Strategy for stereoselective annulation.

as a catalyst (10 mol%) allowed the enantioselective nucleophilic attack of **29** to the C(4) position of **28** to give adduct **30**, which was treated with *p*-toluenesulfonic acid to give bicycle **31** in 56% yield with a high enantioselectivity (94% *ee*).

The present authors' group recently developed a novel concept in a *flavan annulation* (Figure 2.18) (Ito et al. 2014). Two requirements en route to the requisite dioxabicyclic skeleton include (i) design of a suitable precursor **I** to generate the dicationic species **II**, and (ii) regioselective dual bond formation with **III** via mode A, not vice versa (mode B). An additional requirement was the stereoselectivity. Thanks to the C(3) stereocenter in **II**, the annulation would proceed in a stereoselective manner.

As a dication equivalent, two oxygen-based leaving groups X at the C(2) and C(4) positions were envisioned, and the precursor **I** could be available by oxidation of flavan derivatives. Thus, the oxidation of epicatechin derivative **32** with DDQ in the presence of ethylene glycol gave 2,4-ethylenedioxy derivative **33** in 69% yield (Figure 2.19). The C(8) position of **33** was masked by a bromine atom, affording bromide **34** in excellent yield.

As a model study using the dication precursor **34,** the reaction with phloroglucinol derivative **35** was examined in the presence of $BF_3 \cdot OEt_2$ (Figure 2.20). Upon the reaction started

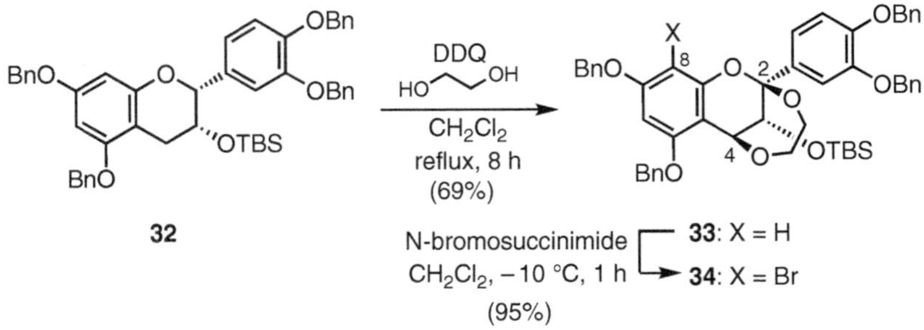

Figure 2.19 Synthesis of a 2,4-dioxy flavan derivative.

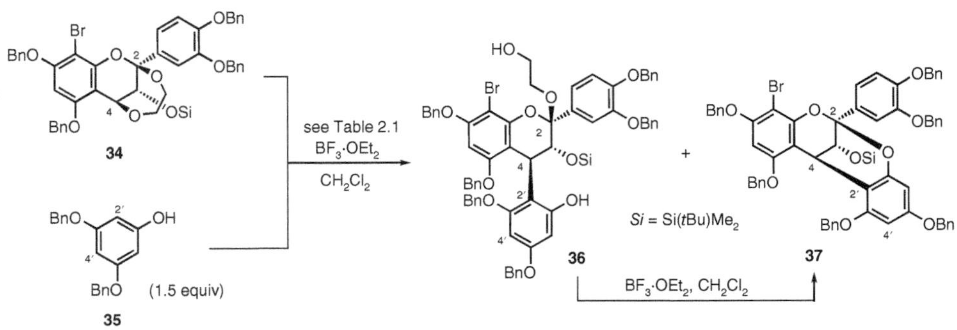

Figure 2.20 Model study for stereoselective flavan annulation.

Table 2.1 Results of flavan annulation.

run	time / h[a]	temperature / °C	product (yield)
1	2	−78 → −40	**36** (66%), **37** (20%)
2	3	−78 → −20	**37** (93%)

[a] Duration of the gradual warming.

at −78 °C followed by gradual warming, the starting material **34** was consumed at −40 °C, and two products were produced (run 1, Table 2.1). The major product was single-linked compound **36** with the C(4)–C(2) bond, whereas the minor product **37** had the desired dioxabicyclic structure. Importantly, these products, **36** and **37**, were the single stereoisomers, respectively. Re-exposure of **36** to the same conditions led to a smooth conversion into **37**. Thus, the annulation proceeded in a stepwise manner, starting with the formation of the C(4)–C(2) bond from the β-side to form **36** followed by the C(2)–O bond formation to give **37**. Indeed, by extending the reaction time and raising the temperature, the annulation of **34** and **35** went to completion, giving **37** in excellent yield (run two).

2.3.6 Total Synthesis

This section describes several completed syntheses of the oligomeric PAs with A-type linkages.

2.3.6.1 Synthesis of Diinsininol Aglycon (45)

Selenski and Pettus (2006) developed an efficient synthetic approach to bicycle **45**, corresponding to the aglycon of diinsininol (**46**), by exploiting the [3+3] annulation approach (Figure 2.21). As an electrophilic unit, flavylium **40** was prepared by the condensation of 2,4,6-triacetoxybenzaldehyde (**38**) and 3,4-dihydroxyacetophenone (**39**). On the other hand, the nucleophilic partner **44** was synthesized from styrene **41** and benzaldehyde **42**. Treatment of **42** with $LiAlH_4$ in the presence of $MgBr_2$ generated *o*-quinonemethide **A**, which underwent a [4+2]-cycloaddition with styrene **41**, giving the corresponding flavan **43** in 45% yield. DDQ-oxidation in the presence of H_2O followed by removal of the protecting groups afforded free flavanone **44** in 72% yield (two steps). The reaction of flavylium salt **40** with nucleophile **44** (6 equiv) proceeded under microwave irradiation at 120 °C, giving the annulated product **45** in 32% yield. The product was obtained as a single diastereomer, albeit in a racemic form.

2.3.6.2 Synthesis of Procyanidin A2 (3)

The protocol described earlier (Figure 2.20) was applied to the synthesis of procyanidin A2 (**3**) (see Figure 2.5), in which a nucleophilic monomer unit needed to be selectively protected (*vide infra*). The present authors have developed a *de novo* synthetic approach to the epi-type catechins (Figure 2.22) (Stadlbauer et al. 2012), relying on the *ortho*-metalation of aryl fluoride **I** (#1) and reaction with epoxy alcohol **II**, followed by an S_NAr oxycyclization of adduct **III** to give flavan **IV** (#2). This protocol enabled access to monomer unit **52** with a free hydroxy group at C(7) (Figure 2.23) (Ito et al. 2014). Starting from 1,3,5-trifluorobenzene (**47**), sequential substitutions with *t*-BuOK and BnOK gave mono-fluoride **48**. Regioselective lithiation of **48** followed by the reaction with epoxy alcohol **49** in the presence of $BF_3{\cdot}OEt_2$ gave adduct **50**. After the protecting group arrangement, the resulting alcohol **51** was subjected to the pyran-ring formation. Treatment of **51** with KH cleanly afforded the cyclized product, which was exposed to acidic conditions to give the monomer

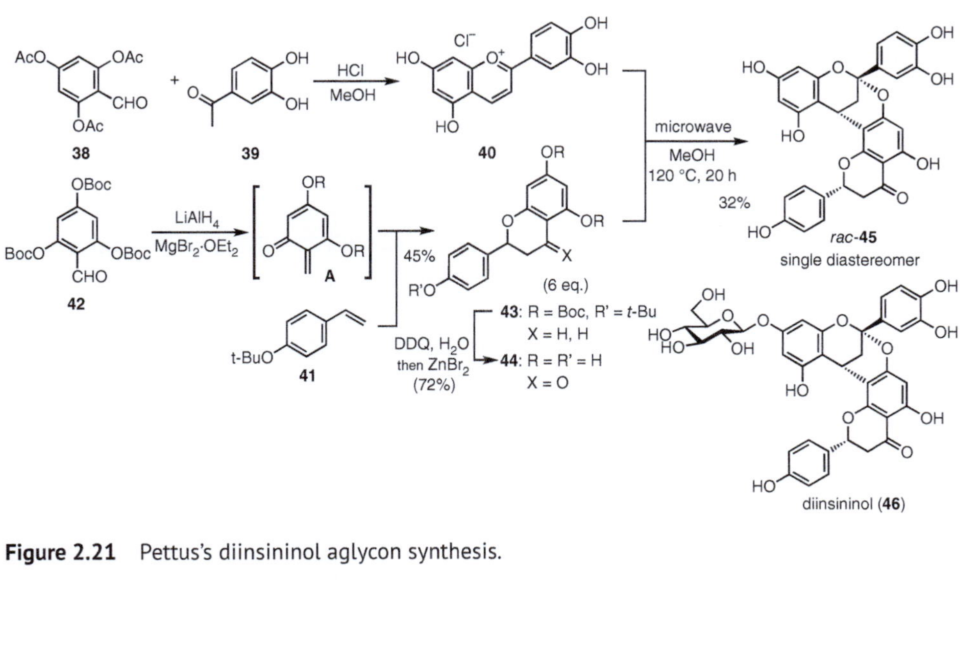

Figure 2.21 Pettus's diinsininol aglycon synthesis.

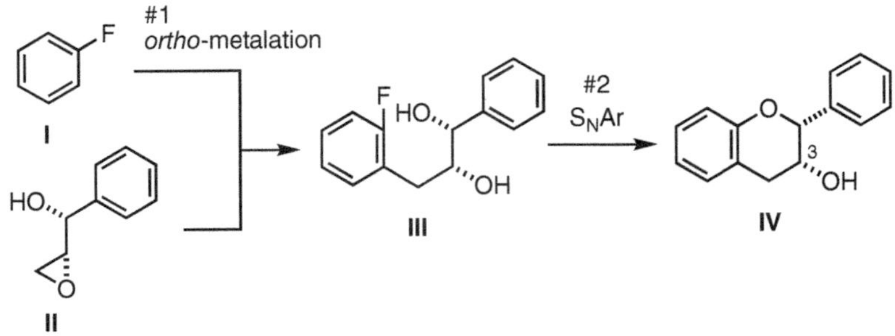

Figure 2.22 Strategy for monomer synthesis.

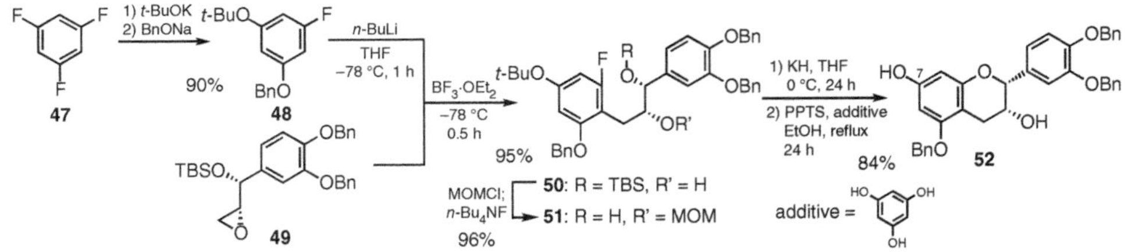

Figure 2.23 *De novo* synthesis of the C(7)-hydroxy monomer unit.

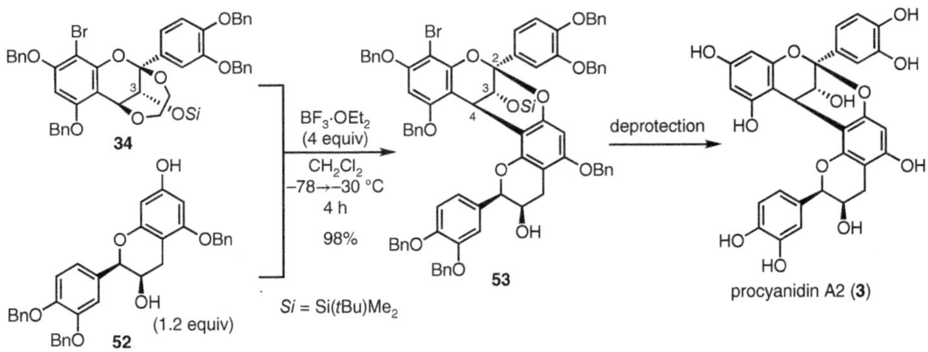

Figure 2.24 Total synthesis of procyanidin A2 via flavan annulation.

unit **52**, which was employed as the nucleophilic flavan unit in the following annulation (Figure 2.24). Upon treatment of a mixture of **34** and **52** with BF_3·OEt_2, the corresponding annulation product **53** was obtained in excellent yield. Notably, none of other regio- and stereoisomers was observed. As the α-face of the electrophilic flavan unit **34** is shielded by the C(3)-substituent, nucleophile **52** approached from the β-side to form the double linkages at the C(2) and C(4) positions. In addition, the nucleophile **52** selectively reacted at its C(8) position rather than the C(6) position. Finally, all protecting groups were removed to give procyanidin A2 (**3**).

More recently, it was found that the annulation could work even with a non-protected flavan unit, e.g. catechin or epicatechin, as a nucleophile (Figure 2.25) (Betkekar et al.

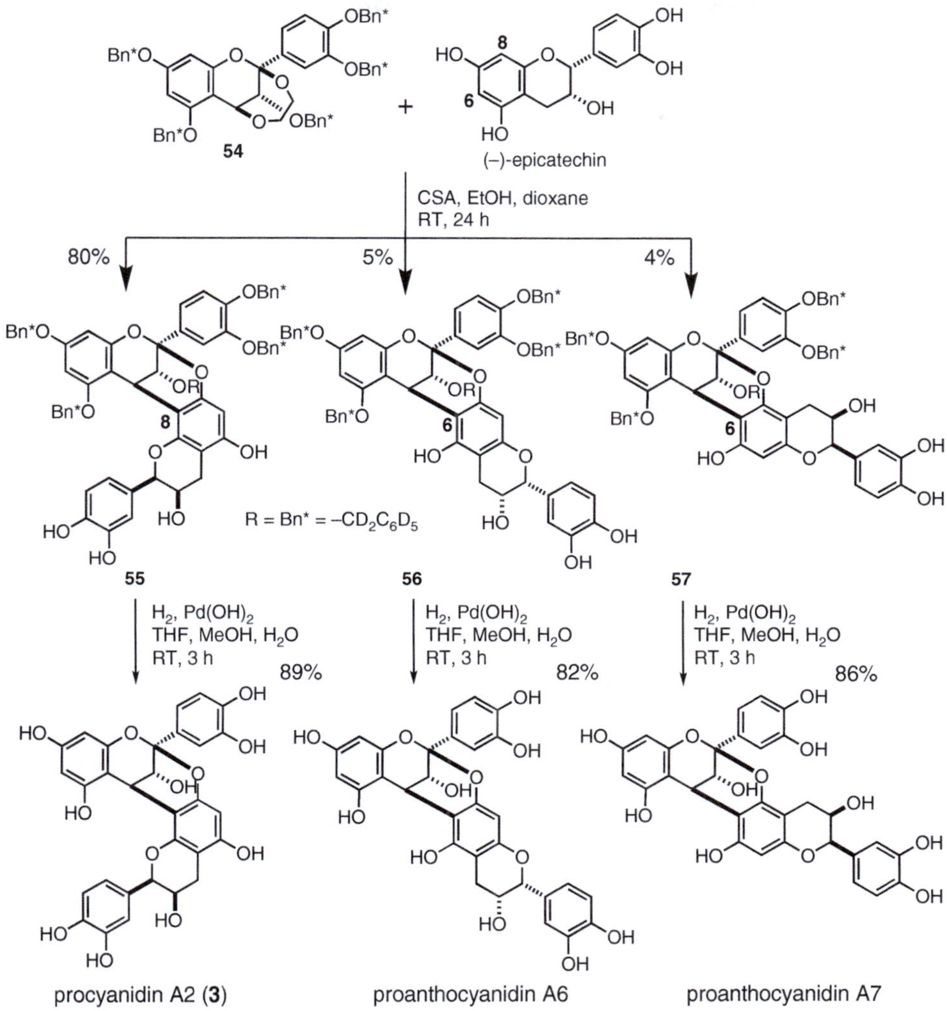

Figure 2.25 Annulation with a free flavan unit and syntheses of procyanidin A2, and proanthocyanidin A6 and A7.

2019). In a solvent mixture of ethanol and 1,4-dioxane, (–)-epicatechin was allowed to react with electrophilic flavan unit **54** in the presence of camphor sulfonic acid, affording annulation product **55** along with minute amounts of its regioisomers **56** and **57**. After chromatographic separation (silica gel) and deprotection, all products were converted to their free-forms, i.e. procyanidin A2 (**3**), proanthocyanidin A6, and proanthocyanidin A7.

These results indicate that the free flavan unit preferentially reacted at its C(8) position rather than the C(6) position. To address the origin of this C8/C6-regioselectivity, a DFT calculation was conducted on the putative Wheland intermediates **II** and **III**, generated by the reaction on the model substrate **I** with CH_3^+ (Figure 2.26). The result suggested that the C8-methyl intermediate **II** is energetically more stable over the C6-counterpart **III**.

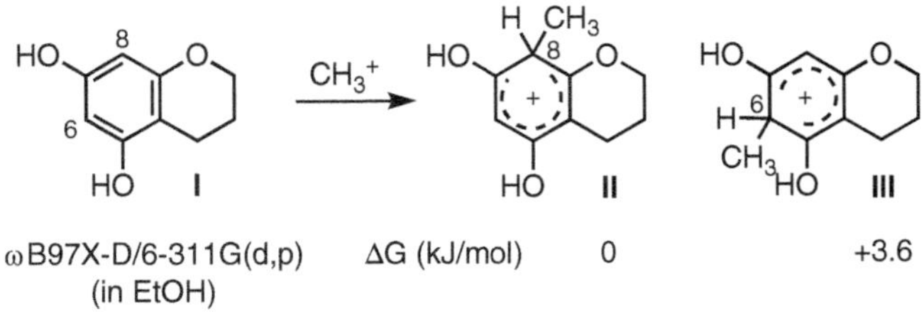

Figure 2.26 DFT calculations of the Wheland intermediates, **II** and **III**, derived from model compound **I**.

2.3.6.3 Synthesis of (+)-Cinnamtannin B1 (62)

This annulation method was also applied to the synthesis of a trimeric natural product, (+)-cinnamtannin B1 (**62**) (Figure 2.27) (Ito et al. 2014), which is known as a rare sweet polyphenol. To realize the convergent assembly of the trimeric scaffold, the orthogonal coupling strategy was adapted by employing monomers **34**, **58**, and **52**. Annulation of dioxy electrophilic unit **34** with sulfide **58** was carried out by using $BF_3{\cdot}OEt_2$ to give the coupling product **59** in 91% yield. The arylthio group remained intact under the hard Lewis acidic conditions. Next, the soft activation of dimeric sulfide **59** was conducted by using I_2 and Ag_2O, enabling the reaction with the nucleophilic epicatechin unit **52** to give trimer **60** in 96% yield. After two-step deprotection via **61**, (+)-cinnamtannin B1 (**62**) was obtained in 70% overall yield.

2.3.6.4 Synthesis of Selligueain A (63)

The coupling protocols were applied to the synthesis of selligueain A (**63**), isolated from the rhizomes of *Selliguea feei* (Figure 2.28) (Noguchi et al. 2018). Although this compound is interesting because of the pronounced sweetness, 35 times sweeter than sucrose, it had been obtained in only minute amounts from nature. A serious issue was that the trimers are comprised of two rare flavan monomer units, i.e. (−)-epiafzelechin (EZ) and (+)-afzelechin (AZ), which manifests a general bottleneck in the chemical synthesis of flavan oligomers – that is, the limited availability of their monomeric constituents. Note that commercially available flavan-3-ol is basically limited to (+)-catechin, and other congeners are scarce and could not be used as the starting materials.

The initial step in the assembly of these monomer units was the annulation of the top unit **64** and the middle unit **65** by using $BF_3{\cdot}OEt_2$ to give **66** as a sole product in 86% yield (Figure 2.29). Importantly, no formation of the C(4)→C(6)-linked products was observed, and the sulfide moiety remained intact under these hard Lewis acidic conditions. After detachment of the silyl group in **66**, the arylthio group in the resulting alcohol **67** was activated under soft Lewis acidic conditions to react with the bottom unit **68**. The annulation reaction proceeded the annulation, giving trimer **69** in 93% yield. It should be noted that the seemingly labile benzylic acetal moiety at the C(2) position of the top unit remained untouched. Finally, all benzyl groups and the bromine atom were removed by hydrogenolysis to give pure **63** in 85% yield.

2.3.6.5 Synthesis of Procyanidin A2 (3)

As another synthetic approach, Sharma et al. (2015) reported the *de novo* synthesis of a series of the dimeric PAs having an A-type structure, which relied on the Friedel–Crafts

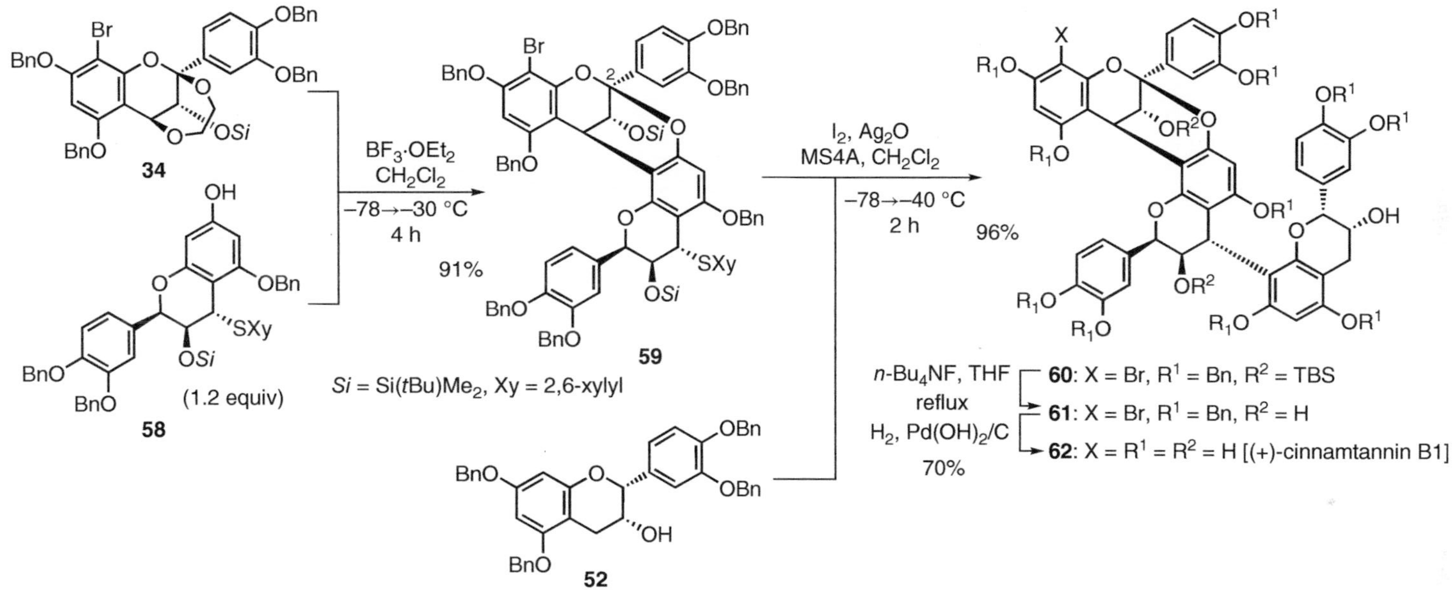

Figure 2.27 Synthesis of cinnamtannin B1 based on the orthogonal activation approach.

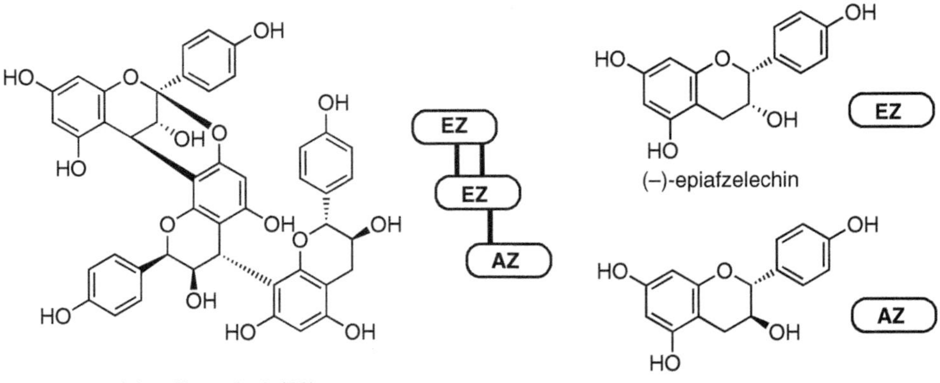

Figure 2.28 Structure of (+)-selligueain A, its monomeric flavan constituents, (+)-afzelechin (AZ) and (−)-epiafzelechin (EZ), and three monomeric synthetic units for orthogonal assembly.

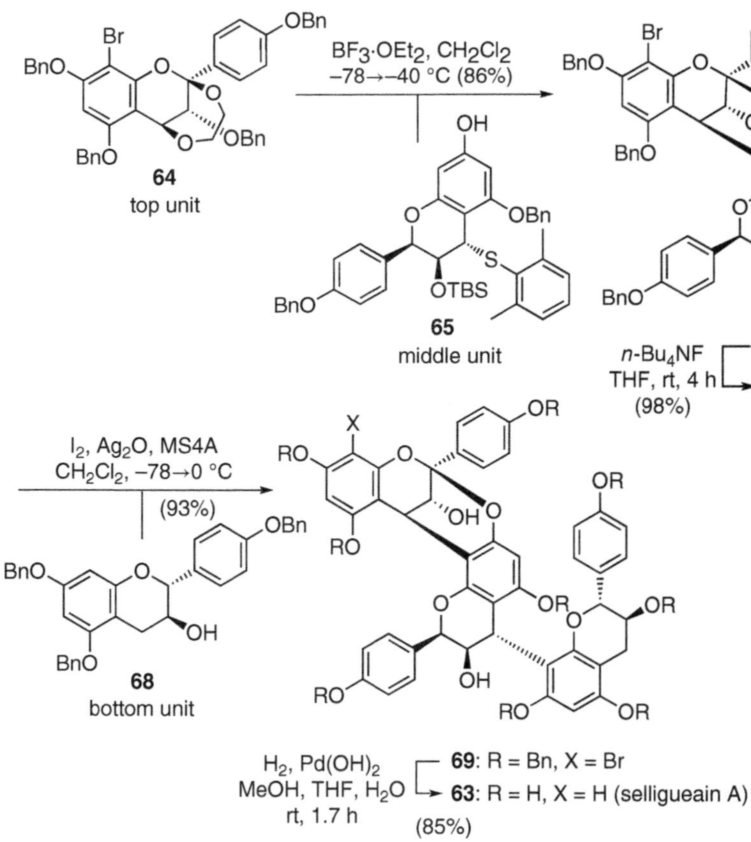

Figure 2.29 Orthogonal activation and synthesis of selligueain A.

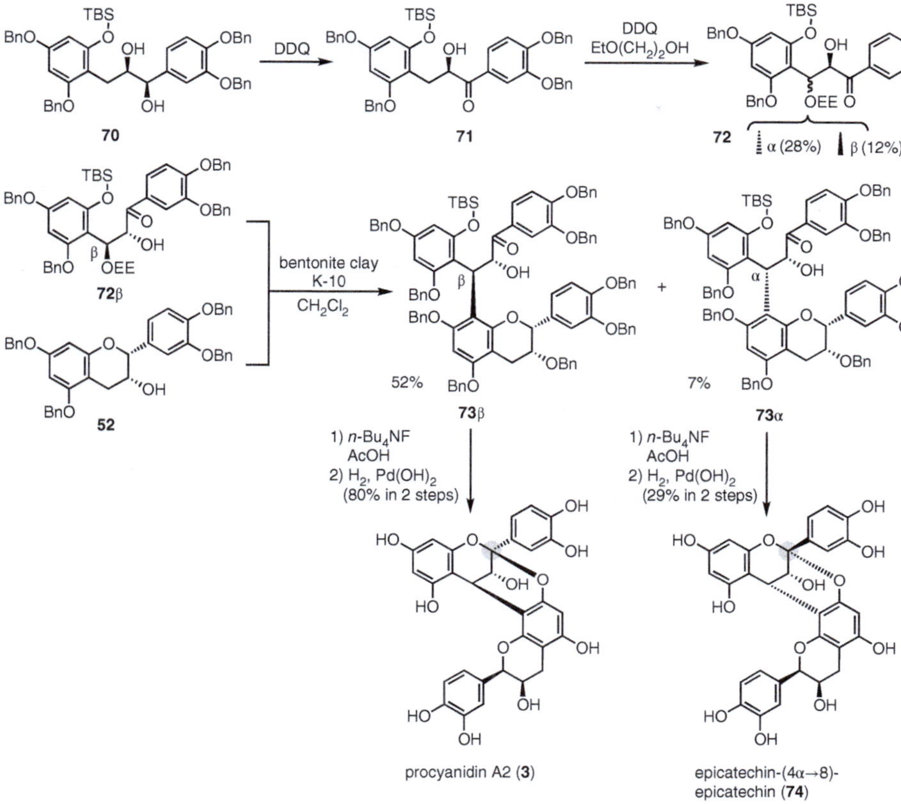

Figure 2.30 Synthesis of a series of dimeric PAs having A-type structure via the Friedel–Crafts reaction.

reaction of two monomeric units followed by internal acetalization (Figure 2.30). Successive DDQ oxidation of diol **70** with/without 2-ethoxyethanol gave the diastereomeric alkoxy ketones, **72α** and **72β**, in 28% and 12% yield, respectively.

The latter diastereomer **72β** was subjected to condensation with nucleophilic flavan unit **52**. Treatment of **72β** with **52** in the presence of bentonite clay K-10 gave the coupling products as a diastereomer mixture, which could be separated by silica gel column chromatography to give β-adduct **73β** and α-adduct **73α** in 52% and 7% yield, respectively. Subsequent two-step deprotection, involving desilylation and hydrogenolysis of the benzyl ethers, allowed a spontaneous intramolecular acetal formation to give procyanidin A2 (**3**) in 80% overall yield from **73β** and its bis-2,4-*epi* isomer **74** in 29% overall yield from **73α**, respectively.

2.4 Conclusion

This chapter has reviewed an emerging class of oligomeric PAs possessing A-type dual interflavan linkages, covering their structures, putative biosynthesis, and chemical synthesis. The research on this class of complex oligomers is still limited by the scarce availability of samples with high purity. For future progress, the role of organic synthesis is quite important, requiring the development of reliable synthetic methods, including (i) the flexible preparative methods of the flavan-3-ol monomeric constituents, and (2) facile means of constructing the [3.3.1]-bicyclic skeleton with control of the stereo- and regiochemistry. The latter process needs to be reiterative, in dealing with the targets having multiple A-type linkages. We hope this review will provide insights for further development and might serve as an invitation to researchers of various fields, attracting interest to this potentially exciting field.

References

Alejo-Armijo, A., Parola, A. J., Pina, F., Altarejos, J., and Salido, S. (2018). Thermodynamic stability of flavylium salts as a valuable tool to design the synthesis of A-type proanthocyanidin analogues. *The Journal of Organic Chemistry* 83: 12297–12304.

Balde, A., de Bruyne, T., Pieters, L., et al. (1995). Tetrameric proanthocyanidins containing a double interflavanoid (A-type) linkage from *Pavetta owariensis*. *Phytochemistry* 40: 933–938.

Betkekar, V.V., Harachi, M., Suzuki, K., and Ohmori, K. (2019). Syntheses of doubly linked proanthocyanidins using free flavan units as nucleophiles: insight into the origin of the high regioselectivity of annulation. *Organic & Biomolecular Chemistry* 17: 9129–9134.

Ito, Y., Ohmori, K. and Suzuki, K. (2014). Annulation approach to doubly linked (A-type) oligocatechins: syntheses of (+)-procyanidin A2 and (+)-cinnamtannin B1. *Angewandte Chemie International Edition* 53: 10129–10133.

Jacques, D., Haslam, E., Bedford, G.R., and Greatbanks, D. (1973). Structure of the dimeric proanthocyanidin A2 and its derivatives. *Journal of the Chemical Society, Chemical Communications* 518–520.

Jacques, D., Haslam, E., Bedford, G.R., and Greatbanks, D. (1974). Plant proanthocyanidins. Part II. Proanthocyanidin A2 and its derivatives. *Journal of the Chemical Society, Perkin Transactions* 1: 2663–2671.

Jurd, L. and Waiss, A.C. (1965). Anthocyanins and related compounds – VI. *Tetrahedron*, 21, 1471–1483.

Kondo, K., Kurihara, M., Fukuhara, K., et al. (2000). Conversion of procyanidin B-type (catechin dimer) to A-type: evidence for abstraction of C-2 hydrogen in catechin during radical oxidation. *Tetrahedron Letters* 41: 485–488.

Kozikowski, A.P. and Tückmantel, W. (2009). Methods for synthesizing the cocoa-derived oligomeric epi-catechins – observations on the anticancer activity of the cocoa polyphenols. In: *Recent Advances in Polyphenol Research* (eds. Daayf, F. and Lattanzio, V., 88–112. Wiley.

Kraus, G.A., and Geraskin, I.M. (2017). Rapid assembly of the procyanidin A skeleton. *Tetrahedron Letters* 58: 4609–4611.

Kraus, G.A., Yuan, Y., and Kempema, A. (2009). A convenient synthesis of type A procyanidins. *Molecules* 14: 807–815.

Makabe, H. (2013). Recent syntheses of proanthocyanidins. *Heterocycles* 87: 2225–2248.

Mayer, W. Goll, L., Arndt, E.M.V., and Mannschreek, A. (1966). Procyanidino-(−)-epicatechin, ein zweiarmig verknüpftes, kondensiertes proanthocyanidin aus *Aesculus hippocastanum*. *Tetrahedron Letters* 7: 429–435.

Morimoto, S., Nonaka, G., and Nishioka, I. (1985). Tannins and related compounds. XXXV. Proanthocyanidins with a doubly linked unit from the root bark of *Cinnamomum sieboldii* MEISNER. *Chemical and Pharmaceutical Bulletin* 33: 4338–4345.

Morimoto, S., Nonaka, G., and Nishioka, I. (1987). Tannins and related compounds. LIX. Aesculitannins, novel proanthocyanidins with doubly-bonded structures from *Aesculus hippocastanum* L. *Chemical and Pharmaceutical Bulletin* 35: 4717–4729.

Nam, J.W., Phansakkar, R.S., Lankin, D.C., et al. (2017). Absolute configuration of native oligomeric proanthocyanidins with dentin biomodification potency. *The Journal of Organic Chemistry* 82: 1316–1329.

Noguchi, Y., Takeda, R., Suzuki, K., and Ohmori, K. (2018). Total synthesis of selligueain A, a sweet flavan trimer. *Organic Letters* 20: 2857–2861.

Nonaka, G.-I., Morimoto, S., Kinjo, J.-E., et al. (1987). Tannins and related compounds. L structures of proanthocyanidin A-1 and related compounds. *Chemical and Pharmaceutical Bulletin* 35: 149–155.

Nonaka, G.-I., Morimoto, S. and Nishioka, I. (1983). Tannins and related compounds. Part 13. Isolation and structures of trimeric, tetrameric, and pentameric proanthocyanidins from cinnamon. *Journal of the Chemical Society, Perkin Transactions* 1: 2139–2145.

Ohmori, K., Shono, T., Hatakoshi, Y., et al. (2011). Integrated synthetic strategy for higher catechin oligomers. *Angewandte Chemie International Edition* 50: 4862–4867.

Ohmori, K., Ushimaru, N. and Suzuki, K. (2004). Oligomeric catechins: an enabling synthetic strategy by orthogonal activation and C(8) protection. *Proceedings of the National Academy of Sciences of the United States of America* 101: 12002–12007.

Oyama, K.-I., Kuwano, M., Ito, M., et al. (2008). Synthesis of procyanidins by stepwise- and self-condensation using 3,4-cis-4-acetoxy-3-*O*-acetyl-4-dehydro-5,7,3',4'-tetra-*O*-benzyl-(+)-catechin and (–)-epicatechin as a key building monomer. *Tetrahedron Letters* 49: 3176–3180.

Pomilio, A., Müller, O., Schilling, G. and Weinges, K. (1977). Zur kenntnis der proanthocyanidine, XXII. Über die konstitution der kondensationsprodukte von phenolen mit flavyliumsalzen. *Justus Liebigs Annalen der Chemie* 1977: 597–601.

Saito, A, Mizushina, Y., Tanaka, A. and Nakajima, N. (2009). Versatile synthesis of epicatechin series procyanidin oligomers, and their antioxidant and DNA polymerase inhibitory activity. *Tetrahedron* 65: 7422–7428.

Selenski, C., and Pettus, T.R. (2006). (+/–)-Diinsininone: made nature's way. *Tetrahedron* 62: 5298–5307.

Sharma, P.K., Romanczyk, L.J., Jr., Kondaveti, L., et al. (2015). Total synthesis of proanthocyanidin A1, A2, and their stereoisomers. *Organic Letters* 17: 2306–2309.

Stadlbauer, S., Ohmori, K., Hattori, F., and Suzuki, K. (2012). A new synthetic strategy for catechin-class polyphenols: concise synthesis of (–)-epicatechin and its 3-*O*-gallate. *Chemical Communications* 48: 8425–8427.

Van Rooyen, P.H., and Redelinghuys, H.J.P. (1983). Crystal structure and molecular conformation of proanthocyanidin-A2, a bitter substance in litchis (*Litchi chinensis* Sonn.). *South African Journal of Chemistry* 36: 49–53.

Weinges, K., Kaltenhäuser, W., Marx, H.-D., et al. (1968). Zur kenntnis der proanthocyanidine, X. Procyanidine aus früchten. *Justus Liebigs Annalen der Chemie* 711: 184–204.

Weinges, K. and Theobald, H. (1971). Synthese des 6-phenyl-12H-6.12-methano-dibenzo[*d*.g][1.3]-dioxocins, einer modellsubstanz für die $C_{30}H_{24}O_{12}$-procyanidine. *Justus Liebigs Annalen der Chemie* 743: 203–206.

Xia, L., Cai, H., and Lee, Y.R. (2014). Catalyst-controlled regio- and stereoselective synthesis of diverse 12H-6,12-methanodibenzo[*d*,g][1.3]dioxocines. *Organic and Biomolecular Chemistry* 12: 4386–4396.

Yang, Z., He, Y. and Toste, F.D. (2016). Biomimetic approach to the catalytic enantioselective synthesis of flavonoids. *Journal of the American Chemical Society* 138: 9775–9778.

Yano, T., Ohmori, K., Takahashi, H., et al. (2012). Unified approach to catechin hetero-oligomers: first total synthesis of trimer EZ-EG-CA isolated from *Ziziphus jujuba*. *Organic and Biomolecular Chemistry* 10: 7685–7688.

3

Answering the Call of the Wild: Polyphenols in Traditional Therapeutic Practice

Mary Ann Lila[1] *and Kriya Dunlap*[2]

[1] *Plants for Human Health Institute, Department of Food Bioprocessing and Nutrition Sciences, North Carolina State University, Kannapolis, USA*
[2] *Department of Biochemistry, University of Alaska Fairbanks, Fairbanks, USA*

3.1 Introduction

Polyphenols, ubiquitous and exceptionally human health–relevant plant-derived compounds, have had a central role in ethnomedicine (traditional herbal medicine) and subsistence diets throughout history, even prerecorded history (Buttriss 2012; Schmidt and Klaser Cheng 2017). Foraging wild edible plants for foods and therapeutic applications is an ancient tradition that was considered obsolescent by most modern urban consumers, but now is enjoying a resurgence in interest. Thanks to some new insights into the composition and resiliency of endemic wild plants, and the quest for novel flavors and textures—especially in niche, fine dining establishments—wild foraging has become "cool" again. Foraged cuisine and indigenous ingredients are coveted, trendy options in high-end restaurants, where the tannic astringency of wildcrafted leaves and fruits can complement blander foods or flavor botanical-infused gins and vodkas, and consumers are keen to experience new foods that they perceive as healthier, natural, and unadulterated (Reyes-Garcia et al. 2015; Schatzker 2015; Soukand 2016; Sebag-Montefiore 2017; Pinela et al. 2017). But what is the real history of human use for polyphenol-rich biologically active edible wild plants? Where were they discovered and which species were preferentially targeted as medicinally relevant?

3.2 The Wildcrafting Tradition

Our hunter-gatherer ancestors were opportunistic omnivores, combining indigenous plant-based foods and ingredients and wild-caught fish and game. Today, wildcrafted plant foods are still part of the cultural and genetic heritage in many nonurbanized regions of the world. They can form the mainstay of the traditional diet, complement staple agricultural

Recent Advances in Polyphenol Research, Volume 7, First Edition. Edited by Jess Dreher Reed, Victor Armando Pereira de Freitas, and Stéphane Quideau.

foods, help to alleviate poverty, and provide food security, especially in rural villages during times when food is scarce (Soukand 2016; Pinela et al. 2017).

Historically, wild plants, including polyphenol-rich berryfruits, were put to use as both food and medicines as well as for dyes and preservatives in the Americas, Europe, Africa, and Asia, and in most cases, responsibility for wildcrafting fell to the female villagers (Densmore 1974; Alarcomicronn et al. 2015; Soukand 2016; Pinela et al. 2017). Some of the earliest historic records from Native Americans and Alaska Natives not only describe the sources and preparation of polyphenol-rich food sources, but also stress the importance of collecting the wild plants in proper season and stage of development, in order to ensure bioactive efficacy, palatability, and resource sustainability (Burns Kraft et al. 2008; Wapner 2012; Allkin 2017). Traditional ecological knowledge and folk medicine provide an ethnobotanical record that guides current discovery efforts (Joseph et al. 2014; Pinela et al. 2017).

Regular consumption of plant-based diets supports health maintenance by regulating the balance between cellular (endogenous) and dietary (exogenous) antioxidants to maintain cell redox homeostasis, (Al-Gubory and Laher 2018) whereas plant medicines are generally prepared by extraction and concentration of the active compounds from medicinal species prior to administration either orally or topically to humans (Moerman 1996; Allkin 2017; Pinela et al. 2017). Herbal teas, for example, were social/recreational beverages used without restriction, but more concentrated decoctions were taken only for a limited time to treat a health condition such as a digestive disorder or respiratory symptoms.

Phytotherapy, the integration of traditional practices and scientific method, supports the biological basis for traditional ecological medicines, and traditional knowledge is increasingly evolving as evidence-based medicine. In the past, the generally potent antioxidant capacities of polyphenolic resources were credited with the multifaceted uses of these plants, however, research now suggests that the diverse health benefits coincident with plant polyphenol intake are more related to their anti-inflammatory, prebiotic, vasodilation and signaling pathways' influences (Cordova and Sumpio 2009; Margina et al. 2015). In each case, specific dosages and controlled periods of intake are recommended, often combined with cultural healing rituals (Pinela et al. 2017). Interestingly, Native Americans used only a small percentage of available polyphenol-containing flora as medicines, and with the exception of several genera (i.e. *Ribes*, *Vaccinium*, *Malus*, *Fragaria*, *Rubus*, *Crataegus*, *Prunus*, and *Arctostaphylos*) the species used for medicines excluded those used for foods. This makes some sense, as medicines can be toxic and proper attention to homeopathy implies that the lowest possible effective doses should be used, as is the case for prescription synthetic drugs (Moerman 1996). Even so, there is some recorded overlap. For example, American tribes used wild strawberry fruits as a medicinal for diarrhea, stomach pain, and irregular menstruation, as well as a dietary item (Schmidt and Klaser Cheng 2017).

Plant medicines were used to treat a wide range of human symptoms, including liver conditions, urinary infection, skin problems, inflammation, toothache, diabetes, flu, and bronchitis. Other plant preparations were valued primarily as disinfectants. Historical records reveal that *single plants* could serve as resources for prescribed treatments against a *wide range of diseases*, because the diverse phytochemical profiles in wild herbal species enable single plant species to have bioactivities relevant against multiple potential disease targets.

Edible medicinal and dietary polyphenolic resources are usually processed/preserved today, for year-round use, by freezing the plant material. In traditional cultures, where freezing is (or in the past, *was*) not an option, materials are preserved and used after

(i) desiccation, (ii) preparation of infusions, (iii) decoction and maceration, or (iv) tincture (Pinela et al. 2017). Dried plant materials were either ground into a powder form or pressed between stones into dried cakes. In some cases, berries were pressed and pounded into wild game meats before co-drying (with a low fire or just hot sun) into a protein/fat/polyphenol-rich high-energy food called pemmican or *wasna*, used in Native American Lakota and Cree cultures. This method not only preserved the polyphenol-rich berries for year-round consumption, but served as a natural antimicrobial additive for the meats. Infusions were prepared by soaking dried leaves or powders in (usually boiling) water so that the water-soluble polyphenols were released (tea-like preparations). Decoction/maceration is applied especially to tougher plant tissues (woody plant materials), which can be cut into small sections and crushed, then boiled in water for 10–20 minutes to fully extract the phytochemicals from the intractable matrix. When the volume of liquid has been reduced and cooled, additional water can be added and the process repeated. The replicated simmering and boiling cycles can concentrate some of the more stable phytoactive components. Tincture involves extraction of plant materials with alcohol, and storage in the same medium. A still-popular preservation method in Alaska and northern Canada is the preparation of Eskimo ice cream or *agutuk*, which is made by whipping fresh wild berries into animal fats (caribou, moose, or walrus tallow, or seal oil). Copious amounts of sugar are typically added (and sometimes fish are added) to the shelf-stable mixture, which is most often stored in a freezer prior to consumption (Wapner 2012).

The Native American traditional diet was one that many nutritionists would consider a healthy gold standard (consisting of fruits/berries, lean meat, fish, and wild vegetables), and the active hunter/fisher/gatherer lifestyle contributed to lean body mass. Polyphenol-rich plant intake contributed to the very low incidence of diabetes and other pathologies of metabolic syndrome in Native communities. The forced relocation of Native American tribal communities to reservations in the 1800s separated people from their traditional food sources, and made them dependent on government rations of high-carbohydrate commodity foods (Goetz 2012). Consequently, the modern incidence of obesity is now 1.6 times higher in Native communities than in the general population, and health conditions including diabetes and heart disease are running rampant (American Diabetes Association 2018). Poverty, lack of access to healthy food options (food deserts), and increasing dependence on highly processed, nutritionally devoid staple foods have contributed to the conditions. Declining Native health status is the impetus behind the resurgence of interest in re-examining wildcrafted subsistence foods, and re-educating tribal youth on the health-relevant attributes of traditional diets and plant-based medicines (Burns Kraft et al. 2008; Kellogg et al. 2010; Schreckinger et al. 2010; Flint et al. 2011; Wapner 2012; Joseph et al. 2014; Kellogg et al. 2016).

3.3 How Wildcrafted Edible Plants Differ from Agricultural Commodities

Obviously, wildcrafted foods are more time consuming to source and to harvest as opposed to simply stopping by a grocery store. Fruits and vegetables foraged in the wild *may,* to some consumers, be less aesthetically appealing than typical commercial market produce

(Soukand 2016; Pinela et al. 2017). For example, wild berryfruits may be smaller and less sweet, have larger, harder, and denser seeds, and provide lower yield per plant; root crops grown in unprepared soil will not be uniformly shaped or sized, fruit or vegetable produce that has not been sprayed routinely with pesticides may exhibit some insect or pathogen damage. However, *it is the adversity of the harsh unprotected wild growing environment that specifically provokes accumulation of redundant, overlapping polyphenolic profiles, which effectively protect the plant from the ravages of nature* (Wynn and Fougere 2007; Wapner 2012; Li et al. 2016; Pinela et al. 2017). The impetus behind a plant's production of antioxidants is to counter the oxidative stressors in the natural habitat (Li et al. 2016). In addition, the unbuffered environmental pressures in the wild cause spatiotemporal variation in the chemical profiles of the plants (Dhami and Mishra 2015). Dar et al. (2017) hypothesized that in the wild, perhaps nature has preselected phytochemicals that influence health (primarily plant health) and then may have specific metabolic roles for all living things that consume or interface with those plants. Farm-cultivated plants are bred to encourage higher yields of the edible plant part, but generally have less energy to expend on production of expansive root systems or generation of secondary compounds. Wild plants maintain a much higher level of genetic diversity than domesticated cultivars, which is in part responsible for the rich phytochemical multiplicity accumulated in the edible fruits and foliage, and is the impetus behind some current efforts to breed back key traits from wild relatives into farmed plants (Zhang et al. 2016).

Wild plants demonstrate pharmacologically unique activities, and a plethora of phytochemical constituents with lesser pharmaceutical activity play roles in augmenting the activity of primary active constituents (Wynn and Fougere 2007; Dhami and Mishra 2015). Traditional medicine has rarely included identification of the specific bioactive components in the whole plant or whole wild fruit extract, in part because these recognized interactions (additive, concomitant, antagonistic, or synergistic) between the myriad phytochemical constituents potentiate their bioactivity once ingested by animals (Phan et al. 2018). The presence of multiple recognized biologically active constituents combined with a diversity of other phytochemicals with lesser (or moderating) bioactivities allows for a wide range of therapeutic coverage from wildcrafted botanical drugs; in herbal medicine, polypharmacy is *de rigueur*. The potency of wildcrafted medicines is linked to these interactive phytochemical effects, as a single isolated compound from the plant will not be as biologically active as a crude or semipurified extract that retains the potentiating interactions. Herbals have nutritional and pharmaceutical elements that interact with one another polyvalently; thus the clinical effects may have greater depth and breadth than those affected by synthetic drugs (Wynn and Fougere 2007; Joseph et al. 2014). Wild polyphenol-enriched fruits have been noted for potent antiviral, antimicrobial, anti-inflammatory, cognition enhancement, and cancer chemoprotective capacities, in addition to robust antioxidant activities, that generally exceed levels in cultivated fruits (Li et al. 2016).

Combination effects (additive or synergistic interactions, as well as antagonisms) that impact human health are widely cited for natural product compounds, although notoriously difficult to precisely characterize (Caesar and Cech 2019). Only recently, the crucial role of the gut microbiome in catabolizing polyphenolic compounds from plant foods, generating biologically active metabolites that return to circulation, has been documented, which has enriched our understanding of polyphenolic metabolite bioavailability (Lila et al. 2016). In terms of the potentiating interactions, *pharmacokinetic synergy* refers to the situation when one phytoactive component may enhance or alter intestinal absorption, metabolism, distribution, or elimination of another component (resulting in a change in concentration of the active component at the therapeutic target in the body). For example, herbal medicines with fibers, mucilage, or tannins can alter the absorption of other phytochemicals. *Pharmacodynamic synergy* refers to the case when two or more compounds interact with a single therapeutic target at a receptor site (Wynn and Fougere 2007).

Three seismic changes in the food we eat have occurred over history: the discovery of cooking, the emergence of agriculture/cropping, and the invention of strategies for processing and preserving foods (Krebs 2013). The shift from hunter-gatherer of wild foods to consumption of agricultural produce allowed the human population to increase and for greater socialization to occur (as foraging for food in hunter-gatherer societies can take on average nearly seven hours of the waking day) (Soukand 2016), but the net effect on human health initially was negative. Early farmers suffered stunting (a form of malnutrition) and higher tooth decay than wild foragers, likely due to the less varied diet higher in farmed grains (carbohydrates), but as agriculture became better established and more efficient, it provided a more reliable and stable, easily stored food supply for humans (Krebs 2013). However, agriculture's emergence has dramatically changed the nutritive and phytoactive content of plant foods.

There are relatively few direct, side by side published compositional (or efficacy) comparisons of wild versus cultivated plants, in part because the genotypes of cultivated plants have for the most part been selected and bred for adaptation to a large-scale cropping regime, and typically diverge widely from parent genotypes endemic in wild settings. A recent study analyzed the differences in mineral composition of wild, naturally growing and cultivated blueberries (*V. myrtillus* versus *V. corymbosum*, respectively) by inductively coupled plasma optical emission spectrometry, and reported that the wild species were higher in five minerals (Ca, Na, Mg, Mn, and Zn) while cultivated plants were higher in Fe and Cd, which was related to the composition of the soil at the farm site (Drozdz et al. 2018). However, it was necessary to examine *two distinct species* in order to compare wild to cultivated blueberries. Another recent study compared mineral composition of wild and cultivated blueberries and cranberries, but had to rely on comparisons of fruits and leaves of two *different* blueberry species (*V. myrtillus* and *V. corymbosum*) and two *different* cranberry species (*V. oxycoccos* and *V. macrocarpon*) to determine their statistically distinct micronutrient compositions in wild versus cultivated, respectively (Karlsons et al. 2018). Wild genotypes of *Vaccinium* species in general demonstrated higher health-relevant bioactive potencies than cultivated selections in other studies (Braga et al. 2013). A wild banana species had higher phenolic and tannin content, and higher antioxidant capacities measured in three distinct assays than a commercial banana variety (Sasipriya et al. 2014), and wild strawberries recorded antioxidant capacity three times higher than cultivated strawberry genotypes (Ozgen et al. 2007). With few exceptions, studies have reported higher phytochemical content and bioactive potency for wild genotypes as compared to cultivated fruit selections, but in nearly all cases the species or genotypes collected in the wild were different than those produced under cultivation (Li et al. 2016).

3.4 Animal Mimickry/Zoopharmacognosy

An evolutionary relationship between plants and mammals has conditioned the unique physiological/biomedical effects of wild plants (Wynn and Fougere 2007). Traditional ecological knowledge (as held in indigenous tribal communities) asserts that at least some animals instinctively know how to treat their own injuries or ills and self-medicate using herbs growing in the wild. Consequently, traditional herbalists routinely use judicious observations of the behavior of sick or distressed wild animals to locate new potential human remedies derived from wild plants (Engel 2007; Wynn and Fougere 2007). Native Americans, as well as tribal groups throughout the world, have observed animal feeding behavior and used these clues to guide their own use of polyphenol-rich plants as medicines for specific ailments. For animals, is it assumed that nutritional wisdom or use of plant resources for healing is a combination of post-ingestive hedonic feedback and individual learning. Animals can use hedonic feedback to find ways of removing the unpleasant sensations of disease and injury. Some research has asserted that there is a purgative effect on the animal's intestines that helps alleviate parasitic infections (see details below), whereas other theories maintain that the natural antimicrobial/wound-healing effects provided by plant polyphenols are impetus for a wild animal's tendency to seek berries or other polyphenolic resources, especially after they've incurred defensive wounds.

Zoopharmacognosy is defined as the use of plant medicines/medicinally active herbs by animals—specifically when an animal seeks out a plant that is not normally part of its diet (and may have no nutritive value) in order to cure injury or disease (Moerman 1996). Mammals and even birds and insects are able to self-medicate using plant resources for both physical and psychological treatments. Native Americans revered the bear as one of the most advanced animals in terms of self-medication, as they use their clawed paws to dig for medicinal root crops (Densmore 1974).

Native American traditional ecological knowledge observed that animals tend to "self-regulate" tannin content via the diet (Moerman 1996; Engel 2007). Feeding deterrents (like tannins) are deliberately sought out by animals when the bioactive benefits outweigh the adverse tastes. For example, wild deer deliberately select plants not with very low tannin content, but those with moderate tannin content, as an adaptive taste preference, presumably because of the impact of the tannins on intestinal parasites. Repeated evidence suggests that animals attempt to self-regulate tannin consumption to optimal levels. Similarly, the Asiatic two-horned rhinoceros periodically will consume copious amounts of mangrove bark (tannin-rich) to the point where their urine turns orange, to self-medicate against endemic dysentery, just as a common medication for human dysentery (clioquinol) has a 50% tannin content (Engel 2007). During the coevolution of plants and animals, animals developed receptors for plant-produced bioactive chemicals (benzodiazepine, opioid, and vanilloid), which had direct effects on the sensations of pain or of consciousness; the animal-to-plant relationships provided a template by which herbalists could treat human patients for seizures, anxiety, and pain.

Natural selection has streamlined a variety of animal behavioral health maintenance strategies. These innate behavioral strategies are well documented for wild animals and although the capacity is partially lost in domesticated animals, the latter have also been observed to self-medicate. It has been noted that Native Americans who depended on their domesticated animals and observed their behaviors (e.g. Plains tribes and horses) had greater knowledge of plant medicine than other tribes (Stowe 1976). Farmers in New Zealand have noted that when captive deer grazed on forage containing tannin-rich plants such as chicory, there was significantly less need to administer chemical de-wormer. Parasitized deer and lambs selectively choose bitter and astringent chicory, which reduces their parasite loads. Consumption of condensed tannins increases sheep lactation, wool growth, and live weight gain by reducing the burden of internal parasites (Wynn and Fougere 2007).

Enthusiastic consumption of other polyphenol-rich plants, including wild berries, is a typical behavior for dogs (Sueda et al. 2008). Anecdotal evidence suggests that dogs will mimic humans when foraging for berries and harvesting from gardens. Dogs apparently love all types of berries—blueberries, strawberries, blackberries, and raspberries have all been scavenged, often to the chagrin of their owners, as well as peas, fallen apples, and corn on the cob from gardens. Sled dogs routinely accompany their owners during wild berry harvesting in Alaska, as a safeguard to keep foraging bears at bay, and these dogs routinely scavenge the berries from the lower branches in a berry stand. In a formal study on the antioxidant status of sled dogs fed wild blueberries, berries were initially mixed into the kibble and fed to the treated animals; however, it soon became apparent that the dogs preferred the berries and if placed on top of the kibble, they'd eat the berries first (Dunlap et al. 2006).

While it has been suggested that domesticated dogs are more omnivorous than wild animals, scat analysis has shown that wolves consume plants as well (Engel 2007) (Figure 3.1). As facultative carnivores, dogs and wolves do not have a physiological requirement for plant carbohydrates, but plants in the canine diet can play a role in health maintenance. Fruits are a vitamin and mineral source for the animals as well as a fiber source. While canines lack

Figure 3.1 Wild wolf feeding on Alaskan salmonberries. *Source:* Illustration by Ginger Dunlap.

salivary amylase (to break down plant carbohydrates) they do produce the enzyme through the pancreas and small intestine. Scat analysis and anecdotal observations of hunters and trappers indicate that fox, coyote, moose, and wolf typically have plant fiber hulls and berry seeds/skins in their feces. Fiber, both soluble and insoluble, may be a primary benefit to the animal of plant consumption, and is an important additive in commercial companion-animal feed. Plant-eating behavior is coincident with meat consumption, and does not occur only in the absence of animal prey.

Grizzly and black bears in Alaska are also notorious for copious consumption of wild berries, necessitating that Alaska Native and aboriginal Canadian berry harvesters must carry firearms for protection (as well as their domesticated sled dogs). Hunters have reported that during berry ripening season, grizzly bears actually have blue coloration from the berries interlaced in their fat deposits, and the meat acquires a sweeter flavor (Dr. Arleigh J. Reynold, DVM, PhD, University of Alaska Fairbanks, 2018, personal communication, 10 August).

Nonhuman primates frequently make use of plants and other natural resources to self-medicate. The most frequent observations are ingestion of natural products to purge intestinal parasites. Chimpanzees in Tanzania have been observed to fold, and then swallow, whole *Aspilia* spp. leaves without mastication; there is no known nutritional value but the undigested rough textured leaves help to expel parasitic worms into the feces. Similar cases where monkeys have ingested certain plants to dislodge parasites have been recorded. Primate behavior of this kind has helped to guide scientists to previously unscreened plant material that may have antimicrobial or antiparasitic properties. One very intriguing recent discovery with primates involves lemurs in Madagascar who deliberately chew on and vigorously rub millipedes, causing them to release secreted toxins in defense. Only then, once the toxins containing benzoquinones are released, the lemurs eat the crushed insects and rub them on their genitalia in order to kill parasitic worms (Bittel 2018).

3.5 Probing the Mechanisms Behind Polyphenol-rich Traditional Medicines Bioactivity

Traditional ecological knowledge concerning the medicinal use of polyphenol-rich plants reaches back many hundreds and even thousands of years, but the naturopathic doctors and healers who routinely rely on wildcrafted plants are frequently loath to share their knowledge with anyone from outside the native communities. The reason for this reluctance has been the exploitation (and resultant overharvesting and/or theft) of native resources by outsiders in the past, which has undermined the authority and historical access of traditional healers (Andrae-Marobela et al. 2012). The reductionist paradigm of Western science marginalizes the holistic approaches of wildcrafted medicine, and was imposed on top of indigenous knowledge systems during colonialism, which has made "bioprospecting" a disrespected term in the opinion of many elders in native tribal groups (Kellogg et al. 2010).

Harvest, attempted commercialization, and even in-depth analytical research on indigenous polyphenol resources carries obligations related to ethics, equitable benefit sharing, and intellectual property, covered in part by the CBD (Convention on Biological Diversity, a multilateral international treaty opened for signature in 1992). Close collaboration between scientists and members of indigenous communities is critical for thorough identification and validation of local, endemic plant resources (Schmidt and Cheng 2017), but these trust issues frequently prevent outsiders from gaining access into traditional ecological knowledge about culturally important species. In order to bridge these concerns, our collaborating laboratory teams from three US universities (www.gibex.org) joined together to produce a portfolio of mobile biodiscovery kits. The interactive experiments in the kits enable participation of both elders and youth in bioexploration and therapeutic lead discovery, focusing on indigenous local plants with a history of medicinal use in traditional ecological knowledge (Kellogg et al. 2010; Flint et al. 2011; McOliver et al. 2015). This unconventional approach provided a sustainable and equitable entrée into traditional cultures and local healing practices, and allowed us to interface with tribal groups in Africa, Asia, and the Americas to create a bridge between traditional knowledge and chemical/physical science discovery experiments.

The mobile discovery workshops have been added into high school and junior college curricula in several participating countries including the USA (Alaska and North Dakota). The mobile biodiscovery kits are simple, straightforward, rapidly deployed bioassays. Experiential learning is the key to resolving the bioactivity of the natural resources, and all of the demonstration/instruction takes place in a team-centered environment. Initial scouting in the forest or field is led by traditional healers, identified plantings in the wild are coded with GPS coordinates, and samples of identified local medicinal species undergo botanical authentication and are vouchered as herbarium specimens. Next, small quantities of plant material (leaves, bark, roots, stems, fruits) are ground, and crude ethanolic extracts are prepared and tested in a series of screens (antifungal, antibacterial, antiprotozoal, enzyme inhibitory, etc.) (Table 3.1). Assays can be conducted using only a few grams of plant material, and allow the participants to identify bioactive properties of plants used in traditional medicines that have been underappreciated by modern pharma-based medicine (Figure 3.2). The strategy has successfully engaged youth from junior high school to college age in science discovery that is hands on, and utilizes plants with intrinsic cultural significance (Kellogg et al. 2010, 2016). The traditional ecological knowledge of ethnic groups regarding harvest and use of polyphenol-rich plants can be lost due to habitat destruction (wars, overharvesting, development or clearing of wild lands for agriculture), but it can also be sidelined or lost when younger generations lose interest in the traditional knowledge of their elders (Schmidt and Klaser Cheng 2017). Community elders report a loss of interest in traditions and subsistence foods among tribal youth, and with that disconnect, a shift to commodity and fast-food diets which increases incidence of diabetes and obesity. Due to these trends, the mobile biodiscovery approach has been widely accepted and encouraged by tribal elders as a means to refocus on traditional values (Kellogg et al. 2010).

Following the leads provided by naturopathic healers in various global settings, the mobile biodiscovery approach was used to shed light on the mechanistic properties of some of the traditionally wildcrafted polyphenol-rich plants.

3.5.1 Phlorotannins in Alaskan Seaweeds/Marine Algae

Alaska's Unangan (Aleut) name is Alaxsxix, which means "place the sea crashes against"; as such, it is a rich harvest site for seaweeds all along its 10,690 km coastline. Seaweeds are

Table 3.1 Portfolio of mobile biodiscovery training modules used in workshops with both elders and youth in participating native communities.

Module ID	Approximate workshop time required to conduct the bioassays/procedures	How long before results can be evaluated	Relevant-disease targets
Field collection*	2–3 hours	n/a	n/a
Plant material extraction	1–2 hours	n/a	n/a
Antibacterial assay	2–3 hours	24–48 hours	Bacterial pathogens
Antioxidant assay	1 hour	Immediate	General health
Glucosidase and glucosidase inhibitor assays	2 hours	Immediate	Metabolic syndrome type 2 diabetes
α-amylase inhibition	1–2 hours	Immediate	Metabolic syndrome type 2 diabetes
Anthocyanin detection	1 hour	Immediate	General health
Protease and protease inhibitor assay	30 minutes–1 hour	10 minutes	Digestive diseases Viral pathogens
Planaria lethality assay	1–2 hours	8–24 hours	Parasitic worms
Planaria regeneration assay	1–2 hours	5 days	Parasitic worms
Nematode lethality assay	1 hour	4 hours	Parasitic worms

*Field identification of medicinal plants and sample collection are conducted in concert with local elders from the community.

Figure 3.2 Alaska Native youth engage in a workshop featuring simple mobile biodiscovery assays during a 2019 fish camp on the Yukon River in Alaska. *Source:* Mary Ann Lila. (*See insert for color representation of the figure.*)

utilized for food, livestock fodder, and traditional medicines for Pacific Rim Native American, Alaska Native, and First Nation tribal communities. As foods, they provide a limitless resource for macro- and micronutrients, and are prepared by drying, toasting, fermenting, and brewing in soups (Kellogg et al. 2015). Recent research has cited the utility of seaweed extracts to significantly increase insulin sensitivity and diminish hyperglycemia (Paradis et al. 2011; Nuno et al. 2013). In partnership with the Southeast Alaska Regional Consortium and Alaska Native elders, we were able to engage in a deeper analysis of the phytoactive constituents of seaweeds and their relevant bioactivities. Extracts from six species of seaweed and one tidal plant were screened for antioxidant capacity using first the mobile biodiscovery kits (described previously) to establish "first hits," then in both chemical and *in vitro* bioassays, and total phenolic content was gauged. A brown seaweed (*Fucus distichus*) proved to have both the highest total phenolic content (557 μg/mg extract, measured in phloroglucinol equivalents) and one of the highest radical scavenging activities (Kellogg and Lila 2013). Follow-up work established that the phlorotannin oligomers (fucophloroethol structures with DP from 3-18 monomer units) of *F. distichus* (Figure 3.3) were involved in potent α-amylase and α-glucosidase inhibition, suggesting a mechanism for their traditionally recognized antidiabetic properties (Kellogg et al. 2014).

Crude extracts from the same six Alaskan seaweeds were assayed for inhibition against five common inflammatory markers (COX2, iNOS, TNFα, IL-10, and MCP-1). While red and green seaweed species exhibited some activity, the brown seaweeds demonstrated significant anti-inflammatory capacity against all five markers. Subfractionation of the brown seaweed F. *distichus* revealed modulation in a dose-dependent manner by some fractions of lipid metabolism gene expression in an adipocyte model. Multiple anti-inflammatory properties in both adipose and macrophage cells were revealed, and two natural product classes (long chain mono- and polyunsaturated fatty acids and phlorotannin oligomers) were linked to these biomarkers (Kellogg et al. 2015). The multifunctional activities of edible seaweeds towards inhibition of the cardiometabolic risk factors associated with metabolic syndrome were linked to high phenolic content and coincident angiotensin converting enzyme I inhibitor activity and anti-inflammatory activities (Rico et al. 2018).

3.5.2 Isolating Phytoactive Principles from the Mediterranean Region

The mobile bioassay approach featured prominently in a joint ENPI CBCMED (European Neighbourhood and Partnership Instrument [ENPI] Cross-border Cooperation [CBC] Mediterranean [MED] Program) grant (funded by the European Union) which linked scientists from Israel and the Palestinian Authority, along with Greece and Spain. The program conducted an efficient yet comprehensive survey of medicinal plants in the region, and identified primary leads with high potency bioactivities that held promise for development and further commercialization (Joseph et al. 2014). The mobile biodiscovery assays enabled rapid screening over a wide diversity of climate zones and habitats, and cataloguing of plant materials that were used in traditional medicine, but had not been assayed previously (Joseph et al. 2014; Ali-Shtayeh et al. 2015). Natural areas in the wild were screened systematically over multiple territories across Palestine's West Bank and Israel. Both in Palestine (Ali-Shtayeh et al. 2015) and in Israel (Joseph et al. 2014), the research teams confirmed that the plants growing in the most hostile wild environments

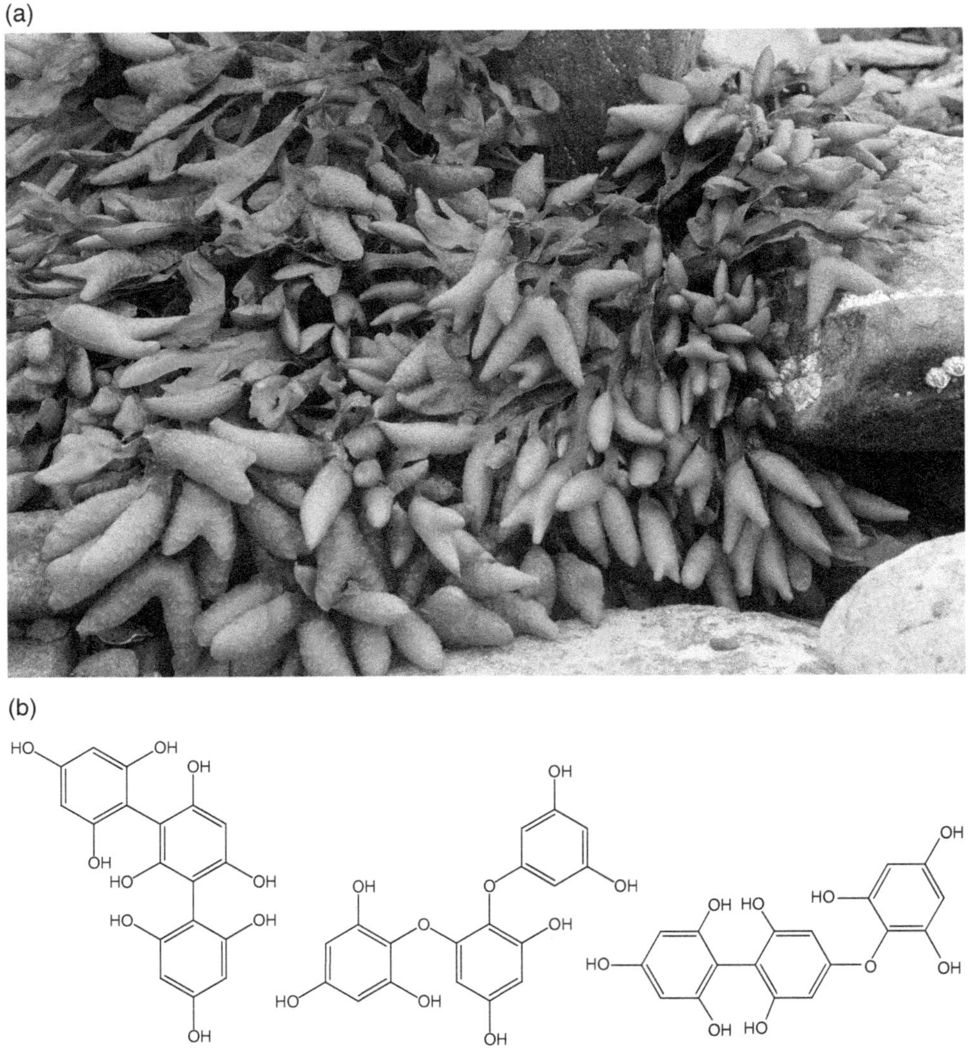

Trimeric fucol Trimeric phloroethol Trimeric fucophloroethol

Figure 3.3 (a) *Fucus distichus* (bladder wrack), a traditionally used phlorotannin-rich seaweed harvested in Whittier, AK. (b) Structural units of phlorotannins from the Alaskan brown algae *F. distichus*. *Source:* (a) Mary Ann Lila. (*See insert for color representation of the figure.*)

demonstrated more potent bioactive properties than plants in moderate environments. Regions with high solar radiation levels hosted plant materials with the highest recorded antioxidant levels and the most concentrated and diverse profiles of flavonoids and phenylpropanoids. Polyphenol-rich plants were particularly identified as active against carbohydrate hydrolyzing enzymes (important in diabetes management). Commercial α-glucosidase-inhibiting drugs in general have been associated with gastrointestinal side effects, whereas the polyphenolic phytoactives were identified as alternatives that did not cause side reactions (Ali-Shtayeh et al. 2015).

3.5.3 Drug Discovery in Cooperation with Traditional Healers in Botswana

In Botswana, the population at large, both in rural areas and in towns, subscribes to a mixture of traditional medicine and Western medicine. To the chagrin of traditional healers, however, the traditional practices are slowly falling out of fashion, and are not regarded in as high esteem as Western medicine, especially among younger citizens. In an attempt to bridge this gap, a series of mobile discovery workshops was conducted in two regions of the country for traditional healers and other community members, who were invited to provide their own indigenous medicinal plants for use in the bioassay screens (Andrae-Marobela et al. 2012). The mobile biodiscovery approach was used to establish an indigenous knowledge-guided drug-discovery platform. Because opportunistic infections such as tuberculosis and sexually transmitted infections are relatively high in Botswana, mobile screens were specifically tailored to include pathogenic bacterial strains with high potential to detect promising new drug candidates (Andrae-Marobela et al. 2012). The ethnomedical approaches (workshops) were conducted in hot spots of biodiversity as well as ethnic diversity. The bioassays revealed "first hits" of bioactivity to medicinal plants that had previously never been examined using these tools, and provided a way to conserve traditional knowledge in qualitative and semiquantitative data formats. One of the most appreciated features of this approach was that the indigenous healers were true research partners given all rights to transparency and benefits, and were never undermined as informant sources only.

3.5.4 Antidiabetic Mechanisms of Wild Tundra Berries

Wild berries have traditionally been integral dietary components for Native Americans and Alaska Natives, are used ceremonially and medicinally, and remain a treasured resource today (Burns Kraft et al. 2008; Kellogg et al. 2010). In Alaska and northern Canadian territories, lands with a high proportion of indigenous people, wild berry species often are the *only* edible terrestrial plants endemic to the environments. Site-specific variation in anthocyanin and proanthocyanidin levels in these wild berries, and more concentrated phytoactive constituents in the more spartan, high-stress environments of the arctic have been documented. Initial on-site mobile bioassays conducted with Alaska Natives identified a wealth of primary bioactivities relevant to human health, including higher antioxidant potentials (relative to commercially harvested domestic berries), more concentrated and diverse phytochemical profiles, and significant effects of climatic fluctuations on berry abundance and quality. The strong partnerships forged in workshops that put scientific inquiry in the context of traditional knowledge (Flint et al. 2011) were followed by more extensive laboratory bioassays of wild berry species.

The incredibly diverse phenolic profiles of wild tundra berries (Figure 3.4), in particular, were found to include abundant A-type proanthocyanidins, and were linked to the berries' efficacy against diabetes and metabolic syndrome biomarkers, including expression of preadipocyte factor 1. Berry-derived flavonoids and, more specifically, their gut-derived phenolic microbial metabolites have demonstrated efficacy in modulation of serum glucose levels (Grace et al. 2009; Kellogg et al. 2010), in neuroprotective properties (Gustafson et al. 2012; Lila et al. 2014), and in wound healing and other anti-inflammatory bioactivities (Grace et al. 2014; Nieman et al. 2018; Esposito et al. 2019; Nieman et al. 2019).

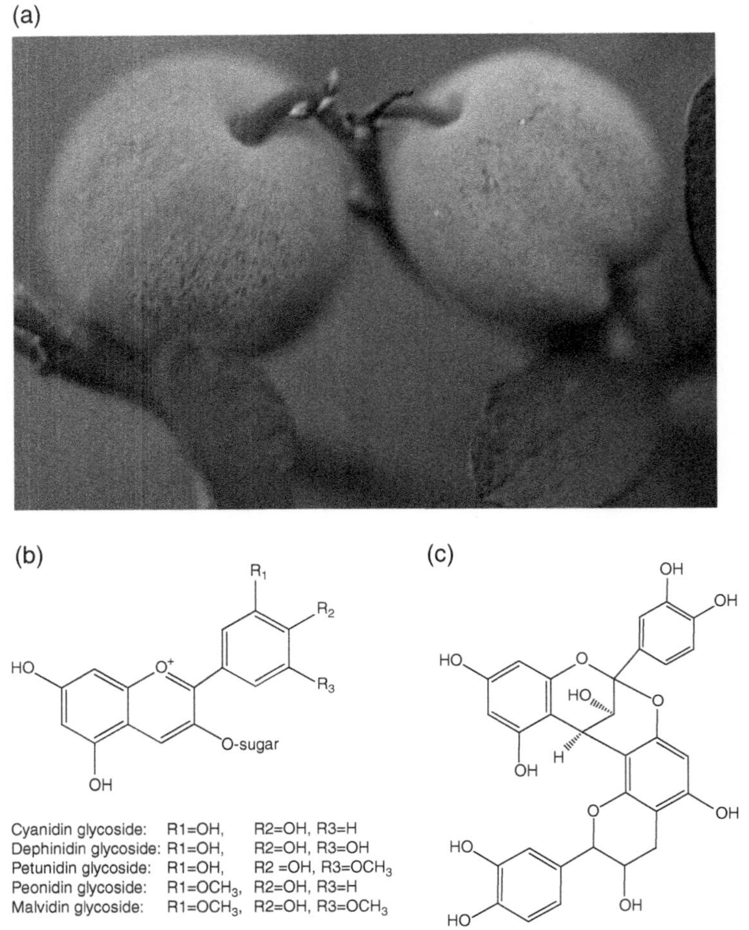

Figure 3.4 (a) Wild *Vaccinium uliginosum* (bog blueberry) growing on the Alaskan tundra. (b) Representative anthocyanins from Alaskan berries. (c) Proanthocyanidin A-type structure characteristic of many Alaskan berry genotypes. *Source:* (a) Mary Ann Lila. (*See insert for color representation of the figure.*)

3.6 Commercialization Prospects for Wildcrafted Polyphenol-rich Plants

The exceptionally high polyphenol content and diverse profiles found in wild plants have led to their incorporation in new functional food and cosmeceutical formulations. There is a global trend favoring clean labels on food and cosmetic products, which has been the impetus behind industrial interest in using wild-harvested plants as natural sources for antioxidants, flavorings, and antimicrobials in unique yogurt and cheese products and topical moisturizers (Pinela et al. 2017; Jacobsen et al. 2019).

For example, as a consequence of the current consumer trend towards natural, healthy foods from clean environments, edible macroalgae or seaweeds are increasingly consumed as trendsetting foods and condiments. Seaweeds, which were until recently harvested only by native groups, feature a range of antioxidant compounds (phlorotannins, pigments, tocopherols, and polysaccharides) which can now be efficiently, economically extracted from wild species, and/or wild seaweeds have been brought into industrial cultivation (Jacobsen et al. 2019). These advances have increased the impetus for the pharmaceutical industry to develop novel antidiabetic, anticarcinogenic, and anti-inflammatory drugs from extracted phytoactives from seaweeds, particularly in Asia (Abirami et al. 2018). The polyphenolic (phlorotannin) content of seaweeds has been the impetus for the launch of new cosmeceutical and pharmacological products based on radiation protective, prebiotic, antibiotic, and antidiabetic attributes. The resiliency of native plants thriving under harsh environmental extremes has provided a novel marketing angle for these derived products. Biorefineries worldwide are currently using seaweeds in bulk as feedstock based on functional (prebiotic and anti-inflammatory) bioactive properties (Holdt and Kraan 2011). Similarly, arctic berries which flourish in the extended 23 h photoperiod of the tundra regions and survive harsh conditions have become niche ingredients for novel edible products as well as cosmeceutical formulations (Kellogg et al. 2011).

Despite escalating investments in pharmaceutical research, the innovation capacity of industrial R and D models has appeared to stagnate. Safety issues are generally addressed rather late in the typical drug development pipeline, and can easily be responsible for failure of drug approval. In contrast, the alternative and time-tested "reverse pharmacology" approach, based on the leads from traditional ecological medicines and a long history of human use, allows safety issues to be addressed early in the discovery process, and can significantly reduce time and costs while introducing novel new treatment approaches to modern consumers (Andrae-Marobela et al. 2012).

3.7 Acknowledgements

We are grateful to the Southeast Alaska Regional Health Consortium (SEARHC) for their help in identifying selected natural resources. We gratefully acknowledge support from USDA NIFA HEC Grant 10790074, entitled In-Field Biodiscovery Framework—a Catalyst for Science Education and Validation of Traditional Knowledge, from the Global Institute for BioExploration (GIBEX), and from USDA NIFA ANNH grant 561939-02689, entitled Back to the River: the Science Behind Alaska's Traditional Subsistence Lifestyle.

References

Abirami, A., Uthra, S., and Arumugam, M. (2018). Nutritional value of seaweeds and their potential pharmacological role of polyphenolic substances. *Journal of Emerging Technologies and Innovative Research (JETIR)* 5: 930–939.

Alarcomicronn, R., Pardo-de-Santayana, M., Priestley, C., et al. (2015). Medicinal and local food plants in the south of Alava (Basque country, Spain). *Journal of Ethnopharmacology* 176: 207–224.

Al-Gubory, K., and Laher, I. (eds.). (2018). *Nutritional Antioxidant Therapies: Treatments and Perspectives*. NY: Springer.

Ali-Shtayeh, M., Jamous, R., and Abu Zaitoun, S. (2015). A comprehensive science-based field assessment of bioactive properties of the native plants of Palestine. *Journal of Biodiversity, Bioprospecting and Development* 2: 151.

Allkin, B. (2017). Useful plants—medicines: At least 28,187 plant species are currently recorded as being of medicinal use. In: *State of the World's Plants 2017* (ed. K.J. Willis). London (UK).

American Diabetes Association (2018). http.www.diabetes.org.

Andrae-Marobela, K., Ntumy, A., Makobela, M., et al. (2012). "Now I heal with pride"—the application of screens-to-nature technology to indigenous knowledge systems research in Botswana: Implications for drug discovery. In: *Drug Discovery in Africa. Impacts of Genomics, Natural Products, Traditional Medicines, Insights into Medicinal Chemistry, and Technology Platforms in Pursuit of New Drugs* (eds. K. Chibale, M. Davies-Coleman and C. Masimirembwa), 239–264. Heidelberg: Springer-Verlag.

Bittel, J. (2018) Lemurs may be making medicine out of millipedes. *National Geographic*, 03 August 2018.

Braga, P.C., Antonacci, R., Wang, Y.Y., et al. (2013). Comparative antioxidant activity of cultivated and wild *Vaccinium* species investigated by EPR, human neutrophil burst and COMET assay. *European Review for Medical and Pharmacological Sciences* 17 (15): 1987–1999.

Burns Kraft, T.F., Dey, M., Rogers, et al. (2008). Phytochemical composition and metabolic performance-enhancing activity of dietary berries traditionally used by native North Americans. *Journal of Agricultural and Food Chemistry* 56 (3): 654–660.

Buttriss, J. (2012). Plant Foods and Health. In: *Phytonutrients* (eds. A. Salter, H. Wiseman and G. Tucker), 1–51. Blackwell Publishing Ltd.

Caesar, L. and Cech, N.B. (2019). Synergy and antagonism in natural product extracts: when 1+1 does not equal two. *Natural Product Reports* 36, 869–888. doi: 10.1039/C9NP00011A.

Cordova, A.C., and Sumpio, B.E. (2009). Polyphenols are medicine: Is it time to prescribe red wine for our patients? *The International Journal of Angiology: Official Publication of the International College of Angiology, Inc*, 18 (3): 111–117.

Dar, R., Shahnawaz, M., Rasool, S., and Qazi, P. (2017). Natural product medicines: A literature update. *J. Phytopharmacol.* 6: 340–342.

Densmore, F. (1974). *How Indians Use Wild Plants for Food, Medicine and Crafts*. New York, New York: Dover Publications.

Dhami, N., and Mishra, A. (2015). Phytochemical variation: How to resolve the quality controversies of herbal medicinal products? *Journal of Herbal Medicine* 5: 118–127.

Drozdz, P., Seziene, V., and Pyrzynska, K. (2018). Mineral composition of wild and cultivated blueberries. *Biological Trace Element Research* 181 (1): 173–177.

Dunlap, K.L., Reynolds, A.J., and Duffy, L.K. (2006). Total antioxidant power in sled dogs supplemented with blueberries and the comparison of blood parameters associated with exercise. *Comparative Biochemistry and Physiology.Part A, Molecular and Integrative Physiology* 143 (4): 429–434.

Engel, C. (2007). Zoopharmacognosy. In: *Veterinary Herbal Medicine* (eds. S. Wynn, and B. Fougere), 7–15. St. Louis Missouri: Mosby Elsevier.

Esposito, D., Overall, J., Grace, M., et al. (2019). Alaskan berry extracts promote dermal wound repair through modulation of bioenergetics and integrin signaling. *Frontiers in Pharmacology* 10: 1058. doi: 10.3389/fphar.2019.01058.

Flint, C.G., Robinson, E.S., Kellogg, J., et al. (2011). Promoting wellness in Alaskan villages: Integrating traditional knowledge and science of wild berries. *EcoHealth,* 8 (2): 199–209.

Goetz, G. (2012). Nutrition a pressing concern for American Indians. *Food Safety News 5,* March 2012.

Grace, M., Ribnicky, D., Kuhn, P. et al. (2009). Hypoglycemic activity of a novel anthocyanin-rich formulation from lowbush blueberry, *Vaccinium angustifolium* Aiton. *Phytomedicine* 16: 406–415.

Grace, M., Esposito, D., Dunlap, K.L., and Lila, M.A. (2014). Comparative analysis of phenolic content and profile, antioxidant capacity, and anti-inflammatory bioactivity in wild Alaskan and commercial *vaccinium* berries. *Journal of Agricultural and Food Chemistry* 62: 4007–4017.

Gustafson, S.J., Yousef, G.G., Grusak, M.A., and Lila, M.A. (2012). Effect of postharvest handling practices on phytochemical concentrations and bioactive potential in wild blueberry fruit. *Journal of Berry Research* 2: 215–227.

Holdt, S., and Kraan, S. (2011). Bioactive compounds in seaweed: Functional food applications and legislation. *Journal of Applied Phycology* 23: 543–597.

Jacobsen, C., Sorensen, A., Holdt, S., et al. (2019). Novel antioxidants from seaweed. *Annual Review of Food Science and Technology* 10: 541–568.

Joseph, G., Faran, M., Raskin, I., et al. (2014). Medicinal plants of Israel: A model approach to enable an efficient, extensive, and comprehensive field survey. *Journal of Biodiversity, Bioprospecting and Development* 1: 134.

Karlsons, A., Osvalde, A., Ceksere, G., and Pormale, J. (2018). Research on the mineral composition of cultivated and wild blueberries and cranberries. *Agronomy Research* 6: 454–463.

Kellogg, J., Croom, B., Plundrich, N., et al. (2016). Engaging American Indian/Alaska Native (AI/AN) students with participatory bioexploration assays. *NACTA Journal* 60: 42–50.

Kellogg, J., Esposito, D., Grace, M., Komarnytsky, S., and Lila, M. (2015). Alaskan seaweeds lower inflammation in RAW 264.7 macrophages and decrease lipid accumulation in 3T3-L1 adipocytes. *Journal of Functional Foods* 15: 396–407.

Kellogg, J., Grace, M.H., and Lila, M.A. (2014). Phlorotannins from Alaskan seaweed inhibit carbolytic enzyme activity. *Marine Drugs* 12 (10): 5277–5294.

Kellogg, J., Higgs, C., and Lila, M.A. (2011). Prospects for commercialization of an Alaska native wild resource as a commodity crop. *Journal of Entrepreneurship* 20: 77–101.

Kellogg, J., and Lila, M.A. (2013). Chemical and in vitro assessment of Alaskan coastal vegetation antioxidant capacity. *Journal of Agricultural and Food Chemistry* 61 (46): 11025–11032.

Kellogg, J., Wang, J., Flint, C. et al. (2010). Alaskan wild berry resources and human health under the cloud of climate change. *Journal of Agricultural and Food Chemistry* 58 (7): 3884–3900.

Krebs, J. (2013). *Food. A very short introduction.* Oxford, UK: Oxford University Press.

Li, Y., Zhang, J.J., Xu, D.P. et al. (2016). Bioactivities and health benefits of wild fruits. *International Journal of Molecular Sciences* 17 (8): 1258–1285.

Lila, M.A., Burton-Freeman, B., Grace, M., and Kalt, W. (2016). Unraveling anthocyanin bioavailability for human health. *Annual Reviews Food Science and Technology* 7: 17.1–17.19.

Lila, M.A., Kellogg, J., Grace, M.H., Yousef, G.G., Kraft, T.B., and Rogers, R.B. (2014). Stressed for success: How the berry's wild origins result in multifaceted health protections. *Acta Horticulturae (ISHS)* 1017: 23–43.

Margina, D., Ilie, M., Gradinaru, D., et al. (2015). Natural products—friends or foes? *Toxicology Letters*, 236 (3): 154–167.

McOliver, C.A., Camper, A.K., Doyle, et al. (2015). Community-based research as a mechanism to reduce environmental health disparities in American Indian and Alaska Native communities. *International Journal of Environmental Research and Public Health* 12 (4): 4076–4100.

Moerman, D.E. (1996). An analysis of the food plants and drug plants of Native North America. *Journal of Ethnopharmacology* 52 (1): 1–22.

Nieman, D.C., Kay, C.D., Rathore, A. et al. (2018). Increased plasma levels of gut-derived phenolics linked to walking and running following 2-weeks flavonoid supplementation. *Nutrients* 10: 1718.

Nieman, D.C., Lila, M.A., and Gillett, N. (2019). Immunometabolism: A multi-omics approach to interpreting the influence of exercise and diet on the immune system. *Annual Reviews Food Science and Technology* 10: 6.1–6.23.

Nuno, K., Villarruel-Lopez, A., Puebla-Perez, A., et al. (2013). Effects of the marine microalgae *Isochrysis galbana* and *Nannochloropsis ocula* in diabetic rats. *Journal of Functional Foods* 5: 106–115.

Ozgen, M., Serce, S., Gunduz, K., et al. (2007). Determining total phenolics and antioxidant activity of selected *Fragaria* genotypes. *Asian Journal of Chemistry* 19: 5573–5581.

Paradis, M.E., Couture, P., and Lamarche, B. (2011). A randomised crossover placebo-controlled trial investigating the effect of brown seaweed (*Ascophyllum nodosum* and *Fucus vesiculosus*) on postchallenge plasma glucose and insulin levels in men and women. *Applied Physiology, Nutrition, and Metabolism = Physiologie Appliquee, Nutrition Et Metabolisme* 36 (6): 913–919.

Phan, M., Paterson, J., Bucknall, M., and Arcot, J. (2018). Interactions between phytochemicals from fruits and vegetables: Effects on bioactivities and bioavailability. *Critical Reviews in Food Science and Nutrition* 58 (8): 1310–1329.

Pinela, J., Carvalho, A.M., and Ferreira, I.C.F.R. (2017). Wild edible plants: Nutritional and toxicological characteristics, retrieval strategies and importance for today's society. *Food and Chemical Toxicology : An International Journal Published for the British Industrial Biological Research Association* 110: 165–188.

Reyes-Garcia, V., Menendez Baceta, G., Aceituno-Mata, L., et al. (2015). From famine foods to delicatessen: Interpreting trends in the use of wild edible plants through cultural ecosystem services. *Ecological Economics* 120: 303–311.

Rico, D., Diana, A., Milton-Laskibar, I., et al. (2018). Characterization and in vitro evaluation of seaweed species as potential functional ingredients to ameliorate metabolic syndrome. *Journal of Functional Foods* 46: 185–194.

Sasipriya, G., Maria, C.L., and Siddhuraju, P. (2014). Influence of pressure cooking on antioxidant activity of wild (ensete superbum) and commercial banana (musa paradisiaca var. monthan) unripe fruit and flower. *Journal of Food Science and Technology* 51 (10): 2517–2525.

Schatzker, M. (2015). *The Dorito Effect*. Simon and Schuster.

Schmidt, B.M., and Klaser Cheng, D.M. (eds.). (2017). *Ethnobotany. A phytochemical perspective*. Hoboken, New Jersey: Wiley Blackwell.

Schreckinger, M.E., Lotton, J., Lila, M.A., and de Mejia, E.G. (2010). Berries from South America: A comprehensive review on chemistry, health potential, and commercialization. *Journal of Medicinal Food* 13 (2): 233–246.

Sebag-Montefiore, C. (2017). The wild bunch. *American Way, March Issue*, 58.

Soukand, R. (2016). Perceived reasons for changes in the use of wild food plants in Saaremaa, Estonia. *Appetite* 107: 231–241.

Stowe, C. (1976). History of veterinary pharmacotherapeutics in the United States. *Journal of the Veterinary Medicine Association* 169: 83–89.

Sueda, K., Hart, B., and Cliff, K. (2008). Characterisation of plant eating in dogs. *Applied Animal Behaviour Science* 111: 120–132.

Wapner, J. (2012). Where the wild berries grow. *Cosmos* 42: 74–78.

Wynn, S., and Fougere, B. (2007). *Veterinary Herbal Medicine. St.* Louis Missouri: Mosby Elsevier.

Zhang, H., Mittal, N., Leamy, L.J., et al. (2016). Back into the wild-apply untapped genetic diversity of wild relatives for crop improvement. *Evolutionary Applications* 10 (1): 5–24.

4

Causes and Consequences of Condensed Tannin Variation in *Populus*

A Molecules to Ecosystems Perspective

Kennedy F. Rubert-Nason[1] *and Richard L. Lindroth*[2]

[1] *Natural Sciences, University of Maine–Fort Kent, Fort Kent, USA*
[2] *Department of Entomology, University of Wisconsin–Madison, Madison, USA*

4.1 Introduction

Condensed tannins (CTs) are flavonol polymers found within the leaf, stem, and root tissues of many plants (Bate-Smith and Metcalfe 1957; Mole 1993; Ayres et al. 1997). These compounds are synthesized via the phenylpropanoid and malonic acid pathways (Marles et al. 2003; Tsai et al. 2006; Philippe and Bohlmann 2007) (Figure 4.1). The macromolecules share common chemical properties, including the capacity to bind to proteins (Ayres et al. 1997; Barbehenn and Constabel 2011) and incorporate into soil organic matter (Adamczyk et al. 2018), and they influence ecological processes at diverse scales (Schweitzer et al. 2008).

Condensed tannins occur at low to high concentrations in *Populus*, a fast-growing, abundant tree genus that is native to, or cultivated in, large expanses of the northern hemisphere (Sibley 2009; Stettler 2009). Generally, CT concentrations are highest in leaves (up to 40% d.m.) followed by stems (up to 10% d.m.) and roots (up to 5% d.m.) (Table 4.1). *Populus* CT polymers tend to be unbranched, consisting predominantly of subunits of various enantiomers, epimers, and diastereoisomers of catechin (procyanidins) and gallocatechin (prodelphinidins) (Salminen and Karonen 2011) (Figure 4.2). The concentration (Table 4.1) and polymer composition (Ayres et al. 1997; Scioneaux et al. 2011; Rubert-Nason and Lindroth 2019) vary substantially among and within species, with variation determined by genetic (Table 4.1), developmental (Donaldson et al. 2006b; Rehill et al. 2006; Holeski et al. 2012), and environmental factors (Tables 4.2 and 4.3). An important further consideration is that this CT-producing genus is abundant in geographical areas that are experiencing substantial human-mediated environmental changes (IPCC 2013).

Given the substantial concentrations of CTs in many *Populus* species, and the widespread distribution of this genus in the northern hemisphere, *Populus* CTs are likely among the most abundant secondary metabolites in the terrestrial biosphere, perhaps second to only lignin. Moreover, *Populus* CTs influence a variety of biological and ecological interactions (Schweitzer et al. 2008; Stettler 2009) at levels ranging from individual organisms to entire

Recent Advances in Polyphenol Research, Volume 7, First Edition. Edited by Jess Dreher Reed, Victor Armando Pereira de Freitas, and Stéphane Quideau.

ecosystems (Tables 4.4–4.6). For example, CTs govern herbivore feeding and performance, tritrophic interactions, pathogen infection, and soil microbial community structure and function. These associations have been linked to differences in CT quantity and quality that are mediated by genetic, developmental, and environmental factors.

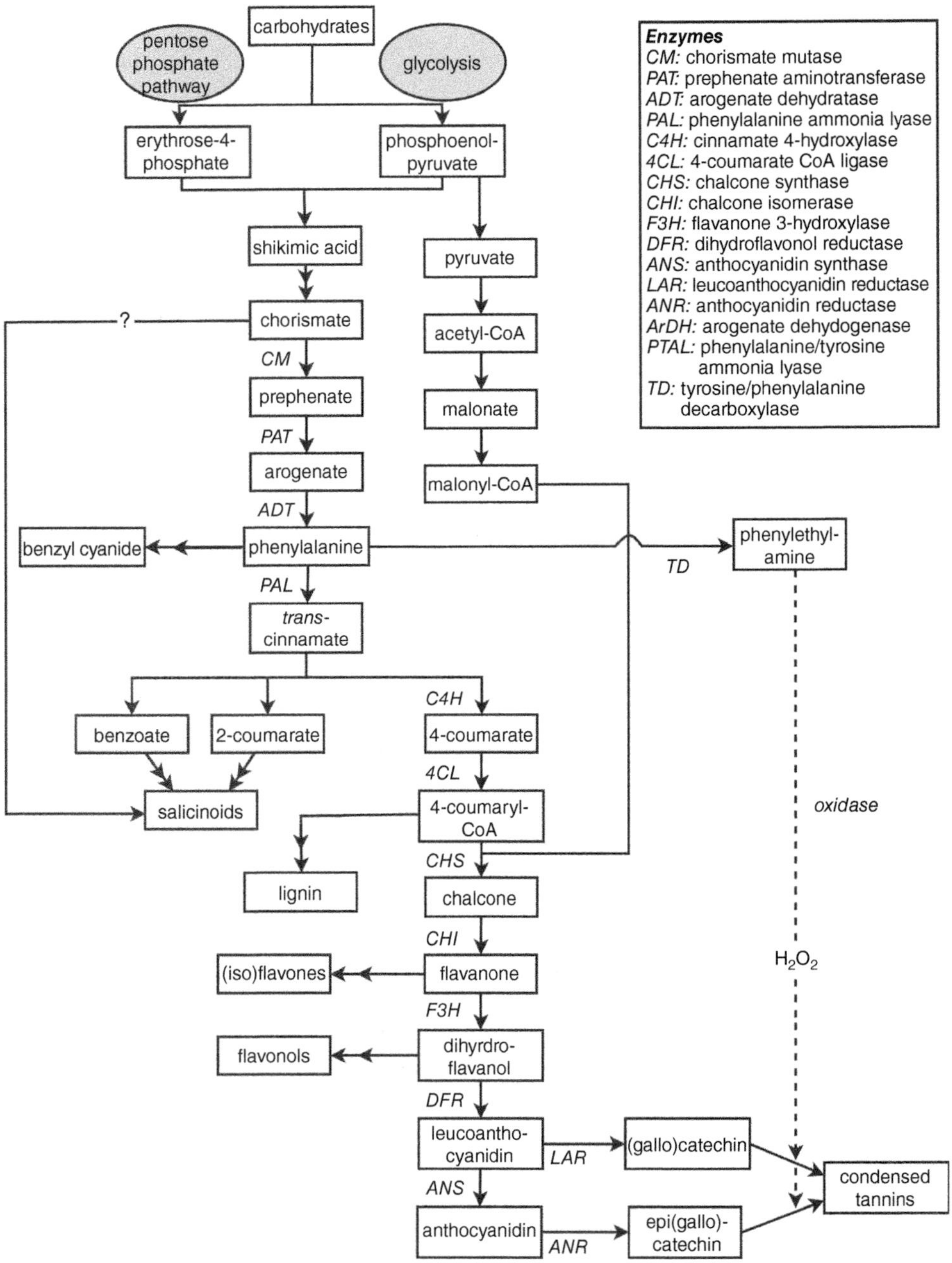

Figure 4.1 Biosynthetic pathways leading to production of condensed tannins. *Sources:* Marles et al. (2003), Tsai et al. (2006), Taiz and Zeiger (2010), Harding et al. (2014), Irmisch et al. (2014), and Cole and Ingvarsson (2018).

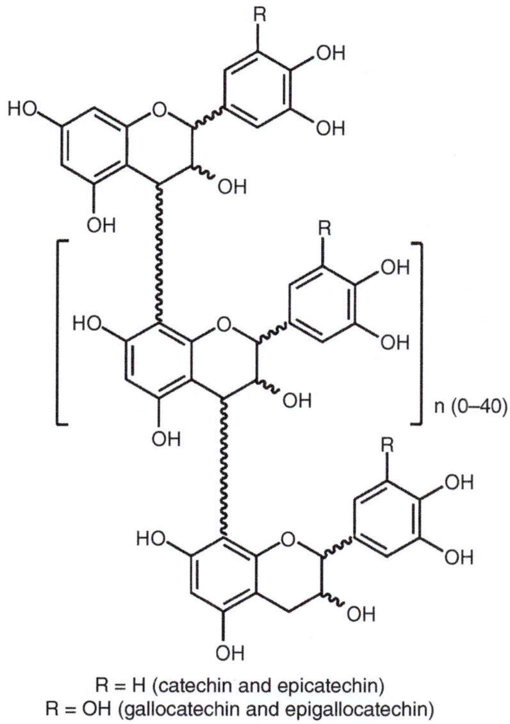

Figure 4.2 Molecular structure of a condensed tannin polymer.

The rich literature documenting causes and consequences of variation in CT levels reveals that the compounds play prominent roles in forest ecosystems. Because CT concentrations and molecular structures change in response to environmental conditions, the functional roles of CTs may change during the course of the Anthropocene – a proposed geological era characterized by rapid and extensive human impacts on biomes and climate (Crutzen and Stoermer 2000; Edwards 2015). Moreover, the interactive effects of genotype and environment on CTs, and subsequent effects on ecological interactions, have the potential to alter evolutionary trajectories of *Populus* communities.

Herein, we briefly review the biosynthesis of CTs, then focus on the genetic, developmental, and environmental determinants of CT variation in *Populus* species, and the consequences thereof for ecological interactions. We also briefly consider the roles of CTs in the context of global changes that characterize the Anthropocene. This work is not a comprehensive review of all relevant literature, as such would require monographic treatment. Rather, we draw on a sufficiently large body of literature to generate informative conclusions about the causes, consequences, and importance of variation in CTs in the genus *Populus*. We focus on *Populus* as it is the foremost model system for tree research (Jansson and Douglas 2007), and because more is known about CTs in the biology of *Populus* species than for any other genus of trees.

4.2 Condensed Tannin Biosynthesis

Biochemical pathways of CT synthesis in *Populus* have been elucidated in increasing detail over the past two decades, with informative descriptions appearing in Tsai et al. (2006), Philippe and Bohlmann (2007), Salminen and Karonen (2011), and Cole and Ingvarsson

(2018). Figure 4.1 summarizes the CT biosynthesis pathways. The existence of multiple, parallel biosynthetic pathways and a large number of regulatory genes (Peters and Constabel 2002; Harding et al. 2005; Ralph et al. 2006; Tsai et al. 2006; Tuskan et al. 2006) is presumably responsible for the substantial quantitative and qualitative variation in CTs evident in *Populus*.

Although the pathways for CT biosynthesis in *Populus* have been generally elucidated, understanding of the mechanisms regulating CT production is limited. Elevated CT production results from increased expression of multiple genes within the phenylpropanoid pathway (Decker et al. 2017); however, most studies of gene expression are restricted to specific genes or gene collections, and all of the regulatory mechanisms are probably not known. To date, several MYB transcription factors (i.e. MYB115, MYB134, MYB182) have been shown to interactively regulate distribution of the CT precursor molecules among different biosynthetic pathways, influencing allocation to CTs, anthocyanins, and flavonoids (Yoshida et al. 2015; James et al. 2017). In addition, signaling molecules such as gibberellic acid and salicylic acid have been shown to positively associate with CT production (Buhl et al. 2019; Ullah et al. 2019). Roles of additional genes, and their associated enzymes, in regulating CT production may be inferred from studies of other tree species (e.g. phenylalanine ammonia lyase [PAL] in *Quercus* [Kariñho-Betancourt et al. 2019], and PAL2 in *Salix* [de Jong et al. 2015]). The functions of the genes, however, can vary considerably among species even within the same family (de Jong et al. 2015).

While many studies report overall increases or decreases in CT concentrations in response to environmental factors, individual genes associated with the CT biosynthetic pathway may respond differently to environmental elicitors. For example, soil nitrogen amendment increased expression of genes (e.g. PAL3, 4CL, and CHS1) early in the phenylpropanoid pathway, while decreasing expression of genes (e.g. DFR1, LAR3, and ANR2) downstream in the pathway, thereby diverting CT precursor molecules to other compounds such as anthocyanins, and leading to a net decline in CTs (Decker et al. 2017). In *Quercus*, subtle differences among environmental variables (e.g. species of galling insects) can exert substantially different impacts on expression of specific genes regulating CT biosynthesis (Kariñho-Betancourt et al. 2019). This scenario seems likely for *Populus* as well.

4.3 Allocational Tradeoffs Influence CT Production

Condensed tannin production is influenced by tradeoffs among biochemical pathways that compete for common precursors. Phenylalanine is a key point of entry into the phenylpropanoid pathway. Irmisch et al. (2014) demonstrated that feeding by gypsy moths (*Lymantria dispar*) induces production of benzyl cyanide from phenylalanine. Benzyl cyanide in turn is repellent to gypsy moths, but at a cost of nitrogen loss to the plant. A more complex tradeoff was reported by Harding et al. (2014) for cottonwoods (*P. angustifolia* × *P. fremontii*). Phenylalanine can be metabolized to phenylethylamine, removing substrate from the phenylpropanoid pathway. Phenylethylamine can subsequently be deaminated, however, in a process that releases peroxides (e.g. H_2O_2) that contribute to CT polymerization. The most compelling evidence for biochemical tradeoffs was discovered, inadvertently, by Boeckler et al. (2014). When they genetically modified hybrid aspen (*P. tremula* × *tremuloides*) to increase CT production, levels of salicinoid phenolic glycosides declined. Strong inverse correlations between levels of foliar CTs and these salicinoid compounds have been documented in studies with both wild and experimental aspen (*P. tremuloides*; [Donaldson et al. 2006a; Cole et al. 2016; Barker et al. 2018]). Competition among biosynthetic pathways for the substrates for CT production appears to change with plant maturation, with preferential allocation to salicinoids in very young trees and allocation to CTs in older trees (Donaldson et al. 2006b; Cope et al. 2019). The studies described here contribute to the developing perspective that CT biosynthesis is tightly integrated and regulated at the genetic level, and not largely a passive consequence of mass-flow source-sink carbon dynamics in plants.

Allocational tradeoffs occur not only among biochemical pathways, but also among the physiological processes that compete for resources (e.g. carbohydrates) required for growth, reproduction, and defense (Bazzaz et al. 1987; Züst and Agrawal 2017). Cole et al. (2016) reported that CT concentrations and growth are inversely proportional in wild *P. tremuloides* trees. In that study, regression models revealed that trees with 25% foliar CTs grew at less than half the rate of trees with 5% foliar CTs. Concentrations of CTs and root growth are positively correlated in cottonwoods, suggesting that increased root growth is necessary to provide the water and nutrients required for high CT production or to facilitate uptake of soil N when soil CT concentrations are high (Fischer et al. 2006).

A 14-year longitudinal study by Cope et al. (2019) revealed that growth-defense tradeoffs vary both genetically and ontogenetically in *P. tremuloides*. Thus, not only can CT *biosynthesis* be regulated to enhance tree fitness in diverse environments, but biochemical and physiological *tradeoffs* themselves may be subject to natural selection and thus influence evolutionary trajectories of aspen populations.

4.4 Causes of Quantitative and Qualitative Variation in Populus CTs

4.4.1 Genetic Variation in CT Quantity and Quality

Condensed tannin concentrations vary substantially among and within *Populus* species and their hybrids (Table 4.1). Comparisons among species and genotypes in common garden experiments (Schweitzer et al. 2004; Barker et al. 2018), along with high broad-sense heritabilities in aspen and cottonwoods (Stevens and Lindroth 2005; Bailey et al. 2006), reveal that genes play a major role in determining variation in CT concentrations. Variation in CT concentrations is linked to differential gene expression in the CT biosynthesis pathways (Figure 4.1) (Peters and Constabel 2002; Harding et al. 2005; Ralph et al. 2006; Tsai et al. 2006; Tuskan et al. 2006).

Most information about genetically mediated variation in CT concentrations derives from studies of a relatively small number of *Populus* species (i.e. *P. tremula*, *P. tremuloides*, *P. angustifolia*, *P. fremontii*, and their hybrids). In general, *P. tremula*, *P. tremuloides*, and *P. angustifolia* tend to have high foliar CT concentrations (up to 40% d.m.), while *P. fremontii* tend to have low CT concentrations (0–3% d.m.), and hybrids (e.g. *P. fremontii* × *angustifolia*, *P. tremula* × *tremuloides*) have concentrations intermediate between the parental species. With the exception of *P. angustifolia*, *P. fremontii*, and their hybrids, CT concentrations are

Table 4.1 Representative examples of variation in CT concentration among *Populus* genotypes, species and their hybrids, and among tissue types

Populus species or hybrid	Tissue type	Number of genotypes	CT Conc. range[a] (% d.m.)	Reference
P. angustifolia	Leaf	5–10	2–24	Schweitzer et al. (2004), LeRoy et al. (2007), Holeski et al. (2012)
P. balsamifera	Leaf	n.r.[b]	0–12[c]	Ayres et al. (1997), Schimel et al. (1998)
P. deltoides	Leaf	n.r.	0.2–0.3[c]	Liu et al. (2015)
P. fremontii	Leaf	5–10	0.04–2	Driebe and Whitham (2000), Schweitzer et al. (2004), LeRoy et al. (2007), Holeski et al. (2012)
P. fremontii x angustifolia	Leaf	5–10	1–21	Schweitzer et al. (2004), Harding et al. (2005), Schweitzer et al. (2005a), LeRoy et al. (2007), Holeski et al. (2012)
P. tremula	Leaf	8–12	0.5–25	Robinson et al. (2012), Randriamanana et al. (2014), Bandau et al. (2015), Randriamanana et al. (2015), Bandau et al. (2017), Decker et al. (2017), Nissinen et al. (2017)
P. tremula x tremuloides	Leaf	2–5	1–18	Haikio et al. (2009), Boeckler et al. (2014)
P. tremuloides	Leaf	3–328	0–40	Hemming and Lindroth (1995), Ayres et al. (1997), Hwang and Lindroth (1997), Mansfield et al. (1999), Osier and Lindroth (2001), Donaldson and Lindroth (2004), Stevens and Lindroth (2005), Lindroth and Hwang (1996), Stevens et al. (2007), Vigue and Lindroth (2010), Cole et al. (2016), Call and St. Clair (2017), Barker et al. (2018), Lämke and Unsicker (2018), Rubert-Nason and Lindroth (2019)
P. trichocarpa	Leaf	n.r.	7–17[c]	Warren et al. (2002)
P. nigra	Stem	n.r.	3–10[c]	Luo et al. (2008)
P. tremula	Stem	12	1–3	Randriamanana et al. (2014)
P. tremuloides	Stem	4–12	0.2–3	Lindroth et al. (2007), Keefover-Ring et al. (2016)
Populus × *canadensis*	Root	1	0.5–2[c]	Kosola et al. (2006)
P. deltoides	Root	12	1–3[c]	Kosola et al. (2004)
P. tremula	Root	12	4–7	Randriamanana et al. (2014)
P. tremuloides	Root	12	2–5	Stevens et al. (2014)

[a] Concentration range among all genotypes reported in study.
[b] n.r. = not reported.
[c] No genotypic variation reported for this entry.

rarely if ever compared across different species within the same study. Compared with foliage, less is known about how CT concentrations vary within/among species in stem and root tissues. For example, stem concentrations have been evaluated in only four studies and three species, and root concentrations have been evaluated in only three studies and species.

Several studies have demonstrated that CT quality also varies substantially among and within *Populus* species and their hybrids (Ayres et al. 1997; Scioneaux et al. 2011; Rubert-Nason and Lindroth 2019). Qualitative variation is exhibited in terms of mean degree of polymerization (MDP), representation of catechin and gallocatechin subunits, and stereochemistry. Mean degree of polymerization varies at least 10-fold among species (ranging from 2 in *P. fremontii* to 22 in *P. angustifolia* [Scioneaux et al. 2011]), and at least twofold within species (ranging from 2.6–6.0 within *P. tremuloides* [Rubert-Nason and Lindroth 2019]). The mean degree of polymerization, relative abundance of hydroxyl groups (catechin:gallocatechin ratio, *aka* procyanidin:prodelphinidin or PC:PD ratio), and the proportions of different stereoisomers also vary substantially among species (e.g. *P. tremuloides* and *P. balsamifera*) (Ayres et al. 1997). Current knowledge of CT structural variation in *Populus* is limited to a small number of studies involving foliage of *P. tremuloides*, *P. angustifolia*, and *P. fremontii*. To our knowledge, the structures of CTs in root or stem tissue have not been evaluated.

4.4.2 Developmental Variation in CT Quantity and Quality

4.4.2.1 Seasonal Variation

Condensed tannins vary in concentration during the course of a growing season, and that variation itself differs among genotypes and species (Hwang and Lindroth 1998; Rehill et al. 2006; Holeski et al. 2012; Dettlaff et al. 2018). In *P. tremuloides*, foliar CT concentrations increase from budbreak through leaf expansion and maturation, and then decline slightly toward autumn (Lindroth et al. 2002; Dettlaff et al. 2018). Root CT concentrations follow a similar trajectory, being highest in early summer and declining thereafter (Dettlaff et al. 2018). In *P. fremontii*, *P. angustifolia* and their hybrids, foliar CTs sometimes increased between early and late summer, with substantial variation among genotypes and hybrids (Rehill et al. 2006; Holeski et al. 2012). The seasonal trajectories may be further complicated by tree ontogeny (Rehill et al. 2006) and environmental factors (e.g. light availability, herbivore damage), all of which can substantially influence CT concentrations (Stevens and Lindroth 2005; Call and St. Clair 2017). As of this time, seasonal changes in CT concentrations in stem tissues have not been evaluated.

Condensed tannin polymer composition also varies during the course of a growing season, although there are no definitive trends. Scioneaux et al. (2011) showed that the relative representation of catechin in CT polymers increased then decreased in *P. angustifolia* and *P. angustifolia* × *P. fremontii* hybrids during a growing season. In contrast, the catechin representation in *P. fremontii* initially increased and then remained relatively stable. Preliminary findings (K. Rubert-Nason and R. Lindroth unpublished data) revealed increases in polymer size, decreases in representation of catechin, and changes in CT stereochemistry between budburst and autumn senescence.

4.4.2.2 Ontogenetic Variation

Condensed tannins vary ontogenetically, and tree maturation is generally (but not universally) associated with increases in foliar CT concentrations (Donaldson et al. 2006a; Rehill et al. 2006; Barbehenn and Constabel 2011; Holeski et al. 2012; Dettlaff et al. 2018). In a cross-sectional study of different-aged ramets of multiple *P. tremuloides* genotypes, Donaldson et al. (2006b) found that foliar CT concentrations increased rapidly during the first several years of a ramet's life. Thereafter, levels stabilized, as also reported by Cope et al. (2019). Foliar CT concentrations also increased during maturation of *P. angustifolia* trees and their hybrids with *P. fremontii* (Rehill et al. 2006; Holeski et al. 2012). In contrast, foliar CTs in *P. fremontii* remained relatively constant across age classes, although overall levels are quite low in that species (Rehill et al. 2006; Holeski et al. 2012). Root CT concentrations did not vary in relation to tree size (an ontogenetic covariate) in *P. tremuloides* (Dettlaff et al. 2018).

Comparatively little is known about ontogenetic changes in CT molecular structures. Thus far, Scioneaux et al. (2011) demonstrated that the relative proportion of catechin in CT polymers increased with maturation in *P. angustifolia*. And preliminary observations suggest minimal qualitative changes in foliar CTs between 3- and 7-year-old *P. tremuloides* (K. Rubert-Nason and R. Lindroth unpublished data).

4.4.3 Environmental Variation in CT Quantity and Quality

4.4.3.1 Abiotic Environment

Numerous abiotic environmental variables, such as soil nutrients, light intensity, and atmospheric conditions, influence CT concentrations (Table 4.2; Figure 4.3). Foliar CT concentrations generally increase as soil nutrient availability decreases. Based on the limited number of studies to date, there is no clear trend regarding the effect of elevated soil nutrients on CT concentrations in stem and root tissues (Kosola et al. 2006; Luo et al. 2008; Randriamanana et al. 2014; Stevens et al. 2014). Condensed tannin concentrations are highly responsive to light intensity; increased exposure to photosynthetically active radiation (PAR) leads to elevated CT levels. Elevated atmospheric CO_2 and O_3 levels, as predicted by climate models to occur within the next 100 years (IPCC 2013), generally increase foliar CTs, and do so independently, interactively, and over time (Kopper and Lindroth 2003a; Couture et al. 2014). Increases in CT concentrations in trees grown under enriched CO_2 are consistent with an increased supply of carbon substrate for photosynthesis. In contrast, increases in CT concentrations in trees exposed to O_3 are likely due to an alternative mechanism: O_3 is phytotoxic, causing cell death and activation of transduction pathways for production of phenolic-based defense compounds (Lindroth 2010). Finally, increases in air temperatures at magnitudes predicted to occur within this century generally decrease foliar CT concentrations. No published work has evaluated the effects of environmental variation on CT polymer composition, although preliminary findings (K. Rubert-Nason and R. Lindroth unpublished data) indicate that soil nutrients, atmospheric composition, and temperature all alter the mean degree of polymerization, relative proportions of catechin and gallocatechin, and stereochemistry of CTs in *P. tremuloides*.

The responses of *Populus* CT concentrations to changes in the availability of required resources such as nutrients, light, and CO_2 align well with the predictions of the Carbon-Nutrient Balance (CNB) (Bryant et al. 1983; Tuomi et al. 1988) and Growth-Differentiation Balance (GDB) (Herms and Mattson 1992) theories. Indeed, changes in *Populus* CT levels in response to those environmental factors provide some of the strongest empirical support for the CNB and GDB theories. While differing in detail (comparison of which is beyond the scope of this paper), both theories posit that levels of carbon-based plant secondary metabolites are determined by the relative balance of resources (nutrients and light) required for growth and defense. When the availability of carbon is high (e.g. high light, enriched CO_2) relative to that of other nutrients, fixed carbon that cannot be used to support plant growth (because of nutrient constraints) will be allocated to production of carbon-based defense compounds, such as CTs. As reviewed in Section 3, however, these resource-mediated shifts in CT production are likely not simple consequences of shifting carbon mass flow balances, but rather are tightly regulated at the genetic level.

Table 4.2 Effects of abiotic environmental factors on CT concentrations in *Populus*

Factor	*Populus* species	Tissue type	Effect on CTs	Reference
Increased soil nutrients[a]				
P	*P. tremula*	Leaf	–	Randriamanana et al. (2014)
N	*P. tremuloides*	Leaf	–	Kinney et al. (1997)
N	*P. tremula*	Leaf	–	Randriamanana et al. (2014)
N	*P. tremula*	Leaf	–	Bandau et al. (2015)
N	*P. tremula*	Leaf	–	Decker et al. (2017)
N, K, micronutrients	*P. tremuloides*	Leaf	+	Calder et al. (2011)
N, P	*P. tremula*	Leaf	–	Randriamanana et al. (2014)
N, P, K	*P. tremuloides*	Leaf	–	Hemming and Lindroth (1999)
N, P, K, micronutrients	*P. tremuloides*	Leaf	+/0/–	Lindroth et al. (2007)
N, P, K, micronutrients	*P. tremuloides*	Leaf	–	Osier and Lindroth (2001)
N, P, K, micronutrients	*P. tremuloides*	Leaf	–	Stevens and Lindroth (2005)
N, P, K, micronutrients	*P. tremuloides*	Leaf	–	Rubert-Nason et al. (2015)
N, P, K	*P. tremuloides*	Litter	–	LeRoy et al. (2012)
N, P, K	*P. nigra*	Stem	+	Luo et al. (2008)
P	*P. tremula*	Stem	–/0	Randriamanana et al. (2014)
N	*P. tremula*	Stem	–	Randriamanana et al. (2014)
N, P	*P. tremula*	Stem	–	Randriamanana et al. (2014)
N	*Populus* × *canadensis*	Root	+/–	Kosola et al. (2006)

(*Continued*)

Table 4.2 (Continued)

Factor	*Populus* species	Tissue type	Effect on CTs	Reference
P	*P. tremula*	Root	+/0	Randriamanana et al. (2014)
N	*P. tremula*	Root	+/–	Randriamanana et al. (2014)
N, P	*P. tremula*	Root	+/0/–	Randriamanana et al. (2014)
N, P, K, micronutrients	*P. tremuloides*	Root	–	Stevens et al. (2014)
Increased light				
	P. tremula	Leaf	+/0	Randriamanana et al. (2015)
	P. tremuloides	Leaf	+	Hemming and Lindroth (1999)
	P. tremuloides	Leaf	+	McDonald et al. (1999)
	P. tremuloides	Leaf	+	Osier and Lindroth (2006)
	P. tremuloides	Leaf	+	Call and St. Clair (2017)
	P. tremuloides	Leaf	–	Calder et al. (2011)
	P. trichocarpa	Leaf	0	Warren et al. (2002)
Elevated CO_2				
	P. tremula	Leaf	–	Randriamanana et al. (2018)
	P. tremuloides	Leaf	+	Roth et al. (1997)
	P. tremuloides	Leaf	+	Kinney et al. (1997)
	P. tremuloides	Leaf	+	Roth et al. (1998)
	P. tremuloides	Leaf	+	McDonald et al. (1999)
	P. tremuloides	Leaf	+	Couture et al. (2012)
	P. tremuloides	Leaf	+	Couture et al. (2014)
	P. tremuloides	Leaf	+	Couture and Lindroth (2014)
	P. tremuloides	Leaf	+	Couture et al. (2017)
	P. tremuloides	Leaf	+/0	Roth et al. (1998)
	P. tremuloides	Leaf	+/0	Agrell et al. (2000)
	P. tremuloides	Leaf	0	Holton et al. (2003)
	P. tremuloides	Leaf	0	Kopper and Lindroth (2003a)
	P. tremuloides	Leaf	0	Kopper and Lindroth (2003b)
	P. tremuloides	Leaf	0	Vigue and Lindroth (2010)

Table 4.2 (Continued)

Factor	*Populus* species	Tissue type	Effect on CTs	Reference
	P. tremuloides	Leaf	0	Couture and Lindroth (2012)
	P. tremuloides	Litter	0	Liu et al. (2005)
	P. nigra	Stem	0	Luo et al. (2008)
Elevated O_3				
	P. tremuloides	Leaf	+	Couture et al. (2012)
	P. tremuloides	Leaf	+	Couture et al. (2014)
	P. tremuloides	Leaf	+	Couture and Lindroth (2014)
	P. tremuloides	Leaf	+/0	Couture and Lindroth (2012)
	P. tremuloides	Leaf	+/−	Kopper and Lindroth (2003a)
	P. tremuloides	Leaf	0	Lindroth et al. (2001a)
	P. tremuloides	Leaf	0	Kopper and Lindroth (2003b)
	P. tremuloides	Leaf	0	Vigue and Lindroth (2010)
Elevated temperature				
	P. tremula	Bud	0	Sivadasan (2018)
	P. tremula	Leaf	−/0	Randriamanana et al. (2015)
	P. tremula	Leaf	−	Nissinen et al. (2017)
	P. tremula × tremuloides	Leaf	−	Kosonen et al. (2012)
	P. tremuloides	Leaf	−	Jamieson et al. (2015)
	P. tremuloides	Root	−	Z. Li and R. Lindroth (unpublished data)
Vernal frost				
	P. tremuloides	Leaf	+/0	St. Clair et al. (2009)
	P. tremuloides	Leaf	−	Rubert-Nason et al. (2017)
Fire damage				
	P. tremuloides	Leaf	+	Wan et al. (2014)
Drought				
	P. tremuloides	Leaf	0	Roth et al. (1997)
	P. yunnanensis	Leaf	0/−	Zhang et al. (2019)

[a] macronutrients: N, nitrogen; P, phosphorus; K, potassium.

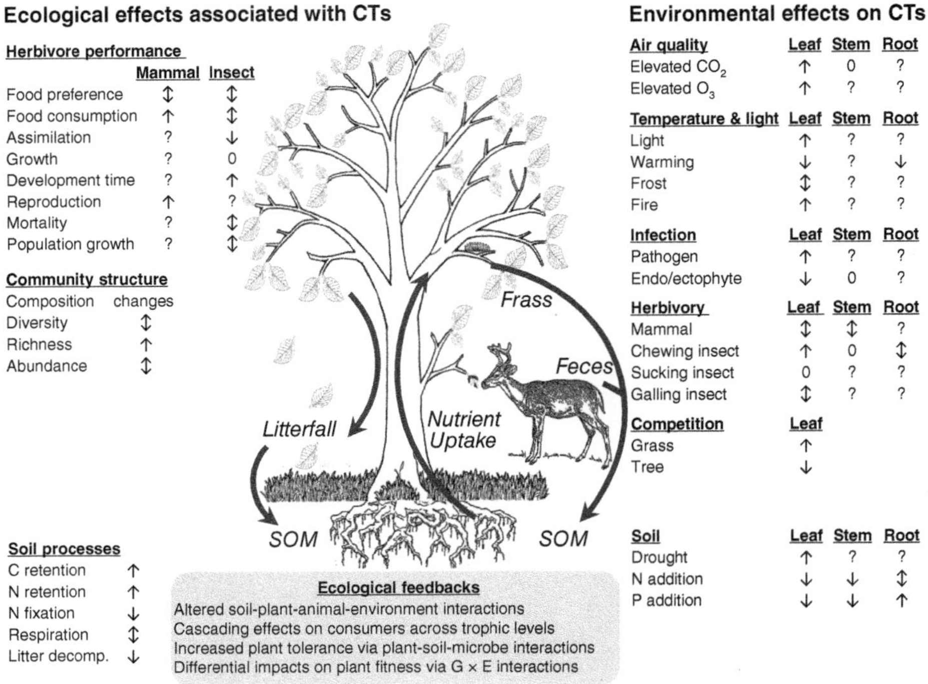

Figure.4.3 Condensed tannins (CTs) influence many aspects of *Populus* ecology and mediate plant-soil-herbivore/pathogen feedbacks that determine plant fitness (see Tables 4.2–4.6). Arrows indicate the full range of effect directions (positive, ↑; negative, ↓; variable, ↕; 0, absent; ?, unknown); consult Tables 4.2–4.6 for predominant effects. SOM = soil organic matter.

Abiotic factors influence *Populus* CT concentrations not only independently, but also interactively. For example, light intensity and CO_2 concentration have much stronger effects on CT levels at low soil nutrient availability than at high nutrient availability (Hemming and Lindroth 1999; Lindroth et al. 2001b; Osier and Lindroth 2006). Similarly, the effects of enriched CO_2 on CT concentrations can be greater at high light levels than at low light levels (McDonald et al. 1999). Elevated levels of CO_2 and temperature at magnitudes predicted for this century (IPCC 2013) have opposing effects on CT concentrations (Veteli et al. 2007). However, given that CO_2 and temperature will increase at different rates, their net effects on CT levels are difficult to predict.

Extreme events that may become more frequent in the Anthropocene (IPCC 2013) also affect CT concentrations (Table 4.2; Figure 4.3). Damaging vernal frosts increase or decrease CTs depending on tree ontogeny and severity of damage (St. Clair et al. 2009; Rubert-Nason et al. 2017). *Populus tremuloides* recovering from forest fire damage have elevated foliar CTs (Wan et al. 2014). Drought during a single growing season has minimal impact on *Populus* CTs, with no effect on *P. tremuloides* foliar CTs (Roth et al. 1997) and a weak negative effect on female but not male *P. yunnanensis* foliar CTs (Zhang et al. 2019). No studies to date have documented how extreme temperatures (i.e. "heat waves") influence CT concentrations. In general, although few studies have evaluated the impacts of extreme events on *Populus* CTs, the findings summarized here suggest that both the direction and magnitude of effects will vary among environmental factors, and will be context-dependent.

4.4.3.2 Biotic Environment

Numerous biotic factors, including herbivory, disease, and competition also contribute to variation in *Populus* CTs (Table 4.3). Effects can occur by induction or repression of various genes in the phenylpropanoid and malonic acid biosynthesis pathways (Figure 4.1) (Peters and Constabel 2002; Miranda et al. 2007; Jelinkova et al. 2012; Decker et al. 2017), and are influenced by the abiotic variables and resource tradeoffs described previously. In general, the magnitudes and directions of CT variation are determined by the type of stress agent; its intensity, location and timing; as well as by plant genotypic and environmental variation.

Herbivore damage generally increases foliar CT concentrations (Table 4.3; Figure 4.3). The positive effect of herbivory on CTs is most evident with folivorous insects (e.g. coleopterans and lepidopterans). Effects are more variable for mammal (e.g. ungulate) browsing that damages both leaves and stems, perhaps because substantial stem loss stimulates compensatory growth and reversion to developmentally younger leaves. A single study showed that the sap sucking aphid *Chaitophorus stevensis* had no effect on CT levels (Wong 2013), perhaps because aphid stylets are adapted to curtail defense induction by minimizing physical damage to the plant (Walling 2008). Galling insects had variable effects on CTs, which may have occurred because of the different feeding strategies of various gallers, gall densities, plant species or leaf ages (Schweitzer et al. 2005a; Frost et al. 2012).

Condensed tannin induction in response to herbivore damage is mediated by complex interactions among genetic, environmental, spatial, and temporal factors. For example, both short-term (Osier and Lindroth 2001) and long-term (Osier and Lindroth 2004) induction of CTs in *P. tremuloides* is highly genetically variable, and responses for the former are stronger under low soil nutrient conditions than under high nutrient conditions. *Populus tremuloides* exhibits systemic-level induction of CTs in response to foliar damage. But the magnitude of induction is greater in damaged leaves than in nearby undamaged leaves (Rubert-Nason et al. 2015) and induction does not occur in adjacent ramets that share a common root system with damaged trees (Cope and Lindroth 2018). Although *Populus* genotypes vary in their constitutive and herbivore-induced expression of CTs, the magnitude of induction appears to be largely independent of constitutive expression (Osier and Lindroth 2001). Finally, herbivore-induced variation in CT expression changes seasonally during an insect outbreak. Donaldson and Lindroth (2008) reported that CT concentrations increased in *P. tremuloides* during an active outbreak of *Malacosoma disstria*, but declined in late-season, reflushed leaves.

In contrast to herbivory, microbial infection of *Populus* has variable effects on CT levels (Table 4.3). Limited evidence suggests that responses of CTs to infection vary depending on whether the microbe is pathogenic (no effect or increase in CTs) or symbiotic (no effect or decrease in CTs).

Intraspecific and interspecific competitive interactions can also influence expression of CTs in *Populus*. Foliar CT levels are lower in high-density than low-density stands of *Populus tremuloides*, most likely because of reduced light availability (O. Cope and R. Lindroth unpublished data). When competition is predominantly belowground, however, opposite effects can occur. Competition of *P. tremuloides* with *Calamagrostis canadensis* grass resulted in elevated foliar CT levels, likely a consequence of reduced nutrient availability (Donaldson et al. 2006a).

Table 4.3 Effects of biotic environmental factors on CT concentrations in *Populus*

Category	Biotic factor	*Populus* species	Plant tissue type	Effect on CTs	Reference
Mammal browsing					
	Artificial + *Odocoileus virginianus* saliva	*P. tremuloides*	Leaf	–	Keefover-Ring et al. (2016)
	Various ungulates	*P. tremuloides*	Leaf	+	Rhodes et al. (2017)
	Artificial + *Odocoileus virginianus* saliva	*P. tremuloides*	Stem	+/0/–	Keefover-Ring et al. (2016)
Leaf-chewing insect					
	Chrysomela confluens	*P. angustifolia*	Leaf	0	Holeski et al. (2012)
	Various	*P. euphratica*	Leaf	0	Zhao et al. (2014)
	Chrysomela confluens	*P. fremontii*	Leaf	+	Holeski et al. (2012)
	Artificial	*P. fremontii* × *angustifolia*	Leaf	+	Tsai et al. (2006)
	Chrysomela confluens	*P. fremontii* × *angustifolia*	Leaf	+	Holeski et al. (2012)
	Artificial	*P. tremuloides*	Leaf	+	Osier and Lindroth (2001)
	Artificial	*P. tremuloides*	Leaf	+	Kao et al. (2002)
	Artificial	*P. tremuloides*	Leaf	+	Peters and Constabel (2002)
	Artificial	*P. tremuloides*	Leaf	+	Osier and Lindroth (2004)
	Leucoma salicis	*P. tremuloides*	Leaf	+	Peters and Constabel (2002)
	Lymantria dispar	*P. tremuloides*	Leaf	+	Rubert-Nason et al. (2015)
	Malacosoma disstria	*P. tremuloides*	Leaf	+	Peters and Constabel (2002)
	Malacosoma disstria + artificial	*P. tremuloides*	Leaf	+/–/0	Stevens et al. (2012)
	Malacosoma disstria + artificial	*P. tremuloides*	Leaf	–/0	Stevens and Lindroth (2005)

	Malacosoma disstria + artificial	*P. tremuloides*	Stem	0	Lindroth et al. (2007)
	Lymantria dispar	*P. canadensis*	Root	+/0	Kosola et al. (2006)
	Malacosoma disstria + artificial	*P. tremuloides*	Root	+/–	Stevens et al. (2014)
Sap-sucking insect					
	Chaitophorus stevensis	*P. tremuloides*	Leaf	0	Wong (2013)
Galling insect					
	Pemphigus betae	*P. fremontii* × *angustifolia*	Leaf litter	+	Schweitzer et al. (2005b)
	Ectoedemia populella	*P. grandidentata*	Leaf	–	Frost et al. (2012)
	Ectoedemia populella	*P. tremuloides*	Leaf	–	Frost et al. (2012)
Infection					
Fungal pathogen	*Drepanopeziza* spp.	*P. tremuloides*	Leaf	0	Call and St. Clair (2017)
	Melampsora medusae	*P. trichocarpa* × *deltoides*	Leaf	+	Miranda et al. (2007)
	Melampsora larici-populina	*P. nigra*	Leaf	+	Ullah et al. (2017)
	Melampsora larici-populina	*P. nigra*	Leaf	+	Ullah et al. (2019)
	Melampsora larici-populina	*P. trichocarpa* × *deltoides*	Leaf	0	Pfabel et al. (2012)
Fungal endophyte	Consortium	*P. fremontii*	Stem	–	Bailey et al. (2005)
Ectomycorrhizae	*Hebeloma esophaeum*	*P. trichocarpa* × *deltoides*	Leaf	0	Pfabel et al. (2012)
Competition					
Grass	*Calamagrostis canadensis*	*P. tremuloides*	Leaf	+	Donaldson et al. (2006a)
Trees	*P. tremuloides*	*P. tremuloides*	Leaf	–	O. Cope and R. Lindroth (unpublished data)

In contrast to leaves, CTs in stem and root tissues exhibit no consistent responses to biological factors. However, relatively little research has been conducted on these tissues. Similarly, no published studies describe the effects of herbivores on CT molecular structures, although preliminary findings (K. Rubert-Nason and R. Lindroth unpublished data) show that herbivory can alter CT polymer size (by interaction with genotype) and increase the PC:PD ratio.

4.5 Roles of CT Variation in *Populus*-Environment Interactions

Effects of CTs on *Populus* ecology can be deduced from two basic categories of information: (i) (bio)chemical evidence relating CTs to biological effects such as oxidative stress and nutrient absorption; and (ii) correlations between CT concentrations and biological functions at organismal (Table 4.4), community (Table 4.5), and ecosystem scales (Table 4.6). In general, the biological and ecological effects of CTs depend on the target organism, the CT concentration (Table 4.4), and (by inference from non-*Populus* species) CT molecular structure (Zucker 1983; Clausen et al. 1990; Boettcher and Targett 1993). An important limitation of most studies linking CTs to biological effects is that they do so based on associative rather than mechanistic evidence. In the remainder of this section we summarize the evidence for, and the implications of, the major putative biological roles of CTs.

4.5.1 Biological Activity of CTs

4.5.1.1 Mode of Action

The observed biological effects of CTs follow from their chemical properties. Condensed tannins from a variety of plant species form complexes with ions and proteins (Quideau et al. 2011; Constabel et al. 2014), and influence oxidation-reduction reactions (Hagerman et al. 1998) by shuttling electrons, affording protection from oxidative stress (Gourlay and Constabel 2019). The capacity of CTs to associate with ions, proteins, and fungal chitin (Ayres et al. 1997; Barbehenn and Constabel 2011; Quideau et al. 2011; Constabel et al. 2014; Adamczyk et al. 2019; Hättenschwiler et al. 2019) is attributed to hydrogen bonding and ionic forces resulting from the abundance of carboxyl and hydroxyl functional groups. The chemical properties of CTs vary based on molecular structure – e.g. proportions of procyanidin and prodelphinidin (PC:PD) units, degree of polymerization, and polymer stereochemistry (Ayres et al. 1997; Quideau et al. 2011; Salminen and Karonen 2011). Increased hydroxylation of CT polymers (i.e. greater prodelphinidin content) is generally associated with increased antiherbivore activity – a probable consequence of the more hydroxylated molecules having higher affinities for proteins and ions and/or different oxidation-reduction potentials (Ayres et al. 1997; Quideau et al. 2011; Constabel et al. 2014; Marsh et al. 2019). Increased molecular size (degree of polymerization) is also associated with increased antiherbivore activity, although that generalization is based on only three species (Ayres et al. 1997). The effect of molecular size could be due to different modes of action; for example, larger molecules bind proteins more strongly (Marsh et al. 2019) whereas smaller molecules may more easily translocate across membranes where they can engage in oxidation-reduction reactions. Overall, the wide range of CT molecular structures observed in *Populus* suggests the potential for myriad, context-dependent biological effects.

4.5.1.2 Trends in Observed CT Biological Activity

Based on the relationships between CT chemical structures and properties, the biological activity of *Populus* CTs is generally expected to (i) increase with CT concentration, (ii) increase with hydroxylation (decreasing PC:PD ratio), (iii) increase with degree of polymerization (with caveats), and (iv) be influenced by stereochemistry. Numerous studies demonstrating correlations between CT concentrations and biological effects (Tables 4.4 and 4.5) suggest that CT effects are dose dependent. The impact of chemical structure variations on the biological activity of *Populus* CTs must be inferred from studies *in vitro* and in other plant species, as this topic has received little attention in *Populus*. Another major limitation in our understanding of the biological effects of *Populus* CTs is a general paucity of genetic, biochemical, and physiological evidence that explains correlations of CTs with organism performance (e.g. mortality, feeding). Nevertheless, numerous studies demonstrate associations between CTs and biological effects (Table 4.4). The plethora of such associations suggests that CTs enhance *Populus* fitness by influencing interactions with herbivores and pathogens, tolerance to environmental stress, and ecosystem function.

4.5.2 Relationships of Populus CTs to Organism Performance: CTs as Agents of Defense

Differences in CT concentrations are associated with a variety of changes in the performance of mammal and insect herbivores (Table 4.4; Figure 4.3), as well as microbial pathogens utilizing *Populus* as hosts (Holeski et al. 2009; Ullah et al. 2017). Decades ago, negative relationships between phenolic compounds and herbivore performance in a variety of plant taxa (Feeny 1970; Cates and Rhoades 1977; Bernays et al. 1989; Forkner et al. 2004) led to the hypothesis that CTs defend plants against herbivory. Condensed tannins in *Populus* (Ullah et al. 2017) and other plant taxa also have *in vitro* activity against microbes (McAllister et al. 2005), suggesting that CTs may inhibit pathogen infection.

Evidence for the efficacy of *Populus* CTs as defenses against herbivores varies among herbivore species, and is complicated by several factors. First, nearly all studies are correlative, showing relationships, but not causation, between herbivore performance and CT content. Second, the effects of CTs are context-specific, varying in relation to other dietary constituents and the type of animal performance metric. Given those caveats, some patterns can still be identified (Table 4.4). *Populus* CTs reduce food preferences and consumption in beaver (*Castor canadensis*) (Bailey et al. 2004) and some chrysomelid beetle species (Gruppe et al. 1999), but generally not in ungulates (Wooley et al. 2008; Holeski et al. 2016) and Lepidoptera (Couture et al. 2012; Boeckler et al. 2014). Condensed tannins have been associated with negative and no effects on food digestion/assimilation (Hemming and Lindroth 1999; Compson et al. 2018). Numerous studies have assessed the relationship of invertebrate growth and development to *Populus* CTs, and most have revealed no effect (Table 4.4). Two studies have reported a positive association between CT consumption and reproduction in sheep (*Ovis aries*) (McWilliam et al. 2004, 2005), although whether this pattern generalizes to other ruminant species is unknown.

The surprisingly modest effects of CTs on herbivore performance can be attributed to two major factors. First, CTs are less labile and reactive in biological contexts than are many other phenolic constituents, including hydrolysable tannins (Barbehenn and Constabel 2011). Second, herbivores exhibit multiple defenses against CTs, including tannin-binding salivary proteins in many mammals, and high gut pH and peritrophic membranes in insects (Barbehenn and Constabel 2011).

Table 4.4 Effects of *Populus* CTs on mammals and invertebrates

Performance metric	Organism affected	Effect	Tree species	Tissue type	CT conc. (% d.m.)[a]	Reference
Food preference	*Castor canadensis*	–	*P. angustifolia*	Bark	0–4	Bailey et al. (2004)
	Castor canadensis	–	*P. fremontii*	Bark	0–4	Bailey et al. (2004)
	Castor canadensis	–	*P. fremontii* × *angustifolia*	Bark	0–4	Bailey et al. (2004)
	Cervus canadensis	0	*P. tremuloides*	Leaf	0–2	Wooley et al. (2008)
	Cervus elaphus	–/0	*P. euramericana*	Leaf + stem	~3	Semiadi et al. (1995)
	Cervus unicolor	+	*P. euramericana*	Leaf + stem	~3	Semiadi et al. (1995)
	Erethizon dorsatum	+	*P. tremuloides*	Leaf	15–17	Diner et al. (2009)
	Microtus agrestis	0	*P. tremula* × *tremuloides*	Leaf	0.1–0.2	Hjältén et al. (2008)
	Clethrionomys glareolus	0	*P. tremula* × *tremuloides*	Leaf	0.1–0.2	Hjältén et al. (2008)
	Odocoileus virginianus	0/–	*P. tremuloides*	Leaf + stem	0.5–5	Holeski et al. (2016)
	Byctiscus populi	–/0	*P. trichocarpa*	Leaf	0.1–3	Gruppe et al. (1999)
	Byctiscus populi	–/0	*P. tremula* × *tremuloides*	Leaf	0.1–1	Gruppe et al. (1999)
	Chrysomela spp.	–/0	*P. trichocarpa*	Leaf	0.1–3	Gruppe et al. (1999)
	Chrysomela spp.	–/0	*P. tremula* × *tremuloides*	Leaf	0.1–1	Gruppe et al. (1999)
	Chaitophorus stevensis	–	*P. tremula* × *tremuloides*	Leaf	0.3–9	Wong (2013)
	Hesperophylax magnus	+/–	*P. angustifolia*	Litter	n.r.[b]	Compson et al. (2018)
	Hesperophylax magnus	+/–	*P. fremontii*	Litter	n.r.	Compson et al. (2018)
	Hesperophylax magnus	+/–	*P. fremontii* × *angustifolia*	Litter	n.r.	Compson et al. (2018)
	Lymantria dispar	+	*P. tremula* × *tremuloides*	Leaf	7–18	Boeckler et al. (2014)
	Malacosoma disstria	+	*P. tremula* × *tremuloides*	Leaf	7–18	Boeckler et al. (2014)

Food consumption	*Ovis aries*	+/0	*P. tremuloides*	Leaf	2–8	Heroy et al. (2018)
	Ovis aries	+/0/–	*P tremuloides*	Leaf	2–11	Villalba et al. (2014)
	Lumbricus terrestris	0	*P. tremuloides*	Litter	0.6–2	Meehan et al. (2010)
	Lymantria dispar	+	*P. tremuloides*	Leaf	2–27	Osier and Lindroth (2001)
	Lymantria dispar	+	*P. tremuloides*	Leaf	1–7	Hemming and Lindroth (1999)
	Lymantria dispar	0	*P. tremuloides*	Leaf	3–15	Hwang and Lindroth (1997)
	Lymantria dispar	0	*P. tremuloides*	Leaf	13–18	Couture et al. (2012)
	Malacosma disstria	0	*P. tremuloides*	Leaf	3–15	Hwang and Lindroth (1997)
	Malacosoma disstria	0	*P. tremuloides*	Leaf	14–32	Donaldson and Lindroth (2008)
	Malacosoma disstria	0	*P. tremuloides*	Leaf	13–18	Couture et al. (2012)
	Malacosoma disstria	–	*P. tremuloides*	Leaf	1–7	Hemming and Lindroth (1999)
	Papilio canadensis	0	*P. tremuloides*	Leaf	5–20	Hwang and Lindroth (1998)
Digestion/ assimilation	*Hesperophylax magnus*	–/0	*P. angustifolia*	Litter	n.r.	Compson et al. (2018)
	Hesperophylax magnus	–/0	*P. fremontii*	Litter	n.r.	Compson et al. (2018)
	Hesperophylax magnus	–/0	*P. fremontii* × *angustifolia*	Litter	n.r.	Compson et al. (2018)
	Lymantria dispar	–	*P. tremula* × *tremuloides*	Leaf	7–18	Boeckler et al. (2014)

(*Continued*)

Table 4.4 (Continued)

Performance metric	Organism affected	Effect	Tree species	Tissue type	CT conc. (% d.m.)[a]	Reference
	Lymantria dispar	–/0	*P. tremuloides*	Leaf	3–15	Hwang and Lindroth (1997)
	Lymantria dispar	–	*P. tremuloides*	Leaf	1–7	Hemming and Lindroth (1999)
	Malacosoma disstria	0	*P. tremuloides*	Leaf	3–15	Hwang and Lindroth (1997)
	Malacosoma disstria	–	*P. tremuloides*	Leaf	1–7	Hemming and Lindroth (1999)
	Papilio canadensis	0	*P. tremuloides*	Leaf	5–20	Hwang and Lindroth (1998)
Growth	*Lumbricus terrestris*	0	*P. tremuloides*	Litter	0.6–2	Meehan et al. (2010)
	Chrysomela falsa	0	*P. balsamifera*	Leaf	~6	Ayres et al. (1997)
	Chrysomela falsa	0	*P. tremuloides*	Leaf	0–12	Ayres et al. (1997)
	Chrysomela scripta	–	*P. tremuloides*	Leaf	2–9	Donaldson and Lindroth (2004)
	Hesperophylax magnus	0	*P. angustifolia*	Litter	n.r.	Compson et al. (2018)
	Hesperophylax magnus	0	*P. fremontii*	Litter	n.r.	Compson et al. (2018)
	Hesperophylax magnus	0	*P. fremontii* × *angustifolia*	Litter	n.r.	Compson et al. (2018)
	Lymantria dispar	+	*P. tremula* × *tremuloides*	Leaf	7–18	Boeckler et al. (2014)
	Lymantria dispar	+	*P. tremuloides*	Leaf	11–16	Couture and Lindroth (2012)
	Lymantria dispar	0	*P. tremuloides*	Leaf	13–18	Couture et al. (2012)
	Lymantria dispar	0	*P. tremuloides*	Leaf	10–20	Rubert-Nason et al. (2015)

Lymantria dispar	0	*P. tremuloides*	Leaf	3–11	Hemming and Lindroth (1995)
Lymntria dispar	0	*P. tremuloides*	Leaf	3–15	Hwang and Lindroth (1997)
Lymantria dispar	0	*P. tremuloides*	Leaf	1–7	Hemming and Lindroth (1999)
Lymantria dispar	0	*P. tremuloides*	Leaf	2–27	Osier and Lindroth (2001)
Lymantria dispar	0	*P. tremuloides*	Leaf	2–20	Donaldson and Lindroth (2007)
Malacosoma disstria	0	*P. tremuloides*	Leaf	13–18	Couture et al. (2012)
Malacosoma disstria	–/0	*P. tremuloides*	Leaf	3–15	Hwang and Lindroth (1997)
Malacosoma disstria	0	*P. tremuloides*	Leaf	7–15	Jamieson et al. (2015)
Malacosoma disstria	0	*P. tremuloides*	Leaf	3–11	Hemming and Lindroth (1995)
Malacosoma disstria	–/0	*P. tremuloides*	Leaf	1–7	Hemming and Lindroth (1999)
Pachysphinx modesta	0	*P. tremuloides*	Leaf	5–19	Hwang and Lindroth (1998)
Papilio canadensis	0	*P. tremuloides*	Leaf	5–19	Hwang and Lindroth (1998)
Pemphigus betae	+	*P. fremontii × angustifolia*	Leaf	2–16	Bailey et al. (2006)
Pyrrhalta luteola	–	*P. tremuloides*	Leaf	0–12	Ayres et al. (1997)

(*Continued*)

Table 4.4 (Continued)

Performance metric	Organism affected	Effect	Tree species	Tissue type	CT conc. (% d.m.)[a]	Reference
Development time	*Chrysomela scripta*	0	*P. tremuloides*	Leaf	2–9	Donaldson and Lindroth (2004)
	Lymantria dispar	+	*P. tremula × tremuloides*	Leaf	7–18	Boeckler et al. (2014)
	Lymantria dispar	+	*P. tremuloides*	Leaf	1–7	Hemming and Lindroth (1999)
	Lymantria dispar	+/0	*P. tremuloides*	Leaf	3–15	Hwang and Lindroth (1997)
	Lymantria dispar	0	*P. tremuloides*	Leaf	11–16	Couture and Lindroth (2012)
	Lymantria dispar	0	*P. tremuloides*	Leaf	3–11	Hemming and Lindroth (1995)
	Lymantria dispar	–	*P. tremuloides*	Leaf	2–27	Osier and Lindroth (2001)
	Malacosoma disstria	+/0	*P. tremuloides*	Leaf	1–7	Hemming and Lindroth (1999)
	Malacosoma disstria	0	*P. tremuloides*	Leaf	3–11	Hemming and Lindroth (1995)
	Malacosoma disstria	0	*P. tremuloides*	Leaf	3–15	Hwang and Lindroth (1997)
	Malacosoma disstria	0	*P. tremuloides*	Leaf	7–15	Jamieson et al. (2015)
	Pachysphinx modesta	0	*P. tremuloides*	Leaf	5–19	Hwang and Lindroth (1998)
	Papilio canadensis	0	*P. tremuloides*	Leaf	5–19	Hwang and Lindroth (1998)

Reproduction	*Ovis aries*	+	*Populus deltoides* × *nigra*	Leaf + stem	~7	McWilliam et al. (2004)
	Ovis aries	+/0	*Populus* spp.	Leaf + stem	~19	McWilliam et al. (2005)
Mortality	*Chrysomela falsa*	0	*P. tremuloides*	Leaf	0–12	Ayres et al. (1997)
	Lymantria dispar	0	*P. tremuloides*	Leaf	1–9	Hwang and Lindroth (1997)
	Lymantria dispar	0	*P. tremuloides*	Leaf	11–16	Couture and Lindroth (2012)
	Lymantria dispar	–	*P. tremula* × *tremuloides*	Leaf	7–18	Boeckler et al. (2014)
	Malacosoma disstria	0	*P. tremuloides*	Leaf	1–9	Hwang and Lindroth (1997)
	Pyrrhalta luteola	+	*P. tremuloides*	Leaf	0–12	Ayres et al. (1997)
Population growth	*Chaitophorus stevensis*	+	*P. tremula* × *tremuloides*	Leaf	0.3–8	Wong (2013)
	Chaitophorus stevensis	–	*P. tremuloides*	Leaf	1–7	Rubert-Nason et al. (2017)

[a] When multiple means are reported, the range between smallest and largest means is presented. When individual data points, rather than means, are reported, the range between the smallest and largest data points is presented.
[b] n.r., not reported.

Finally, we note that the apparent positive associations between *Populus* CTs and herbivore performance (Table 4.4) must be interpreted with caution. In many cases (e.g. Hemming and Lindroth 1999; Osier and Lindroth 2001; Boeckler et al. 2014), the putative positive effect of CTs on performance was a statistical artifact: CTs were inversely correlated with salicinoid phenolic glycosides, the plant constituents that more strongly governed the interactions. These results highlight the need for experimental feeding studies with purified CTs, to truly assess the effects of CTs on herbivore performance.

The efficacy of CTs as defense agents against pathogens has been explored in only a few studies. Holeski et al. (2009) reported a strong negative correlation between foliar CT concentrations and *Venturia* infection in *P. tremuloides*. Barker et al. (2016) found substantial environmental variation in susceptibility of *P. tremuloides* to powdery mildew (*Erysiphe adunca*), and implicated (but did not evaluate) CTs as the causative factor. In the most thorough investigations to date, Ullah et al. (2017, 2019) provide strong evidence that CTs protect *Populus* from infection by the rust fungus *Melampsora larici-populina*, corroborating the negative association between CTs and infection initially reported by McKown et al. (2014).

4.5.3 Relationships of Populus CTs to Community Structure

Populus CTs influence the structure of aboveground and belowground communities through differential effects on the abundance, richness, and diversity of organisms (Table 4.5; Figure 4.3). Aboveground, CTs influence the species composition of arthropod communities associated with *P. tremula*, *P. tremuloides*, *P. fremontii*, *P. angustifolia* and their hybrids (Bangert et al. 2006; Robinson et al. 2012; Barker et al. 2018). These effects are presumably determined by how CT concentration (and potentially structure) differentially affect the performance of individual arthropod species. Condensed tannins are also an important driver of understory plant species richness (Lamit et al. 2011), presumably due to the effects of CTs on soil nutrient cycling and microbes. In soil, CTs impact the abundance, richness, and diversity of bacteria and fungi (Table 4.5). One particularly notable consequence of increased CT inputs to soil is decreased abundance of N-fixing bacteria (Winder et al. 2013), which may in turn influence the structure of plant communities. Condensed tannins ingested by insects also alter the abundance of bacterial taxa in the gut microbiome, potentially influencing the performance of those insects (Mason et al. 2015).

A limited number of studies suggest that CTs influence multitrophic interactions in *Populus* communities (Table 4.5). Bailey et al. (2006) reported that cottonwood CT concentrations are positively associated with *Pemphigus* gall densities, which in turn are positively associated with levels of gall predation by birds. In controlled feeding studies, Müller et al. (2006) found that predation by chickadees on gypsy moths is inversely related to levels of aspen CTs in the insects' guts. Finally, Holton et al. (2003) showed that survivorship of the parasitoid *Compsilura concinnata* in forest tent caterpillars was inversely related to concentrations of CTs in the caterpillars' diets. These examples illustrate that the effects of CTs are not limited to plant-animal or plant-microbe interactions, and support the notion that CTs may be involved in broader community- and ecosystem-level functions.

Table 4.5 Effects of *Populus* CTs on community structure

Parameter	Effect of increasing CT conc.	CT-producing species	Reference
Community composition (location)			
Arthropod (tree canopy)	Altered	*P. tremula*	Robinson et al. (2012)
Arthropod (tree canopy)	Altered	*P. fremontii, P. angustifolia and hybrids*	Bangert et al. (2006)
Insects	Altered	*P. tremuloides*	Barker et al. (2018)
Microbe (soil)	Altered	*P. angustifolia and P. angustifolia* × *fremontii*	Schweitzer et al. (2007)
Community diversity (location)			
Bacteria (soil)	0	*P. tremula* × *tremuloides*	Winder et al. (2013)
Bacteria (leaf litter)	+	*P. fremontii, P. angustifolia,* and *hybrids*	Marks et al. (2009)
Bacteria (leaf litter)	+	*P. tremuloides*	Kelly et al. (2010)
Fungi (soil)	–/0[a]	*P. tremula* × *tremuloides*	Winder et al. (2013)
Fungi (leaf litter)	+	*P. fremontii, P. angustifolia,* and *hybrids*	Marks et al. (2009)
Fungi (on leaf litter)	0	*P. tremuloides*	Kelly et al. (2010)
Community richness (location)			
Plants (forest understory)	+	*P. fremontii, P. angustifolia* and *hybrids*	Lamit et al. (2011)
Fungi (soil)	0	*P. tremula* × *tremuloides*	Winder et al. (2013)
Community abundance (location)			
Ralstonia bacteria (insect gut)	+	*P. tremuloides*	Mason et al. (2015)
Acinetobacter bacteria (insect gut)	–	*P. tremuloides*	Mason et al. (2015)
Bacteria (leaf litter)	–	*P. tremuloides*	Kelly et al. (2010)
β-proteobacteria (soil)	+	*P. tremula* × *tremuloides*	Winder et al. (2013)

(*Continued*)

Table 4.5 (Continued)

Parameter	Effect of increasing CT conc.	CT-producing species	Reference
Bacteroides (leaf litter)	+	*P. tremuloides*	Kelly et al. (2010)
Firmicutes bacteria (leaf litter)	–	*P. tremuloides*	Kelly et al. (2010)
Gram+ bacteria (leaf litter)	–	*P. tremuloides*	Kelly et al. (2010)
N-fixing soil bacteria	–	*P. tremula* × *tremuloides*	Winder et al. (2013)
Ammonia-oxidizing bacteria (soil)	0	*P. fremontii*, *P. angustifolia* and *hybrids*	Selmants et al. (2019)
Ammonia-oxidizing Archaea (soil)	–	*P. fremontii*, *P. angustifolia* and *hybrids*	Selmants et al. (2019)
Fungi (leaf litter)	–	*P. tremuloides*	Kelly et al. (2010)
Endophytic fungi (twig)	–	*P. fremontii* × *angustifolia*	Bailey et al. (2005)
Multitrophic interactions			
	More CTs correlated with increased gall density and increased gall predation by birds	*P. fremontii* × *angustifolia*	Bailey et al. (2006)
	More CTs in *Lymantria dispar* diet decreased feeding on *L. dispar* by birds (*Poecile atricapillus*)	*P. tremuloides*	Müller et al. (2006)
	More CTs in *Malacosoma disstria* diet decreased parasitoid (*Compsilura concinnata*) survivorship	*P. tremuloides*	Holton et al. (2003)

[a] Effect direction is context-dependent.

4.5.4 Relationships of Populus CTs to Ecosystem Function

At the ecosystem level, CTs are key drivers of soil processes such as litter decomposition and nutrient cycling (Schweitzer et al. 2008; Chomel et al. 2016; Hunter 2016). In general, *Populus* CTs alter soil respiration and decrease decomposition and N mineralization (Table 4.6; Figure 4.3). These changes can be attributed to differential effects of CTs on soil microbes (Hunter 2016). For example, higher molecular weight CTs tend to inhibit microbial growth while lower molecular weight CTs can stimulate microbial growth by being preferentially utilized as substrates (Bhat et al. 1998; Madritch et al. 2007b; Schweitzer et al. 2008). By altering the availability of soil nutrients (Constabel et al. 2014) and serving as substrates that increase or inhibit the growth of specific soil microbial taxa (Bhat et al. 1998; Madritch et al. 2007a; Constabel et al. 2014), CTs can alter the abundance of different taxa and change microbial community structure (Table 4.5).

Table 4.6 Effects of *Populus* CTs on ecosystem function

Parameter	Effect	CT-producing species	Reference
Soil respiration	+/–	*P. balsamifera*	Fierer et al. (2001)
	–	*P. balsamifera*	Schimel et al. (1996)
	+/0/–	*P. tremuloides*	Madritch et al. (2007b)
	–	*P. balsamifera*	Schimel et al. (1998)
Soil nutrient cycling	Decrease N availability	*P. balsamifera*	Schimel et al. (1996)
	Decrease N availability, alter C mineralization	*P. balsamifera*	Fierer et al. (2001)
	Decrease N mineralization	*P. balsamifera*	Schimel et al. (1998)
	Decrease nitrification	*P. fremontii*, *P. angustifolia* and *hybrids*	Selmants et al. (2019)
	Alter N mineralization	*P. fremontii*, *P. angustifolia* and *hybrids*	Schweitzer et al. (2004)
	Alter rates of C, N, S mineralization	*P. tremuloides*	Madritch et al. (2006)
	Increase nitrogen recovery by plants after insect defoliation	*P. tremuloides*	Madritch and Lindroth (2015)
	Stabilize C (in fungal necromass)		Adamczyk et al. (2019)
Nitrogen fixation	–	*P. balsamifera*	Cates and Rhoades (1977)
	–	*P. balsamifera*	Schimel et al. (1998)
Litter decomposition	–	*P. angustifolia*	Shay et al. (2018)
	–	*P. balsamifera*	Schimel et al. (1998)
	–	*P. fremontii*, *P. angustifolia*	LeRoy et al. (2006)

(*Continued*)

Table 4.6 (Continued)

Parameter	Effect	CT-producing species	Reference
	–	*P. fremontii, P. angustifolia*	LeRoy et al. (2007)
	–	*P. fremontii* × *angustifolia*	Driebe and Whitham (2000)
	–	*P. fremontii, P. angustifolia* and *hybrids*	Schweitzer et al. (2004)
	–	*P. fremontii, P. angustifolia, and hybrids*	Marks et al. (2009)
	–	*P. tremuloides*	Madritch et al. (2006)
	–	*P. tremuloides*	Parsons et al. (2008)
	–	*P. tremuloides*	Liu et al. (2009)
	–	*P. tremuloides*	Preston et al. (2009)
	–	*P. tremuloides*	Frost et al. (2012)

Tannin-mediated feedbacks link ecosystem processes in areas with abundant *Populus*. Condensed tannins may enhance *Populus* tolerance to nutrient stress and defoliation, through effects on plant-soil and plant-soil-herbivore interactions. Schweitzer et al. (2008) argued that CTs, through their negative effects on N mineralization and differential effects on soil microbial communities, may facilitate conservation of soil N when it is in short supply. They further suggested that CTs could enhance uptake of soil N by plants through their effects on root symbionts. These hypotheses have broad implications – that CTs can facilitate the growth of *Populus* in N-limited environments, and that CTs can help *Populus* recover N lost through defoliation (caused by herbivory or an extreme climate event). Although evidence supporting these hypotheses remains limited, Madritch and Lindroth (2015) successfully demonstrated that CTs deposited to soil via insect frass can increase N recovery by *P. tremuloides* after defoliation. This discovery agrees with the broader hypothesis presented by Hunter (2016), that CT-mediated feedbacks can increase plant fitness through differential effects on soil microbes, nutrient uptake, and insect herbivores. This work also provides some of the strongest evidence to date that CTs influence both the slow recycling of nutrients from senesced litter as well as the rapid recycling of nutrients from herbivore feces and residues, which are typically less recalcitrant to decomposition than leaf litter.

Finally, because expression of CTs is strongly genetically determined, genotypic variation in clonal *Populus* species such as *P. tremuloides* can create complex mosaics of ecosystem function at the landscape level (Madritch et al. 2009). Genotypic and environmental variation interact to determine CT levels, which in turn directly or indirectly influence trophic interactions, herbivore and soil microbial communities, and nutrient cycling. These effects feed back over ecological time to affect tree fitness, and over evolutionary time to affect population genetic structure. This perspective of the complex causes and consequences of variation in CT expression supports the notion of the "phytochemical landscape" advanced by Hunter (2016).

4.6 Importance of CTs in *Populus*-dominated Ecosystems of the Anthropocene

4.6.1 Condensed Tannins and Populus Ecology

Environmental conditions are projected to change more dynamically and in novel ways during the Anthropocene, compared with recent climate history. For example, atmospheric CO_2, and O_3, as well as temperature, are projected to increase substantially in this century, and extreme events such as droughts and heat waves are likely to become more frequent (IPCC 2013). Because these environmental factors can alter the geographical distribution of *Populus* and/or the quantity and quality of the CTs that they produce, environmental changes associated with the Anthropocene are likely to cause unique shifts in the many ecological interactions mediated by CTs.

Global change may influence CT-mediated ecological interactions directly, by changing tissue CT levels, or indirectly, by changing the genetic composition of *Populus* populations. Both processes were demonstrated at the large Aspen FACE (Free Air CO_2 and O_3 Enrichment) experiment in northern Wisconsin, USA. Carbon dioxide and O_3 differentially affected (i) CT production in multiple *P. tremuloides* genotypes (Couture et al. 2014), (ii) canopy insect community composition and defoliation (Hillstrom and Lindroth 2008; Couture et al. 2015), (iii) insect-mediated transfer of frass, litter and CTs from the canopy to the forest floor (Meehan et al. 2014), (iv) litter decomposition (Parsons et al. 2008), and (v) soil C and N mineralization (Hillstrom et al. 2010). McDonald et al. (2002) reported that the same *P. tremuloides* genotypes exhibited differential growth responses to atmospheric CO_2 and O_3, and that those differences interacted with competition to influence forest stand structure. More recently, Ware et al. (2019) reported that temperature increases reduce genetic variation in phenology of riparian *P. angustifolia*, with repercussions for soil microbial communities and nutrient pools. Whether CTs play a role in those effects was not determined, but is likely given results from previous experiments in the same system (Schweitzer et al. 2004).

Specific predictions about how the ecological impacts of *Populus* CTs may shift in environments of the future are difficult to make because different environmental factors produce different effects, which are often context-dependent (Tables 4.2 and 4.3). For example, both atmospheric CO_2 and O_3 typically increase foliar CT concentrations, but the magnitudes of increase vary among genotypes and with tree age (Couture et al. 2014). In contrast, elevated temperatures decrease foliar CT concentrations (Table 4.2), while the effects of drought are poorly known. To summarize, the combined effects of multiple global change factors on CT production, and consequences thereof for *Populus* ecology, are expected to be highly context-dependent.

Global change may also influence CT-mediated ecological interactions at much larger scales by altering distributional ranges of *Populus* species. Range shifts may occur naturally, due to changing environmental conditions that make current geographical locations more or less suitable to *Populus* fitness (McKenney et al. 2007; Morin et al. 2008), or via forest management practices aimed at increasing *Populus* cultivation for bioenergy, wood products, and carbon sequestration. Population declines of *P. tremuloides* throughout much of North America have been linked to warmer and drier climates, and amplified by insect and pathogen damage (Huang and Anderegg 2012; Worrall et al. 2013). Because CTs influence herbivore performance (Table 4.4), aboveground and belowground communities (Table 4.5), and soil processes (Table 4.6), altered CT inputs to ecosystems due to range shifts of *Populus* spp. are likely to produce a variety of effects at different biological scales. In addition, exportation of *Populus* species and their associated herbivores outside of native ecosystems will present new opportunities for herbivore proliferation, as exemplified by Asian longhorn beetle outbreaks in North America (Hu et al. 2009).

4.6.2 Condensed Tannins and Populus Evolution

The extraordinary genetic diversity of *Populus* spp. (Cole 2005; Slavov and Zhelev 2010) is reflected in the range of CT concentrations and molecular structures observed within and among species in this genus (Table 4.1) (Scioneaux et al. 2011; Lin et al. 2014; Rubert-Nason and Lindroth 2019), which are linked to plant fitness through CT-mediated ecological interactions (Schweitzer et al. 2008). Plasticity of CTs in response to environmental factors (Tables 4.2 and 4.3) suggests that environment, via differential interactions with genotypes, may drive the evolutionary trajectories of *Populus* populations over time. Moreover, these interactions are expected to change as environments shift in the Anthropocene, exerting new selective pressures on *Populus* populations. Just how they will change, however, is difficult to predict. Not only do different environmental change factors have opposing effects on CT production (e.g. elevated CO_2 and elevated temperature), but CT expression is influenced by biochemical and physiological tradeoffs, and organismal interactions, that will themselves be altered by environmental change. In short, while mechanistic details are presently speculative, the strong genetic determination of CT responses to environmental change factors suggests that novel suites of selective pressures in the Anthropocene will influence the evolutionary trajectories of *Populus* populations.

4.7 Conclusions and Challenges

Condensed tannins are a major class of secondary metabolites with diverse biological and ecological functions in *Populus* species (Figure 4.3). The metabolic pathways for CT biosynthesis are complex, with multiple branch points and modes of control. These divergent pathways and control mechanisms may explain tradeoffs between production of CTs and other secondary metabolites, or with physiological processes such as growth. Condensed tannin expression is highly variable and determined by a complex of interacting genetic, ontogenetic and environmental factors. This variation has repercussions for diverse biological functions at levels ranging from individual organisms to entire ecosystems. Current evidence suggests that environmentally mediated tradeoffs among CT production, other secondary metabolite production, and plant growth differentially impact *Populus* fitness, both directly and via interaction with genotype (G×E and G×E×E interactions).

Although CTs have a limited role in defending *Populus* against many herbivores, emerging evidence suggests that the compounds may confer protection against fungal pathogens and influence the structure and function of associated communities and ecosystems. Condensed tannins appear to have prominent belowground roles, with effects on soil microbial communities that influence ecosystem-level processes such as litter decomposition and nutrient cycling. Accruing evidence also suggests that CTs affect the dynamics of nutrient transfer in plant-soil and insect-plant-soil complexes. Indeed, CTs can facilitate nutrient uptake after defoliation, and thus may function more as agents of plant tolerance, rather than resistance, to herbivore damage. As suggested by Hunter (2016), these effects are anticipated to translate to the landscape scale.

On the basis of the literature reviewed herein, we advance the notion that CTs are a pivotally important phenotypic trait of *Populus*, which can mediate multitrophic interactions and inform understanding of ecological and evolutionary processes across scales of organization, space, and time (Abdala-Roberts et al. 2019). Moreover, quantitative and qualitative changes in *Populus* CTs, as effected by changing environmental conditions, are likely to alter organismal interactions, community and ecosystem functions, and evolutionary processes in novel ways in the Anthropocene.

Our effort to identify emergent patterns in the literature regarding the causes and consequences of CT variation in *Populus* has revealed numerous knowledge gaps, which should be addressed in future research:

1) Although the fundamental CT biosynthetic pathways have been well delineated, understanding of how those pathways are regulated at the genetic and biochemical levels is incomplete. Future research should examine how production and distribution of CTs is regulated, how that regulation results in tradeoffs at both the biochemical and physiological levels, and how regulation is influenced by biotic and abiotic environmental factors. Such research will strengthen understanding of the causal relationships between environmental factors and CT plasticity and provide insights into the evolution of CT expression.
2) Present understanding of CT dynamics in *Populus* is derived from measuring changes in foliar CT concentrations in relatively few species, small populations, and limited geographical areas. Studies that evaluate the plasticity of CT quantity and quality in more species and at larger spatial scales will enhance understanding of the roles of *Populus* CTs in diverse contexts across ecological landscapes.
3) The efficacy of *Populus* CTs as defense compounds appears to be limited to specific combinations of species and environmental conditions. To ascertain whether the effects of CTs are truly idiosyncratic, or whether they exhibit generalizable patterns among major taxonomic groups (e.g. insects vs. fungi; chrysomelids vs. lepidopterans), future research should examine a broader range of taxa, especially in combination.
4) Present understanding of the effects of CTs on individual organisms, communities, and ecosystems is based almost entirely on correlative studies. Future research should consider how CTs, either in purified form or in genetically up- or down-regulated *Populus* trees, affect ecological interactions. This research should take into account the effects of variation in both CT concentration and molecular structure.
5) *Populus* CTs provide the opportunity to explore how key plant phenotypic traits influence multitrophic interactions and link to community and ecosystem structure and function. Research integrated across multiple scales of organization, space, and time will help reveal the emergent properties of communities and ecosystems (Abdala-Roberts et al. 2019).

6) Although *Populus* CTs appear to play more important roles in ecosystem function than in plant defense against herbivores, how the ecological roles of CTs may confer a fitness advantage to individual trees is unclear. Future research should elucidate how a complex of CT-mediated direct effects, indirect effects, and feedback loops are tied to individual tree fitness, and can thereby drive the evolutionary trajectories of *Populus* populations.
7) We have only nascent understanding of how environments of the future will influence CT production and CT-mediated interactions. Given the plethora of environmental factors that are changing simultaneously, research should investigate multiple factors in concert, and over sufficiently long periods of time to account for ontogenetic effects. Both large-scale experiments (e.g. FACE) and landscape-level environmental simulations (e.g. where altitude is used as a proxy for climate) are needed.

4.8 Acknowledgements

We thank the organizers of the XXIX International Conference on Polyphenols and 9th Tannin Conference for the invitation to write this paper. We are indebted to the large number of undergraduate, graduate, and postgraduate students over 30 years in the Lindroth Lab, whose research provided the basis for much of the literature reviewed here. We thank Chris Cole for creative insights and conversations about tannin biosynthesis and tradeoffs, and Alex Guenther for clerical assistance. Funding to support the preparation of this manuscript was provided by the United States Department of Agriculture (McIntire-Stennis project WIS01842, Hatch project WIS01934, and AFRI grant 2016-67013-25088), by the National Science Foundation (grant DEB-1456592), and by the University of Maine at Fort Kent.

References

Abdala-Roberts, L., Puentes, A., Finke, D.L. et al. (2019). Tri-trophic interactions: bridging species, communities and ecosystems. *Ecology Letters.* 22: 2151–2167.

Adamczyk, B., Adamczyk, S., Smolander, A. et al. (2018). Plant secondary metabolites – missing pieces in the soil organic matter puzzle of boreal forests. *Soil Systems* 2: 2–10.

Adamczyk, B., Sietiö, O., Biasi, C., and Heinonsalo, J. (2019). Interaction between tannins and fungal necromass stabilizes fungal residues in boreal forest soils. *New Phytologist* 223: 16–21.

Agrell, J., McDonald, E.P., and Lindroth, R.L. (2000). Effects of CO_2 and light on tree phytochemistry and insect performance. *Oikos* 88: 259–272.

Ayres, M.P., Clausen, T.P., Maclean Jr. et al. (1997). Diversity of structure and antiherbivore activity in condensed tannins. *Ecology* 78: 1696–1712.

Bailey, A., Wooley, S.C., Lindroth, R.L., and Whitham, T G. (2006). Importance of species interactions to community heritability: a genetic basis to trophic-level interactions. *Ecology Letters* 9: 78–85.

Bailey, J., Schweitzer, J., Rehill, B. et al. (2004). Beavers as molecular geneticists: a genetic basis to the foraging of an ecosystem engineer. *Ecology* 85: 603–608.

Bailey, J.K., Deckert, R., Schweitzer, J.A. et al. (2005). Host plant genetics affect hidden ecological players: links among *Populus*, condensed tannins, and fungal endophyte infection. *Canadian Journal of Botany* 83: 356–361.

Bandau, F., Albrectsen, B.R., Julkunen-Tiitto, R., and Gundale, M.J. (2017). Genotypic variability in *Populus tremula* L. affects how anthropogenic nitrogen enrichment influences litter decomposition. *Plant and Soil* 410: 467–481.

Bandau, F., Decker, V.H.G., Gundale, M.J., and Albrectsen, B.R. (2015). Genotypic tannin levels in *Populus tremula* impact the way nitrogen enrichment affects growth and allocation responses for some traits and not for others. *PLoS One* 10 (10): e0140971.

Bangert, R.K., Turek, R.J., Rehill, B.J. et al. (2006). A genetic similarity rule determines arthropod community structure. *Molecular Ecology* 15: 1379–1391.

Barbehenn, R.V. and Constabel, C.P. (2011). Tannins in plant–herbivore interactions. *Phytochemistry* 72: 1551–1565.

Barker, H.L., Holeski, L.M., and Lindroth, R.L. (2018). Genotypic variation in plant traits shapes herbivorous insect and ant communities on a foundation tree species. *PLoS One* 13 (7): e0200954.

Barker, H.L., Smith, D., Stanosz, G., and Lindroth, R.L. (2016). Host genetics and environment shape fungal pathogen incidence on a foundation forest tree species, *Populus tremuloides*. *Canadian Journal of Forest Research* 46: 1167–1172.

Bate-Smith, E.C. and Metcalfe, C.R. (1957). Leucoanthocyanins. 3. The nature and systematic distribution of tannins in dicotyledonous plants. *Journal of the Linnean Society* 55: 669–705.

Bazzaz, F.A., Chiariello, N.R., Coley, P.D., and Pitelka, L.F. (1987). Allocating resources to reproduction and defense. *BioScience* 37: 58–67.

Bernays, E., Driver, G., and Bilgener, M. (1989). Herbivores and plant tannins. *Advances in Ecological Research* 19: 263–302.

Bhat, T., Singh, B., and Sharma, O. (1998). Microbial degradation of tannins – a current perspective. *Biodegradation* 9: 343–357.

Boeckler, G.A., Towns, M., Unsicker, S.B. et al. (2014). Transgenic upregulation of the condensed tannin pathway in poplar leads to a dramatic shift in leaf palatability for two tree-feeding Lepidoptera. *Journal of Chemical Ecology* 40: 150–158.

Boettcher, A. and Targett, N. (1993). Role of polyphenolic molecular size in reduction of assimilation efficiency in *Xiphister mucosus*. *Ecology* 74: 891–903.

Bryant, J., Chapin III, S., and Klein, D. (1983). Carbon/nutrient balance of boreal plants in relation to vertebrate herbivory. *Oikos* 40: 357–368.

Buhl, C., Strauss, S.H., and Lindroth, R.L. (2019). Genetic down-regulation of gibberellin results in semi-dwarf poplar but few non-target effects on chemical resistance and tolerance to defoliation. *Journal of Plant Ecology* 12: 124–136.

Calder, W.J., Horn, K.J., and St. Clair, S.B. (2011). Conifer expansion reduces the competitive ability and herbivore defense of aspen by modifying light environment and soil chemistry. *Tree Physiology* 31: 582–591.

Call, A.C. and St. Clair, S.B. (2017). Outbreak of Drepanopeziza fungus in aspen forests and variation in stand susceptibility: leaf functional traits, compensatory growth and phenology. *Tree Physiology* 37: 1198–1207.

Cates, R. and Rhoades, D. (1977). Patterns in production of anti-herbivore chemical defenses in plant communities. *Biochemical Systematics of Ecology* 5: 185–193.

Chomel, M., Guittonny-Larcheveque, M., Fernandez, C. et al. (2016). Plant secondary metabolites: a key driver of litter decomposition and soil nutrient cycling. *Journal of Ecology* 104: 1527–1541.

Clausen, T., Provensa, F., Burritt, E., Reichardt, P., and Bryant, J. (1990). Ecological implications of condensed tannin structure – a case study. *Journal of Chemical Ecology* 16: 2381–2392.

Cole, C.T. (2005). Allelic and population variation of microsatellite loci in aspen (*Populus tremuloides*). *New Phytologist* 167: 155–164.

Cole, C.T. and Ingvarsson, P.K. (2018). Pathway position constrains the evolution of an ecologically important pathway in aspens (*Populus tremula* L.). *Molecular Ecology* 27: 3317–3330.

Cole, C.T., Stevens, M.T., Anderson, J.E., and Lindroth, R.L. (2016). Heterozygosity, gender, and the growth-defense trade-off in quaking aspen. *Oecologia* 181: 381–390.

Compson, Z., Hungate, B., Whitham, T. et al. (2018). Linking tree genetics and stream consumers: isotopic tracers elucidate controls on carbon and nitrogen assimilation. *Ecology* 99: 1759–1770.

Constabel, C.P., Yoshida, K., and Walker, V. (2014). Diverse ecological roles of plant tannins: plant defense and beyond. In: *Recent Advances in Polyphenol Research*, 4 (eds. A. Romani, V. Lattanzio and S. Quideau), 115–142. New Jersey, USA: Wiley-Blackwell.

Cope, O.L., Kruger, E.L., Rubert-Nason, K.F., and Lindroth, R.L. (2019). Chemical defense over decadal scales: Ontogenetic allocation trajectories and consequences for fitness in a foundation tree species. *Functional Ecology* 33: 2105–2115.

Cope, O.L. and Lindroth, R.L. (2018). Clonal saplings of trembling aspen do not coordinate defense induction. *Journal of Chemical Ecology* 44: 1045–1050.

Couture, J.J., Holeski, L.M., and Lindroth, R.L. (2014). Long-term exposure to elevated CO_2 and O_3 alters aspen foliar chemistry across developmental stages. *Plant, Cell and Environment* 37: 758–765.

Couture, J.J. and Lindroth, R.L. (2012). Atmospheric change alters performance of an invasive forest insect. *Global Change Biology* 18: 3543–3557.

Couture, J.J. and Lindroth, R.L. (2014). Atmospheric change alters frass quality of forest canopy herbivores. *Arthropod Plant Interactions* 8: 33–47.

Couture, J.J., Meehan, T.D., Kruger, E.L., and Lindroth, R.L. (2015). Insect herbivory alters impact of atmospheric change on northern temperate forests. *Nature Plants* 1: 15016.

Couture, J.J., Meehan, T.D., and Lindroth, R.L. (2012). Atmospheric change alters foliar quality of host trees and performance of two outbreak insect species. *Oecologia* 168: 863–876.

Couture, J.J., Meehan, T.D., Rubert-Nason, K.F., and Lindroth, R.L. (2017). Effects of elevated atmospheric carbon dioxide and tropospheric ozone on phytochemical composition of trembling aspen (*Populus tremuloides*) and paper birch (*Betula papyrifera*). *Journal of Chemical Ecology* 43: 26–38.

Crutzen, P. and Stoermer, E. (2000). The "Anthropocene." *Global Change Newsletter* 41: 17–18.

Decker, V.H.G., Bandau, F., Gundale, M.J. et al. (2017). Aspen phenylpropanoid genes' expression levels correlate with genets' tannin richness and vary both in responses to soil nitrogen and associations with phenolic profiles. *Tree Physiology* 37: 270–279.

de Jong, F., Hanley, S.J., Beale, M.H., and Karp, A. (2015). Characterisation of the willow phenylalanine ammonia-lyase (PAL) gene family reveals expression differences compared with poplar. *Phytochemistry* 117: 90–97.

Dettlaff, M.A., Marshall, V., Erbilgin, N., and Cahill, J.F. (2018). Root condensed tannins vary over time, but are unrelated to leaf tannins. *AOB Plants* 10: ply044.
Diner, B., Berteaux, D., Fyles, J., and Lindroth, R.L. (2009). Behavioral archives link the chemistry and clonal structure of trembling aspen to the food choice of North American porcupine. *Oecologia* 160: 687–695.
Donaldson, J.R., Kruger, E.L., and Lindroth, R.L. (2006a). Competition- and resource-mediated tradeoffs between growth and defensive chemistry in trembling aspen (*Populus tremuloides*). *New Phytologist* 169: 561–570.
Donaldson, J.R. and Lindroth, R.L. (2004). Cottonwood leaf beetle (Coleoptera: Chrysomelidae) performance in relation to variable phytochemistry in juvenile aspen (*Populus tremuloides* Michx.). *Environmental Entomology* 33: 1505–1511.
Donaldson, J.R. and Lindroth, R.L. (2007). Genetics, environment, and their interaction determine efficacy of chemical defense in trembling aspen. *Ecology* 88: 729–739.
Donaldson, J.R. and Lindroth, R.L. (2008). Effects of variable phytochemistry and budbreak phenology on defoliation of aspen during a forest tent caterpillar outbreak. *Agricultural and Forest Entomology* 10: 399–410.
Donaldson, J.R., Stevens, M.T., Barnhill, H.R., and Lindroth, R.L. (2006b). Age-related shifts in leaf chemistry of clonal aspen (*Populus tremuloides*). *Journal of Chemical Ecology* 32: 1415–1429.
Driebe, E.M. and Whitham, T.G. (2000). Cottonwood hybridization affects tannin and nitrogen content of leaf litter and alters decomposition. *Oecologia* 123: 99–107.
Edwards, L.E. (2015). What is the Anthropocene? *EOS* 96. doi: 10.1029/2015EO040297.
Feeny, P. (1970). Seasonal changes in oak leaf tannins and nutrients as a cause of spring feeding by winter moth caterpillars. *Ecology* 51: 565–581.
Fierer, N., Schimel, J., Cates, R., and Zou, J. (2001). Influence of balsam poplar tannin fractions on carbon and nitrogen dynamics in Alaskan taiga floodplain soils. *Soil Biology and Biochemistry* 33: 1827–1839.
Fischer, D.G., Hart, S.C., Rehill, B.J. et al. (2006). Do high-tannin leaves require more roots? *Oecologia* 149: 668–675.
Forkner, R., Marquis, R.J., and Lill, J. (2004). Feeny revisited: condensed tannins as anti-herbivore defences in leaf-chewing herbivore communities of *Quercus*. *Ecological Entomology* 29: 174–187.
Frost, C.J., Dean, J.M., Smyers, E.C. et al. (2012). A petiole-galling insect herbivore decelerates leaf lamina litter decomposition rates. *Functional Ecology* 26: 628–636.
Gourlay, G. and Constabel, C.P. (2019). Condensed tannins are inducible antioxidants and protect hybrid poplar against oxidative stress. *Tree Physiology* 39: 345–355.
Gruppe, A., Fusseder, M., and Schopf, R. (1999). Short rotation plantations of aspen and balsam poplar on former arable land in Germany: defoliating insects and leaf constituents. *Forest Ecology and Management* 121: 113–122.
Hagerman, A., Riedl, K., Jones, G. et al. (1998). High molecular weight plant polyphenolics (tannins) as biological antioxidants. *Journal of Agricultural and Food Chemistry* 46: 1887–1892.
Haikio, E., Makkonen, M., Julkunen-Tiitto, R. et al. (2009). Performance and secondary chemistry of two hybrid aspen (*Populus tremula* L. × *Populus tremuloides* Michx.) clones in long-term elevated ozone exposure. *Journal of Chemical Ecology* 35: 664–678.
Harding, S.A., Jiang, H.Y., Jeong, M.L. et al. (2005). Functional genomics analysis of foliar condensed tannin and phenolic glycoside regulation in natural cottonwood hybrids. *Tree Physiology* 25: 1475–1486.

Harding, S.A., Xue, L.J., Du, L. et al. (2014). Condensed tannin biosynthesis and polymerization synergistically condition carbon use, defense and growth in *Populus*. *Tree Physiology* 34: 1240–1251.

Hättenschwiler, S., Sun, T., and Coq, S. (2019). The chitin connection of polyphenols and its ecosystem consequences. *New Phytologist* 223: 5–7.

Hemming, J. and Lindroth, R.L. (1995). Intraspecific variation in aspen phytochemistry: effects on performance of gypsy moths and forest tent caterpillars. *Oecologia* 103: 79–88.

Hemming, J.D.C. and Lindroth, R.L. (1999). Effects of light and nutrient availability on aspen: growth, phytochemistry, and insect performance. *Journal of Chemical Ecology* 25: 1687–1714.

Herms, D.A. and Mattson, W.J. (1992). The dilemma of plants: to grow or defend. *Quarterly Review of Biology* 67: 283–335.

Heroy, K.Y., St. Clair, S.B., Burritt, E.A. et al. (2018). Nutritional state and secondary compounds influence trembling aspen (*Populus tremuloides*) intake by lambs. *Applied Animal Behaviour Science* 206: 39–51.

Hillstrom, M.L. and Lindroth, R.L. (2008). Elevated atmospheric carbon dioxide and ozone alter forest insect abundance and community composition. *Insect Conservation and Diversity* 1: 233–241.

Hillstrom, M.L., Meehan, T.D., Kelly, K., and Lindroth, R.L. (2010). Soil carbon and nitrogen mineralization following deposition of insect frass and greenfall from forests under elevated CO_2 and O_3. *Plant Soil* 336: 75–85.

Hjältén, J., Lindau, A., Wennstrom, A. et al. (2008). Vole response to unintentional changes in the chemistry of GM poplars. *Chemoecology* 18: 227–231.

Holeski, L.M., Hillstrom, M.L., Whitham, T.G., and Lindroth, R.L. (2012). Relative importance of genetic, ontogenetic, induction, and seasonal variation in producing a multivariate defense phenotype in a foundation tree species. *Oecologia* 170: 695–707.

Holeski, L.M., McKenzie, S.C., Kruger, E.L. et al. (2016). Phytochemical traits underlie genotypic variation in susceptibility of quaking aspen (*Populus tremuloides*) to browsing by a keystone forest ungulate. *Journal of Ecology* 104: 850–863.

Holeski, L.M., Vogelzang, A., Stanosz, G., and Lindroth, R.L. (2009). Incidence of *Venturia* shoot blight in aspen (*Populus tremuloides* Michx.) varies with tree chemistry and genotype. *Biochemical Systematics and Ecology* 37: 139–145.

Holton, M.K., Lindroth, R.L., and Nordheim, E.V. (2003). Foliar quality influences tree-herbivore-parasitoid interactions: effects of elevated CO_2, O_3, and plant genotype. *Oecologia* 137: 233–244.

Hu, J.F., Angeli, S., Schuetz, S., Luo, Y.Q., and Hajek, A.E. (2009). Ecology and management of exotic and endemic Asian longhorned beetle *Anoplophora glabripennis*. *Agricultural and Forest Entomology* 11: 359–375.

Huang, C.Y. and Anderegg, W.R.L. (2012). Large drought-induced aboveground live biomass losses in southern Rocky Mountain aspen forests. *Global Change Biology* 18: 1016–1027.

Hunter, M.D. (2016). *The Phytochemical Landscape*. (eds. S.A. Levin and H.S. Horn). Princeton, New Jersey, USA: Princeton University Press.

Hwang, S.Y. and Lindroth, R.L. (1997). Clonal variation in foliar chemistry of aspen: effects on gypsy moths and forest tent caterpillars. *Oecologia* 111: 99–108.

Hwang, S.Y. and Lindroth, R.L. (1998). Consequences of clonal variation in aspen phytochemistry for late season folivores. *Ecoscience* 5: 508–516.

IPCC (2013). *Climate Change 2013: The Physical Science Basis. Contribution of Working Group I to the Fifth Assessment Report of the Intergovernmental Panel on Climate Change.* Cambridge, UK and New York, New York, USA: Cambridge University Press.

Irmisch, S., McCormick, A.C., Günther, J. et al. (2014). Herbivore-induced poplar cytochrome P450 enzymes of the CYP71 family convert aldoximes to nitriles which repel a generalist caterpillar. *The Plant Journal* 80: 1095–1107.

James, A.M., Ma, D.W., Mellway, R. et al. (2017). Poplar MYB115 and MYB134 Transcription factors regulate proanthocyanidin synthesis and structure. *Plant Physiology* 174: 154–171.

Jamieson, M.A., Schwartzberg, E.G. Raffa, K.F. et al. (2015). Experimental climate warming alters aspen and birch phytochemistry and performance traits for an outbreak insect herbivore. *Global Change Biology* 21: 2698–2710.

Jansson, S. and Douglas, C. (2007). *Populus*: a model system for plant biology. *Annual Review of Plant Biology* 58: 435–458.

Jelinkova, H., Tremblay, F., and Desrochers, A. (2012). Herbivore-simulated induction of defenses in clonal networks of trembling aspen (*Populus tremuloides*). *Tree Physiology* 32: 1348–1356.

Kao, Y., Harding, S.A., and Tsai, C. (2002). Differential expression of two distinct phenylalanine ammonia-lyase genes in condensed tannin-accumulating and lignifying cells of quaking aspen. *Plant Physiology* 130: 796–807.

Kariñho-Betancourt, E., Hernández-Soto, P., Rendón-Anaya, M. et al. (2019). Differential expression of genes associated with phenolic compounds in galls of *Quercus castanea* induced by *Amphibolips michoacaensis*. *Journal of Plant Interactions* 14: 177–186.

Keefover-Ring, K.K., Rubert-Nason, K.F., Bennett, A., and Lindroth, R.L. (2016). Growth and chemical responses of trembling aspen to simulated browsing and ungulate saliva. *Journal of Plant Ecology* 9: 474–484.

Kelly, J.J., Bansal, A., Winkelman, J. et al. (2010). Alteration of microbial communities colonizing leaf litter in a temperate woodland stream by growth of trees under conditions of elevated atmospheric CO_2. *Applied and Environmental Microbiology* 76: 4950–4959.

Kinney, K.K., Lindroth, R.L., Jung, S.M., and Nordheim, E.V. (1997). Effects of CO_2 and NO_3^- availability on deciduous trees: phytochemistry and insect performance. *Ecology* 78: 215–230.

Kopper, B.J. and Lindroth, R.L. (2003a). Effects of elevated carbon dioxide and ozone on the phytochemistry of aspen and performance of an herbivore. *Oecologia* 134: 95–103.

Kopper, B.J. and Lindroth, R.L. (2003b). Responses of trembling aspen (*Populus tremuloides*) phytochemistry and aspen blotch leafminer (*Phyllonorycter tremuloidiella*) performance to elevated levels of atmospheric CO_2 and O_3. *Agricultural and Forest Entomology* 5: 17–26.

Kosola, K., Parry, D., and Workmaster, B.A.A. (2006). Responses of condensed tannins in poplar roots to fertilization and gypsy moth defoliation. *Tree Physiology* 26: 1607–1611.

Kosola, K.R., Dickmann, D.I., Hall, R.B., and Workmaster, B.A.A. (2004). Cottonwood growth rate and fine root condensed tannin concentration. *Tree Physiology* 24: 1063–1068.

Kosonen, M., Keski-Saari, S., Ruuhola, T. et al. (2012). Effects of overproduction of condensed tannins and elevated temperature on chemical and ecological traits of genetically modified hybrid aspens (*Populus tremula* × *P. tremuloides*). *Journal of Chemical Ecology* 38: 1235–1246.

Lamit, L.J., Wojtowicz, T., Kovacs, Z. et al. (2011). Hybridization among foundation tree species influences the structure of associated understory plant communities. *Botany-Botanique* 89: 165–174.

Lämke, J.S. and Unsicker, S.B. (2018). Phytochemical variation in treetops: causes and consequences for tree-insect herbivore interactions. *Oecologia* 187: 377–388.

LeRoy, C.J., Whitham, T.G., Keim, P., and Marks, J.C. (2006). Plant genes link forests and streams. *Ecology* 87: 255–261.

LeRoy, C.J., Whitham, T.G., Wooley, S.C., and Marks, J.C. (2007). Within-species variation in foliar chemistry influences leaf-litter decomposition in a Utah river. *Journal of the North American Benthological Society* 26: 426–438.

LeRoy, C.J., Wooley, S.C., and Lindroth, R.L. (2012). Genotype and soil nutrient environment influence aspen litter chemistry and in-stream decomposition. Freshwater Science 31: 1244–1253.

Lin, L., Sun, S., Chen, P. et al. (2014). UHPLC-PDA-ESI/HRMSn profiling method to identify and quantify oligomeric proanthocyanidins in plant products. *Journal of Agricultural and Food Chemistry* 62: 9387–9400.

Lindroth, R.L. (2010). Impacts of elevated atmospheric CO_2 and O_3 on forests: phytochemistry, trophic interactions, and ecosystem dynamics. *Journal of Chemical Ecology* 36: 2–21.

Lindroth, R.L., Donaldson, J.R., Stevens, M.T., and Gusse, A. (2007). Browse quality in quaking aspen (*Populus tremuloides*): effects of genotype, nutrients, defoliation, and coppicing. *Journal of Chemical Ecology* 33: 1049–1064.

Lindroth, R.L. and Hwang, S.Y. (1996). Clonal variation in foliar chemistry of quaking aspen (*Populus tremuloides* Michx). *Biochemical Systematics of Ecology* 24: 357–364.

Lindroth, R.L., Kopper, B.J., Parsons, W.F.J. et al. (2001a). Consequences of elevated carbon dioxide and ozone for foliar chemical composition and dynamics in trembling aspen (*Populus tremuloides*) and paper birch (*Betula papyrifera*). *Environmental Pollution* 115: 395–404.

Lindroth, R.L., Osier, T.L., Barnhill, H.R., and Wood, S.A. (2002). Effects of genotype and nutrient availability on phytochemistry of trembling aspen (*Populus tremuloides* Michx.) during leaf senescence. *Biochemical Systematics of Ecology* 30: 297–307.

Lindroth, R.L., Roth, S., and Nordheim, E.V. (2001b). Genotypic variation in response of quaking aspen (*Populus tremuloides*) to atmospheric CO_2 enrichment. *Oecologia* 126: 371–379.

Liu, L.L., King, J.S., and Giardina, C.P. (2005). Effects of elevated concentrations of atmospheric CO_2 and tropospheric O_3 on leaf litter production and chemistry in trembling aspen and paper birch communities. *Tree Physiology* 25: 1511–1522.

Liu, L.L., King, J.S., Giardina, C.P., and Booker, F.L. (2009). The influence of chemistry, production and community composition on leaf litter decomposition under elevated atmospheric CO_2 and tropospheric O_3 in a northern hardwood ecosystem. *Ecosystems* 12: 401–416.

Liu, Q., Zhou, Y.Q., Chen, J., and Hao, D.J. (2015). Defensive responses of *Populus deltoides* 895 seedlings against exogenous methyl jasmonate. *Pakistan Journal of Botany* 47: 177–188.

Luo, Z.B., Calfapietra, C., Scarascia-Mugnozza, G., Liberloo, M., and Polle, A. (2008). Carbon-based secondary metabolites and internal nitrogen pools in *Populus nigra* under Free Air CO_2 Enrichment (FACE) and nitrogen fertilization. *Plant and Soil* 304: 45–57.

Madritch, M.D., Donaldson, J.R., and Lindroth, R.L. (2006). Genetic identity of *Populus tremuloides* litter influences decomposition and nutrient release in a mixed forest stand. *Ecosystems* 9: 528–537.

Madritch, M.D., Donaldson, J.R., and Lindroth, R.L. (2007a). Canopy herbivory can mediate the influence of plant genotype on soil processes through frass deposition. *Soil Biology and Biochemistry* 39: 1192–1201.

Madritch, M.D., Jordan, L.M., and Lindroth, R.L. (2007b). Interactive effects of condensed tannin and cellulose additions on soil respiration. *Canadian Journal of Forest Research* 37: 2063–2067.

Madritch, M.D., Greene, S.G., and Lindroth, R.L. (2009). Genetic mosaics of ecosystem functioning across aspen-dominated landscapes. *Oecologia* 160: 119–127.

Madritch, M.D. and Lindroth, R.L. (2015). Condensed tannins increase nitrogen recovery by trees following insect defoliation. *New Phytologist* 208: 410–420.

Mansfield, J.L., Curtis, P.S., Zak, D.R., and Pregitzer, K.S. (1999). Genotypic variation for condensed tannin production in trembling aspen (*Populus tremuloides*, Salicaceae) under elevated CO_2 and in high- and low-fertility soil. *American Journal of Botany* 86: 1154–1159.

Marks, J.C., Haden, G.A., Harrop, B.L. et al. (2009). Genetic and environmental controls of microbial communities on leaf litter in streams. *Freshwater Biology* 54: 2616–2627.

Marles, M., Ray, H., and Gruber, M. (2003). New perspectives on proanthocyanidin biochemistry and molecular regulation. *Phytochemistry* 64: 367–383.

Marsh, K.J., Wallis, I.R., Kulheim, C. et al. (2019). New approaches to tannin analysis of leaves can be used to explain in vitro biological activities associated with herbivore defence. *New Phytologist* 225: 488–498.

Mason, C.J., Rubert-Nason, K.F., Lindroth, R.L., and Raffa, K.F. (2015). Aspen defense chemicals influence midgut bacterial community composition of gypsy moth. *Journal of Chemical Ecology* 41: 75–84.

McAllister, T., Martinez, T., Bae, H. et al. (2005). Characterization of condensed tannins purified from legume forages: chromophore production, protein precipitation, and inhibitory effects on cellulose digestion. *Journal of Chemical Ecology* 31: 2049–2068.

McDonald, E.P., Agrell, J., and Lindroth, R.L. (1999). CO_2 and light effects on deciduous trees: growth, foliar chemistry, and insect performance. *Journal of Chemical Ecology* 119: 389–399.

McDonald, E.P., Kruger, E.L., Riemenschneider, D.E., and Isebrands, J.G. (2002). Competitive status influences tree-growth responses to elevated CO_2 and O_3 in aggrading aspen stands. *Functional Ecology* 16: 792–801.

McKenney, D., Pedlar, J., Lawrence, K. et al. (2007). Potential impacts of climate change on the distribution of North American trees. *BioScience* 57: 939–948.

McKown, A.D., Guy, R.D., Quamme, L. et al. (2014). Association genetics, geography and ecophysiology link stomatal patterning in *Populus trichocarpa* with carbon gain and disease resistance trade-offs. *Molecular Ecology* 23: 5771–5790.

McWilliam, E.L., Barry, T.N., Lopez-Villalobos, N. et al. (2004). The effect of different levels of poplar (*Populus*) supplementation on the reproductive performance of ewes grazing low quality drought pasture during mating. *Animal Feed Science and Technology* 115: 1–18.

McWilliam, E.L., Barry, T.N., Lopez-Villalobos et al. (2005). Effects of willow (*Salix*) versus poplar (*Populus*) supplementation on the reproductive performance of ewes grazing low quality drought pasture during mating. *Animal Feed Science and Technology* 119: 69–86.

Meehan, T.D., Couture, J.J., Bennett, A.E., and Lindroth, R.L. (2014). Herbivore-mediated material fluxes in a northern deciduous forest under elevated carbon dioxide and ozone concentrations. *New Phytologist* 204: 397–407.

Meehan, T.D., Crossley, M.S., and Lindroth, R.L. (2010). Impacts of elevated CO_2 and O_3 on aspen leaf litter chemistry and earthworm and springtail productivity. *Soil Biology and Biochemistry* 42: 1132–1137.

Miranda, M., Ralph, S., Mellway, R.D. et al. (2007). The transcriptional response of hybrid poplar (*Populus trichocarpa* × *P. deltoides*) to infection by *Melampsora medusae* leaf rust involves induction of flavonoid pathway genes leading to the accumulation of proanthocyanidins. *Molecular Plant-Microbe Interactions* 20: 816–831.

Mole, S. (1993). The systematic distribution of tannins in the leaves of angiosperms: a tool for ecological studies. *Biochemical Systematics and Ecology* 21: 833–846.

Morin, X., Viner, D., and Chuine, I. (2008). Tree species range shifts at a continental scale: new predictive insights from a process-based model. *Journal of Ecology* 96: 784–794.

Müller, M.S., McWilliams, S.R., Podlesak, D. et al. (2006). Tri-trophic effects of plant defenses: chickadees consume caterpillars based on host leaf chemistry. *Oikos* 114: 507–517.

Nissinen, K., Virjamo, V., Randriamanana, T. et al. (2017). Responses of growth and leaf phenolics in European aspen (*Populus tremula*) to climate change during juvenile phase change. *Canadian Journal of Forest Research* 47: 1350–1363.

Osier, T.L. and Lindroth, R.L. (2001). Effects of genotype, nutrient availability, and defoliation on aspen phytochemistry and insect performance. *Journal of Chemical Ecology* 27: 1289–1313.

Osier, T.L. and Lindroth, R.L. (2004). Long-term effects of defoliation on quaking aspen in relation to genotype and nutrient availability: plant growth, phytochemistry and insect performance. *Oecologia* 139: 55–65.

Osier, T.L. and Lindroth, R.L. (2006). Genotype and environment determine allocation to and costs of resistance in quaking aspen. *Oecologia* 148: 293–303.

Parsons, W., Bockheim, J., and Lindroth, R.L. (2008). Independent, interactive, and species-specific responses of leaf litter decomposition to elevated CO_2 and O_3 in a northern hardwood forest. *Ecosystems* 11: 505–519.

Peters, D. and Constabel, C.P. (2002). Molecular analysis of herbivore-induced condensed tannin synthesis: cloning and expression of dihydroflavonol reductase from trembling aspen (*Populus tremuloides*). *Plant Journal* 32: 701–712.

Pfabel, C., Eckhardt, K.U., Baum, C. et al. (2012). Impact of ectomycorrhizal colonization and rust infection on the secondary metabolism of poplar (*Populus trichocarpa* × *deltoides*). *Tree Physiology* 32: 1357–1364.

Philippe, R.N. and Bohlmann, J. (2007). Poplar defense against herbivores. *Canadian Journal of Botany* 85: 1111–1126.

Preston, C.M., Nault, J.R., Trofymow, J.A. et al. (2009). Chemical changes during 6 years of decomposition of 11 litters in some Canadian forest sites. Part 1. Elemental composition, tannins, phenolics, and proximate fractions. *Ecosystems* 12: 1053–1077.

Quideau, S., Deffieux, D., Douat-Casassus, C., and Pouysegu, L. (2011). Plant polyphenols: chemical properties, biological activities, and synthesis. *Angewandte Chemie – International Edition* 50: 586–621.

Ralph, S., Cooper, D., Yueh, H. et al. (2006). Genomics of hybrid poplar (*Populus trichocarpa* × *deltoides*) interacting with forest tent caterpillars (*Malacosoma disstria*): normalized and full-length cDNA libraries, expressed sequence tags, and a cDNA microarray for the study of insect-induced defences in poplar. *Molecular Ecology* 15: 1275–1297.

Randriamanana, T., Nybakken, L., Lavola, A. et al. (2014). Sex-related differences in growth and carbon allocation to defence in *Populus tremula* as explained by current plant defence theories. *Tree Physiology* 34: 471–487.

Randriamanana, T.R., Lavola, A., and Julkunen-Tiitto, R. (2015). Interactive effects of supplemental UV-B and temperature in European aspen seedlings: implications for growth, leaf traits, phenolic defense and associated organisms. *Plant Physiology and Biochemistry* 93: 84–93.

Randriamanana, T.R., Nissinen, K., Ovaskainen, A. et al. (2018). Does fungal endophyte inoculation affect the responses of aspen seedlings to carbon dioxide enrichment? *Fungal Ecology* 33: 24–31.

Rehill, B.J., Whitham, T.G., Martinsen, G.D. et al. (2006). Developmental trajectories in cottonwood phytochemistry. *Journal of Chemical Ecology* 32: 2269–2285.

Rhodes, A.C., Anderson, V., and St. Clair, S.B. (2017). Ungulate herbivory alters leaf functional traits and recruitment of regenerating aspen. *Tree Physiology* 37: 402–413.

Robinson, K.M., Ingvarsson, P.K., Jansson, S., and Albrectsen, B.R. (2012). Genetic variation in functional traits influences arthropod community composition in aspen (*Populus tremula* L.). *PLoS One.* 7 (5): e37679.

Roth, S., Lindroth, R.L., Volin, J.C., and Kruger, E.L. (1998). Enriched atmospheric CO_2 and defoliation: effects on tree chemistry and insect performance. *Global Change Biology* 4: 419–430.

Roth, S., McDonald, E.P., and Lindroth, R.L. (1997). Atmospheric CO_2 and soil water availability: consequences for tree–insect interactions. *Canadian Journal of Forest Research* 27: 1281–1290.

Rubert-Nason, K.F., Couture, J.J., Gryzmala, E.A. et al. (2017). Vernal freeze damage and genetic variation alter tree growth, chemistry, and insect interactions. *Plant, Cell and Environment* 40: 743–2753.

Rubert-Nason, K.F., Couture, J.J., Major, I.T. et al. (2015). Influence of genotype, environment, and gypsy moth herbivory on local and systemic chemical defenses in trembling aspen (*Populus tremuloides*). *Journal of Chemical Ecology* 41: 651–661.

Rubert-Nason, K.F. and Lindroth, R.L. (2019). Analysis of condensed tannins in *Populus* spp. using reversed phase UPLC-PDA-(-)esiMS following thiolytic depolymerization *Phytochemical Analysis* 30: 257–267.

Salminen, J.P. and Karonen, M. (2011). Chemical ecology of tannins and other phenolics: we need a change in approach. *Functional Ecology* 25: 325–338.

Schimel, J., Vancleve, K., Cates, R. et al. (1996). Effects of balsam poplar (*Populus balsamifera*) tannins and low molecular weight phenolics on microbial activity in taiga floodplain soil: implications for changes in N cycling during succession. *Canadian Journal of Botany* 74: 84–94.

Schimel, J.P., Cates, R.G., and Ruess, R. (1998). The role of balsam poplar secondary chemicals in controlling soil nutrient dynamics through succession in the Alaskan taiga. *Biogeochemistry* 42: 221–234.

Schweitzer, J.A., Bailey, J., Rehill, B. et al. (2004). Genetically based trait in a dominant tree affects ecosystem processes. *Ecology Letters* 7: 127–134.

Schweitzer, J.A., Bailey, A., Bangert, R.K. et al. (2007). The role of plant genetic variation in determining above- and belowground microbial communities. In: Microbial Ecology of Aerial Plant Surfaces (eds. M.J. Bailey, A.K. Lilley, T.M. Timms-Wilson and P.T.N. Spencer-Phillips). Wallingford, UK: CABI Publishing.

Schweitzer, J.A., Bailey, J.K., Hart, S.C., and Whitham, T.G. (2005a). Nonadditive effects of mixing cottonwood genotypes on litter decomposition and nutrient dynamics. *Ecology* 86: 2834–2840.

Schweitzer, J.A., Bailey, J.K., Hart, S.C. et al. (2005b). The interaction of plant genotype and herbivory decelerate leaf litter decomposition and alter nutrient dynamics. *Oikos* 110: 133–145.

Schweitzer, J.A., Madritch, M.D., Bailey, J.K. et al. (2008). From genes to ecosystems: the genetic basis of condensed tannins and their role in nutrient regulation in a *Populus* model system. *Ecosystems* 11: 1005–1020.

Scioneaux, A.N., Schmidt, M.A., Moore, M.A. et al. (2011). Qualitative variation in proanthocyanidin composition of *Populus* species and hybrids: genetics is the key. *Journal of Chemical Ecology* 37: 57–70.

Selmants, P.C., Schweitzer, J.A., Adair, K.L. et al. (2019). Genetic variation in tree leaf chemistry predicts the abundance and activity of autotrophic soil microorganisms. *Ecosphere* 10: e02795.

Semiadi, G., Barry, T.N., Muir, P.D., and Hodgson, J. (1995). Dietary preferences of sambar (*Cervus unicolor*) and red deer (*Cervus elaphus*) offered browse, forage legume and grass species. *Journal of Agricultural Science* 125: 99–107.

Shay, P.E., Constabel, C.P., and Trofymow, J.A. (2018). Evidence for the role and fate of water-insoluble condensed tannins in the short-term reduction of carbon loss during litter decay. *Biogeochemistry* 137: 127–141.

Sibley, D.A. (2009). *The Sibley Guide to Trees*. New York, USA: Alfred A. Knopf.

Sivadasan, U. (2018). Climate change effects on growth and metabolism of the dioecious *Salix myrsinifolia* and *Populus tremula*. Ph.D. thesis. University of Eastern Finland.

Slavov, T.T. and Zhelev, P. (2010). Salient biological features, systematics, and genetic variation of *Populus*. In: *Genetics and Genomics of Populus* (eds. S. Jansson, R.P. Bhalerao and A.T. Groover). New York, USA: Springer.

St. Clair, S.B., Monson, S.D., Smith, E.A. et al. (2009). Altered leaf morphology, leaf resource dilution and defense chemistry induction in frost-defoliated aspen (*Populus tremuloides*). *Tree Physiology* 29: 1259–1268.

Stettler, R.F. (2009). *Cottonwood and the River of Time: On Trees, Evolution, and Society*. Seattle, Washington, USA: University of Washington Press.

Stevens, M.T., Gusse, A.C., and Lindroth, R.L. (2012). Genotypic differences and prior defoliation affect re-growth and phytochemistry after coppicing in *Populus tremuloides*. *Journal of Chemical Ecology* 38: 306–314.

Stevens, M.T., Gusse, A.C., and Lindroth, R.L. (2014). Root chemistry in *Populus tremuloides*: effects of soil nutrients, defoliation, and genotype. *Journal of Chemical Ecology* 40: 31–38.

Stevens, M.T. and Lindroth, R.L. (2005). Induced resistance in the indeterminate growth of aspen (*Populus tremuloides*). *Oecologia* 145: 298–306.

Stevens, M.T., Waller, D.M., and Lindroth, R.L. (2007). Resistance and tolerance in *Populus tremuloides*: genetic variation, costs, and environmental dependency. *Evolutionary Ecology* 21: 829–847.

Taiz, L. and Zeiger, E. (2010). *Plant Physiology*. Oxford, UK: Sinauer Associates.

Tsai, C., Harding, S.A., Tschaplinski, T.J. et al. (2006). Genome-wide analysis of the structural genes regulating defense phenylpropanoid metabolism in *Populus. New Phytologist* 172: 47–62.

Tuomi, J., Niemela, P., Chapin III, S. et al. (1988). Defensive responses of trees in relation to their carbon/nutrient balance. In: *Mechanisms of Woody Plant Defenses Against Insects: Search for Pattern* (eds. W.J. Mattson, J. Levieux and C. Bernard-Dagan). Heidelberg: Springer-Verlag.

Tuskan, G., Difazio, S., Jansson, S. et al. (2006). The genome of black cottonwood, *Populus trichocarpa* (Torr. and Gray). *Science* 313: 1596–1604.

Ullah, C., Tsai, C., Unsicker, S.B. et al. (2019). Salicylic acid activates poplar defense against the biotrophic rust fungus *Melampsora larici-populina* via increased biosynthesis of catechin and proanthocyanidins. *New Phytologist* 221: 960–975.

Ullah, C., Unsicker, S.B., Fellenberg, C. et al. (2017). Flavan-3-ols are an effective chemical defense against rust infection. *Plant Physiology* 175: 1560–1578.

Veteli, T., Mattson, W., Niemela, P. et al. (2007). Do elevated temperature and CO_2 generally have counteracting effects on phenolic phytochemistry of boreal trees? *Journal of Chemical Ecology* 33: 287–296.

Vigue, L.M. and Lindroth, R.L. (2010). Effects of genotype, elevated CO_2 and elevated O_3 on aspen phytochemistry and aspen leaf beetle *Chrysomela crotchi* performance. *Agricultural and Forest Entomology* 12: 267–276.

Villalba, J.J., Burritt, E.A., and St. Clair, S.B. (2014). Aspen (*Populus tremuloides* Michx.) intake and preference by mammalian herbivores: the role of plant secondary compounds and nutritional context. *Journal of Chemical Ecology* 40: 1135–1145.

Walling, L.L. (2008). Avoiding effective defenses: strategies employed by phloem-feeding insects. *Plant Physiology* 146: 859–866.

Wan, H.Y., Rhodes, A.C., and St. Clair, S.B. (2014). Fire severity alters plant regeneration patterns and defense against herbivores in mixed aspen forests. *Oikos* 123: 1479–1488.

Ware, I.M., Fitzpatrick, C.R., Senthilnathan, A. et al. (2019). Feedbacks link ecosystem ecology and evolution across spatial and temporal scales: empirical evidence and future directions. *Functional Ecology* 33: 31–42.

Warren, J. M., Bassman, J. H., and Eigenbrode, S. (2002). Leaf chemical changes induced in *Populus trichocarpa* by enhanced UV-B radiation and concomitant effects on herbivory by *Chrysomela scripta* (Coleoptera: Chrysomelidae). *Tree Physiology* 22: 1137–1146.

Winder, R.S., Lamarche, J., Constabel, C.P., and Hamelin, R.C. (2013). The effects of high-tannin leaf litter from transgenic poplars on microbial communities in microcosm soils. *Frontiers in Microbiology* 4: 1–12.

Wong, A.R. (2013). Effects of poplar phenolics on the fitness and behaviour of *Chaitophorus* aphids. MS thesis. University of British Columbia.

Wooley, S.C., Walker, S., Vernon, J., and Lindroth, R.L. (2008). Aspen decline, aspen chemistry, and elk herbivory: are they linked? *Society of Range Management*, 17–21.

Worrall, J.J., Rehfeldt, G.E., Hamann, A. et al. (2013). Recent declines of *Populus tremuloides* in North America linked to climate. *Forest Ecology and Management* 299: 35–51.

Yoshida, K., Ma, D.W., and Constabel, C.P. (2015). The MYB182 protein down-regulates proanthocyanidin and anthocyanin biosynthesis in poplar by repressing both structural and regulatory flavonoid genes. *Plant Physiology* 167: 693–710.

Zhang, R., Liu, J., Liu, Q. et al. (2019). Sexual differences in growth and defence of *Populus yunnanensis* under drought stress. *Canadian Journal of Forest Research-Revue Canadienne De Recherche Forestiere* 49: 491–499.

Zhao, S., Li, J., Wang, Y. et al. (2014). Effects of herbivory and defoliation on foliar phenolic compounds in *Populus euphratica* in the Ejina Oasis, Inner Mongolia, China. *Allelopathy Journal* 34: 49–57.

Zucker, W. (1983). Tannins: does structure determine function? An ecological perspective. *American Naturalist* 121: 335–365.

Züst, T. and Agrawal, A. (2017). Trade-offs between plant growth and defense against insect herbivory: an emerging mechanistic synthesis. *Annual Review of Plant Biology* 68: 513–543.

5

Matrix-Assisted Laser Desorption/Ionization Time-of-Flight Mass Spectrometry (MALDI-TOF MS) of Proanthocyanidins to Determine Authenticity of Functional Foods and Dietary Supplements

Daniel Esquivel-Alvarado[1], Jess Dreher Reed[1,2], and Christian G. Krueger[1,2]

[1] *Department of Animal Sciences, University of Wisconsin–Madison, Madison, USA*
[2] *Complete Phytochemical Solutions LLC, Cambridge, Wisconsin, USA*

5.1 Introduction

There is a significant need within the functional food and dietary supplement industries and regulatory agencies for rapid, accurate, and reliable analytical methods to authenticate bioactive ingredients (Clydesdale 2004). The health benefits conferred by the bioactive compounds found within many functional foods (e.g. fruits, juices, purees, and dried fruits/vegetables) depend on the quantity, type, source, and ratio of bioactive ingredients within these products. The functional food market is substantial; in the United States fruits and vegetables are an \$80.2B/year market, and botanicals represent an ~\$37B/year market (Leclair 2016). Cranberries alone, for example, are the second most consumed supplement in the US herbal ingredient industry, with a \$800M/year market, and cranberry value-added products are a \$54M/year market (Smith et al. 2015; Polito 2015).

Attempts have been made to characterize and authenticate the presence and type(s) of PAC within natural products and dietary supplements. However, current methods that are used by the industry are obsolete and do not rapidly identify, characterize, and distinguish PAC. For instance, high-performance thin layer chromatography (HPTLC) and high-performance liquid chromatography (HPLC) can identify and quantify individual, low molecular weight compounds, and distinguish genuine compounds from adulterants (Craciun et al. 2015; Rumalla et al. 2012). However, they are incapable of characterizing complex classes of oligomeric polyphenolic compounds, such as PAC or hydrolysable tannins. Genomic methods can identify the presence of plant species and distinguish genuine plant materials from some types of adulterants (Wallace et al. 2012). However, nucleic acids are not the bioactive compounds within natural products. The simple detection of a species' DNA within a sample does not indicate whether the bioactive compound from that species is present within the product, nor does it rule out the presence of other compounds (bioactive or otherwise) within the product. For example, roots and leaves could test positive for the plant of interest, but could fail to contain the bioactive components of interest if those

Recent Advances in Polyphenol Research, Volume 7, First Edition. Edited by Jess Dreher Reed, Victor Armando Pereira de Freitas, and Stéphane Quideau.

compounds were found primarily in the fruit. These shortcomings have not prevented DNA testing from being promoted and adopted by some in the industry as a standard method for the authentication of natural products (Twilley 2015).

A lack of standardized data and information on bioactive ingredients, coupled with an inability to quickly and accurately characterize and quantify the bioactive ingredients within functional foods, undermines the market because these problems prevent verification of authenticity. As a result, many products are improperly labeled, contain adulterants (including allergens), or lack the bioactive ingredients of interest. This hinders: (i) research on their health benefits and the establishment of dosage and/or dietary guidelines; (ii) the ability of the industry (growers, processors, distributors, and sellers) to improve crops and food products based on bioactive compounds, and; (iii) the ability of consumers to gauge which functional foods are most effective when used to improve their health.

The United States Department of Agriculture (USDA) Economic Research Service report on Food and Drug Administration (FDA) refusals of imported food products from 2005 through 2013 indicated that 41% of the 142,679 total violations reported were for misbranding, which may include untruthful or misleading labels (Bovay 2016). As stated in the report, "Although adulteration generally poses a greater risk to human health than misbranding, improper labeling, such as failure to identify an allergen, may lead to illness and fatalities in some cases" (Bovay 2016). Bioactive ingredients may degrade or be lost as the result of processing or improper handling. Regulatory bodies are unable to assess whether a product is properly labeled or if it complies with FDA regulations, such as the Food Modernization Safety Act (FMSA), current Good Manufacturing Practices (cGMP), Generally Recognized As Safe (GRAS), and the New Dietary Ingredient (NDI) requirements (US Congress 1938, 2011).

Thus, many products currently being marketed to consumers as dietary supplements may lack the bioactive ingredients of interest, contain subclinical quantities, or contain unsafe adulterants, such as allergens and toxins. For example, some unethical manufacturers substitute peanut skins as an inexpensive source of PAC in products purported to contain cranberry or grape seed PAC (Kupina and Gafner 2016), introducing the risk of unknowing consumption of peanut allergens by consumers (Villani et al. 2015). Improper handling or processing may also degrade or eliminate the bioactive compounds within products. The lack of efficacy of the components in these products only serves to erode consumer confidence and the overall perception of the health benefits.

There are greater than 10000 phenolic compounds that have been identified in plants (Cheynier et al. 2013). The main classes found in fruits and fruit products are hydroxycinnamic acids, flavonoids (including flavonol glycosides, anthocyanins, and proanthocyanidins), and hydrolysable tannins (gallotannins and ellagitannins). All species of plants contain a unique "fingerprint" of these polyphenolic compounds and research is required to demonstrate that technology to characterize the unique polyphenolic composition of plants may be used to determine authenticity. This "fingerprint" may be determined by liquid chromatography/mass spectrometry (LC/MS) methods to produce a data set (i.e. the "polyphenol fingerprint") that is unique to each species. This information may then be used to determine the presence or absence of the plant species through analysis of the mass spectral data sets. In the case of PAC and oligomeric ellagitannins, we have previously demonstrated that MALDI-TOF MS has good potential for rapid development of fingerprints for authentication of whole fruit and fruit products (Reed et al. 2005; Feliciano et al. 2014).

5.2 Introduction to Matrix-Assisted Laser Desorption/Ionization Time-of-Flight Mass Spectrometry (MALDI-TOF MS)

In MALDI-TOF MS (Figure 5.1), an analyte is combined with a matrix and placed on a target plate within the instrument. A pulsed laser irradiates and ablates the compounds, ionizing them in the process. The ionized particles are drawn into a vacuum tube to a detector. The molecules arrive in waves based on their transit time, which is determined by their molecular weight (MW). Lower MW particles arrive first, followed by higher MW particles. The detector's response (output) is then converted to a mass spectrum with high mass resolution (1000 to 60000), which detects compounds that differ by one mass unit or less. Since elements have both relative abundances and specific isotopic masses, mass spectrometers are able to differentiate masses of less than one mass unit because compounds with the same nominal masses have different proportions of elements. For instance, 4-methyl-1-pentene (C_6H_{12}), 2-methyl-2-butenal (C_5H_8O), and 3-(methylamino)propanenitrile ($C_4H_8N_2$) all have the same nominal mass of 84. However, C_6H_{12} has an exact mass of 84.0939, C_5H_8O has an exact mass of 84.0575, and $C_4H_8N_2$ has an exact mass of 84.0688.

MALDI-TOF MS was approved by the FDA in 2013 as "the first mass spectrometry system for automated identification of bacteria and yeasts that are known to cause serious illness in humans"; the method was approved for clinical applications (Laine 2013). The method depends on matching peptide mass fingerprints (PMF) from direct MALDI-TOF MS analysis of complex mixtures of microorganisms with PMF databases (Singhal et al. 2015). Thus, MALDI-TOF MS offers the potential for rapid, inexpensive analysis that avoids complex chromatography, giving it a competitive advantage over other approaches when analyzing the presence and type of polyphenolic compounds to determine authenticity of functional foods.

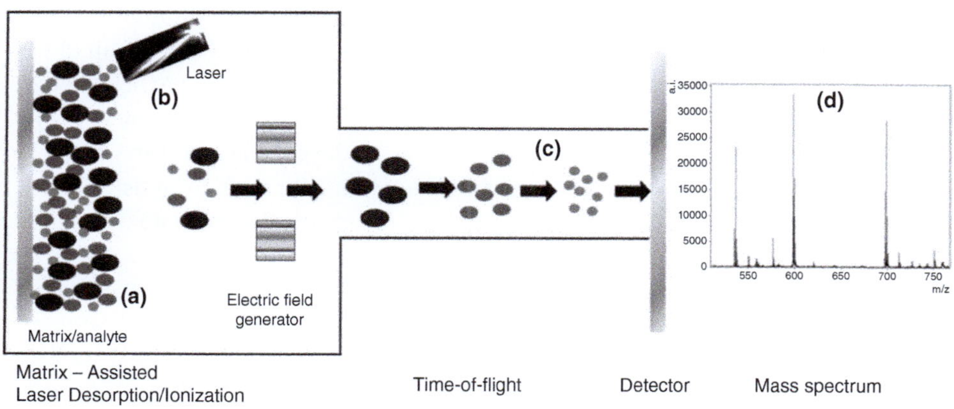

Figure 5.1 Representative schematic of the MALDI-TOF MS process. From left to right: (a) matrix and analyte are deposited on a target; (b) a laser beam irradiates the target, desorbing and ionizing the compounds; (c) compounds are separated by molecular weight (MW) in response to their transit time (time of flight) down a vacuum tube, with lower MWs arriving at the detector before heavier MWs; and (d) the detector response is converted to a mass spectrum with unit mass resolution.

5.3 Mass Spectrometry of Proanthocyanidins

Research on the application of MALDI-TOF MS for analysis of PAC oligomers in foods has been developed over the last 20 years (Ohnishi-Kameyama et al. 1997; Krueger et al. 2000; Foo et al. 2000a, 2000b; Reed et al. 2005). The first publication demonstrated that MALDI-TOF MS could determine the degree of polymerization (DP) of PAC oligomers in apples (Ohnishi-Kameyama et al. 1997). Oligomers were detected from the dimer to the pentadecamer, whereas previous research with fast atom bombardment (FAB) MS had only given a range up to the pentamer.

Krueger et al. (2000) used MALDI-TOF MS to characterize the structural heterogeneity of galloylated procyanidins from grape seed extract (GSE) and were able to detect a series of PAC oligomers up to the nonamer in the positive reflectron mode and up to the undecamer in the positive linear mode. Additional masses corresponding to a series of galloyl procyanidins were also detected; the highest degree of galloylation observed was six. Yang and Chien (2000) subsequently published similar results and indicated that the method has potential as a sensitive and quantitative technique. Both electrospray ionization (ESI) and MALDI-TOF MS were used to characterize the DP and structure of PAC oligomers in cranberries that are associated with the *in vitro* inhibition of adhesion of *Escherichia coli* to uroepithelial cells (Foo et al. 2000b, 2000a). The methods detected oligomers up to a DP of 7 in the PAC fraction that had the greatest antiadhesion activity. The application of MALDI-TOF MS clearly demonstrated the presence of A-type linkages in the tetramer through hexamer.

Porter et al. (2001) used MALDI-TOF MS to characterize the PAC oligomers that are associated with decreased susceptibility of human low density lipoproteins (LDL) to *in vitro* Cu^{2+} induced oxidation. The identified masses closely resembled the epicatechin structures reported by Foo et al. (2000b) to be present in cranberries. However, the spectra also had masses that were 16 Dalton (Da) greater than the masses previously reported, that may correspond to oligomers with one epigallocatechin unit. Although Foo et al. (2000b) reported the presence of epigallocatechin subunits in their cranberry juice fraction, they did not detect heteropolymers with both epicatechin and epigallocatechin units by mass spectrometry. Differences in the detection of these heteropolymers may be a result of the types of instrumentation that were used and methods of isolating the PAC (Foo et al. 2000b). MALDI-TOF MS results also showed the presence of more complex heteropolymers for PAC with a DP greater than four. These results indicated that cranberry PAC oligomers with a DP greater than four have more than one A-type interflavan bond. These pioneering results on the application of MALDI-TOF MS clearly demonstrate its value to characterize the nature of PAC oligomers in functional foods.

Rush et al. (2018) used MALDI-TOF MS in negative reflectron mode to characterize procyanidins from cocoa, peanut skins, cinnamon, and crab apples. The negative reflectron mode was preferred over the positive reflectron mode to avoid splitting the PAC signals between $[M+Na]^+$ and potassium $[M+K]^+$ ion adducts. Results indicate that MALDI-TOF MS in negative reflectron mode demonstrated the structural heterogeneity of procyanidins. Although Rush et al. (2018) described a rapid method to determine structures of procyanidins, this work used isolated procyanidins and not complex mixtures of PAC from botanical sources.

Recently, Esquivel-Alvarado et al. (2020) used MALDI-TOF MS in positive reflectron mode to characterize PAC in three *Vaccinium* species from Costa Rica. Masses that correspond to sodium adducts $[M+Na]^+$ based on repeating (epi)catechin monomers from trimers to tetradecamers were observed. MALDI-TOF MS spectra indicated that the *Vaccinium* species from Costa Rica have different isotopic patterns compared to cranberry and blueberry. Results also indicated that the majority of PAC in these *Vaccinium* species from Costa Rica have at least one A-type interflavan bond, which suggest that PAC from these *Vaccinium* species are closer to cranberry PAC than blueberry PAC. Although Esquivel-Alvarado et al. (2020) mentioned the ability of MALDI-TOF to show different isotopic patterns of PAC, this work split the PAC signals as both $[M+Na]^+$ and potassium $[M+K]^+$ ion adducts.

Currently, our laboratory uses a protocol to obtain high-resolution MALDI-TOF MS spectra in positive reflectron mode. This protocol eliminates the signal splitting between sodium $[M+Na]^+$ and potassium $[M+K]^+$ ion adducts by using cesium trifluoroacetate. As a result the PAC signal is detected only as $[M+Cs]^+$ ion adducts. In our protocol, 2,5-dihydroxybenzoic acid (DHB) at a concentration of 1.3 mol/L in methanol and samples at a concentration of 1.3 mg/mL in methanol are subjected to deionization using PureSpeed IEX strong cation resin tips. After deionization, cesium trifluoroacetate at a concentration of 0.01 mol/L in methanol and the deionized DHB are mixed at a ratio of 1:1 (v/v). Each deionized sample (0.5 μL) is spotted on the stainless-steel target, followed by the addition of the matrix (0.6 μL), which is pipetted a minimum of five times up and down in the pipette tip and allowed to dry on the target. The samples are analyzed in positive reflectron mode using a Bruker UltraFlex® III MALDI-TOF MS equipped with a SmartBeam laser. The settings are modified to a pulsed ion extraction of 130 ns, detector gain of 2.64x and 1548kV, preamplifier filter bandwidth small, and digitizer sampling frequency of 100 Hz.

5.4 Deconvolution of Isotope Patterns of A- to B-type Interflavan Bonds in Proanthocyanidins

Deconvolution is defined as a process of resolving something into its constituent elements or removing complication in order to understand composition. MALDI-TOF MS analysis of PAC that contains both A- and B-type interflavan bonds results in mass spectra with complex overlapping isotope patterns. Based on the known structures of PAC and the known relative abundance of naturally occurring isotopes, equations can be written and solved by methods of matrix algebra to deconvolute the overlapping isotope patterns, resulting in the ability to quantify relative ratios of A- to B-type interflavan bonds within each DP of PAC (Feliciano et al. 2012). The deconvolution data sets can be analyzed by multivariate analysis (see Section 5.5) and used to determine authenticity of PAC.

Previous research has shown that B-type interflavan bonds can be converted to A-type interflavan bonds as a result of oxidative intramolecular reactions (Kondo et al. 2000; Poupard et al. 2011). These oxidative reactions lead to loss of two hydrogens and may be mistakenly identified as A-type interflavan bonds in the MS spectra. Kondo et al. (2000) demonstrated that procyanidins B1 and B2 were converted into procyanidin A1 and A2 by radical oxidation using 1,1-diphenyl-2-picrylhydrazyl at neutral pH. Similarly, Poupard et al. (2011) described oxidation of procyanidin B2 in an apple juice model solution using caffeoylquinic acid o-quinone. Results indicate that no extensive polymerization occurred when procyanidin B2 was oxidized with caffeoylquinic acid o-quinone. Moreover, procyanidin A2 did not accumulate, but rather continued to react and form dimer end-products with three linkages ($[M\text{-}H]^-$ at m/z 573). Previous results from our laboratory showed that when MALDI-TOF MS was performed on samples of apple juice, unexplained masses corresponding to trimers with more than two A-type interflavan bonds and tetramers with more than 3 A-type interflavan bonds were observed. These masses do not correspond to the suggested structures of trimers and tetramers with all A-type interflavan bonds because there would be more interflavan bonds present in the oligomer than what is possible. However, when MALDI-TOF was performed on apple, masses indicative of intramolecular oxidations were not detected. We hypothesize that intramolecular oxidation in PAC that are extracted from the apple fruit does not occur but oxidation occurs during the production of apple juice (Poupard et al. 2011). Currently, no evidence of intramolecular oxidation has been reported on cranberry PAC.

5.4.1 Isotope Distributions

Mass calculating programs such as IsoPro 3.0 can be used to predict the masses and abundance of the isotopes for a given formula and allow for comparison between predicted and observed isotopic distributions (Senko 2002). For example; the predicted monoisotope of procyanidin A2 dimer is $m/z = 599$ (Figure 5.2a), detected as a sodium ion adduct $[M+Na]^+$, which represents the contribution of ^{12}C, ^{16}O, and ^{1}H, to the compound. The mass at $m/z = 600$ represents one ^{13}C, or one ^{17}O, or one ^{2}H. The mass at $m/z = 601$ represents two ^{13}C, or one ^{13}C and one ^{2}H, or one ^{18}O, or two ^{2}H.

Similarly, the isotope distribution of procyanidin B2 can be predicted (Figure 5.2b). The monoisotopic mass of B2 is $m/z = 601$. The 2 atomic mass unit (amu) difference between the monoisotope of the B2 and A2 dimers is due to the loss of two H in the formation of the C-O-C bond that defines the A-type interflavan linkage in the procyanidin A2 dimer. A series of PAC, which vary only in the ratio of A- to B-type interflavan bonds, produces a mass spectrum with overlapping isotope patterns for each individual oligomer. As an example, a 1:1 (w/w) mixture of procyanidin A2 and B2 has predictable overlapping isotope patterns at $m/z = 601$ to $m/z = 604$ (Figure 5.2c). At $m/z = 601$, the monoisotopic mass of B2 overlaps with the mass of A2 that contains the third most abundant combination of heavier isotopes of C, O, and H that are 2 amu heavier than monoisotopic mass of A2. Thus, in a 1:1 (w/w) mixture of these dimers, the relative intensity ($r_i = 0.3739$) at $m/z = 601$ is higher than the r_i at m/z = 599 ($r_i = 0.3474$) because the third isotopic peak of A2 as well as the monoisotope of B2 contribute to the intensity of m/z = 601 (Figure 5.2c).

5.4.2 Precision and Accuracy Validation for Binary Mixtures of Procyanidin A2 and B2

Feliciano et al. (2012) used the MALDI-TOF MS spectra of procyanidin A2 and B2 dimers to deconvolute overlapping isotope patterns and determine relative ratios of A- to B-type interflavan bonds. In a binary system, 11 known ratios of procyanidin A2 and B2 were combined and subjected to positive reflectron mode MALDI-TOF MS. The data sets were then subjected to deconvolution. All observed ratios were within 3.6% of the predicted ratios for both compounds (Table 5.1), showing that the method is accurate.

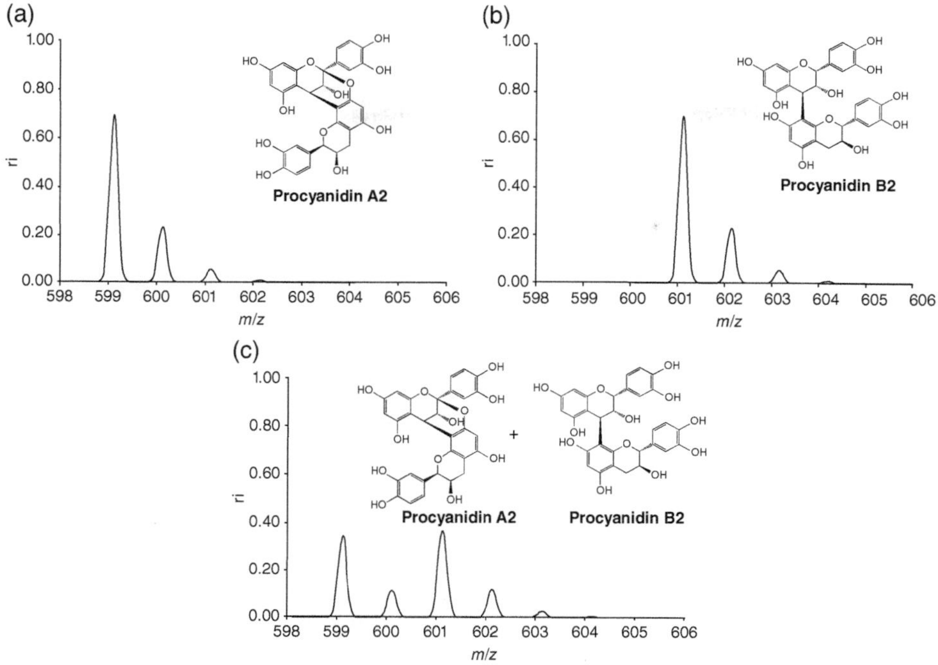

Figure 5.2 Natural isotope distribution of procyanidin A2 (a) and procyanidin B2 (b). Complex overlapping isotope patterns of procyanidin A2 and B2 mixed at a 1:1 (w/w) ratio (c).

Table 5.1 Predicted and observed percentages of binary mixtures of procyanidins A2 and B2 after deconvolution.

Procyanidin B2		Procyanidin A2	
Predicted	**Observed**	**Predicted**	**Observed**
95.0	93.3	5.0	6.7
90.0	88.2	10.0	11.8
80.0	79.0	20.0	21.0
70.0	67.5	30.0	32.5
60.0	56.9	40.0	43.1
50.0	49.0	50.0	51.0
40.0	39.9	60.0	60.1
30.0	28.8	70.0	71.2
20.0	16.5	80.0	83.6
10.0	8.6	90.0	91.4
5.0	5.1	95.2	94.9

The standard deviation among each series of replicates was ≤3.5% and the standard error of the mean (SE) was <0.8%, which indicates that the method is also precise. The results using the A2 and B2 dimers indicated that the deconvolution method could be used to accurately and precisely predict ratios of A- to B-type bonds in PAC oligomers.

5.4.3 Deconvolution of Cranberry PAC Ratios of A- to B-type Bonds within Each Degree of Polymerization

Obviously, cranberry PAC have a greater diversity in polymer distribution than A- and B-type dimers and as the DP of PAC increases, complexity of overlapping isotope patterns increases. For example, cranberry PAC trimers contain three molecules that differ in the number of A-type interflavan bonds: two A-type and zero B-type ($m/z = 885$), one A-type and one B-type ($m/z = 887$), and zero A-type and two B-type ($m/z = 889$) (Figure 5.3) (Foo et al. 2000b). The spectra in this mass range will have overlapping isotope patterns if trimers with one or two A-type interflavan bonds are present in the sample, as previously demonstrated for the A2 and B2 dimers. For example, the intensity of the signal at $m/z = 889$ is the combination of three PAC trimers: (i) the monoisotopic signal of the trimer with 2 B-type bonds, (ii) lower signals due to heavier isotopes present in the trimer with one A-type and one B-type interflavan bond, and (iii) the trimer with two A-type interflavan bonds. Again, matrix algebra (solving simultaneous unknown equations) enables deconvolution to be performed on increasingly complex overlapping isotope patterns at each DP within a PAC polymer distribution.

Procyanidin dimers showed high correlation between predicted natural isotope abundance and observed isotope distributions, suggesting good precision. By extension, positive reflectron mode MALDI-TOF MS spectra were collected on isolated purified cranberry PAC and subjected to deconvolution for PAC oligomers from DP3 to DP11 (Feliciano et al. 2012). The relative ratios of A-type to B-type interflavan bonds was determined for each DP. Results showed that PAC oligomers with one or more A-type interflavan bonds occurred at higher proportion than PAC oligomers with all B-type interflavan bonds. Furthermore, >90% of the

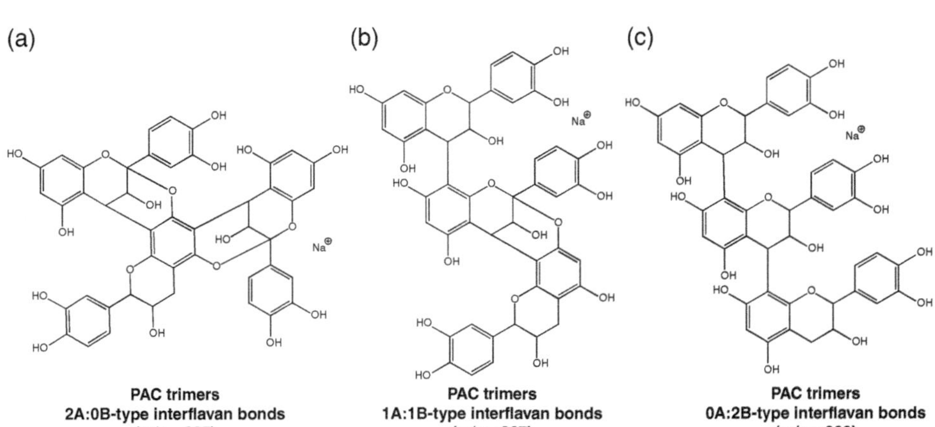

Figure 5.3 Chemical structures of PAC trimers, which show 2A:0B-type interflavan bonds (a), 1A:1B-type interflavan bonds (b), and 0A:2B-type interflavan bonds (c).

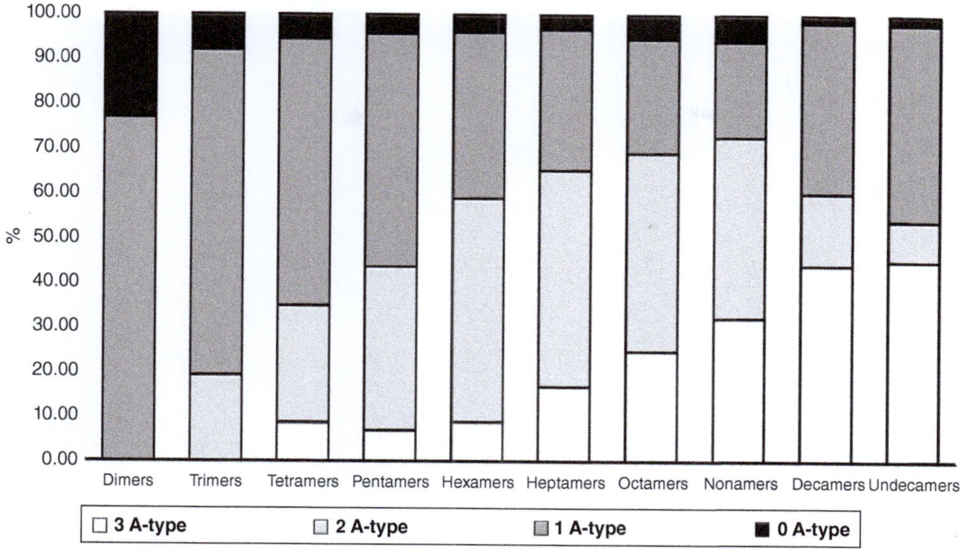

Figure 5.4 Percentage of A-type interflavan bonds in cranberry PAC oligomers from dimers to undecamers analyzed by deconvolution for isotope patterns in MALDI-TOF MS.

cranberry PAC were shown to contain at least one A-type interflavan bond and the number of A-type interflavan bonds within each oligomer increased as DP increased (Figure 5.4).

5.4.4 Deconvolution Application for PAC Structure in Studies of Bioactivity

The unique structural attributes of PAC (i.e. nature of interflavan bonds) are associated with bioactivity. Putative health benefits of PAC include a decreased risk of heart disease and dyslipidemia for PAC in grape seed extract or pine bark (Schoonees et al. 2012; Sahebkar 2014; Zhang et al. 2016), and a decreased risk of urinary tract infections (UTIs) for PAC in cranberry. However, meta-analysis results of clinical trials on UTI prevention are often conflicted, with associations between PAC intake and health benefits in the trials being confounded due to use of unstandardized product forms that lack compositional analysis, doses that may be subefficacious, and little to no mechanistic guidance for selection of subjects and study design (Jepson and Craig 2007, Jepson et al. 2012). The MALDI-TOF MS may be used to correlate PAC structure to bioactivity.

Feliciano et al. (2014) demonstrated the utility of MALDI-TOF MS to determine the relative ratios of A- to B-type PAC in cranberry and apples and correlated PAC structure to the ability of PAC to agglutinate *E. coli*, and inhibit *E. coli* invasion of gut epithelial cells. Results from this work indicated that >94.5% of the isolated cranberry PAC oligomers had at least one A-type bond, while >88.3% of the isolated apple PAC oligomers had no A-type bonds (Figure 5.5). A-type bonds found in apple PAC are likely due to oxidative intramolecular linkages produced during the extraction, isolation, and characterization of PAC. Furthermore, the cranberry A-type PAC had greater bioactivity in *E. coli* agglutination and inhibition of *E. coli* invasion of epithelial cells in comparison to the apple B-type PAC.

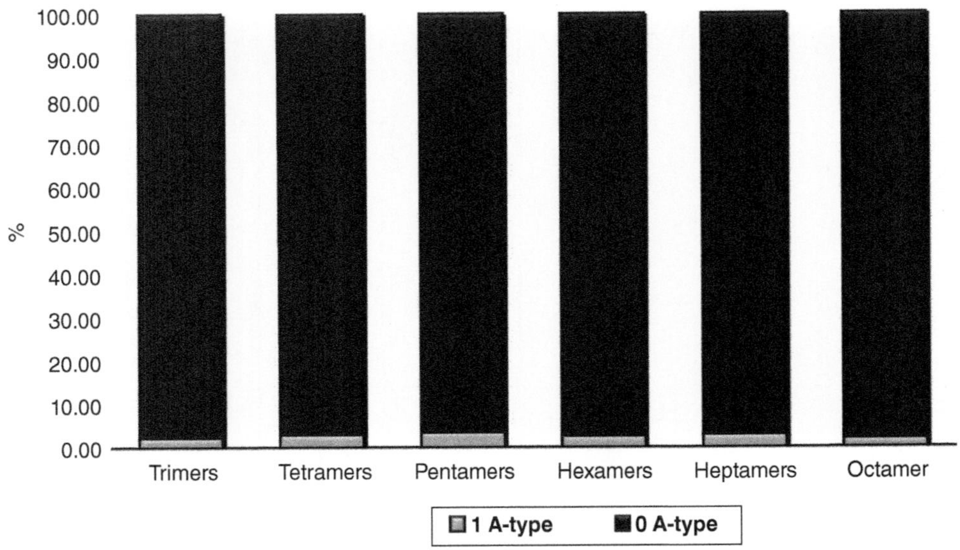

Figure 5.5 Percentage of A-type interflavan bonds in apple PAC oligomers from trimers to octamers analyzed by deconvolution of isotope patterns in MALDI-TOF MS.

5.4.5 MALDI-TOF MS for Mixtures of Isolated Cranberry and Apple Proanthocyanidins

We used isolated cranberry and apple PAC to further validate the precision and accuracy for deconvolution of overlapping isotope patterns in complex mixtures across a wide range of PAC polymers (DP 3–7). In a binary system, 21 known ratios of isolated cranberry and apple PAC were combined and subjected to positive reflectron mode MALDI-TOF MS. The data sets were then subjected to deconvolution. Observed ratios were within 3.9% of the predicted ratios, showing that the method is accurate (Figure 5.6).

This research demonstrated that MALDI-TOF MS and deconvolution of overlapping isotope patterns may be used to determine the authenticity of PAC from a single source and to accurately determine the ratios at which two known PAC sources are combined. By extension, these methods could be used to determine the source of PAC in blends of whole fruits, juices, and dietary ingredients and to confirm authenticity of the products.

5.5 Multivariate Analysis of MALDI-TOF MS Spectra Data

The rapid development of high-throughput analytical techniques such as MALDI-TOF MS allow generation of a large amount of data. Multivariate analysis is a useful tool to discover trends and reveal information that is not obvious through visual inspection of large data sets. MALDI-TOF spectral data combined with multivariate analysis can therefore be used to determine the authenticity of functional foods and dietary supplements that contain PAC.

Figure 5.6 Deconvolution of MALDI-TOF MS of 21 different ratios of isolated cranberry and apple PAC shows predicted percentage cranberry PAC is within 3.9% of actual mixed ratios. *Source:* Christian G Krueger. (*See insert for color representation of the figure.*)

Multivariate analysis may be divided into two categories: unsupervised and supervised. Unsupervised methods use unlabeled data and are categorized as exploratory techniques. Unsupervised methods include PCA, self-organizing maps (SOM), and hierarchical clustering analysis (HCA). Supervised methods use labeled data and are categorized as predictive techniques (Abney 2007). Supervised methods include LDA, partial least squares discriminant analysis (PLSDA), k-nearest neighbor (KNN), and soft independent modeling of class analogy (SIMCA).

PCA uses an orthogonal transformation to convert a large number of correlated variables into a small number of uncorrelated variables called principal components (PCs), which are useful for data reduction and cluster visualization (Johnson 1998). PCA is composed of matrices of direction (eigenvectors) and scale (eigenvalues). PCA may be represented by two plots: loading and score. The loading plot shows how strongly each variable influences a PC, whereas the score plot shows the clustering of samples based on their similarity (Lucci et al. 2017).

Deconvoluted or autoscaled MALDI-TOF spectra data may be used to classify botanical sources of PAC. The deconvolution method only uses the monoisotopic peaks for each predicted A- or B-type interflavan bonds. Deconvoluted MALDI-TOF spectra data differ from the autoscaled MALDI-TOF spectra data because the latter divides the absolute intensity of each A- or B-type interflavan bond by the sum of all absolute intensities of each A- or B-type interflavan bond for each individual PAC oligomer.

Three approaches that use deconvoluted or autoscaled MALDI-TOF MS spectra data combined with PCA have been evaluated. In our first approach, MALDI-TOF MS spectra of 19 samples (5 blueberries and 14 cranberries) were subjected to PCA. PCA differentiated samples by the presence of A- or B-type interflavan bonds in PAC oligomers. Score plot based on the deconvoluted MALDI-TOF spectra data indicated that PC1 and PC2 explained 87.6% of the total variance (Figure 5.7). In addition, samples with all B-type interflavan bonds (blueberries) were located on the negative part of PC1, whereas samples with one A-type interflavan bond (cranberries) were located on the positive part of PC1. This preliminary result showed that PCA can differentiate botanical sources with different PAC profiles.

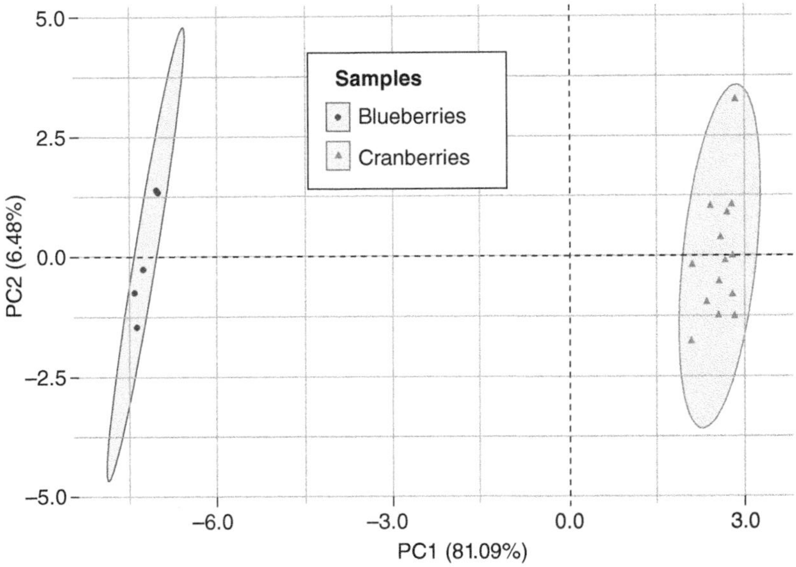

Figure 5.7 Principal component analysis score plot of proanthocyanidins from blueberries and cranberries based on the deconvoluted MALDI-TOF MS spectra. PC1 (x-axis) and PC2 (y-axis) accounts for 87.6% of the total variance. The colored shadows correspond to the 95% confidence ellipse, based on the distribution of A-type interflavan bonds in the PAC oligomers. (*See insert for color representation of the figure.*)

In our second approach, a more robust design was evaluated. MALDI-TOF spectra of 75 samples (30 cranberries, 5 cranberry supplements, 25 apples, 6 grape skins, 4 cocoa powders, and 5 blueberries) were subjected to deconvolution of PAC isotope patterns prior to performing PCA. Score plot based on the deconvoluted MALDI-TOF spectra data indicates that PC1 and PC2 explained 86.7% of the total variance (Figure 5.8). Furthermore, samples were grouped according their PAC profile. Samples in which the most abundant peak corresponds to a PAC structure with all B-type interflavan bonds (apples, grape skins, cocoa powders, and blueberries) were located on the negative part of PC1, whereas samples in which the most abundant peak corresponds to a PAC structure with one A-type interflavan bond (cranberries and cranberry supplements) were located on the positive part of PC1. Results of this study indicate that the deconvolution method is robust, precise, and accurate and also contributes to a better understanding of differences in PAC structure when PCA is used.

Lastly, in our third approach, MALDI-TOF MS data of 86 samples (28 apples, 34 cranberries, and 24 peanut skins) were subjected to both data selection and autoscaling. As previously described by Feliciano et al. (2014), PAC were isolated from the botanical samples prior to MALDI-TOF MS analysis. Data selection was used instead of deconvolution to obtain more information from the MALDI-TOF spectra data. The deconvolution method only uses the relative intensity of the overlapping peaks for each predicted A- or B-type interflavan bond and ignores the signal of the intermediate isotope signals. Data selection uses these intermediate isotope signals. Therefore, the selection of all predicted isotope patterns of PAC from DP3 to DP8 were used. The isotope patterns of PAC were subjected to autoscaling. Autoscaling was chosen instead of other

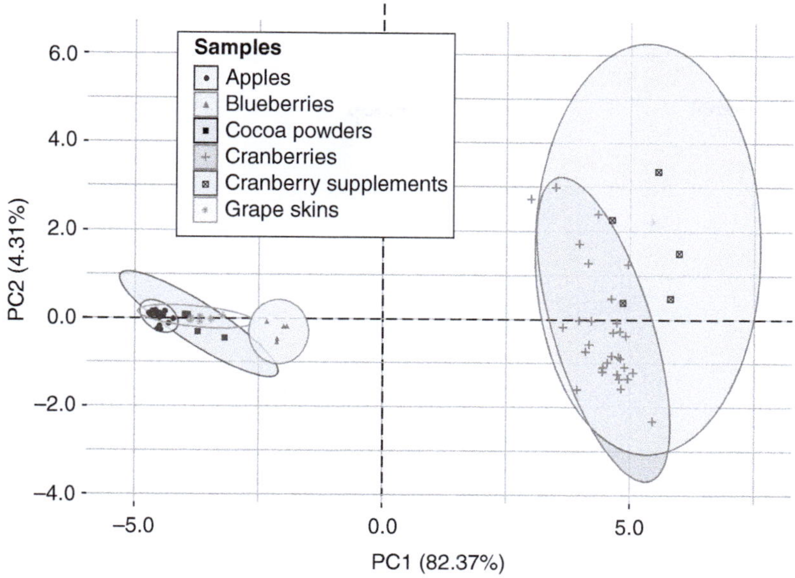

Figure 5.8 Principal component analysis score plot of proanthocyanidins from apples, blueberries, cocoa, powders, cranberries, cranberry supplements, and grape skins based on the deconvoluted MALDI-TOF MS spectra. PC1 (x-axis) and PC2 (y-axis) accounts for 86.7% of the total variance. The colored shadows correspond to the 95% confidence ellipse, based on the distribution of A-type interflavan bonds in the PAC oligomers. (*See insert for color representation of the figure.*)

pretreatment methods to maintain equality amongst all the isotopic peaks (Van Den Berg et al. 2006). Loading plot indicated that PC1 separated all the samples (apples) with no A-type interflavan bonds, which were located on the negative part of PC1, from all the samples (cranberries and peanut skins) with one, two, and three A-type interflavan bonds, which were located on the positive part of PC1. In addition, the loading plot indicated that PC2 separated all the samples (cranberries) with one A-type interflavan bond, which were located on the negative part of PC2, from all the samples (peanut skins) with two and three A-type interflavan bonds, which were located on the positive part of PC2 (Figure 5.9a). Score plot based on the autoscaled MALDI-TOF spectra data indicated that PC1 and PC2 explained 92.0% of the total variance (Figure 5.9b). PCA was able to differentiate PAC from apples, cranberries, and peanut. This approach is an improvement over our previous studies because it introduces PAC from peanut skins, which also contains A-type interflavan bonds (Zhang et al. 2013). Cranberry and peanut skin PAC were differentiated from one another because the latter contained a higher percentage of PAC with two or three A-type interflavan bonds. The separation between cranberries and peanut skins was reflected in the higher percentage variation explained by PC2 (23.3%) compared to the two previous approaches. Results support our previous studies that showed that PCA can accurately predict the PAC profile of products that are used in dietary supplements. In addition, this latter approach allows the inclusion of other features that are present in samples such as the galloyl units in grape skins, the 3-deoxyproanthocyanidins in sorghum, and ellagitannins in pomegranate.

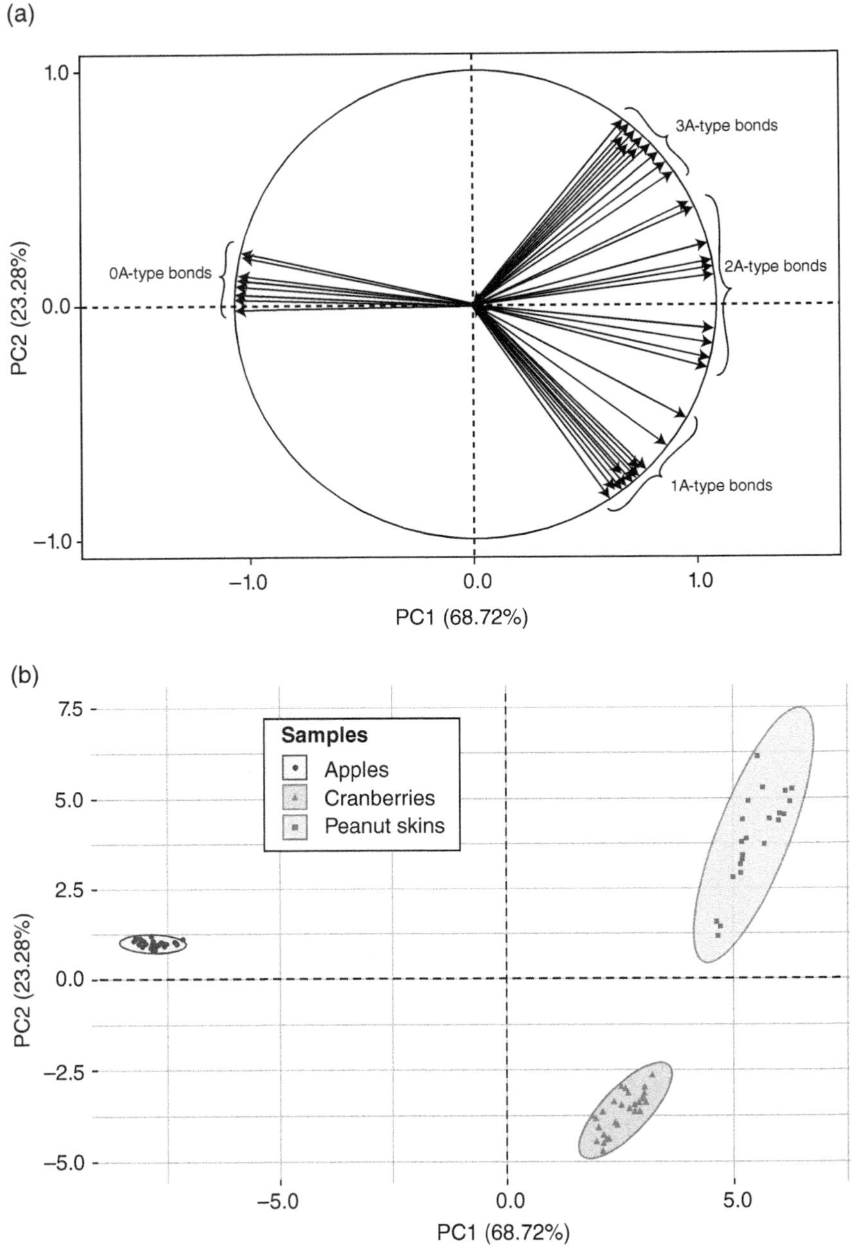

Figure 5.9 Principal component analysis of proanthocyanidins from apples, cranberries, and peanut skins based on the autoscaled MALDI-TOF MS spectral data. PC1 (x-axis) and PC2 (y-axis) accounts for 92.0% of the total variance. (a) PCA loading plot. The arrows indicate the location of all the PAC oligomers with the number of A-type interflavan bonds, as represented by the brackets, for oligomers from DP3 to DP8. (b) PCA score plot. The colored shadows correspond to the 95% confidence ellipse, based on the distribution of A-type interflavan bonds in the PAC oligomers. (*See insert for color representation of the figure.*)

LDA is a reduction technique used to create prediction models to assign unknown samples to each class (Callao and Ruisánchez 2018). LDA maximizes the ratio of between-class variance and minimizes the ratio of within-class variance (Berrueta et al. 2007). LDA was used to evaluate the potential of the autoscaled MALDI-TOF MS data to correctly classify samples. LDA was evaluated using our previous data set (n = 86) of individual PAC samples, a data set (n = 70) in which 10 cranberry PAC samples were combined with 10 apple PAC samples at each of the following ratios [5, 10, 15, 20, 25, 50, and 75% (w/w)], and a data set (n = 70) in which 10 cranberry PAC samples were combined with 10 peanut skin PAC samples at each of the following ratios [5, 10, 15, 20, 25, 50, and 75% (w/w)]. The complete data set (n = 226) was split into two subsets: training set and testing set. The training set composed of 159 samples (70% of the total samples) was used to build the predictive algorithm. The testing set composed of 67 samples (30% of the total samples) was used to provide an unbiased evaluation of the model on the training set. Score plot based on the autoscaled MALDI-TOF MS data indicated that LD1 and LD2 explained 93.9% of the total variance. The classification performance was determined using the training set, the testing set, and the leave-one-out cross validation (LOOCV) method. The LOOCV method uses one sample from the total samples as the testing data, and the remaining samples as the training data. This process is repeated such that each sample is used once as the validation data. Results indicated that the recognition ability (training set) was 99.4%, predictive ability (testing set) was 100%, and the LOOCV method was 94.2%. The high performance obtained by the LDA model indicated that autoscaled MALDI-TOF spectral data were able to correctly classify unknown samples based on PAC structure.

5.6 Conclusion

The use of MALDI-TOF MS analysis of PAC has potential as a method to determine authenticity of functional food and dietary supplements such as cranberries, grape seed extracts, and many other ingredients. In this study, we demonstrate that MALDI-TOF MS accurately distinguishes PAC from different foods and ingredients that are used in dietary supplements. We also demonstrated that MALDI-TOF MS accurately determines the relative proportion of PAC in mixtures of PAC from different plant species. However, more research is needed to develop MALDI-TOF MS as a practical tool for authenticity. For instance, the impact of oxidation on the robustness of the method and its validation when applied to food samples prone to oxidation needs to be studied. The most challenging task will be the development of spectral libraries from samples of authentic dietary supplements and functional foods for use in building predictive capabilities of multivariate analysis.

References

Abney, S. (2007). *Semisupervised learning for computational linguistics*, Chapman and Hall/CRC.

Berrueta, L., Alonso-Salces, R., and Heberger, K. (2007). Supervised pattern recognition in food analysis. *Journal of Chromatography A* 1158: 196–214.

Bovay, J. 2016. FDA refusals of imported food products by country and category, 2005–2013.

Callao, P. and Ruisánchez, I. (2018). An overview of multivariate qualitative methods for food fraud detection. *Food Control* 86: 283–293.

Cheynier, V., Comte, G., Davies, K.M., et al. (2013). Plant phenolics: recent advances on their biosynthesis, genetics, and ecophysiology. *Plant Physiology and Biochemistry* 72: 1–20.

Clydesdale, F. (2004). Functional foods: opportunities and challenges. *Food Technology Magazine* 58.

Craciun, M., Cretu, G., Mirea, C., et al. (2015). Anthocyanins identification, separation and measurement from cranberry, blueberry, bilberry, chokeberry and acai berry extracts by HPTLC. *Revista de Chimie* 66: 929–936.

Esquivel-Alvarado, D., Muñoz-Arrieta, R., Alfaro-Viquez, E., et al. (2020). Composition of anthocyanins and proanthocyanidins in three tropical *Vaccinium* species from Costa Rica. *Journal of Agricultural and Food Chemistry* 68: 2872–2879.

Feliciano, R., Krueger, C., Shanmuganayagam, D., Vestling, M., and Reed, J. (2012). Deconvolution of matrix-assisted laser desorption/ionization time-of-flight mass spectrometry isotope patterns to determine ratios of A-type to B-type interflavan bonds in cranberry proanthocyanidins. *Food Chemistry* 135: 1485–1493.

Feliciano, R., Meudt, J., Shanmuganayagam, D., Krueger, C., and Reed, J. (2014). Ratio of "A-type" to "B-type" proanthocyanidin interflavan bonds affects extra-intestinal pathogenic *Escherichia coli* invasion of gut epithelial cells. *Journal of Agricultural and Food Chemistry* 62: 3919–3925.

Foo, L., Lu, Y., Howell, A., and Vorsa, N. (2000a). The structure of cranberry proanthocyanidins which inhibit adherence of uropathogenic P-fimbriated *Escherichia coli in vitro*. *Phytochemistry* 54: 173–181.

Foo, L., Lu, Y., Howell, A., and Vorsa, N. (2000b). A-Type proanthocyanidin trimers from cranberry that inhibit adherence of uropathogenic P-fimbriated *Escherichia coli*. *Journal of Natural Products* 63: 1225–1228.

Jepson, R. and Craig, J. (2007). A systematic review of the evidence for cranberries and blueberries in UTI prevention. *Molecular Nutrition and Food Research* 51: 738–745.

Jepson, R., Williams, G., and Craig, J. (2012). Cranberries for preventing urinary tract infections. *Cochrane Database of Systematic Reviews* 10 (Art. No.: CD001321). doi: 10.1002/14651858.CD001321.pub5.

Johnson, D. (1998). *Applied multivariate methods for data analysts*. Pacific Grove, CA.: Duxbury Press.

Kondo, K., Kurihara, M., Fukuhara, K., et al. (2000). Conversion of procyanidin B-type (catechin dimer) to A-type: evidence for abstraction of C-2 hydrogen in catechin during radical oxidation. *Tetrahedron Letters* 41: 485–488.

Krueger, C., Dopke, N., Treichel, P., et al. (2000). Matrix-assisted laser desorption/ionization time-of-flight mass spectrometry of polygalloyl polyflavan-3-ols in grape seed extract. *Journal of Agricultural and Food Chemistry* 48: 1663–1667.

Kupina, S. and Gafner, S. (2016). Adulteration of grape seed extract (Vitis vinifera). *Botanical Adulterants Bulletin*, April.

Laine. (2013). New test system identifies 193 different yeasts and bacteria known to cause illnes. *Infection Control Today* (22 August).

Leclair, M. 2016. Fruit and Vegetable Markets in the US. *IBSWorld Industry Report*. IBISWorld.

Lucci, P., Saurina, J., and Núñez, O. (2017). Trends in LC-MS and LC-HRMS analysis and characterization of polyphenols in food. *Trends in Analytical Chemistry* 88: 1–24.

Ohnishi-Kameyama, M., Yanagida, A., Kanda, T., and Nagata, T. (1997). Identification of catechin oligomers from apple (*Malus pumila* cv. Fuji) in matrix-assisted laser desorption/

ionization time-of-flight mass spectrometry and fast-atom bombardment mass spectrometry. *Rapid Communications in Mass Spectrometry* 11: 31–36.

Polito, R. (2015). The regulatory shakeup. *Nutritional Business Journal*.

Porter, M., Krueger, C., Wiebe, D., et al. (2001). Cranberry proanthocyanidins associate with low-density lipoprotein and inhibit in vitro Cu^{2+}-induced oxidation. *Journal of the Science of Food and Agriculture* 81: 1306–1313.

Poupard, P., Sanoner, P., Baron, A., et al. (2011). Characterization of procyanidin B2 oxidation products in an apple juice model solution and confirmation of their presence in apple juice by high-performance liquid chromatography coupled to electrospray ion trap mass spectrometry. *Journal of Mass Spectrometry* 46: 1186–1197.

Reed, J., Krueger, C., and Vestling, M. (2005). MALDI-TOF mass spectrometry of oligomeric food polyphenols. *Phytochemistry* 66: 2248–2263.

Rumalla, C., Avula, B., Wang, Y., et al. (2012). Densitometric-HPTLC method development and analysis of anthocyanins from acai (*Euterpe oleracea* Mart.) berries and commercial products. *Journal of Planar Chromatography* 25: 409–414.

Rush, M., Rue, E., Wong, A., et al. (2018). Rapid determination of procyanidins using MALDI-ToF/ToF mass spectrometry. *Journal of Agricultural and Food Chemistry* 66: 11355–11361.

Sahebkar, A. (2014). A systematic review and meta-analysis of the effects of pycnogenol on plasma lipids. *Journal of Cardiovascular Pharmacology and Therapeutics* 19: 244–255.

Schoonees, A., Visser, J., Musekiwa, A., and Volmink, J. (2012). Pycnogenol® for the treatment of chronic disorders. *Cochrane Database of Systematic Reviews* 2 (Art. No.: CD008294). doi:10.1002/14651858.CD008294.pub3.

Senko, M. (2002). IsoPro 3.0 MS/MS software. *Isotopic abundance simulator version* 3.

Singhal, N., Kumar, M., Kanaujia, P., and Virdi, J. (2015). MALDI-TOF mass spectrometry: an emerging technology for microbial identification and diagnosis. *Frontiers in Microbiology* 6: 791.

Smith, T., Lynch, M., Johnson, J., et al. (2015). Herbal dietary supplement sales in US increases 6.8 in 2014. *The Journal of the American Botanical Council*, 52–59.

Twilley, N. (2015). How not to test a dietary supplement. *The New Yorker* (10 February).

US Congress (1938). Federal Food, Drug, and Cosmetic Act (FD&C Act). Chapter V § 351, Adulterated drugs and devices.

US Congress (2011). Food Safety Modernization Act (FSMA).

Van Den Berg, R., Hoefsloot, H., et al. (2006). Centering, scaling, and transformations: improving the biological information content of metabolomics data. *BMC Genomics* 7: 142.

Villani, T., Reichert, W., Ferruzzi, M., et al. (2015). Chemical investigation of commercial grape seed derived products to assess quality and detect adulteration. *Food Chemistry* 170: 271–280.

Wallace, L., Boilard, S., Eagle, S., et al. (2012). DNA barcodes for everyday life: routine authentication of natural health products. *Food Research International* 49: 446–452.

Yang, Y. and Chien, M. (2000). Characterization of grape procyanidins using high-performance liquid chromatography/mass spectrometry and matrix-assisted laser desorption/ionization time-of-flight mass spectrometry. *Journal of Agricultural and Food Chemistry* 48: 3990–3996.

Zhang, H., Liu, S., Li, L., et al. (2016). The impact of grape seed extract treatment on blood pressure changes: A meta-analysis of 16 randomized controlled trials. *Medicine* 95: e4247.

Zhang, H., Yang, Y., and Ma, C. (2013). Structures and antioxidant and intestinal disaccharidase inhibitory activities of A-type proanthocyanidins from peanut skin. *Journal of Agricultural and Food Chemistry* 61: 8814–8820.

6

Challenges in Analyzing Bioactive Proanthocyanidins

Wayne E. Zeller[1] *and Irene Mueller-Harvey*[2]

[1] *ARS-USDA, U.S. Dairy Forage Research Center, Madison, USA*
[2] *School of Agriculture, Policy and Development, University of Reading, UK*

6.1 Introduction

The leather, wine, wood adhesive, and anticorrosive primer production represent traditional markets for all tannin types, with proanthocyanidins (PAs, also referred to as condensed tannins) accounting for most (86%) and hydrolysable and phlorotannins for the rest of the one million ton annual sales (https://www.grandviewresearch.com/industry-analysis/tannin-market). While the traditional markets for PAs rely largely on 5-deoxy proanthocyanidins (i.e. profisetinidins and prorobinetinidins [Figure 6.1]) that originate from the wood or bark of southern hemisphere trees (Ferreira and Coleman 2011), other PA classes, such as the propelargonidins, procyanidins, or prodelphinidins, are found more widely in plants and fruits that are important for human or animal nutrition and health.

Foods containing PAs have attracted interest due to their purported potential for improving human health through modulating various detrimental processes in the body (i.e. inflammation, cancer, cardiovascular, immune responses) (Selmi et al. 2006; Holderness et al. 2008; de Pascual-Teresa et al. 2010; Park et al. 2012; EFSA 2012; Li et al. 2015; Salaritabar et al. 2017; Hamza et al. 2018; Toden et al. 2018), through their direct effects on the gut biome and the indirect health effects of the resulting PA metabolites (Mena et al. 2015).

In addition, PA-containing forages and other plant materials offer potential for enhancing animal health, production efficiency, and farming sustainability. It is generally accepted that many of these effects are derived from interaction of PAs with forage or microbially generated proteins. PA-containing forages have been shown to conserve protein during both the ensiling process (Albrecht and Muck 1991; Lorenz et al. 2010; Hymes-Fecht et al. 2013) and rumen digestion, elevating levels of protein escaping into the hindgut, where it can bolster milk, wool, and meat production (Jones and Mangan 1977; Waghorn et al. 1998; Barry and McNabb 1999; Min et al. 2003; Mueller-Harvey et al. 2019). PA-containing forages or total mixed rations supplemented with commercially available PAs have also shown potential to reduce greenhouse gas emissions (methane) from ruminants (Patra and Saxena 2010; Pellikaan et al. 2011; Naumann et al. 2013, 2014) and to reduce urea and ammonia excretion (Aguerre et al. 2010).

Recent Advances in Polyphenol Research, Volume 7, First Edition. Edited by Jess Dreher Reed, Victor Armando Pereira de Freitas, and Stéphane Quideau.

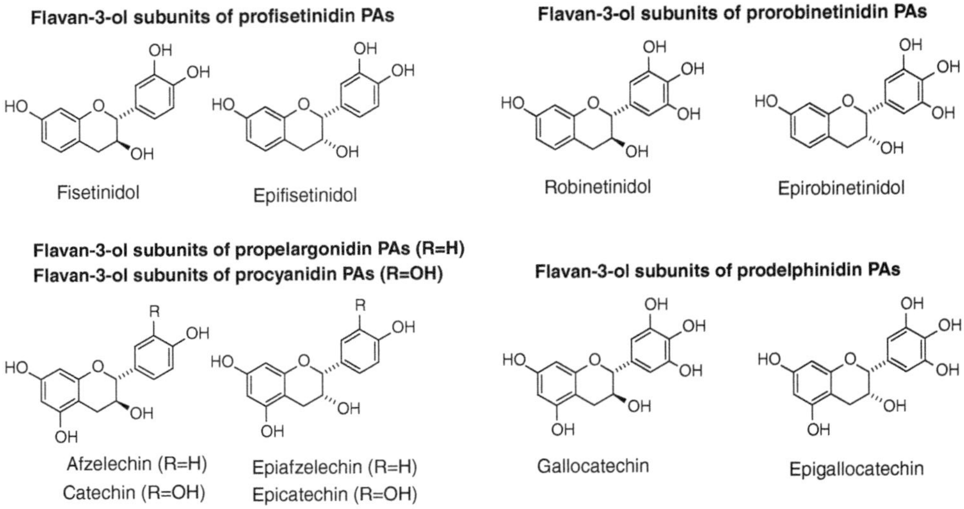

Figure 6.1 Structures of flavan-3-ol subunits that give rise to profisetinidin, prorobinetinidin, propelargonidin, procyanidin, and prodelphinidin proanthocyanidins. These flavan-3-ol subunits may also be galloylated or glycosylated at C3 or the phenolic groups.

In addition to their impact on protein and nitrogen conservation in animal production, PA-containing feedstuffs can act as antiparasitic agents (Hoste et al. 2006, 2015; Quijada et al. 2015; Mueller-Harvey et al. 2019) and prevent bloat (Wang et al. 2012), which impacts positively on farm economic and environmental sustainability (Aerts et al. 1999; Piluzza et al. 2013). However, at high PA concentrations antinutritional effects have been observed leading to lower intakes, while toxic effects have been reported with 5-deoxy-PAs (Hervás et al. 2003; Deaville et al. 2010). Thus, it is imperative that PA measurements are based on robust analytical tools to underpin the research that is driving the development of new PA products and uses, such as medicinal plants, functional foods, and animal feeds.

However, these studies have been hampered from the use of analytical methods which were later shown to be less rigorous in PA determination than once thought. For example, Makkar et al. (1999) showed by cross polarization–magic angle spinning (CPMAS) ^{13}C NMR spectroscopy (see Section 6.9.3 for descriptive uses of this method) that the traditional HCl-butanol assay conditions failed to detect all the PAs in samples of *Acacia saligna* leaves, even after multiple treatments. It was also shown that the traditional HCl-butanol assay does not react with all the nonextractable PAs in *Lotus* species and as a result can lead to considerable underestimation of PA concentrations (Grabber et al. 2013). Moreover, agreement between different methods is often unsatisfactory (see Section 6.6) and can cast doubt on many reports of total PA concentrations of plant or food materials.

This chapter focuses on current procedures that are suitable for the isolation, characterization, and analysis of PAs by a wide number of laboratories and that can serve as a basic set of techniques for PA biological research, whether in plant breeding, human and animal nutrition, or health. The selection is also based on the general instrumental accessibility for the majority of PA researchers. This chapter will focus on techniques that are applicable to PAs, although some (near infrared reflectance spectroscopy [NIRS], NMR, UPLC-MS/MS) are also suitable for the analysis of hydrolysable tannins. In an effort to provide as complete a summary as possible, it was necessary to omit some PA-macro-molecular interaction and state-of-the-art physicochemical techniques which are only available to selected laboratories, such as small-angle neutron scattering (SANS), X-ray, small-angle X-ray scattering (SAXS), and synchroton studies, in this update and the reader is, therefore, referred to recent articles on these techniques (e.g. Zanchi et al. 2008; Yu et al. 2009; Vernhet et al. 2011; Vidal et al. 2016; Dudek et al. 2017; Zhou et al. 2018).

6.2 Structural Diversity of Proanthocyanidins

The terminology used for classifying PAs into propelargonidins, procyanidins, prodelphinidins, etc (Figure 6.1) is somewhat misleading as PAs consisting of predominantly one flavan-3-ol subunit are rarely found in nature. A few plant species synthesize PAs with one preponderant set of subunits (i.e. apple leaves, black or red currant leaves, cocoa seeds, lespedeza plants, sorghum seeds) but for the most part PAs occur in complex mixtures in plants (Porter 1988). The structural diversity of these PA mixtures increases exponentially with polymer size (Table 6.1) due to the numerous combinatorial options of the different variables: flavan-3-ol subunits, interflavan-3-ol links (4-8 B-type, 4-6 B-type, A-type), and relative stereochemistry (at C3 and C4) (Naumann et al. 2017; Phansalkar et al. 2018) (Figure 6.2). The structural diversity of PAs expands even more if substituents at C3 or the phenolic groups (e.g. galloylation, glycosylation, etc) are included (Krueger et al. 2000, 2003).

This complexity poses challenges that limit the applicability of many analytical techniques across the whole PA spectrum, with PAs larger than 3000 Daltons being particularly difficult to purify and analyze (Reed et al. 2005; Stringano et al. 2011; Feliciano et al. 2012; Tanaka et al. 2014).

The biosynthesis of PAs is not only plant species–specific but also leads to distinct fingerprints, even at the accession level (Karonen et al. 2011; Salminen 2018; Mueller-Harvey et al. 2019). This is an often-overlooked fact that is likely to impact their nutritional and health contributions. In general PA concentrations and composition appear to be influenced greatly by plant species, accession, and plant part (Porter 1988; Huang et al. 2012; Azuhnwi et al. 2013), although harvest time (Hagerman 1988; Theodoridou et al. 2010) and growing conditions or environment (Acuña et al. 2008; Grabber et al. 2014) can also have an impact.

Table 6.1 Number of PA compounds in mixtures containing two, three, or four different flavan-3-ol subunits linked via two different bond types.

Number of flavan-3-ol subunits	2 subunits, 2 bond types	3 subunits, 2 bond types	4 subunits, 2 bond types
2 (dimer)	8	18	32
3 (trimer)	32	108	256
4 (tetramer)	128	648	2048
5 (pentamer)	512	3888	16384
6 (hexamer)	2048	23328	131072
7 (heptamer)	8192	139968	1048576
8 (octamer)	32768	839808	8388608
9 (nonamer)	131072	5038848	67108864
10 (decamer)	524288	30233088	536870912

Source: Naumann et al. 2017.

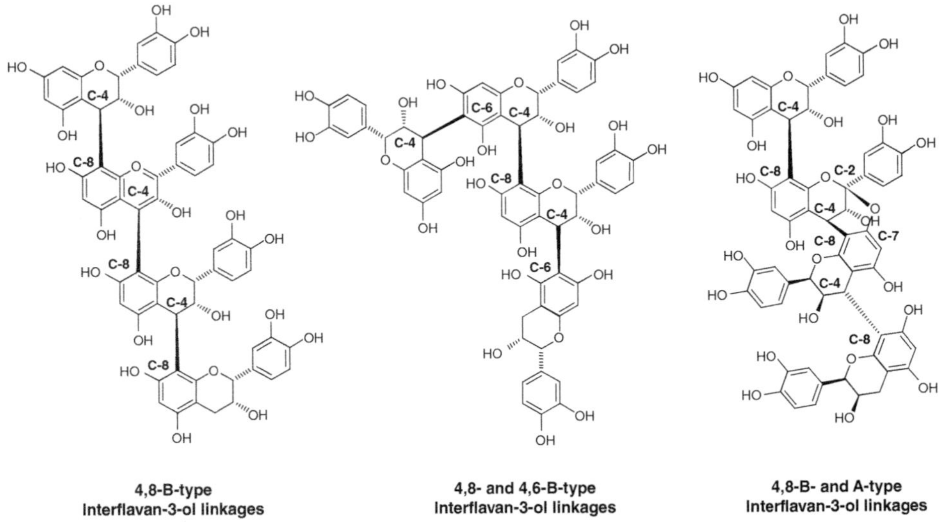

Figure 6.2 Common structural features of PAs: different flavan-3-ol subunits, interflavan-3-ol links (4-8 B-type, 4-6 B-type, or A-type), *cis* or *trans* stereochemistry (at C3 and C4), substitution at C3 or the phenolic groups (e.g. galloylation, glycosylation, etc.).

6.3 Noted Challenges in Proanthocyanidin Analysis

The main challenge to date is the need to identify robust analytical tools that can predict which PA compounds, or which generic structural features, are important for conferring their diverse benefits and for monitoring the consistent delivery of the requisite PA concentration.

Up to now, some structural parameters have been linked to biological responses. The mean degree of polymerization appears to be one of the most important and affects, for example, the interaction with and the precipitation of proteins (Zeller, Sullivan et al. 2015; Ropiak et al. 2017). It also impacts on dentin biomaterial via the formation of collagen crosslinks (Phalsankar et al. 2018), and on antiinflammatory immune responses (Williams et al. 2016). Prodelphinidins have been linked to more potent antiparasitic activities – but also to antinutritional effects – in contrast to procyanidins (reviewed by Mueller-Harvey et al. 2019). Galloylation generates potent in vitro antiparasitic and antioxidant activities (Plumb et al. 1998; da Silva Porto et al. 2003; Ramsay et al. 2016), which may also be linked to antiinflammatory activities (Terra et al. 2007). There is some evidence that A-type rather than B-type inter-flavan-3-ol links enhance activity against biofilm formation (Alejo-Armijo et al. 2018; Wojnicz et al. 2012). Both A-type and B-type linked PAs showed hypoglycemic activities (Lu et al. 2011; Sun et al. 2019). Given that oligomeric and polymeric *cis*-flavan-3-ols are much more widespread than *trans*-flavan-3-ols, it is surprising how little comparative information exists on their bioactivities, as these stereochemical differences are likely to affect their three-dimensional structures and thus their interaction with proteins and other biomolecules.

In addition to the aforementioned structural features, there are some generic features that apply to all PAs, with their antioxidant activities increasing progressively from monomeric to oligomeric and polymeric flavan-3-ols (Zhou et al. 2014). This could be important for the inhibition of lipid oxidation and for reducing chronic inflammation (Selmi et al. 2006; Terra et al. 2007; da Silva Porto et al. 2003), as PAs do not require prior absorption in order to exert antioxidant effects *in vivo*. Lopez-Andres et al. (2013) found that supplementing lambs with quebracho tannins improved the antioxidant capacity of tissues even though no PAs could be detected in plasma. They postulated an indirect effect that possibly enhanced the tissue endogenous antioxidant system or contributed to the regeneration of other antioxidant compounds.

PA size affects bioavailability, as the ability of dietary flavan-3-ol monomers, dimers, and trimers to transfer from the gastrointestinal tract to the blood plasma is rapidly reduced by polymerization (Déprez et al. 2001; Ou and Gu 2014). However, it is useful to differentiate between the bioavailability and bioactivity of PAs, because size influences the urinary or fecal excretion routes of PAs and thus their metabolites. For example, current research is exploring the microbial fermentation products from the colon, as these can be bioavailable – in contrast to their native, bioactive parent PAs – and can undergo further transformations in tissues. Both the PA metabolites and/or their derivatives may contribute to interesting health benefits (Sauro-Calixto et al. 2010; Ou and Gu 2014; Gonzalez-Sarrias et al. 2017). For these reasons, the bioactivity of PAs that are nonextractable (see Sections 6.4 and 6.5) is attracting considerable attention despite their analytical challenges (Mateos-Martin et al. 2012; Mena et al. 2015).

6.4 Fate of Proanthocyanidins in the Digestive Tract and During Plant Fermentation

A substantial proportion of PAs appears to reach the colon possibly in the form of dietary protein or fiber complexes or as nonextractable PAs (NEPAs) (Terrill et al. 1994; Le Bourvellec et al. 2009; Hagerman 2012; Mateos-Martin et al. 2012; Li and Hagerman 2013; Mena et al. 2015). However, there is also evidence for considerable PA losses or structural modifications across the digestive tract, since the analysis of dietary and fecal samples by the HCl-butanol assay (Terrill et al. 1992; see Section 6.7 for a description of depolymerization assays), after correction for organic matter losses, revealed that 62% of the PAs from four forage legumes had disappeared during digestion in sheep (Table 6.2; Waghorn et al. personal communication, 23 March 2011, UK-NZ Ag/Bio/Health Workshop held at Sky City Hotel, Auckland, NZ). Even larger PA losses (85% to >96%) from diet to feces were observed when ^{14}C-labeled PAs were added to sheep diets (Terrill et al. 1994) or portions of a *Medicago sativa* hay diet fed to sheep were substituted with lespedeza pellets (Kronberg et al. 2018).

More recent studies that used the improved HCl-butanol-acetone assay (Grabber et al. 2013) found similar losses. According to a meta-analysis (Riaz et al. 2014) average dry matter digestibilities in sheep are 58%. Therefore, if PAs were inert, measured concentrations in the feces of sainfoin-fed sheep should have been close to 15 g/kg DM. However, measured concentrations were only 2 g/kg DM, which suggests that 87% of the PAs could no longer be detected or had disappeared (Quijada et al. 2018). A cattle study that used the same sainfoin diet and assay estimated a loss of c. 50% of the PAs (Desrues et al. 2017).

Table 6.2 Proanthocyanidin (PA) concentrations and related products in feeds and sheep feces (g/kg dry organic matter).

Forage legume	Feed PA	Feces PA (measured)	Feces PA (calculated)*	Calculated Losses (%)
Hedysarum coronarium	100	130	350	62.9
Lotus corniculatus	50	60	160	62.5
Lotus pedunculatus (a)	80	120	250	52.0
Lotus pedunculatus (b)	110	140	380	63.2
Average	**85**	**113**	**285**	**62.2**

* Calculated from PA feed concentrations and measured digestibilities assuming no loss of PA during digestion.
Source: Waghorn, personal communication 2011.

Further examples of PA losses have been observed after the fermentation of silages – i.e. 17% by the HCl-butanol-acetone assay (Ramsay et al. 2015) – and in the human digestive tract, where larger PAs can be metabolized by intestinal microflora and converted into bioavailable metabolites (Sauro-Calixto et al. 2010; Mateos-Martin et al. 2012; Li and Hagerman 2013; Ou and Gu 2014; Mena et al. 2015; Gonzalez-Sarrias et al. 2017). Although some of these PA losses could be due to an inability to detect native or modified PAs in digesta or feces by currently available techniques, degradation by intestinal enzymes and/or colonic microbiota appears to be important and PAs are not as biologically inert as often suggested.

Traditionally, NEPAs have been viewed as being inaccessible and therefore having little or no effect on ruminant digestion (Makkar 2000), and similarly they have been thought of as a fiber portion in human nutrition. However, PAs may not have to be soluble or absorbed *per se* from the digestive tract in order to produce physiological effects.

Interesting research by Mateos-Martin et al. (2012) has found that the *in vivo* transformation of NEPAs in rats differs in part from that of the smaller extractable PAs (EPAs). Larger PAs can be depolymerized in the intestinal tract into flavan-3-ol monomers and possibly also oligomers. These authors suggested that the "posterior degradation by the intestinal microbiota into small units may also differ between NEPAs and EPAs. As NEPAs are associated with the food matrix in foodstuffs, particularly with other insoluble polymers constitutive of dietary fiber, their conversion may be slow compared to that of EPAs. This deferred release would make NEPA metabolites bioavailable for particularly long times after intake and may result in them having health effects for a long time." In fact, after feeding a dietary fiber with 14.8 g NEPAs/kg dry matter to rats, phenolic metabolites were released into the blood stream for 24 hours (Mateos-Martin et al. 2012), whereas Gonzalez-Sarrias et al. (2017) reported their detection in the blood stream for up to four days after a meal. Differences have been reported between the fecal and urinary phenolic metabolites derived from dietary NEPAs (Mateos-Martin et al. 2012) (Figure 6.3).

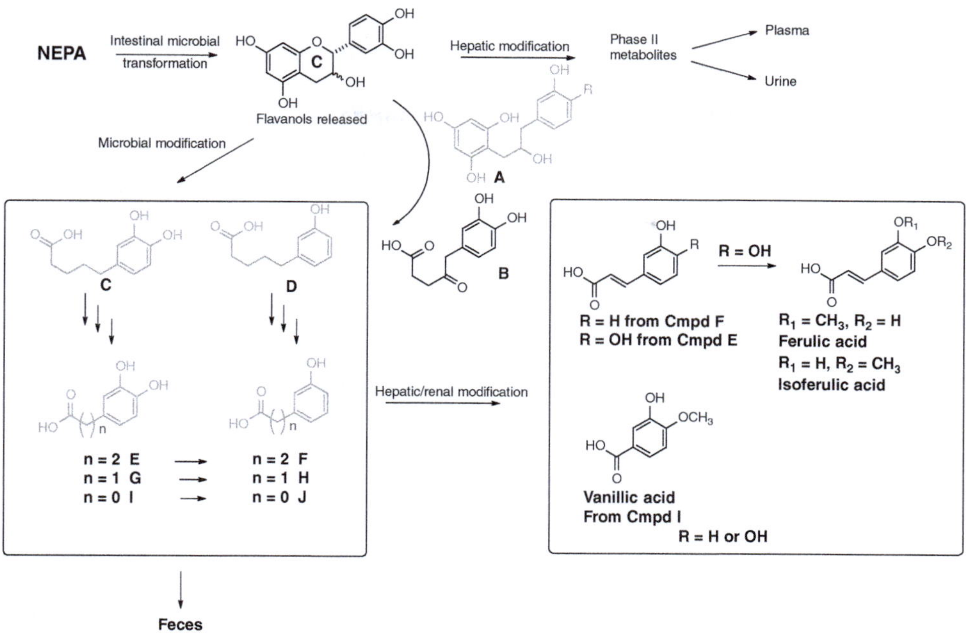

Figure 6.3 Profile of metabolites identified during metabolism of NEPA from dietary fiber fed to rats (Mateos-Martin et al. 2012), human metabolism of procyanidin B2 (Monagas et al. 2010), or after rumen digestion of catechin dosed rumen fluid (Becker et al. 2014). Intermediates common to more than one study are shown in grey. *Source:* Monagas et al. (2010), Mateos-Martin et al. (2012), and Becker et al. (2014).

Common phenolic metabolites have been identified in digestion studies whether examining NEPA digestion in rats (Mateos-Martin et al. 2012), human metabolism of procyanidin B2 (Monagas et al. 2010), or *in vitro* rumen fermentation studies supplemented with catechin (Becker et al. 2014), and parallel compounds isolated from biotransformations of flavan-3-ols by human microflora (Feng 2006) (Figure 6.3). These results point to common biotransformations occurring in metabolic pathways of the digestive microbiota, whether the test subject is rodent, ruminant, or human.

However, it is possible that other fecal fermentation products are also generated, as feeding sheep with PA-containing forages indicated the formation of a tropolone ring (Figure 6.4b; Waghorn et al. unpublished observation, 23 March 2011, UK-NZ Ag/Bio/Health Workshop held at Sky City Hotel, Auckland, NZ), as found previously during fermentation of tea leaves (Nonaka et al. 1986; Hashimoto et al. 1992) and during oxidation of epigallocatechin gallate (Mouls and Fulcrand 2015).

6.5 Definition and Possible Origins of Nonextractable Proanthocyanidins (NEPAs)

Researchers are currently using different extraction and purification conditions for measuring the NEPAs, which means that the NEPA fraction is operationally defined and method-specific. Domínguez-Rodríguez et al. (2017) have summarized the different approaches for isolating and quantifying NEPAs in plants: soluble PAs are first removed with water, aqueous methanol or acetone mixtures (some mixtures are acidified) although some researchers have also used NaOH. The NEPAs are then quantified within the plant residues by depolymerization techniques (see Section 7 below). In contrast, Gonzalez-Sarrías et al. (2017) defined NEPAs as dietary polyphenols that are not soluble in food products or the gastrointestinal content. Reported NEPA concentrations in common foods range from 0.08 g/100 g dry matter in apple to 1.87 g/100 g in bananas (Pérez-Jiménez et al. 2009, 2013, 2015; Zurita et al. 2012) and can be as high as 11.0 g/100 g in processed animal supplements such as lespedeza pellets (Zeller, unpublished observations). In fact, the concentration of NEPAs tends to increase during processing of plant materials, e.g. average NEPA concentrations accounted for 48% and 85% of total PAs in pelleted and ensiled forage legumes compared to 26% in fresh samples (Mueller-Harvey et al. 2019). This could explain why processing of high PA sorghum grain reduced its nutritional value despite the concentrations measured after PA extraction being much lower (Deshpande et al. 1986).

Little is known about what causes the nonextractability of PAs or how NEPAs may be created during processing. There could be several explanations. Firstly, the high molecular weight of some PAs is one likely cause for a lack of solubility. Secondly, decompartmentalization of soluble PAs during the maceration and extraction of plant samples may lead to the formation of a large number of noncovalent bonds between the PAs and other dietary polymers, such as proteins and cell wall constituents (Le Bourvellec et al. 2007, 2009). Thirdly, oxidation may lead to insoluble, larger molecular weight PAs (Mouls and Fulcrand 2015) or to *o*-diphenols being converted to quinones or semiquinones, which in turn form covalent links with thiol or amine groups. And fourthly, covalent bond formation may also occur when the *m*-diphenols such as the A-rings of flavanols react with α,β-unsaturated carbonyl compounds in analogy to amines (Figure 6.4; Zamora and Hidalgo 2018). Fermentation is indeed one possible route to the formation of carbonyl compounds (see also Section 4 above; Monagas et al. 2010), and may include crosslinking of PAs with smaller aldehydes generated from sugars (Totlani et al. 2006). If the adducts originate from large phenol-carbonyl compounds they may well be nonextractable and are also likely to evade PA detection by conventional assays.

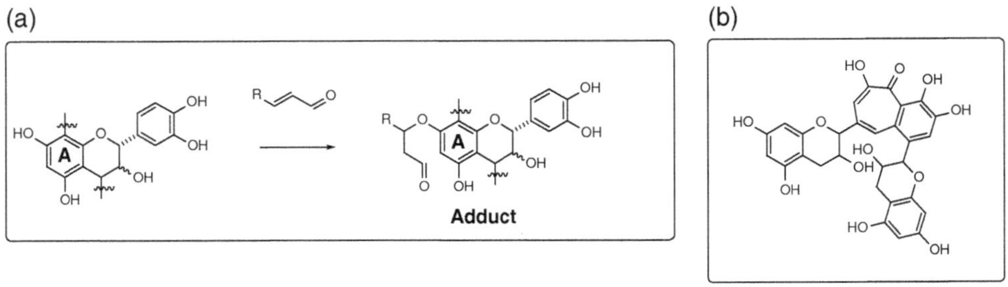

Figure 6.4 Phenolic reactions during fermentation and oxidation. A-ring phenol of a proanthocyanidin subunit reacts with a Michael receptor to form an adduct (a). Tropolone formation through the oxidative coupling of flavan-3-ol B-rings (b). *Source:* Zamora and Hidalgo 2018; Mouls and Fulcrand 2015; Waghorn, personal communication 2011.

6.6 Universal Problems of Proanthocyanidin Analysis

6.6.1 The Need for Reference Materials

The novice analyst is faced with a plethora of methods and choice is difficult as agreement between methods is generally poor. Seddon and Downey (2008) compared protein and methylcellulose precipitation with phloroglucinolysis for grape skin PAs and found no correlations. The reader is referred to Sections 6.7 and 6.10 as these describe some limitations of the different methods, which are likely to explain the observed lack of correlations. The authors concluded that each method detected a different PA fraction. However, it is also important to bear in mind that these techniques measure different PA characteristics; e.g. protein precipitation in particular is affected by PA molecular size (Ropiak et al. 2017) in contrast to phloroglucinolysis. Others found up to 20-fold differences in PA-concentrations between thiolysis, HCl-butanol and solid state ^{13}C-NMR analyses, but some methods provided at least similar rankings of samples (Azuhnwi et al. 2013; Fryganas et al. 2018). It has also been pointed out that the current European and British Pharmacopoeia method underestimates the concentration of procyanidins in *Crataegus* by an order of magnitude compared to the modified HCl-butanol-acetone assay (Leach et al. 2018). There are many possible reasons for these discrepancies.

Firstly, there is a dearth of interlaboratory comparisons of PA analysis (Stodt and Engelhardt 2013). Secondly, there is no universal standard that is suitable for all PA types (Maxon and Rooney 1972; van Beek and Montoro 2009; Gu 2012) because PAs differ greatly in their reactivities towards detection reagents. Thirdly, the preparation of an in-house standard is not straight forward as high purity is essential and far too often not even assessed (Azuhnwi et al. 2013; Grabber et al. 2013; Brown et al. 2017). However, solutions to these problems do exist.

It is generally recommended that an intralaboratory or in-house sample that is matrix matched to the sample being analyzed and its regular use are essential for monitoring data quality (Phillips et al. 2013). In PA analysis, it is not only necessary that the PA-standard is matrix-matched, but it should contain PAs that are as identical as possible to the sample PAs being analyzed and its purity needs to be known; these three aspects are important prerequisites for valid PA data (Stewart et al. 2000; Prior et al. 2010; Gu 2012; Brown et al. 2017). This approach is supported in the literature by several research groups (Giner-Chavez et al. 1997; Schofield et al. 2001; Kraus et al. 2003; Waghorn 2008; Hattas and Julkunen-Tiitto 2012; Grabber et al. 2013) as color and depolymerization yields are greatly dependent on the source of the PAs (del Pino et al. 2005).

In lieu of preparing one's own reference standard, a few samples exist, which are available for use. A purified cocoa procyanidin standard can be requested from Mars Inc. (Mars Botanicals, Rockville, MD) and the composition of this standard has been described (Gu 2012). A PA reference standard purified from apples has also been reported (Li et al. 2010). Interlaboratory method comparisons are also urgently needed, and the first attempts have been made with a cocoa and chocolate product and a cranberry-derived PA standard (Robbins et al. 2009; AOAC 2017). The development of Certified Reference Materials (CRM) for different PAs is long overdue and would greatly advance interlaboratory comparisons and research; a start could be made with CRMs prepared from green tea, grape seed, cranberry, and pine bark and their extracts as large numbers of papers are focused on their health benefits. The availability of a CRM for forage legumes would also greatly advance animal nutrition and health research.

In the absence of a commercially available pure PA standard that is completely free of contaminants, Feliciano et al. (2012) devised a relatively straight-forward gravimetric formaldehyde–HCl precipitation route for assessing the purity of isolated PAs (Spranger et al. 2008). This approach is based on the observation that formaldehyde reacts with flavonoids under acidic conditions and leads to polymerization and precipitation. The freeze-dried supernatant was then weighed in order to take account of nonflavonoid components, such as carbohydrates or phospholipids and any interference from flavonoid monomers could be assessed separately by LC-MS. Other approaches to assessing the purity of isolated PAs are based on ^{1}H-^{13}C HSQC NMR (see Section 6.9.4, Figure 6.18) and on depolymerization techniques, where the sum of the mass of the derived flavanol subunits is compared to the mass of the PA sample. The latter approach requires a knowledge of the UV response factors of the reaction products. Kennedy and Jones (2001) reported response factors for phloroglucinolysis products; however, some response factors of the thiolysis products are still based on estimates (Ropiak et al. 2016).

6.6.2 Extraction and Purification of Proanthocyanidins for Use as In-house Standards

Extraction of PAs has traditionally been performed through shaking or stirring with a protic solvent (methanol, ethanol) or mixtures (acetone/water). However, alternative approaches now include ionic liquids, such as 1-butyl-3-methylimidazolium bromide or the distillable N,N-dimethylammonium N'N'-dimethylcarbamate, coupled to microwave- or ultrasound-assisted extraction (Chowdhury et al. 2010; Liu et al. 2012; Sun et al. 2013; Ćurko et al. 2017; Vu et al. 2017). Brillouet et al. (2017) have published an important new addition to this field. They demonstrated that PAs can be easily oxidized under conventional extraction conditions. By using a buffered ascorbic acid and Triton X-100 pretreatment followed by acetone extraction and precipitation of the PAs from a methanol/chloroform mixture and storing samples at temperatures below −85 °C, these authors succeeded in obtaining white, rather than the typically beige colored, PA samples. These white PAs had a higher degree of polymerization and a lower polydispersity than conventionally extracted PAs. In addition, the high performance liquid chromatography (HPLC) trace after thiolysis was flat and there was no indication of an oxidized PA 'hump'. This study suggests the need for re-evaluating one's sample preparation and extraction techniques.

6.6.3 Gram Scale Isolation and Purification of Proanthocyanidins

Commonly used methods for obtaining purified PA samples involve placing an aqueous, an aqueous alcohol (usually methanol or ethanol) mixture or an ethanolic solution of the dried extract onto Sephadex LH-20 resin packed in water, water/alcohol mixtures or ethanol (Feliciano et al. 2014; Brown et al. 2017). The purification commences with elution of undesired components (mainly carbohydrates) with copious amounts of alcohol or water/alcohol mixtures until baseline UV activity of the eluent is reached followed by elution of tannins with an acetone/water (usually 7:3 v/v) mixture, concentrating and freeze-drying this fraction to afford the 'purified tannins'. In some cases, we have found that this procedure affords

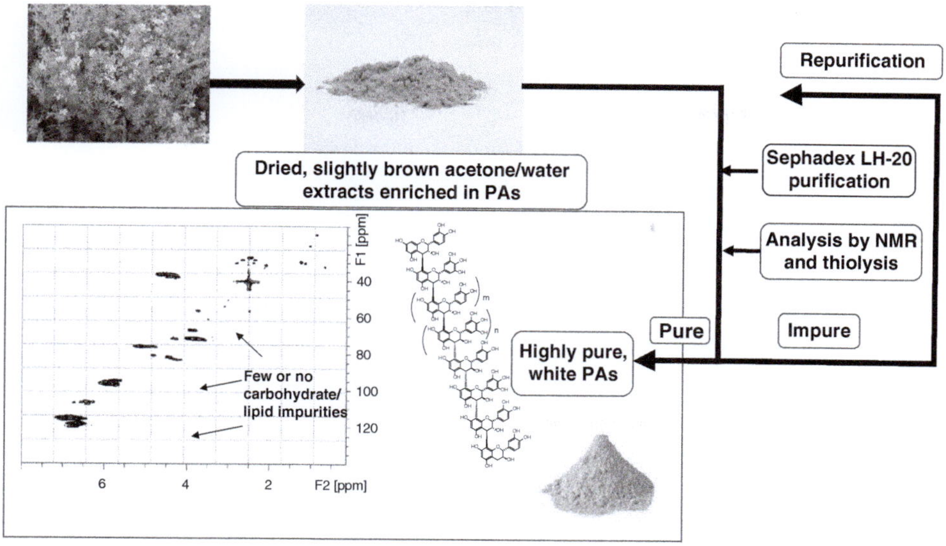

Figure 6.5 Illustration of a rapid approach to proanthocyanidin isolation and purification at the gram scale. *Source:* Adapted from Brown et al. 2017. (*See insert for color representation of the figure.*)

PAs with purity estimates of >90% in this fraction, based on the lack of significant non-PA cross-peaks in the 2D NMR spectrum (Zeller, unpublished observations) and corroborated by quantitative purity analysis by thiolytic degradation. However, in most cases this 'purified PA' fraction is still contaminated with other biomolecules such as carbohydrates and lipids (Brown et al. 2017), especially if lower acetone concentrations have been employed, e.g. with acetone/water (1:1 or 3:7, v/v) solvents. It should also be pointed out that some low molecular weight PAs are eluted in the alcohol fractions together with contaminating biomolecules, whereas the acetone/water (7:3, v/v) mixture tends to elute preferentially the PA oligomers and polymers and generates higher purity PA fractions.

These traditional approaches to purifying PAs have relied on long and narrow Sephadex, Toyopearl HW-40C or HW-50F columns (e.g. 23 cm H × 3 cm W) (Brown et al. 2017). However, purification of PAs at the gram scale can be done much more rapidly in a wide Buchner funnel or on a short and wide Sephadex LH-20 column (7 cm H × 6.5 cm W) (Brown et al. 2017). High flow rates of 15-40 ml/min are important for removing carbohydrate and phospholipid impurities (see Figure 6.5). These approaches yielded gram quantities of PAs with mDP values ranging from 8 to 38 and purities of 64 to 100 g PA/100 g sample (as determined by thiolysis) within 2.5 to 5 hours.

6.6.4 Separation of Proanthocyanidin Mixtures and Preparation of 'DESIGNER' Extract Resources

An initial route to separating oligomers from polymers can be achieved using a simple precipitation scheme with ethanol/*n*-hexane that removes the typical PA 'hump' from extracted oligomers (Esatbeyoglu et al. 2011, 2014).

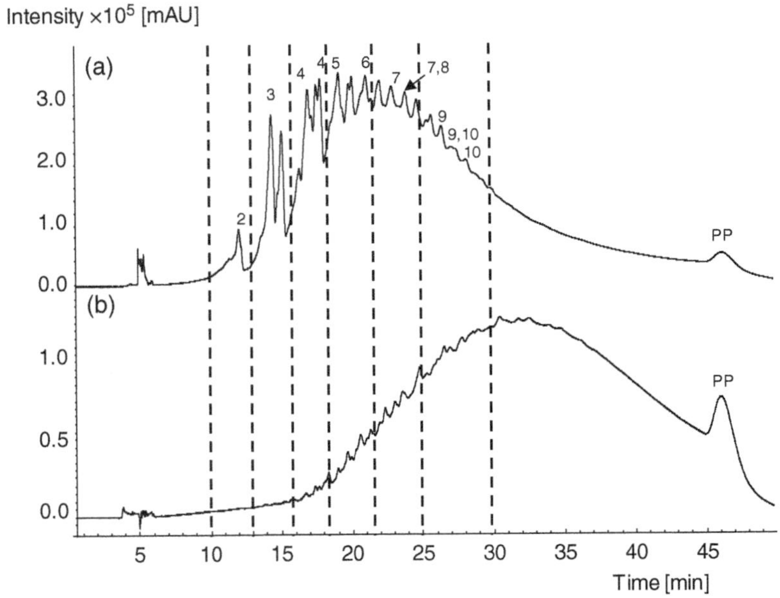

Figure 6.6 HILIC chromatography of (a) pine bark and (b) birch leaf proanthocyanidins (numbers indicate the degree of polymerization, DP). The vertical lines illustrate how fractionation could achieve the isolation of proanthocyanidin subfractions with narrower degrees of polymerization although an overlap between the fractions will be unavoidable. *Source:* Adapted from Karonen et al. 2011. © 2011 John Wiley & Sons.

The above purified PA mixtures (Section 6.3) can be partially separated on HILIC columns as these achieve better peak shapes and size separations than normal phase columns and can cover a wider range of molecular weights (Gu et al. 2002; Karonen et al. 2011; Montero et al. 2013). HILIC chromatography operates on the basis that the hydrophilicity of PAs increases as the degree of polymerization increases. Tanaka et al. (2014) and Karonen et al. (2011) were thus able to separate and detect procyanidins with degrees of polymerization of up to 22 (Figure 6.6). By fractionating these HILIC humps, it should be possible to obtain PAs with different degrees of polymerization and narrower polydispersities that could be useful resources for bioactivity studies.

In addition, several different countercurrent chromatographic techniques exist, which represent an automated, continuous liquid/liquid extraction system (Berthod et al. 2009; Hu and Pan 2012) that can achieve 90% to near 100% recoveries of the applied samples due to the absence of a solid stationary phase (Tanaka et al. 2014). This is unlike Sephadex LH-20 column chromatography, where large molecular weight PAs tend to be retained preferentially. Recent reports demonstrated that oligomeric PA compounds and their galloyl derivatives can be isolated at the 500 mg scale with 80% to 97% purity (Tanaka et al. 2014; Esatbeyoglu et al. 2014, 2015; Peng et al. 2018; Phansalkar et al. 2018). Kurumatani et al. (2005) reported the isolation of high molecular weight PAs with degrees of polymerization of c. 22 and our research (Ramsay et al. 2014) has shown that PAs with mDP values as high as 80 can be separated and isolated by countercurrent chromatography, which would not have been possible by column chromatography.

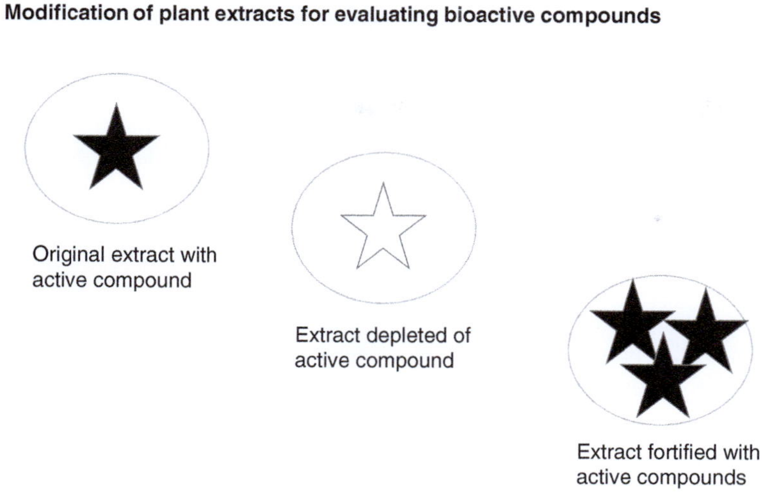

Figure 6.7 Countercurrent chromatography for preparing depleted and fortified plant extracts for bioactivity testing. *Source:* Alvarenga et al. 2014; Phansalkar et al. 2018.

Phansalkar et al. (2018) have described countercurrent chromatography as a novel tool for PA bioactivity studies by applying the 'principle of chemical subtraction or addition' to plant extracts (Figure 6.7; Chen et al. 2008; Alvarenga et al. 2014).

Since material losses in countercurrent chromatography are minimal, this technique can be used for the selective removal (or addition) of target compounds from a complex mixture. Phansalkar et al. (2018) coined the acronym 'DESIGNER' (Deplete and Enrich Select Ingredients to Generate Normalized Extract Resources) for this approach, which allows bioactivity testing within a plant extract, either by removing selected compounds or by enriching the extracts with selected compounds. Depending on the plant source (grape seed, cocoa, pine bark or cinnamon), polymeric PAs can be isolated with methyl acetate/water or ethyl acetate/water solvents and bioactivity studies can then be performed either with PA-depleted or PA-fortified plant extracts.

The approach by Phansalkar et al. (2018) first generated polymeric PA knockout fractions by partitioning between methyl acetate/water. Countercurrent chromatography with hexane/ethyl acetate/methyl acetate/water mixtures (0.8/4/1/4 or 2/4/1/4 v/v) or hexane/ethyl acetate/methanol/water (1/10/1/10 v/v) solvents was used to obtain flavan-3-ol monomer and dimer knockout fractions, which resulted in trimer and tetramer enriched PA fractions. Subsequent bioactivity studies revealed that the trimers and tetramers were the most potent dentin-active PAs from several different extracts and especially the trimers with one A-type and one B-type interflavan-3-ol linkages.

6.7 Proanthocyanidin Characterization by Depolymerization

In the most widely used approach for PA analysis (Porter et al. 1986), the PAs are reacted with HCl/butan-1-ol and converted to anthocyanidins, which can be quantified at their peak maxima that range from c. 490 nm for apigeninidin to c. 560 nm for delphinidin or by

HPLC analysis. However, Zurita et al. (2012) proposed that the sum of absorbances at 555 nm for the anthocyanidin and at 450 nm for the xanthylium reaction products can be used to determine PA concentrations with this reagent. An important recent improvement includes addition of acetone to this reagent, which enhances the anthocyanidin yield (Grabber et al. 2013). This reaction can provide information on PA concentrations if a PA standard of known purity is available (Grabber et al. 2013) and on the flavan-3-ol composition in terms of procyanidin/prodelphinidin ratios (Klongsiriwet 2016).

In addition, several techniques exist that are based on acidic degradation of PAs with *in situ* nucleophilic capture of the electrophilic-generated intermediate for identifying the monomeric subunits either within a PA molecule or within mixtures of PA molecules. Table 6.3 provides examples of different types of nucleophiles and lists the associated techniques. These procedures provide data on the *average* global composition of the whole PA mixture. The 4-8 or 4-6 B-type bonds are the most wide-spread interflavan-3-ol linkage type occurring in PAs (Gu et al. 2003), and are susceptible to depolymerization. In contrast, galloylated PAs and PAs with A-type linkages are more resistant to the acidic depolymerization process; the latter yield PA dimers or trimers with one or two A-type bonds, which can facilitate the detection of the A-type bond position in trimers (Ropiak et al. 2016; Orejola et al. 2017; Gao et al. 2018). However, thiolysis performs poorly for the

Table 6.3 Nucleophiles that have been used for depolymerization of proanthocyanidins.

Nucleophile		Reference	Associated analytical technique
Polyphenols	Phloroglucinol	Kennedy and Jones (2001); Koerner et al. (2009); Orejola et al. (2017)	Phloroglucinolysis
	Pyrogallol	Bordiga et al. (2013); Rebello et al. (2013);	Pyrogallolysis
	2,3-Dimethylpyrazole	Fu et al. (2010)	
	3-Ethyl-2,4-dimethylpyrrole	Fu et al. (2010)	
	3,5-Dimethoxyphenol	Fu et al. (2010)	
	3,5-Dimethoxyaniline	Fu et al. (2010)	
Mercaptans	Benzenethiol	Brown and Shaw (1974)	Thiolysis
	Benzyl mercaptan	Rigaud et al. (1991); Guyot, Marnet, and Drilleau (2001); Gea et al. (2011); Ropiak et al. (2016)	
	Thioglycolic acid	Mouls and Fulcrand (2015)	Thioglycolysis
	2-Mercaptoethanol	Orejola et al. (2017); Zhang et al. (2017)	
	Cysteamine	Gao et al. (2018)	
	o-Thioaniline	Fu et al. (2010)	
	1,2-Dithioethane	Fu et al. (2010)	
Others	Furan, menthofuran, sylvan	Rouméas et al. (2018); Billerach et al. (2019)	Furanolysis, sylvanolysis

chemically resistant 5-deoxy PAs (Steynberg et al. 1994) and thus is not recommended for use with these PAs.

The resulting depolymerization products can be quantified by HPLC, which provides information on the concentration of flavan-3-ols in extension units, which are released as the nucleophilic adducts, and in terminal units, which are released as underivatized flavan-3-ols (Figure 6.8). This allows, for example, the calculation of the i) total PA concentration in a plant sample or the purity of extracted or isolated PAs, ii) average propelargonidin/procyanidin/prodelphinidin ratios, iii) *cis/trans* flavan-3-ol ratio and iv) mean degree of polymerization (Koupai-Abyazani et al. 1993; Guyot, Marnet, and Drilleau 2001; Guyot, Marnet, Sanoner, and Drilleau 2001; Gea et al. 2011; Ropiak et al. 2016). However, one of the obstacles in performing the thiolysis technique is the availability of reference compounds and/or that only a small number of response factors have been published for the nucleophilic adducts that are based on authentic compounds. Moreover, if this reaction is carried out on PAs within plant tissues or extracts rather than on purified PAs, it is also necessary to separately determine the concentration of native, free flavan-3-ol monomers, as they are identical to the released terminal flavan-3-ols and would otherwise interfere with the calculation of the PA composition and concentration.

It is worth noting that several different nucleophiles have now been employed for PA depolymerization reactions as shown in Table 6.3. Thiolysis has been the preferred technique by many researchers as it uses quite strong nucleophiles; however, most thiols have an unpleasant odour. For this reason, the odourless phloroglucinol reagent is preferred by many researchers, although our studies revealed that benzyl mercaptan generated higher reaction yields from *Onobrychis viciifolia* PAs than phloroglucinol (Gea et al. 2011). Recently, several new reagents for PA depolymerization have also been proposed: i.e. the less odorous cysteamine/HCl or mercaptoethanol/HCl reagents (Orejola et al. 2017; Zhang et al. 2017; Gao et al. 2018) plus furanolysis with furan, menthofuran or sylvan (Rouméas et al. 2018; Billerach et al. 2019).

Depolymerization is an evolving field of PA analysis as shown by a surprisingly wide range of reaction conditions even for the same nucleophile. For example, the following conditions have been used for the benzyl mercaptan reaction with different PA types and plant matrices: 40 °C for 30 min to 60 min (Guyot, Marnet, and Drilleau 2001; Guyot, Marnet, Sanoner, and Drilleau 2001; Gea et al. 2011), 70 °C for 22 h (Orejola et al. 2017), 90 °C for 2 min (Labarbe et al. 1999), 105 °C for 4.5 h (Matthews et al. 1997). There appears, therefore, to be a need to understand these reactions better and to work towards a consensus for optimum reaction conditions, especially since little is known about the side reactions during both thiolysis and phloroglucinolysis; e.g. extent of epimerization, heterocyclic ring fission or A-type bond formation with either phloroglucinol or with a neighboring flavan-3-ol A-ring (Gu 2012; Liu et al. 2013; Orejola et al. 2017) (Figure 6.9).

The reaction products after depolymerization are usually separated by HPLC and quantified with UV and MS detection (Ramsay et al. 2016). However, capillary electrophoresis that utilizes micellar electrokinetic chromatography can achieve much faster separation of the reaction products in under five minutes (Qa'dan et al. 2006).

An interesting application of thioglycolysis using thioglycolic acid to the depolymerization of oxidized PAs has been reported by Mouls and Fulcrand (2015). The results

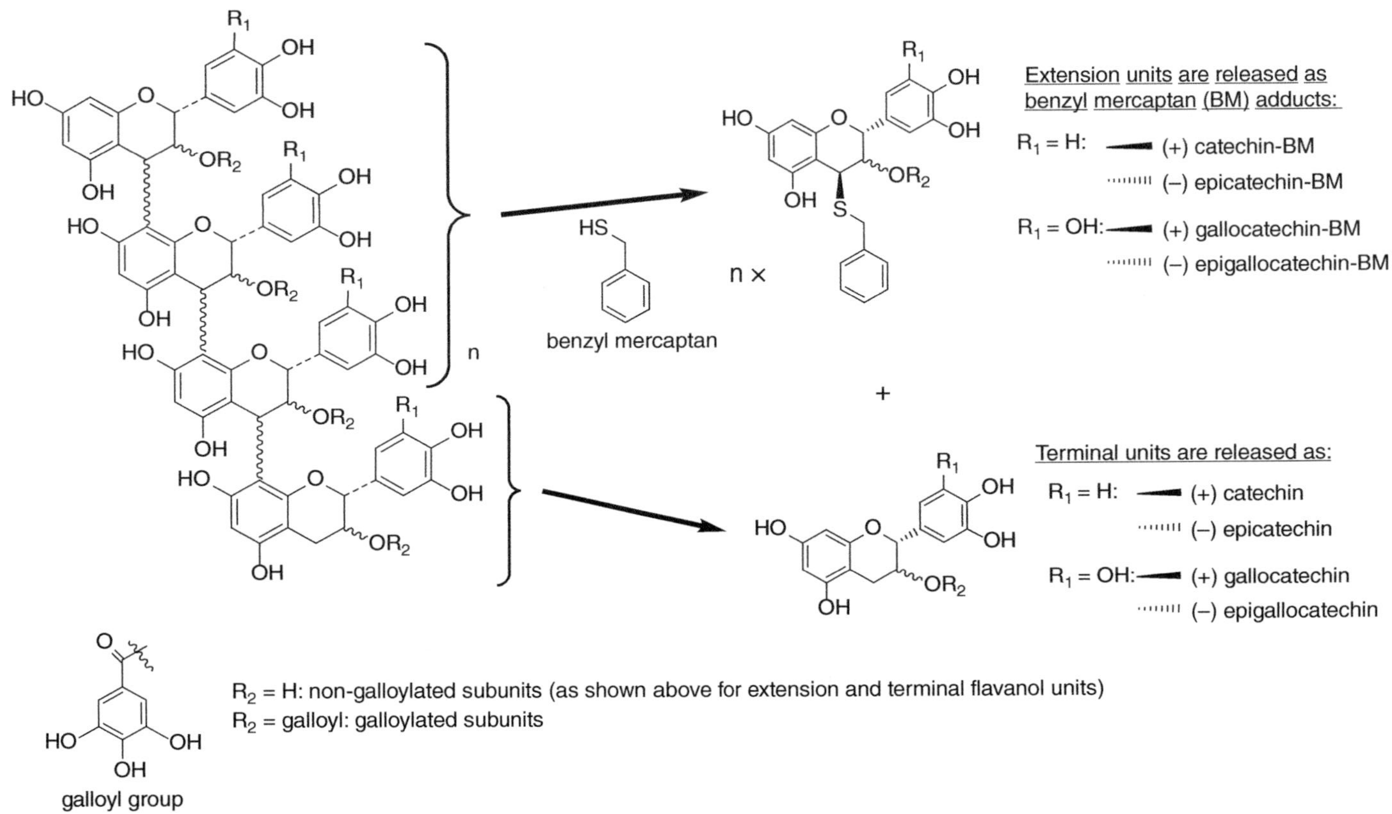

Figure 6.8 Products derived from depolymerization reaction of proanthocyanidins with benzyl mercaptan.

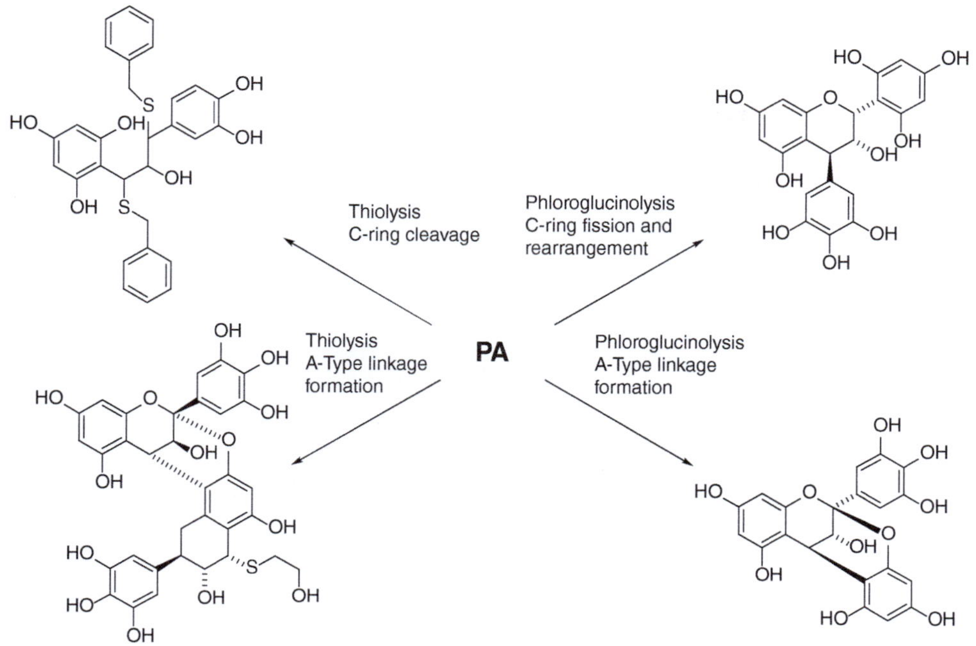

Figure 6.9 Identified or proposed side reaction products from phloroglucinolysis and thiolysis. *Source:* Liu et al. 2013; Orejola et al. 2017.

suggested that there was a difference in the reactivity between the extension and terminal units of procyanidin B2, where the extension unit participated preferentially in the oxidative coupling as a nucleophile through the A ring, while the terminal unit participated via the B ring through an electrophilic *o*-quinone. Their studies revealed 30 oxidation markers in model solutions containing epicatechin, B2 dimer and epicatechin-based PAs from apple (Mouls and Fulcrand 2012) and showed that oxidation of PAs leads to both intra- and intermolecular linkages of PA molecules (Figure 6.10).

6.8 Mass Spectrometry

6.8.1 Matrix-Assisted Laser Desorption Ionization – Time of Flight Mass Spectrometry (MALDI TOF MS)

The depolymerization reactions described previously provide information on the average or global composition of a PA mixture. In contrast, MALDI TOF MS provides information on the degree of polymerization, the number and types of flavan-3-ol subunits, and B- versus A-type bonds per PA molecule – i.e. the molecular composition and complexity within PA mixtures (Reed et al. 2005; Fulcrand et al. 2008; Krueger et al. 2013; Van Huynh and Bevington 2014; Dominguez-Rodriguez et al. 2017). Thus,

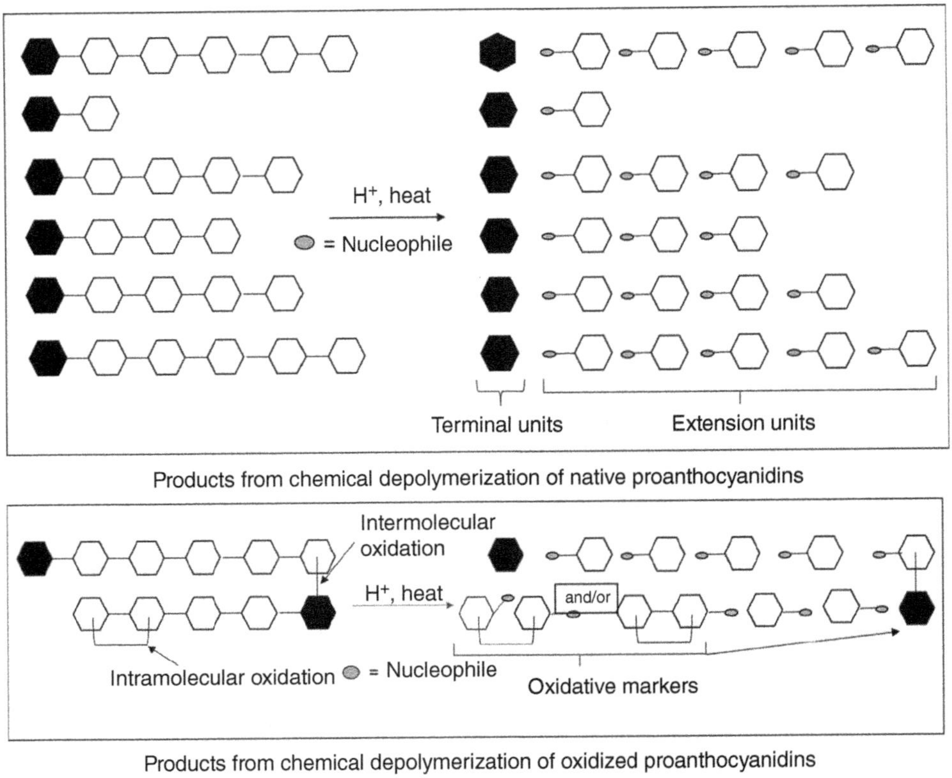

Figure 6.10 Schematic representation of products obtained from chemical depolymerization of native and oxidized proanthocyanidins. *Source:* Adapted from Mouls and Fulcrand 2012.

it highlighted the co-occurrence of homo- and hetero-polymeric PAs in plants (Figure 6.11) (Taylor et al. 2003; Monagas et al. 2010; Stringano et al. 2011; Fraser et al. 2016). Its disadvantages stem from the fact that this technique is not currently suitable; firstly, for quantification purposes as PA concentrations cannot be obtained from peak heights because peak intensity decreases as polymer chain length increases due to ion suppression and/or fragmentation (Taylor et al. 2003) or due to instrumental effects (Chen et al. 2003), and secondly, that most studies only report PAs with degrees of polymerization (DP) of up to c. 12 as large PA polymers suffer from low ionization efficiency (Perret et al. 2003; Gu 2012). Three strategies have been proposed to overcome this problem. If PAs are fractionated the detection of higher molecular weight compounds can be improved after removal of lower molecular weight compounds (Taylor et al. 2003; Mouls et al. 2014). Alternatively, a protein can be used as a charge carrier (Mané et al. 2007) as reviewed by Fulcrand et al. (2008). Thirdly, some researchers have also described difficulties with spectral reproducibility as the PA sample to matrix ratio and the crystallization technique need fine-tuning. However, Feliciano et al. (2012) succeeded in detecting PAs with DP of up to 26, or m/z values of up to c. 7500, after applying a Savitzky-Golay smoothing function to the MALDI-TOF MS spectra.

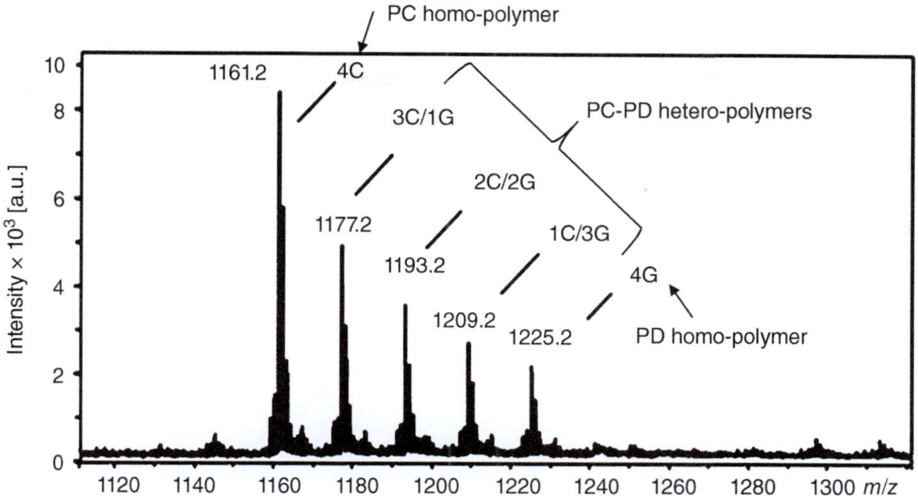

Figure 6.11 MALDI-TOF MS spectrum of sainfoin proanthocyanidins consisting of procyanidin (PC) and prodelphinidin (PD) homo- and hetero-polymers. *Source:* Adapted from Stringano et al. 2011.

6.8.2 LC-MS/MS

Polyphenols including PAs are easily detected in the negative electron ionization mode. Low resolution mass analyzers, such as triple quadrupole MS in the multiple reaction monitoring acquisition mode, are now widely used and have excellent sensitivity (Lucci et al. 2017). Three PA fragmentations are common in mass spectrometry (Gu 2012): retro-Diels-Alder (RDA) fragmentation results in the break-up of the C-ring followed by the loss of the B-ring; heterocyclic ring fission (HRF) also breaks up the C-ring and leads to the loss of the A-ring; and quinone methide (QM) formation causes splitting of interflavanol bonds (Figure 6.12) and leads to different fragments for extension and terminal flavan-3-ol units (Engström et al. 2014).

Mass spectrometry is particularly useful for the quebracho and wattle PAs as the absence of the 5-OH group causes the interflavan-3-ol bonds to be resistant to acid-catalyzed depolymerization. Using a triple-quadrupole mass spectrometer coupled to an ESI source, retro-Diels-Alder fragmentation was identified as the salient fragmentation pattern of wattle dimers to tetramers (Venter et al. 2012a, 2012b). In addition, Venter et al. (2012a) were able to detect up to undecameric PAs and to distinguish between starter/terminal and extension units in wattle PAs (Figure 6.13). Venter et al. (2012b) thus estimated the mDP of quebracho PAs to be c. 3, with catechin as the starter and fisetinidol as the extension units.

Several LC-MS/MS methods have been published for detecting and/or quantifying individual PA compounds, such as cranberry A-type dimers and trimers (van Dooren et al. 2018), B-type procyanidins from *Potentilla erecta* up to decamers (Mari et al. 2013), mixtures of *Theobroma cacao* A- and B-type procyanidins up to decamers (D'Souza et al. 2017), and groups of *Vitis vinifera* procyanidin dimers and trimers and their galloylated derivatives (Luo et al. 2017). Mouls et al. (2011) have reviewed ESI-MS methods for their suitability to determine mDP values and polymer distributions and concluded that this technique can be used for estimating the mDP of fractions that have a narrow polymer distribution provided that the mDP values are below 20. For PAs with mDP values above 20, one of the previously mentioned depolymerization methods needs to be used instead.

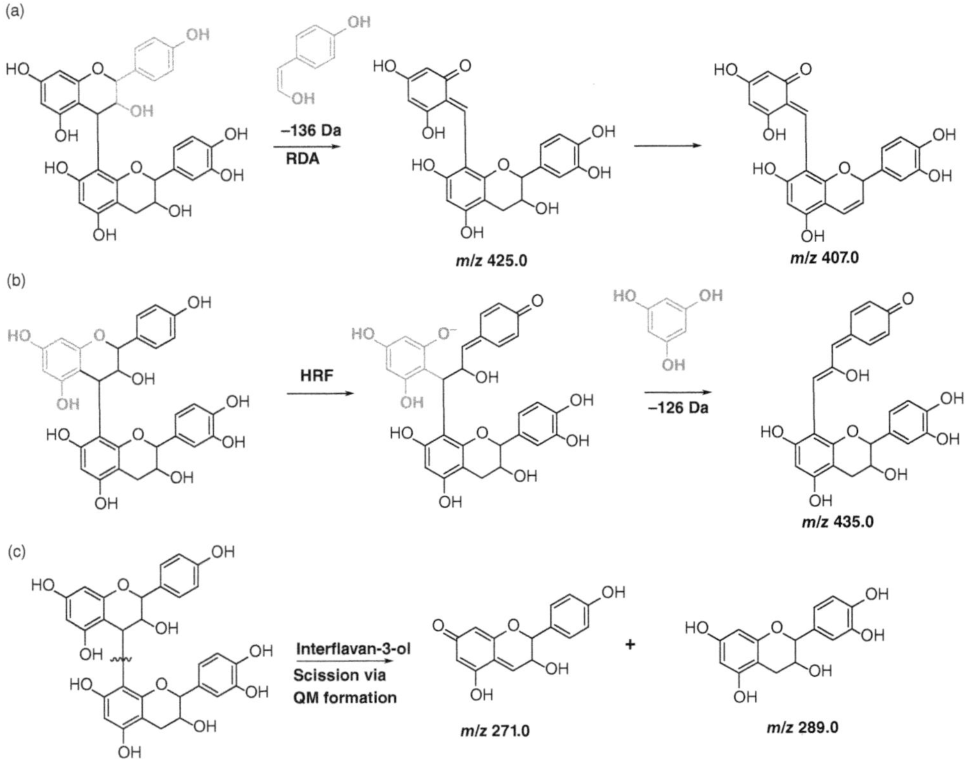

Figure 6.12 Figure showing major fragments observed during mass spectrometric analysis of a proanthocyanidin dimer. (a) Retro-Diels-Alder (RDA) reaction results in loss of a fragment of 136 Da and subsequent loss of water (18 Da), (b) heterocyclic ring fission (HRF) results in loss of phloroglucinol (126 Da), and (c) interflavan linkage cleavage via formation of a quinomethide (QM) with an m/z ratio of 271. *Source:* Adapted from Gu 2012.

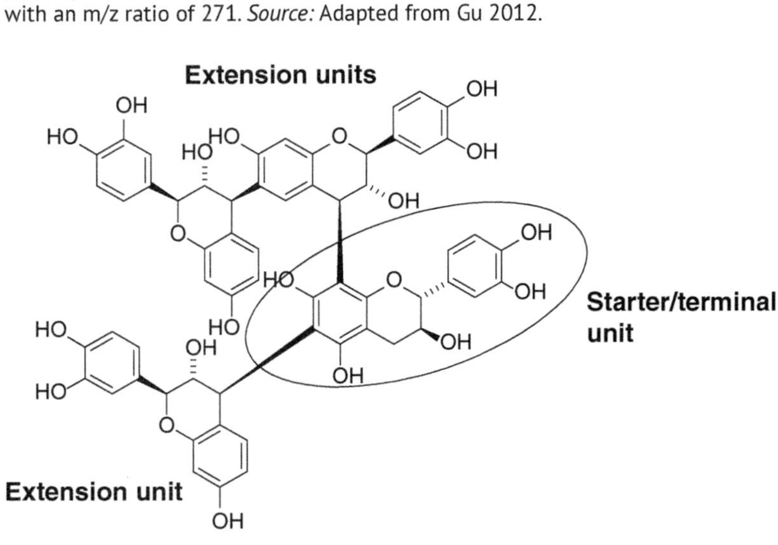

Figure 6.13 Starter/terminal (circled here) and extension units (all others) in 5-deoxy proanthocyanidins.

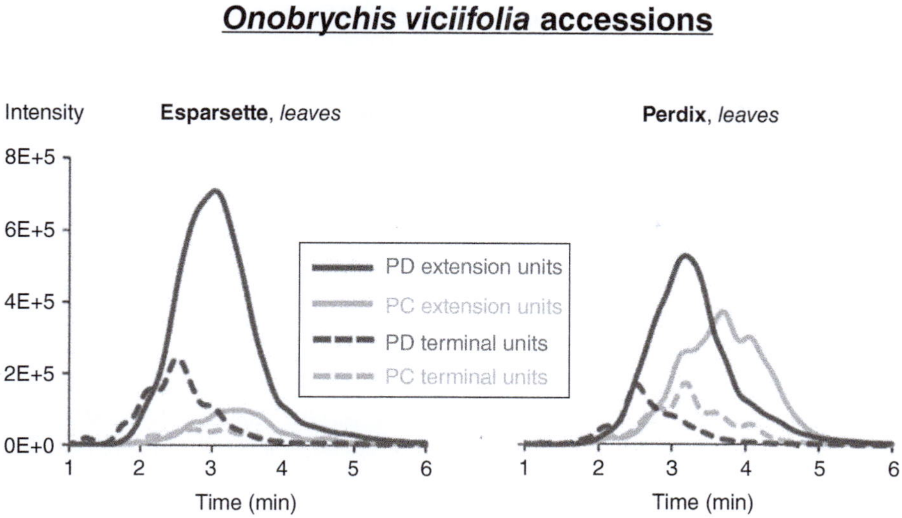

Figure 6.14 Fingerprinting of prodelphinidins and procyanidins in two *Onobrychis viciifolia* accessions by UPLC-MS/MS. *Source:* Engström et al. 2014; Salminen 2018.

The quinone methide fragmentation when coupled to multiple reaction monitoring can be used to quantify extension and terminal units of procyanidins and prodelphinidins (Figure 6.12) (Engström et al. 2014). This method also provides information on the flavan-3-ol connection sequence up to pentameric B-type PAs, and the position of A-type bonds in trimers (Gu 2012; Engström et al. 2014). Careful optimization of cone voltages and collision energies achieved excellent correlations with the thiolysis and the HCl-butanol-acetone assays for total PA, procyanidin and prodelphinidin content for procyanidin/prodelphinidin (PC/PD) ratios and for mDP values (Engström et al. 2014). Unlike the depolymerization methods that can only provide global average data on PA mixtures, this UPLC-MS/MS method can generate distinct 'PA fingerprints' from plant extracts because this technique can detect different PA compounds along the UPLC elution profile (Salminen 2018). Figure 6.14 depicts PAs extracted from two *Onobrychis viciifolia* varieties: the Esparsette variety had mostly prodelphinidins throughout the elution profile, whereas the Perdix variety contained a mixture of prodelphinidins and procyanidins. These data can be collected within a 10 min. chromatogram and, therefore, this UPLC-MS/MS method represents a powerful addition to our toolbox for the qualitative and quantitative analysis of PA mixtures in plant extracts (Salminen 2018).

6.9 Nuclear Magnetic Resonance Spectroscopy

The use of solution-state nuclear magnetic resonance (NMR) spectroscopy has been a cornerstone for the structure determination of PAs ranging from dimers to pentamers. Through use of both one-dimensional (1D) spectroscopic (^{1}H and ^{13}C) and two-dimensional (2D) spectroscopic (HSQC, HMBC, COSY, NOESY) techniques, researchers have determined structures of an increasing number of smaller oligomeric PAs varying in flavan-3-ol subunit structure and interflavan-3-ol connectivity. A compilation of references for these structural determination studies can be found in an internet-available NMR database (Zeller and Schatz 2017).

6.9.1 Use of Variable Temperature ^{1}H NMR Spectroscopy

As the length of the PA increases, ^{1}H NMR spectra become less useful through coalescence of the signals caused by limited rotation about the interflavan-3-ol bonds (atropisomerism). In the cases of small oligomers, variable temperature ^{1}H NMR has been useful in structure determination. For example, Esatbeyoglu et al. (2011) have shown that lowering the temperature transformed the coalesced signals into well resolved entities allowing nuclear Overhauser effects (NOE) to be used to characterize interflavan-3-ol linkages. NMR at low temperature has been used to establish the 3D structures of major conformers of procyanidin dimers and trimers (Tarascou et al. 2006). Low-temperature ^{1}H NMR has also been used to assist in the identification of PAs isolated from unripe *Malus pumila* cv. Fuji fruit (Shoji et al. 2003).

6.9.2 Use of Solution-State ^{13}C NMR Spectroscopy

Detailed examination of solution-state ^{13}C spectra of larger oligomer/polymer mixtures has historically provided a strong method for estimation of procyanidin/prodelphinidin ratios (Czochanska et al. 1979, 1980; Pizzi and Stephanou 1993; Foo et al. 2000; Ossipova et al. 2001; Kraus et al. 2003; Qádan et al. 2010; Zhang et al. 2010; Chai et al. 2012) and *cis/trans* ratios (Czochanska et al. 1979, 1980; Foo et al. 2000; Kraus et al. 2003; Qádan et al. 2010; Zhang et al. 2010; Chai et al. 2012) for identification of different interflavan-3-ol (C4-C6 and C4-C8) linkages (Newman et al. 1987; Pizzi and Stephanou 1993) and for providing estimates of mDP (Czochanska et al. 1979, 1980; Pizzi and Stephanou 1993; Foo et al. 2000; Kraus et al. 2003; Zhang et al. 2010; Chai et al. 2012). It, therefore, remains a staple of the PA researcher's tools in structure determination.

6.9.3 Use of Solid-State ^{13}C NMR Spectroscopy

While first utilized by researchers (Gamble et al. 1996; Parfitt and Newman 2000; Lorenz and Preston 2002; Hill et al. 2003; Wawer et al. 2006) for identification of PAs, renewed interest in recent years in the use of solid-state cross-polarization magic angle spinning (CPMAS) ^{13}C NMR spectroscopy studies of PA-containing plant materials and extracts has emerged (Hoong et al. 2010; Romer et al. 2011; Reid et al. 2013; Fryganas et al. 2018). This technique provides similar analyses to those obtained by solution-state ^{13}C NMR spectroscopy. A distinct advantage of CPMAS ^{13}C NMR spectroscopy is the ability to examine PA-containing materials without derivatization or extraction. However, both solution- and solid-state ^{13}C NMR methods tend to suffer from occasional overlap of non-PA signals and poor resolution. Even in light of these potential shortcomings, these methods have proven useful in PA structure investigations.

A ^{13}C CPMAS NMR method was developed to study procyanidins (PCs) and prodelphinidins (PDs) directly in milled sainfoin (*Onobrychis viciifolia*) plant tissues. These samples were not amenable to gel-state NMR analysis (*vide infra*), presumably because these PAs had a more complex flavan-3-ol composition and matrix peaks prevented PA peak detection (Ramsay et al. unpublished observations, discussion of results on 4 Feb 2014, University of Reading, UK). For this reason, application of solid state ^{13}C CPMAS NMR was used instead.

Using the peak width of the signal at 155 ppm to correct for lignin interference, it was possible to estimate PA concentration from the 155 ppm signal and the PC/PD ratio from the I_{132}/I_{155} ratio (Fryganas et al. 2018). These NMR results were up to 21-fold higher than thiolysis and 1.4- to 2.6-fold higher than HCl-butanol-acetone data and could indicate that some PAs may be difficult to quantify by depolymerization (Fryganas et al. 2018). This approach will, however, require more work as PA concentration was associated with an 8% error and the procyanidin/prodelphinidin ratio with a 16% error. Nevertheless, solid state NMR is likely to be an important tool and worth developing further to ensure that all PAs, i.e. extractable and nonextractable PAs, are detected. A detailed review of NMR methods for the analysis of NEPA components was recently provided by Dominguez-Rodriguez et al. (2017).

Solid state ^{13}C CPMAS NMR has also been applied to monitor the extraction efficiency and approximate quantification of PAs in quebracho heartwood and black wattle bark. The same PA signals at 154, 147, and 131 ppm were used but the presence of lignin caused interferences (Reid et al. 2013).

In an unpublished study, Waghorn et al. (personal communication; 23rd March 2011, UK-NZ Ag/Bio/Health Workshop held at Sky City Hotel, Auckland, NZ) used solid-state ^{13}C-NMR to compare purified PAs, feed PAs, and the resulting fecal products after digestion. The aim of the study was to decipher the nature of the mechanisms by which fecal PA molecules were rendered unresponsive to the HCl-butanol assay. The spectra showed evidence of changes to the A- or B-rings and were consistent with known PA reactions that lead to production of the theaflavins. The chemical shift changes provided evidence for cross-linking reactions, which could help to explain the difficulties in extracting PAs from feces and with detecting these PAs in the HCl-butanol assay. Most notably, a ^{13}C signal at 185 ppm (Figure 6.15a), indicative of the carbonyl carbon of the tropolone moiety (Figure 6.15b) (Nonaka et al. 1986), was identified as a newly generated signal (Waghorn et al. personal communication; 23rd March 2011, UK-NZ Ag/Bio/Health Workshop held at Sky City Hotel, Auckland, NZ). These NMR studies are in agreement with the more recent LC-MS/MS results on the formation of PA crosslinks and the discovery of a theaflavin marker ion after thiolyglycolysis (*m/z* 669) as a PA oxidation product (Figure 6.15d) (Mouls and Fulcrand 2015). This means that tropolone (Figure 6.15c and d) formation from flavan-3-ol reactions has now been reported under both anaerobic microbial fermentation and aerobic conditions (Nonaka et al. 1986; Hashimoto et al. 1992), which makes their formation in the intestinal tract likely, and may account in part for the modification of PA structure during digestion.

6.9.4 Use of ^{1}H–^{13}C HSQC NMR Spectroscopy

Although 2D NMR techniques have rarely been applied in the analysis of oligomeric and polymeric PA samples, recent advances in the use of ^{1}H–^{13}C HSQC NMR spectroscopy can deliver semiquantitative assessment of PA composition and structure (Zeller, Ramsay et al. 2015). Estimation of PC/PD ratios were achieved from comparative integration of the aromatic H/C signals (Figure 6.16) arising from the H/C-2',6' (from prodelphinidin cross-peaks) and the H/C-6' (from procyanidin cross-peaks). The data show excellent correlation between the percentage of procyanidins in these samples determined by thiolysis (Figure 6.17). Furthermore, the H/C-4 cross-peak arising from extension units is segregated into signals for *cis*- and *trans*-flavan-3-ols signals. Relative integration of these cross-peaks also provides a strong correlation ($R^2 = 0.89$) with data obtained by thiolytic degradation.

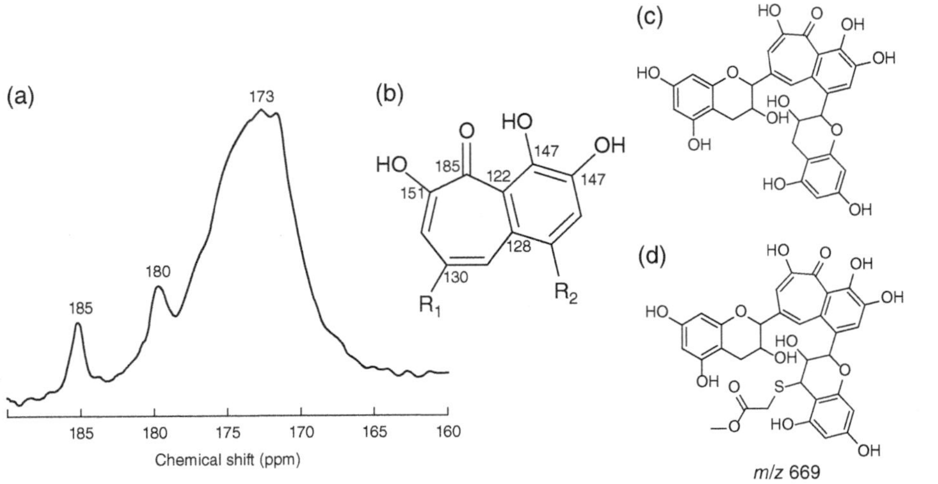

Figure 6.15 Expansion of the carbonyl region of CPMAS ^{13}C NMR spectrum of feces (a) showing the signal at 185 ppm, ^{13}C chemical shifts of tropolone compounds (b), structure of theaflavine (c), and proposed structure of marker ion at m/z 669 (theaflavin-type [d] after thioglycolysis). *Source:* (a) Waghorn, personal communication 2011. (b) Nonaka et al. 1986. (c,d) Mouls and Fulcrand 2015.

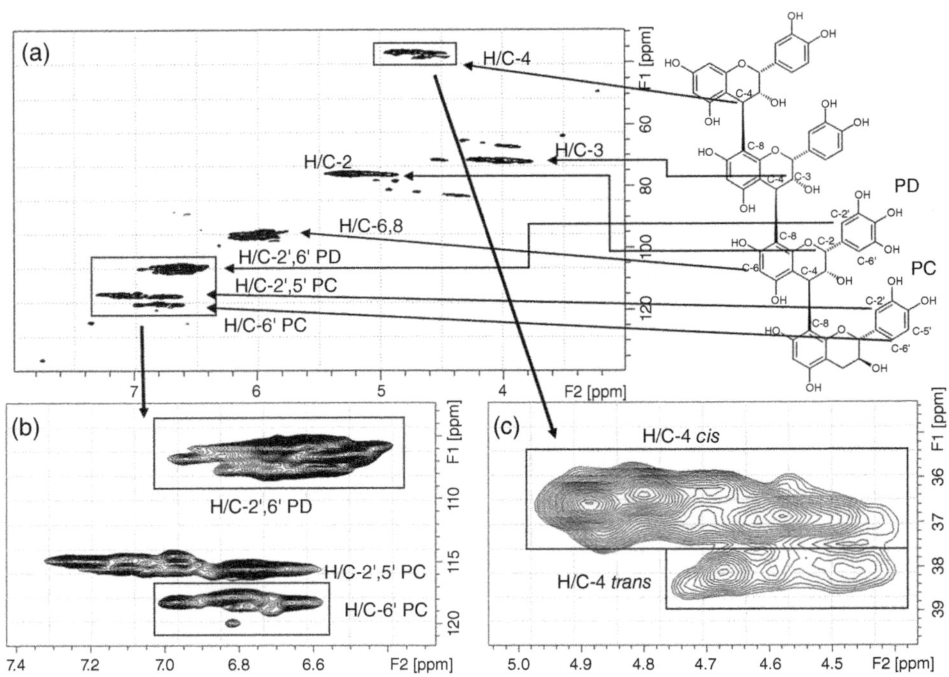

Figure 6.16 ^{1}H-^{13}C HSQC NMR spectrum (a) of purified proanthocyanidins from *Onobrychis viciifolia* (sainfoin) showing assignment of cross-peak signals. Expansion of the aromatic region (b) showing cross-peak signals used in estimating the procyanidin/prodelphinidin ratio of the sample. Expansion of the H/C-4 cross-peaks signals (c) used in estimating the *cis/trans* flavan-3-ol ratio of the sample.

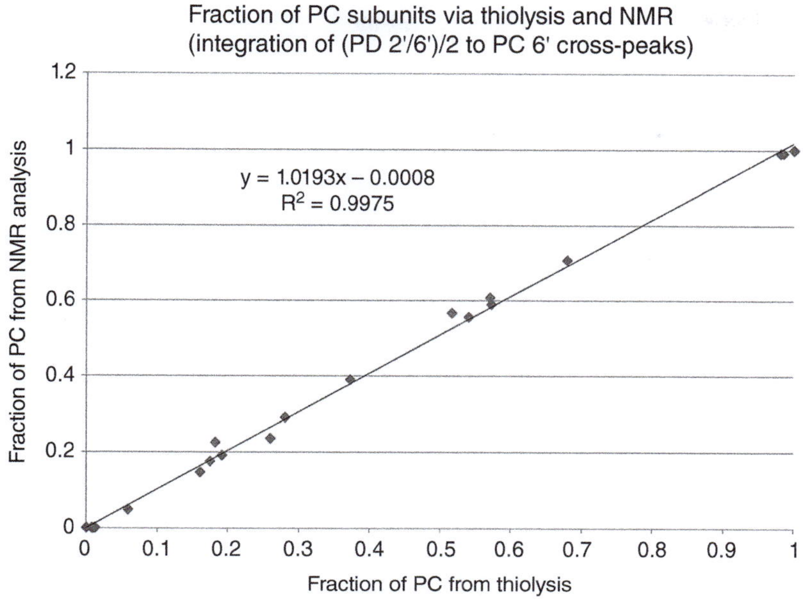

Figure 6.17 Comparison of the results obtained of fraction of procyanidins (PC) determined by 2D NMR and by thiolytic degradation.

In addition, ^{1}H–^{13}C HSQC NMR spectroscopy has an advantage over solution- and solid-phase ^{13}C NMR techniques as the former, adding a second analytical NMR dimension, is more capable of assessing relative levels of impurities and identifying impurity components (Figure 6.18). More recently, an application of this 2D-NMR method has been described for identifying terminal units, the extent of galloylation at C-3, A-/B-type bond ratios, and the mDP of purified PAs (Zeller et al. work in preparation). To facilitate these determinations, a new database was generated (Zeller and Schatz 2017) that collates literature data on solution-state ^{1}H and ^{13}C-NMR chemical shifts of PA oligomers (https://www.ars.usda.gov/mwa/madison/dfrc/tannin).

Most of the same cross-peak signals that had been used from 2D NMR for the determination of procyanidin/prodelphinidin and *cis/trans* ratios in purified PAs are also evident in whole plant samples of *Lotus corniculatus* leaves (Grabber et al. 2013) that had been ball-milled and subjected to gel-state ^{13}C NMR (Figure 6.19) in DMSO-d_6/pyridine-d_5 (Kim et al. 2010). In such cases, data can be obtained to allow assessment of PC/PD and *cis/trans* ratios directly from the whole plant sample.

6.9.5 Use of ^{31}P NMR Spectroscopy

A ^{31}P NMR spectroscopic technique developed for the examination of lignin was recently applied to PAs and involves phosphytilation with 2-chloro-4,4,5,5-tetramethyl-1,3,2-dioxaphospholane (Cl-TMDP). This is based on the observation that Cl-TMDP reacts quantitatively with aliphatic alcohols, phenols, and carboxylic acids moieties of polyphenols in pyridine and was used to assess the purity of PA samples. After determining ^{31}P chemical shifts of P-derivatized model compounds (Melone et al. 2013a), the method was used to

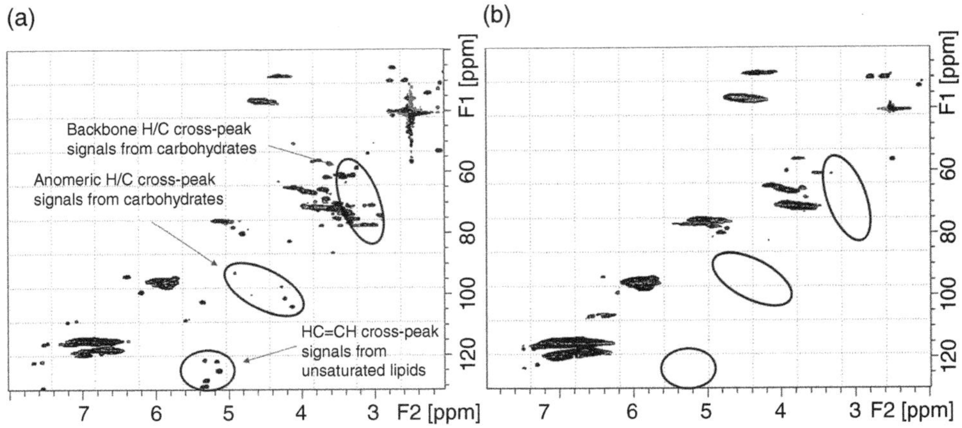

Figure 6.18 $^1H-^{13}C$ HSQC NMR spectra of purified cranberry proanthocyanidin samples showing significant impurities present (a) and absent (b).

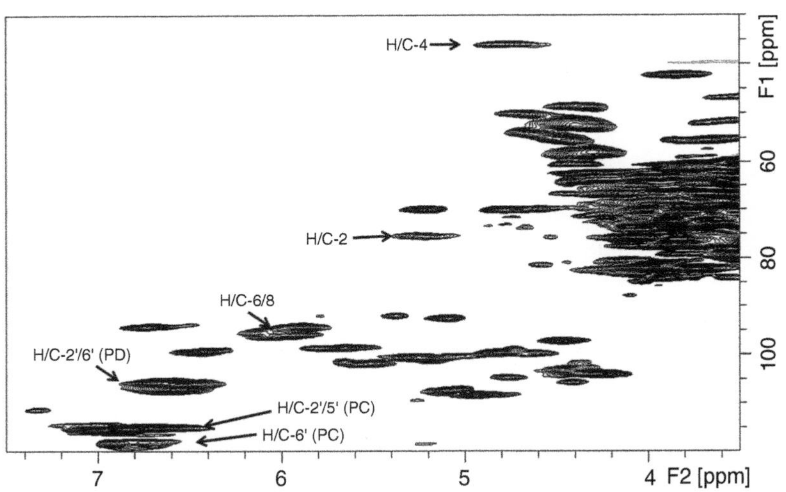

Figure 6.19 Gel-state $^1H-^{13}C$ (500/125 MHz) HSQC NMR spectrum (4:1 DMSO-d_6/pyridine-d_5,) of ball-milled *Lotus corniculatus* leaves. Visible cross-peaks arising from proanthocyanidins present in the sample are labeled.

evaluate structural and quantitative characterization of all OH groups of several hydrolyzable tannins and PAs (Melone et al. 2013b). Recently, this method has also been used in conjunction with $^1H-^{13}C$ HSQC NMR spectroscopy in the analysis of PAs isolated from *Acacia catechu*, *Schinopsis balansae*, and *Acacia mearnsii* (Crestini et al. 2016).

6.10 Colorimetry

Colorimetric techniques suffer from the fact that the various features of PA compounds (e.g. molecular weight, B- vs A-type bonds, 4-8 B-type vs 4-6 B-type bonds, etc) and PA types (e.g. PCs vs PDs, PAs with or without a 5-OH group, galloylation, etc) vary enormously in the color yields generated with their detection reagents (Giner-Chavez et al. 1997; Feliciano et al. 2012; Wang et al. 2016), which poses problems for quantitation. Nevertheless, colorimetry can be useful for certain, clearly defined applications, such as assays adapted into high-throughput screening protocols, as long as the purity and composition of the reference sample have been evaluated.

Colorimetric techniques are particularly useful for quality control purposes on a single product or for screening large collections of samples, whether from countercurrent chromatography (Phansalkar et al. 2018) or from germplasm collections, where the composition of PA structures is either the same or varies only slightly. For example, the DMACA (4-dimethylaminocinnamaldehyde) assay has been optimized to examine PA concentrations in different cranberry sources using procyanidin A2 as a reference standard, which proved reproducible in a multilaboratory validation experiment (Prior et al. 2010). This method was further improved through the use of PAs isolated from cranberry and can be carried out in 96 well plates (Prior et al. 2010; Feliciano et al. 2012). The sulfuric acid assay has also been adapted for use in a continuous flow analyzer (Nitao et al. 2001).

In addition, extract compositions, in terms of PA oligomers versus polymers, can be examined quickly with the Fast Blue Salt B, DMACA, or vanillin/HCl reagents when coupled to HP-TLC-densitometry. This approach has been applied to differentiate between cocoa or chocolate types, varieties, and origins (Pedan et al. 2018) and between commercial cranberry products (Boudesocque-Delaye et al. 2018).

6.11 Infrared Spectroscopy

Infrared spectroscopy techniques, such as near infrared reflectance (NIR) or Fourier transform infrared (FTIR) spectroscopy, benefit from being a fast and nondestructive technique as whole, usually ground, sample tissues but also extracts can be analyzed directly and, due to the absence of reagents, they are environmentally friendly. These techniques are of particular interest to quality control testing; for example, NIRS plus partial least squares discriminant analysis grouped commercial tannin products correctly into PAs and hydrolysable tannins and could be used to detect fraudulent products (Figure 6.20; Grasel and Ferrao 2016).

Other examples can be found in agronomic research and trait-directed plant breeding, as these can profit not only from information on PA concentrations but also on PA compositions. The correlations between the spectral predictions and the primary laboratory methods such as phloroglucinolysis or thiolysis for total PA concentrations or normal phase HPLC for an oligomeric subset concentration typically have correlation coefficients >0.98, which demonstrates the power of this approach. A few examples shall illustrate these points: PA concentrations and compositions (mDP, % galloylation; procyanidin/prodelphinidin and *cis/trans* flavan-3-ol ratios) have been predicted by NIRS for grapeseed, birdsfoot trefoil, sorghum, and sainfoin samples (Mueller-Harvey et al. 2011; Dykes et al. 2014; Pappas et al. 2015) and concentrations of PA oligomers for barley and cocoa samples (Whitacre et al. 2003; Verardo et al. 2015). Examining 14 birdsfoot trefoil cultivars from 10 location-harvest environments by NIRS, Grabber et al. (2014) identified the factors that impacted most on PA concentrations: the maturity score explained 78% of the variation, while genotypes and growth environment had some effect. It has now been established that

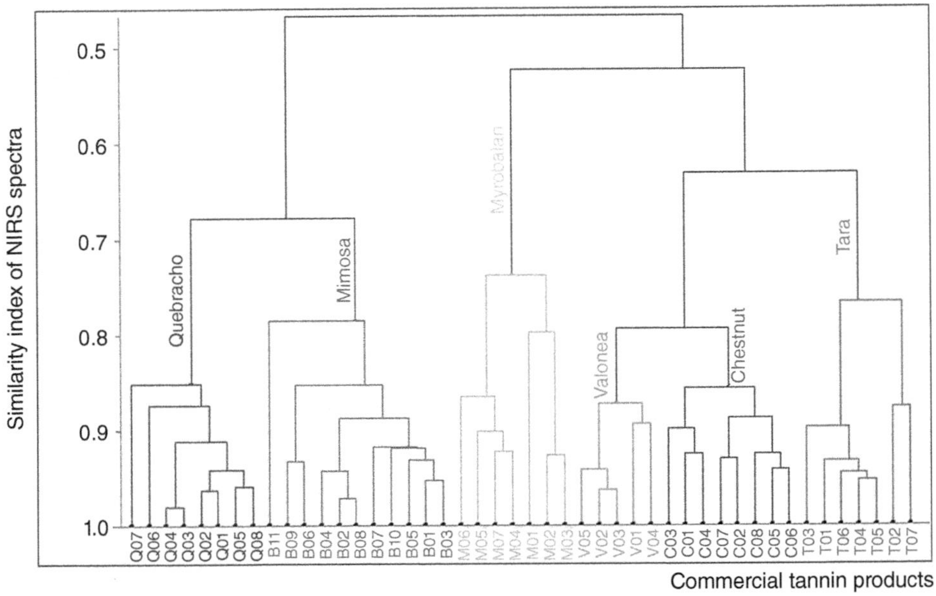

Figure 6.20 Clustering of near-infrared reflectance spectra of commercial tannin products illustrating distinct proanthocyanidin and hydrolyzable tannin groups and tannin-specific subgroups. *Source:* Grasel and Ferrao 2016. (*See insert for color representation of the figure.*)

many germplasm collections possess enormous variation in both PA concentration and composition (Stringano et al. 2012; Grabber et al. 2015; Ferraro et al. 2014). Therefore, the use of fast IR-based techniques will be invaluable for trait-based plant breeding of new chemovars or for identifying cultivars with robust PA expression that are less affected by external factors (Azuhnwi et al. 2013).

6.12 Conclusions

This review has highlighted the need for analytical tools that are sufficiently accurate to underpin research into the numerous nutritional and health benefits of different PA compounds, to unravel the mechanisms that confer desirable bioactivities, and to ensure that PA-containing plants or products deliver consistent and predictable results. Interlaboratory studies are needed to test the validity and robustness of current methods, to support the development of certified PA reference standards, to enable quality control of nutritional and health supplements, and to evaluate new functional food products and novel animal feeds, whether through conventional plant breeding or novel genetic approaches (Hayot Carbonero et al. 2011; Albert 2015). Genome-wide association studies have identified molecular markers and transcription factors for PAs (Rhodes et al. 2014; Mora-Ortiz et al. 2016; Kempf et al. 2016; Constabel 2018) and first reports of stable and heritable PA production in transgenic plants and forage legumes have already appeared (Albert 2015; Fresquet-Corrales et al. 2017).

Due to the large number of techniques required to determine accurate PA concentrations and structures, it is unlikely that the required expertise and equipment are housed in one laboratory. Interdisciplinary and interlaboratory collaborations are, therefore, vital for ensuring progress in the face of the PA complexity that exists in plants, to discover meaningful structure-activity relationships, and to develop a stronger understanding of the function of PAs foods and feeds. This review has attempted to offer an overview of the different techniques, and their advantages and disadvantages for specific applications or research questions. The topics covered PA extraction and purification methods, depolymerization reagents for determining PA concentration and composition, conventional and improved colorimetric, and infrared spectroscopic techniques plus state-of-the-art mass spectrometric and nuclear magnetic resonance techniques.

Relationships are emerging between PA compounds or generic structural features and their biological effects. For example, high molar proportions of prodelphinidins appear to give rise to antinutritional effects in ruminant animals – in contrast to procyanidins (Waghorn 2008; Mueller-Harvey et al. 2019). However, a high percentage of prodelphinidins and increasing mean degrees of polymerization can generate useful anthelmintic (antiparasitic) effects (Terrill et al. 2012; Mueller-Harvey et al. 2019). The mean degree of polymerization appears as a key parameter with relevance for several bioactivities. This may stem from the fact that this parameter strongly affects PA-protein interactions (Figure 6.21; Zeller, Sullivan et al. 2015; Ropiak et al. 2017) and that there appears to be a threshold at a mean degree of polymerization of c. 7 to 8, which corresponds to c. 2000 to 2700 Daltons, at which enhanced protein interaction occurs.

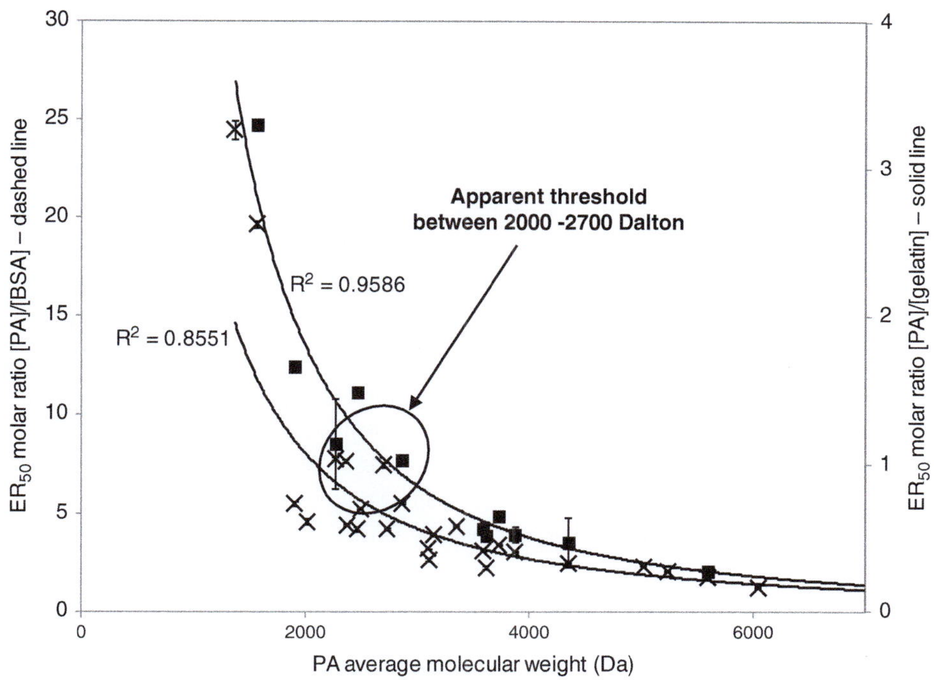

Figure 6.21 Relationship between the average molecular weight of proanthocyanidins and their precipitation capacity (ER_{50}) of bovine serum albumin (BSA, dashed line) and gelatin (solid line). *Source:* Ropiak et al. 2017.

A cellular immune response could be measured with PAs that had a mean degree of polymerization >6 but not below (Williams et al. 2016), while di-, tri- and tetrameric PA oligomers increased the proliferation and expression of genes in dental pulp cells that are necessary for extracellular matrix deposition and biomineralization (Kulakowski et al. 2017). Evidence is also emerging for different physiological effects of extractable versus nonextractable dietary PAs: extractable PAs generated relatively short-term effects in terms of plasma changes, but metabolites could be detected in the blood stream up to four days after consumption of nonextractable PAs (Gonzales-Sarrias et al. 2017).

These examples illustrate the importance of interdisciplinary collaborations that are based on PA analysis that is fit-for-purpose. Unfortunately, it is not uncommon to read publications that purport to report the discovery of important medicinal or nutritional PA activities, but that have not even characterized or quantified the PAs. It is indeed surprising to note how different the pathways have been in, for example, ecological research, which has employed sound tannin analysis tools coupled to robust experimental designs (Agrawal et al. 2012) and the approaches chosen in many animal nutrition studies in the past. We need to progress from *in vitro* to *in vivo* feeding trials that employ distinct PA chemovars for evaluating the potential benefits of different PA types in forages for improved nutrition and health. Information of the role of PAs at the plant and mammalian tissue and cellular level can also be expected from new imaging techniques that are capable of tracking the trafficking of flavan-3-ols, PAs, and their metabolites (Brillouet et al. 2013; Mueller-Harvey et al. 2014).

In summary, this chapter was designed to bring the PA researcher up-to-date on improvements of existing, and the evolution of new, techniques available for determining PA concentrations and composition. Through detailing these techniques, we have highlighted the need for improving the analysis of PAs that are nonextractable or that are modified during sample processing, whether in planta, during storage, extraction, drying, heating, or fermentation.

6.13 Acknowledgments

We would like to extend our sincere gratitude to all of the authors who gave permission and/or forwarded graphs and diagrams for use in this chapter. We also would like to thank all of the individuals responsible for conducting the research outlined in this chapter. Special mention goes out to Yap Foo and Garry Waghorn for sharing their unpublished observations and to Laurie Reinhardt and Honorata M. Ropiak, who assisted with production of some graphs and figures. Mention of trade names or commercial products in this article is solely for the purpose of providing specific information and does not imply recommendation or endorsement by the US Department of Agriculture.

References

Acuña, H., Concha, A., and Figueroa, M. (2008). Condensed tannin concentrations of three Lotus species grown in different environments. *Chilean Journal of Agricultural Research* 68: 31–41.

Aerts, R.J., Barry, T.N., and McNabb, W.C. (1999). Polyphenols and agriculture: beneficial effects of proanthocyanidins in forages. *Agriculture, Ecosystems and Environment* 75: 1–12.

Agrawal, A.A., Hastings, A.P., Johnson, M.T.J., et al. (2012). Insect herbivores drive real-time ecological and evolutionary change in plant populations. *Science* 338: 113–116.

Aguerre, M.J., Wattiaux, M.A., Capozzolo, M.C., et al. (2010). Effect of quebracho-chestnut tannin extracts at two dietary crude protein levels on nitrogen partitioning in lactating dairy cows. *Journal of Dairy Science*, 93 (E-Suppl. 1)., 446 (Abstr.).

Albert, N.W. (2015). Subspecialization of R2R3-MYB repressors for anthocyanin and proanthocyanidin regulation in forage legumes. *Frontiers in Plant Science* 6: 1165.

Albrecht, K.A. and Muck, R.E. (1991). Proteolysis in ensiled forage legumes that vary in tannin concentration. *Crop Science* 31: 464–469.

Alejo-Armijo, A., Glibota, N., María P., et al. (2018). Synthesis and evaluation of antimicrobial and antibiofilm properties of A-type procyanidin analogues against resistant bacteria in food. *Journal of Agricultural and Food Chemistry* 66: 2151–2158.

Alvarenga, R.F.R., Friesen, J.B., Nikolic, D., et al. (2014). K-targeted metabolomic analysis extends chemical subtraction to DESIGNER extracts: selective depletion of extracts of hops (*Humulus lupulus*). *Journal of Natural Products* 77: 2595–2604.

AOAC (2017). AOAC SMPR 2017.003: *Standard Method Performance Requirements* (SMPRs) for quantitation of proanthocyanidin content in cranberry fruit, juice, beverage, dried cranberry, cranberry sauce, ingredients (concentrates, extracts, and powders), and dietary supplement formulations. *AOAC Stakeholder Panel Meeting on Strategic Food Analytical Methods* (SPSFAM), March 13, 2017. http://www.aoac.org/aoac_prod_imis/AOAC_Docs/SMPRs/SMPR%202017_003.pdf (accessed 24 Sep 2018).

Azuhnwi, B.N., Boller, B., Dohme-Meier, F., et al. (2013). Exploring variation in proanthocyanidin composition and content of sainfoin (*Onobrychis viciifolia*). *Journal of the Science of Food and Agriculture* 93: 2102–2109.

Barry, T.N., and McNabb, W.C. (1999). The implications of condensed tannins on the nutritive value of temperate forages fed to ruminants. *British Journal of Nutrition* 81: 263–272.

Becker, P.M., Wikselaar, P.G., Franssen, M.C.R., et al. (2014). Evidence for a hydrogen-sink mechanism of (+)catechin-mediated emission reduction of the ruminant greenhouse gas methane. *Metabolomics* 10: 179–189.

Berthod, A., Ruiz-Ángel, M.J., and Carda-Broch, S. (2009). Countercurrent chromatography: People and applications. *Journal of Chromatography A* 1216: 4206–4217.

Billerach, G., Rouméas, L., Dubreucq, E., and Fulcrand, H. (2020). Furanolysis with menthofuran: a new depolymerization method for analyzing condensed tannins. *Journal of Agricultural and Food Chemistry* 68: 2917–2926.

Bordiga, M., Coïsson, J.D., Locatelli, M., et al. (2013). Pyrogallol: an alternative trapping agent in proanthocyanidins analysis. *Food Analytical Methods* 6, 148–156.

Boudesocque-Delaye, L., Lanoue, A., Dorat, J., et al. (2018). Quality control of commercial cranberry products: HPTLC densitometry a new deal. *Food Control* 86: 214–223.

Brillouet, J.-M., Fulcrand, H., Carrillo, S., et al. (2017). Isolation of native proanthocyanidins from grapevine (*Vitis vinifera*) and other fruits in aqueous buffer. *Journal of Agricultural and Food Chemistry* 65: 2895–2901.

Brillouet, J.-M., Romieu, C., Schoefs, B., et al. (2013). The tannosome is an organelle forming condensed tannins in the chlorophyllous organs of Tracheophyta. *Annals of Botany* 112: 1003–1014.

Brown, B.R., and Shaw, M.R. (1974). Reactions of flavanoids and condensed tannins with sulphur nucleophiles. *Journal of the Chemical Society, Perkin Transactions 1*, 2036–2049.

Brown, R.H., Mueller-Harvey, I., Zeller, W.E., et al. (2017). Facile purification of milligram to gram quantities of condensed tannins according to mean degree of polymerization and flavan-3-ol subunit composition. *Journal of Agricultural and Food Chemistry* 65: 8072–8082.

Chai, W.-M., Shi, Y., Feng, H.-L., et al. (2012). NMR, HPLC-ESI-MS, and MALDITOF MS analysis of condensed tannins from *Delonix regia* (Bojer ex Hook.) Raf. and their bioactivities. *Journal of Agricultural and Food Chemistry* 60: 5013–5022.

Chen, S.-N., Turner, A., Jaki, B.U., et al. (2008). An experimental implementation of chemical subtraction. *Journal of Pharmaceutical and Biomedical Analysis* 46: 692–698.

Chen, W., Fu, C., Qin, Y., and Huang, D. (2009). One-pot depolymerizative extraction of proanthocyanidins from mangosteen pericarps. *Food Chemistry* 114: 874–880.

Chen, X., Westphall, L., and Smith, M. (2003). Mass spectrometric analysis of DNA mixtures: instrumental effects responsible for decreased sensitivity with increasing mass. *Analytical Chemistry* 75: 5944–5952.

Chowdhury, S.A., Vijayaraghavan, R., and MacFarlane, D.R. (2010). Distillable ionic liquid extraction of tannins from plant materials. *Green Chemistry* 12: 1023–1028.

Constabel, C.P. (2018). Molecular controls of proanthocyanidin synthesis and structure: prospects for genetic engineering in crop plants. *Journal of Agricultural and Food Chemistry* 66: 9882–9888.

Crestini, C., Lange, H., and Bianchetti, G. (2016). Detailed chemical composition of condensed tannins via quantitative ^{31}P NMR and HSQC analyses: *Acacia catechu*, *Schinopsis balansae*, and *Acacia mearnsii*. *Journal of Natural Products* 79: 2287–2295.

Ćurko, N., Tomašević, M., Bubalo, M.C., et al. (2017). Extraction of proanthocyanidins and anthocyanins from grape skin by using ionic liquids. *Food Technology and Biotechnology* 55: 429–437.

Czochanska, Z., Foo, L.Y., Newman, R.H., et al. (1979). Direct proof of a homogeneous polyflavan-3-ol structure for polymeric proanthocyanidins. *Journal of the Chemical Society, Chemical Communications*, 375–377.

Czochanska, Z., Foo, L.Y., Newman, R.H., and Porter, L.J. (1980). Polymeric proanthocyanidins. Stereochemistry, structural units, and molecular weight. *Journal of the Chemical Society, Perkin Transactions* 1, 2278–2286.

da Silva Porto, P.A.L., Laranjinha, J.A.N., and de Freitas, V.A.P. (2003). Antioxidant protection of low density lipoprotein by procyanidins: structure/activity relationships. *Biochemical Pharmacology* 66: 947–954.

Deaville, E.R., Givens, D.I., and Mueller-Harvey, I. (2010). Chestnut and mimosa tannin silages: Effects in sheep differ for apparent digestibility, nitrogen utilisation and losses. *Animal Feed Science and Technology* 157: 129–138.

del Pino, M.C.A., Hervas, G., Mantecon, A.R., et al. (2005). Comparison of biological and chemical methods, and internal and external standards, for assaying tannins in Spanish shrub species. *Journal of the Science of Food and Agriculture* 85: 583–590.

de Pascual-Teresa, S., Moreno, D.A., and García-Viguera C. (2010). Flavanols and Anthocyanins in Cardiovascular Health: A Review of Current Evidence. *International Journal of Molecular Sciences* 11: 1679–1703.

Déprez, S., Mila, I., Huneau, J.F., Tome, D., and Scalbert, A. (2001). Transport of proanthocyanidin dimer, trimer, and polymer across monolayers of human intestinal epithelial Caco-2 cells. *Antioxididants and Redox Signaling* 3: 957–967.

Deshpande, S.S., Cheryan, M., and Salunkhe, D.K. (1986). Tannin analysis of food products. *CRC Critical Reviews in Food Science and Nutrition* 24: 401–449.

Desrues, O., Mueller-Harvey, I., Pellikaan, W.F., et al. (2017). Condensed tannins in the gastrointestinal tract of cattle after sainfoin (*Onobrychis viciifolia*) intake and their possible relationship with anthelmintic effects. *Journal of Agricultural and Food Chemistry* 65: 1420–1427.

Domínguez-Rodríguez, G., Marina, M.L., and Plaza, M. (2017). Strategies for the extraction and analysis of non-extractable polyphenols from plants. *Journal of Chromatography A* 1514: 1–15.

D'Souza, R.N., Grimbs, S., Behrends, B., et al. (2017). Origin-based polyphenolic fingerprinting of *Theobroma cacao* in unfermented and fermented beans. *Food Research International* 99: 550–559.

Dudek, M.K., Bujacz, G., and Potrzebowski, M.J. (2017). Experimental tests for quality validation of computationally predicted crystal structures – a case of a conformationally flexible procyanidin A-2 dihydrate. *CrystEngComm* 19: 2903–2913.

Dykes, L., Hoffmann Jr., L., Portillo-Rodriguez, O., et al. (2014). Prediction of total phenols, condensed tannins, and 3-deoxyanthocyanidins in sorghum grain using near-infrared (NIR) spectroscopy. *Journal of Cereal Science* 60: 138–142.

EFSA (European Food Safety Authority) (2012). Scientific opinion on the substantiation of a health claim related to cocoa flavanols and maintenance of normal endothelium-dependent vasodilation pursuant to Article 13(5). of regulation (EC) No 1924/2006. *EFSA Journal* 10: 2809.

Engström, M.T., Pälijärvi, M., Fryganas, C., et al. (2014). Rapid qualitative and quantitative analysis of proanthocyanidin oligomers and polymers by UPLC-MS/MS. *Journal of Agricultural and Food Chemistry* 62: 3390–3399.

Esatbeyoglu, T., Jaschok-Kentner, B., Wray, V., and Winterhalter, P. (2011). Structure elucidation of procyanidin oligomers by low-temperature 1H NMR spectroscopy. *Journal of Agricultural and Food Chemistry* 59: 62–69. doi:10.1021/jf1032334.

Esatbeyoglu, T., Juadjur, A., Wray, V., and Winterhalter, P. (2014). Semisynthetic preparation and isolation of dimeric procyanidins B1–B8 from roasted hazelnut skins (*Corylus avellana* L.) on a large scale using countercurrent chromatography. *Journal of Agricultural and Food Chemistry* 62: 7101–7110.

Esatbeyoglu, T., Wray, V., and Winterhalter, P. (2015). Isolation of dimeric, trimeric, tetrameric and pentameric procyanidins from unroasted cocoa beans (*Theobroma cacao* L.) using countercurrent chromatography. *Food Chemistry* 179: 278–289.

Feliciano, R.P., Meudt, J.J., Shanmuganayagam, D., Krueger, C.G., and Reed, J.D. (2014). Ratio of "A-type" to "B-type" proanthocyanidin interflavan bonds affects extra-intestinal pathogenic Escherichia coli invasion of gut epithelial cells. *Journal of Agricultural and Food Chemistry* 62: 3919–3925.

Feliciano, R.P., Shea, M.P., Shanmuganayagam, D., et al. (2012). Comparison of isolated cranberry (*Vaccinium macrocarpon* Ait.) proanthocyanidins to catechin and procyanidins A2 and B2 for use as standards in the 4-(dimethylamino) cinnamaldehyde assay. *Journal of Agricultural and Food Chemistry* 60: 4578–4585.

Feng, W.Y. (2006). Metabolism of green tea catechins: An overview. *Current Drug Metabolism* 7: 755–809.

Ferraro, K., Jin, A.L., Nguyen, T-D., et al. (2014). Characterization of proanthocyanidin metabolism in pea (*Pisum sativum*) seeds. *BMC Plant Biology* 14: 238.

Ferreira, D., and Coleman, C.M. (2011). Towards the synthesis of proanthocyanidins: half a century of innovation. *Planta Medica* 77: 1071–1085.

Foo, L.Y., Lu, Y., Molan, A.L., et al. (2000). The phenols and prodelphinidins of white clover flowers, *Phytochemistry* 54: 539–548.

Fraser, K., Collette, V., and Hancock, K.R. (2016). Characterization of proanthocyanidins from seeds of perennial ryegrass (*Lolium perenne* L.) and tall fescue (*Festuca arundinacea*) by liquid chromatography—mass spectrometry. *Journal of Agricultural and Food Chemistry* 64: 6676–6684.

Fresquet-Corrales, S., Roque, E., Sarrion-Perdigones, A., et al. (2017). Metabolic engineering to simultaneously activate anthocyanin and proanthocyanidin biosynthetic pathways in *Nicotiana spp. PLoS ONE* 12 (9): e0184839.

Fryganas, C., Drake, C., Ropiak, H.M., et al. (2018). Carbon-13 cross-polarization magic-angle spinning nuclear magnetic resonance (CPMAS NMR) for measuring proanthocyanidin content and procyanidin to prodelphinidin ratio in sainfoin (*Onobrychis viciifolia*) tissues. *Journal of Agricultural and Food Chemistry* 66: 4073–4081.

Fu, C., Chen, W., Quek, Y.L., et al. (2010). Sustainability from agricultural waste: chiral ligands from oligomeric proanthocyanidins via acid-mediated depolymerization. *Tetrahedron Letters* 51: 6322–6324.

Fulcrand, H., Mané, C., Preys, S., et al. (2008). Direct mass spectrometry approaches to characterize polyphenol composition of complex samples. *Phytochemistry* 69: 3131–3138.

Gamble, G.R., Akin, D.E., Makkar, H.P., and Becker, K. (1996). Biological degradation of tannins in sericea lespedeza (*Lespedeza cuneata*) by the white rot fungi *Ceriporiopsis subvermispora* and *Cyathus stercoreus* analyzed by solid-state ^{13}C nuclear magnetic resonance spectroscopy. *Applied and Environmental Microbiology* 62: 3600–3604.

Gao, C., Cunningham, D.G., Liu, H.Y., et al. (2018). Development of a thiolysis HPLC method for the analysis of procyanidins in cranberry products. *Journal of Agricultural and Food Chemistry* 66: 2159–2167.

Gea, A., Stringano, E., Brown, R.H., and Mueller-Harvey, I. (2011). *In situ* analysis and structural elucidation of sainfoin (*Onobrychis viciifolia*) tannins for high throughput germplasm screening. *Journal of Agricultural and Food Chemistry* 59: 495–503.

Giner-Chavez, B., Van Soest, P.J., Robertson, J.B., et al. (1997). A method for isolating condensed tannins from crude plant extracts with trivalent ytterbium. *Journal of the Science of Food and Agriculture* 74: 359–368.

Gonzalez-Sarrías, A., Espín, J.C., and Tomas-Barberan, F.A. (2017). Non-extractable polyphenols produce gut microbiota metabolites that persist in circulation and show antiinflammatory and free radical-scavenging effects. *Trends in Food Science and Technology* 69: 281–288.

Grabber J., Zeller, W.E., and Mueller-Harvey, I. (2013). Acetone enhances the direct analysis of procyanidin- and prodelphinidin-based condensed tannins in *Lotus* species by the butanol-HCl-iron assay. *Journal of Agricultural and Food Chemistry* 61: 2669–2678.

Grabber, J.H., Coblentz, W.K., Riday, H., et al. (2015). Protein and dry-matter degradability of European- and Mediterranean-derived birdsfoot trefoil cultivars grown in the colder continental USA. *Crop Science* 55: 1356–1364.

Grabber, J.H., Riday, H., Cassida, K.A., et al. (2014). Yield, morphological characteristics, and chemical composition of European- and Mediterranean-derived birdsfoot trefoil cultivars grown in the colder continental United States. *Crop Science* 54: 1893–1901.

Grasel, F.S. and Ferrao, M.F. (2016). A rapid and non-invasive method for the classification of natural tannin extracts by near-infrared spectroscopy and PLS-DA. *Analytical Methods* 8: 644–649.

Gu, L. (2012). Analysis Methods of Proanthocyanidins. In: *Analysis of Antioxidant-Rich Phytochemicals* (eds Z. Xu and L.R. Howard), 247–274. Chichester, U.K.: Wiley.

Gu, L, Kelm, M., Hammerstone, J.F., et al. (2002). Fractionation of polymeric procyanidins from lowbush blueberry and quantification of procyanidins in selected foods with an optimized normal-phase HPLC-MS fluorescent detection method. *Journal of Agricultural and Food Chemistry* 50: 4852–4860.

Gu, L., Kelm, M.A., Hammerstone, J.F., et al. (2003). Screening of foods containing proanthocyanidins and their structural characterization using LC-MS/MS and thiolytic degradation. *Journal of Agricultural and Food Chemistry* 51: 7513–7521.

Guyot, S., Marnet, N., and Drilleau, J.-F. (2001). Thiolysis-HPLC characterization of apple procyanidins covering a large range of polymerization states. *Journal of Agricultural and Food Chemistry* 49: 14–20.

Guyot, S., Marnet, N., Sanoner, P., and Drilleau, J.F. (2001). Direct thiolysis on crude apple materials for high-performance liquid chromatography characterization and quantification of polyphenols in cider apple tissues and juices. *Methods Enzymology* 335: 57–70.

Hagerman, A.E. (1988). Extraction of tannins from fresh and preserved leaves. *Journal of Chemical Ecology* 14: 453–461.

Hagerman, A.E. (2012). Fifty years of polyphenol-protein complexes. In: *Recent Advances in Polyphenol Research. Vol 3.* (eds. V. Cheynier, P. Sarni-Manchado, and S. Quideau), 71–97.

Hamza, A.A., Heeba, G.H., Elwy, H.M., et al. (2018). Molecular characterization of the grape seeds extract's effect against chemically induced liver cancer: *In vivo* and *in vitro* analyses. *Scientific Reports* 8: 1270.

Hashimoto, F, Nonaka, G-i., and Nishioka, I. (1992). Tannins and related-compounds. CXIV. structures of novel fermentation products, theogallinin, theaflavonin and desgalloyl theaflavonin from black tea, and changes of tea leaf polyphenols during fermentation. *Chemical and Pharmaceutical Bulletin* 40: 1383–1389.

Hattas, D., and Julkunen-Tiitto, R. (2012). The quantification of condensed tannins in African savanna tree species. *Phytochemistry Letters* 5: 329–334.

Hayot Carbonero, C., Mueller-Harvey, I., Brown, T.A., and Smith, L. (2011). Sainfoin (*Onobrychis viciifolia*): a beneficial forage legume. *Plant Genetic Resources: Utilisation and Characterisation* 9: 70–85.

Hernández-Jiménez, A., Kennedy, J.A., Bautista-Ortín, A.B., and Gómez-Plaza, E. (2012). Effect of ethanol on grape seed proanthocyanidin extraction. *American Journal of Enology and Viticulture* 63: 57–61.

Hervás, G., Frutos, P., Giráldez, F.J., et al. (2003). Effect of different doses of quebracho tannins extract on rumen fermentation in ewes. *Animal Feed Science and Technology* 109: 65–78.

Hill, S.J., McIntosh, C.D., and Grigsby, W.J. (2003). NMR estimation of extractables from bark: analysis method for quantifying tannin extraction from bark. *Journal of Wood Chemistry and Technology* 23: 179–195.

Holderness, J., Hedges, J.F., Daughenbaugh, K., et al. (2008). Response of γδ T cells to plant-derived tannins. *Critical Reviews in Immunology* 28: 377–402.

Hoong, Y.B., Pizzi, A., Tahir, P.M., and Pasch, H. (2010). Characterization of *Acacia mangium* polyflavonoid tannins by MALDI-TOF mass spectrometry and CP-MAS ^{13}C NMR, *European Polymer Journal* 46: 1268–1277.

Hoste, H., Jackson, F., Athanasiadou, S., Thamsborg, S.M., and Hoskin, S.O. (2006). The effects of tannin-rich plants on parasitic nematodes in ruminants. *Trends in Parasitology* 22: 253–261.

Hoste, H., Torres-Acosta, J.F.J., Sandoval-Castro, C.A., et al. (2015). Tannin containing legumes as a model for nutraceuticals against digestive parasites in livestock. *Veterinary Parasitology* 212: 5–17.

Hu, R. and Pan, Y. (2012). Recent trends in counter-current chromatography. *Trends in Analytical Chemistry* 40: 15–27.

Huang, Y.-F., Doligez, A., Fournier-Level, A., et al. (2012). Dissecting genetic architecture of grape proanthocyanidin composition through quantitative trait locus mapping. *BMC Plant Biology* 12: 30. doi: 10. 1186/1471-2229-12-30.

Hymes-Fecht, U.C., Broderick, G.A., Muck, R.E., and Grabber, J.H. (2013). Replacing alfalfa or red clover silage with birdsfoot trefoil silage in total mixed rations increases production of lactating dairy cows. *Journal of Dairy Science* 96: 460–469.

Jones, W.T., and Mangan, J.L. (1977). Complexes of the condensed tannins of sainfoin (*Onobrychis viciifolia* Scop.) with Fraction 1 leaf protein and with submaxillary mucoprotein, and their reversal by polyethylene glycol and pH. *Journal of the Science of Food and Agriculture* 28: 125–136.

Karonen, M., Liimatainen, J., and Sinkkonen, J. (2011). Birch inner bark procyanidins can be resolved with enhanced sensitivity by hydrophilic interaction HPLC-MS. *Journal of Separation Science* 34: 3158–3165.

Kempf, K., Mora-Ortiz, M., Smith, L.M.J., et al. (2016). Characterization of novel SSR markers in diverse sainfoin (*Onobrychis viciifolia*) germplasm. *BMC Genetics* 17: 124.

Kennedy, J.A. and Jones, G.P. (2001). Analysis of proanthocyanidin cleavage products following acid-catalysis in the presence of excess phloroglucinol. *Journal of Agricultural and Food Chemistry* 49: 1740–1746.

Kim, H., and Ralph, J. (2010). Solution-state 2D NMR of ball-milled plant cell wall gels in DMSOd6/pyridine-d5. *Organic and Biomolecular Chemistry* 8: 576–591.

Klongsiriwet, C. (2016). Phytochemical Analysis of a Willow Germplasm Collection and its Pharmacological Activities. PhD thesis. University of Reading, UK.

Koerner, J.L., Hsu, V.L., Lee, J., and Kennedy, J.A. (2009). Determination of proanthocyanidin A2 content in phenolic polymer isolates by reversed-phase high-performance liquid chromatography. *Journal of Chromatography A* 1216: 1403–1409.

Koupai-Abyazani, M.R., Muir, A.D., Bohm, B.A., et al. (1993). The proanthocyanidin polymers in some species of *Onobrychis*. *Phytochmistry* 34: 113–117.

Kraus, T.E.C., Yu, Z., Preston, C.M., et al. (2003). Linking chemical reactivity and protein precipitation to structural characteristics of foliar tannins. *Journal of Chemical Ecology* 29: 703–730.

Kronberg, S.L., Zeller, W.E., Waghorn, G.C., et al. (2018). Effects of feeding *Lespedeza cuneata* pellets with *Medicago sativa* hay to sheep: Nutritional impact, characterization and degradation of condensed tannin during digestion. *Animal Feed Science and Technology* 245: 41–47.

Krueger, C.G., Dopke, N.C., Treichel, P.M., et al. (2000). Matrix-assisted laser desorption/ ionization time-of-flight mass spectrometry of polygalloyl polyflavan-3-ols in grape seed extract. *Journal of Agricultural and Food Chemistry* 48: 1663–1667.

Krueger, C.G., Reed, J.D., Feliciano, R.P., and Howell, A.B. (2013). Quantifying and characterizing proanthocyanidins in cranberries in relation to urinary tract health. *Analytical and Bioanalytical Chemistry* 405: 4385–4395.

Krueger C.G., Vestling M.M., and Reed J.D. (2003). Matrix-assisted laser desorption/ionization time-of-flight mass spectrometry of heteropolyflavan-3-ols and glucosylated heteropolyflavans in sorghum [Sorghum bicolor (L.) Moench]. *Journal of Agricultural and Food Chemistry* 51, 538–543.

Kulakowski, D., Leme-Kraus, A.A., Nam, J-w., et al. (2017). Oligomeric proanthocyanidins released from dentin induce regenerative dental pulp cell response. *Acta Biomaterialia* 55: 262–270.

Kurumatani, M., Fujita, R., Tagashira, M., et al. (2005). Analysis of polyphenols from hop bract region using CCC. *Journal of Liquid Chromatography and Related Technologies* 28: 1971–1983.

Labarbe, B., Cheynier, V., Brossaud, F., et al. (1999). Quantitative fractionation of grape proanthocyanidins according to their degree of polymerization. *Journal of Agricultural and Food Chemistry* 47: 2719–2723.

Leach, D.N., Fryganas, C., Mueller-Harvey, I., et al. (2018). Current pharmacopoeial method greatly underestimates procyanidins in Hawthorn (*Crataegus* spp.). Poster presented at the 66th Annual Meeting of the Society for Medicinal Plant and Natural Product Research, (GA), Shanghai (26–29 August 2018). Poster-PO-22; P 195.

Le Bourvellec, C., Le Quere, J.-M., and Renard, C.M.G.C. (2007). Impact of noncovalent interactions between apple condensed tannins and cell walls on their transfer from fruit to juice: studies in model suspensions and application. *Journal of Agricultural and Food Chemistry* 55: 7896–7904.

Le Bourvellec, C., Guyot, S., and Renard, C.M.G.C. (2009). Interactions between apple (*Malus x domestica* Borkh.) polyphenols and cell walls modulate the extractability of polysaccharides. *Carbohydrate Polymers* 75: 251–261.

Li, C., Trombley, J.D., Schmidt, M.A., and Hagerman, A.E. (2010). Preparation of an acid butanol standard from fresh apples. *Journal of Chemical Ecology* 36: 453–460.

Li, M., and Hagerman, A.E. (2013). Interactions between plasma proteins and naturally occurring polyphenols. *Current Drug Metabolism* 14: 432–445.

Li, Y.-Y., Feng, J., Zhang, X.-L., and Cui, Y.-Y. (2015). Pine bark extracts: nutraceutical, pharmacological, and toxicological evaluation. *Journal of Pharmacology and Experimental Therapeutics* 353: 9–16.

Liu, Y., Yang, L., Zu, Y., et al. (2012). Development of an ionic liquid-based microwave-assisted method for simultaneous extraction and distillation for determination of proanthocyanidins and essential oil in Cortex cinnamomi. *Food Chemistry* 135: 2514–2521.

Liu, Y., Feng, S., Song, L., et al. (2013). Secondary metabolites in durian seeds: oligomeric proanthocyanidins. *Molecules* 18: 14172–14185. doi: 10.3390/molecules181114172.

Lopez-Andres, P., Luciano, G., Vasta, V., et al. (2013). Dietary quebracho tannins are not absorbed, but increase the antioxidant capacity of liver and plasma in sheep. *British Journal of Nutrition* 110: 632–639.

Lorenz, K., and Preston, C.M. (2002). Characterization of high-tannin fractions from humus by carbon-13 cross-polarization and magic-angle spinning nuclear magnetic resonance. *Journal of Environmental Quality* 31: 431–436.

Lorenz, M.M., Eriksson, T., and Udén, P. (2010). Effect of wilting, silage additive, PEG treatment and tannin content on the distribution of N fractions after ensiling of three different sainfoin (*Onobrychis viciifolia*) varieties. *Grass and Forage Science* 65: 175–184.

Lu, Z., Jia, Q., Wang, R., et al. (2011). Hypoglycemic activities of A- and B-type procyanidin oligomer-rich extracts from different Cinnamon barks. *Phytomedicine* 18, 298–302.

Lucci, P., Saurina, J., and Núnez, O. (2017). Trends in LC-MS and LC-HRMS analysis and characterization of polyphenols in food. *Trends in Analytical Chemistry* 88: 1–24.

Luo, L., Cui, Y., Zhang, S., et al. (2017). Detailed phenolic composition of Vidal grape pomace by ultrahigh performance liquid chromatography–tandem mass spectrometry. *Journal of Chromatography B*, 1068–1069, 201–209.

Makkar, H.P.S., Gamble, G., and Becker, K. (1999). Limitation of the butanol-hydrochloric acid-iron assay for bound condensed tannins. *Food Chemistry* 66: 129–133.

Makkar, H.P.S. (2000). *Quantification of tannins in tree foliage: A laboratory manual.* Vienna, FAO, Int. Atomic Energy Agency.

Mané, C., Sommerer, N., Yalcin, T., et al. (2007). Assessment of the molecular weight distribution of tannin fractions through MALDI-TOF MS analysis of protein-tannin complexes. *Analytical Chemistry* 79: 2239–2249.

Mari, A., Eletto, D., Pizza, C., et al. (2013). Integrated mass spectrometry approach to profile proanthocyanidins occurring in food supplements: analysis of *Potentilla erecta* L. rhizomes. *Food Chemistry* 141: 4171–4178.

Mateos-Martın, M.L., Perez-Jimenez, J., Fuguet, E., and Torres, J.L. (2012). Non-extractable proanthocyanidins from grapes are a source of bioavailable (epi) catechin and derived metabolites in rats. *British Journal of Nutrition* 108: 290–297.

Matthews, S., Mila, I., Scalbert, A., et al. (1997). Method for estimation of proanthocyanidins based on their acid depolymerization in the presence of nucleophiles. *Journal of Agricultural and Food Chemistry* 45: 1195–1201.

Maxon, E.D., and Rooney, L.W. (1972). Evaluation of methods for tannin analysis in sorghum grain. *Cereal Chemistry* 49: 719–729.

Melone, F., Saladino, R., Lange, H., and Crestini, C. (2013a). Tannin structural elucidation and quantitative ^{31}P NMR analysis. 1. Model compounds. *Journal of Agricultural and Food Chemistry* 61: 9307–9315.

Melone, F., Saladino, R., Lange, H., and Crestini, C. (2013b). Tannin structural elucidation and quantitative ^{31}P NMR analysis. 2. Hydrolyzable tannins and proanthocyanidins. *Journal of Agricultural and Food Chemistry* 61: 9316–9324.

Mena, P., Calani, L., Bruni, R., and Del Rio, D. (2015). Bioactivation of high-molecular-weight polyphenols by the gut microbiome. In: *Diet-Microbe Interactions in the Gut* (ed. K.T.D. Rio), 73–101. San Diego: Academic Press.

Min, B.R., Barry, T.N., Attwood, G.T., and McNabb, W.C. (2003). The effect of condensed tannins on the nutrition and health of ruminants fed fresh temperate forages: a review. *Animal Feed Science and Technology* 106: 3–19.

Monagas, M., Urpi-Sarda, M., Sánchez-Patán, F., et al. (2010). Insights into the metabolism and microbial biotransformation of dietary flavan-3-ols and the bioactivity of their metabolites. *Food & Function* 1: 233–253.

Montero, L., Herrero, M., Prodanov, M., et al. (2013). Characterization of grape seed procyanidins by comprehensive two-dimensional hydrophilic interaction x reversed phase liquid chromatography coupled to diode array detection and tandem mass spectrometry. *Analytical and Bioanalytical Chemistry* 405: 4627–4638.

Mora-Ortiz, M., Swain, M.T., Vickers, M.J., et al. (2016). De-novo transcriptome assembly for gene identification, analysis, annotation, and molecular marker discovery in *Onobrychis viciifolia*. *BMC Genomics* 17: 756.

Mouls, L., Hugouvieux, V., Mazauric, J-P., et al. (2014). How to gain insight into the polydispersity of tannins: A combined MS and LC study. *Food Chemistry* 165: 348–353.

Mouls, L., and Fulcrand, H. (2012). UPLC-ESI-MS study of the oxidation markers released from tannin depolymerization: toward a better characterization of the tannin evolution over food and beverage processing. *Journal of Mass Spectrometry* 47: 1450–1457.

Mouls, L., and Fulcrand, H. (2015). Identification of new oxidation markers of grape-condensed tannins by UPLC-MS analysis after chemical depolymerization. *Tetrahedron* 71: 3012–3019.

Mouls, L., Mazauric, J.-P., Sommerer, N., et al. (2011). Comprehensive study of condensed tannins by ESI mass spectrometry: average degree of polymerisation and polymer distribution determination from mass spectra. *Analytical and Bioanalytical Chemistry* 400: 613–623.

Mueller-Harvey, I., Bee, G., Dohme-Meier, F., et al. (2019). Benefits of condensed tannins in forage legumes fed to ruminants: importance of structure, concentration and diet composition. *Crop Science* 59, 861–885.

Mueller-Harvey I., Martin-Lorenzo M, Stringano E., et al. (2011). NIR spectroscopy for predicting the nutritional, anthelmintic and environmental effects of sainfoin. *Proceedings of the 8th International Symposium on the Nutrition of Herbivores (ISNH8)*, Aberystwyth, UK (6–9 Sep 2011). Cambridge University Press.

Mueller-Harvey, I., Osterrieder, A., Botchway, S.W., et al. (2014). Excited state fluorescence lifetimes of flavanols and tannins for imaging purposes. *Proceedings of the XXVIIth International Conference on Polyphenols (ICP2014)*, Nagoya, Japan (2–6 Sep 2014).

Naumann, H.D., Hagerman, A.E., Lambert, B.D., et al. (2014). Molecular weight and protein-precipitating ability of condensed tannins from warm-season perennial legumes. *Journal of Plant Interactions* 9: 212–219.

Naumann, H.D., Tedeschi, L.O., Hagerman, A.E., et al. (2013). Methane emission and protein precipitating ability of condensed tannins from warm-season perennial legumes. In: *Energy and Protein Metabolism and Nutrition in Sustainable Animal Production* (eds. J.W. Oltjen, E. Kebreab, and H. Lapierre), 491–492. European Association of Animal Production, EAAP publication No. 134. Wageningen, The Netherlands: Wageningen Academic Publishers.

Naumann, H.D., Tedeschi, L.O., Zeller, W.E., and Huntley, N.F. (2017). The role of condensed tannins in ruminant animal production: Advances, limitations and future directions. *Revista Brasileira de Zootecnia* 46: 929–949.

Newman, R.H., Porter, L.J., Foo, L.Y., et al. (1987). High-resolution ^{13}C NMR studies of proanthocyanidin polymers (condensed tannins). *Magnetic Resonance in Chemistry* 25: 118–124.

Nitao, J.K., Birr, B.A., Nair, M.G., et al. (2001). Rapid quantification of proanthocyanidins (condensed tannins) with a continuous flow analyzer. *Journal of Agricultural and Food Chemistry* 49: 2207–2214.

Nonaka, G., Hashimoto, F., and Nishioka, I. (1986). Tannins and related compounds. XXXVI. Isolation and structures of theaflagallins, new red pigments from black tea. *Chemical and Pharmaceutical Bulletin* 34: 61–65.

Orejola, J., Matsuo, Y., Saito, Y., and Tanaka, T. (2017). Characterization of proanthocyanidin oligomers of *Ephedra sinica*. *Molecules* 22: 1308.

Ossipova, S., Ossipov, V., Haukioja, E., et al. (2001). Proanthocyanidins of mountain birch leaves: quantification and properties. *Phytochemical Analysis* 12: 128–133.

Ou, K., and Gu, L., (2014). Absorption and metabolism of proanthocyanidins. *Journal of Functional Foods* 7: 43–53.

Pappas, C., Kyraleou, M., Voskidi, E., et al. (2015). Direct and simultaneous quantification of tannin mean degree of polymerization and percentage of galloylation in grape seeds using Diffuse Reflectance Fourier Transform-Infrared Spectroscopy. *Journal of Food Science* 80: C298–C306.

Parfitt, R.L., and Newman, R.H. (2000). ^{13}C NMR study of pine needle decomposition. *Plant and Soil* 219: 273–278.

Park, J.-S., Park, M.-K., Oh, H.-J., et al. (2012). Grape-seed proanthocyanidin extract as suppressors of bone destruction in inflammatory autoimmune arthritis. *PLoS ONE*, 7 (12), e51377.

Patra, A.K. and Saxena, J. (2010). A new perspective on the use of plant secondary metabolites to inhibit methanogenesis in the rumen. *Phytochemistry* 71: 1198–1222.

Pedan, V., Weber, C., Do, T., et al. (2018). HPTLC fingerprint profile analysis of cocoa proanthocyanidins depending on origin and genotype. *Food Chemistry* 267: 277–287.

Pellikaan, W.F., Stringano, E., Leenaars, J., et al. (2011). Evaluating effects of tannins on extent and rate of *in vitro* gas and CH_4 production using an automated pressure evaluation system (APES). *Animal Feed Science and Technology*, 166–167, 377–390.

Peng, J., Li, K., Zhu, W., et al. (2018). Separation and purification of four phenolic compounds from persimmon by high-speed counter-current chromatography. *Journal of Chromatography B* 1072: 78–85.

Pérez-Ramírez, I.F., Reynoso-Camacho, R., Saura-Calixto, F., and Pérez-Jiménez, J. (2018). Comprehensive characterization of extractable and nonextractable phenolic compounds by high-performance liquid chromatography–electrospray ionization–quadrupole time-of-flight of a grape/pomegranate pomace dietary supplement. *Journal of Agricultural and Food Chemistry* 66: 661–673.

Pérez-Jiménez, J., Arranz, S., and Saura-Calixto, F. (2009). Proanthocyanidin content in foods is largely underestimated in the literature data: An approach to quantification of the missing proanthocyanidins. *Food Research International* 42: 1381–1388.

Pérez-Jiménez, J., M. Díaz-Rubio, E., and Saura-Calixto, F. (2013). Non-extractable polyphenols, a major dietary antioxidant: occurrence, metabolic fate and health effects. *Nutrition Research Reviews* 26: 118–129.

Pérez-Jiménez, J., and Saura-Calixto, F. (2015). Macromolecular antioxidants or non-extractable polyphenols in fruit and vegetables: Intake in four European countries. *Food Research International* 74: 315–323.

Perret, C., Roger Pezet, R., and Tabacchi, R. (2003). Fractionation of grape tannins and analysis by matrix-assisted laser desorption/ionisation time-of-flight mass spectrometry. *Phytochemical Analysis* 14: 202–208.

Phansalkar, R.S., Nam, J.-W., Chen, S.-N., et al. (2018). Centrifugal partition chromatography enables selective enrichment of trimeric and tetrameric proanthocyanidins for biomaterial development. *Journal of Chromatography A* 1535: 55–62.

Phillips, M.M., Sharpless, K.E., and Wise, S.A. (2013). Standard reference materials for food analysis. *Analytical and Bioanalytical Chemistry* 405: 4325–4335.

Piluzza, G., Sulas, L., and Bullitta, S. (2013). Tannins in forage plants and their role in animal husbandry and environmental sustainability: a review. *Grass and Forage Science* 69: 32–48.

Pizzi, A., and Stephanou, A. (1993). A comparative ^{13}C NMR study of polyflavonoid tannin extracts for phenolic polycondensates. *Journal of Applied Polymer Science* 50: 2105–2113.

Plumb, G.W., de Pascual-Teresa, S., Santos-Buelga, C., et al. (1998). Antioxidant properties of catechins and proanthocyanidins: Effect of polymerization, galloylation and glycosylation. *Free Radical Research* 19: 351–358.

Porter, L.J. (1988). Flavans and proanthocyanidins. In: *The Flavonoids: Advances in Research since 1980* (ed. J.B. Harborne), 21–62. Boston: Springer.

Porter, L.J., Hristich, L.N., and Chan, B.G. (1986). The conversion of procyanidins and prodelphinidins to cyanidin and delphinidin *Phytochemistry* 25: 223–230.

Prior, R.L.; Fan, E.; Ji, H., et al. (2010). Multi-laboratory validation of a standard method for quantifying proanthocyanidins in cranberry powders. *Journal of the Science of Food and Agriculture* 90: 1473–1478.

Qa'dan, F., Petereit, F., Lechtenberg, M., and Nahrstedt, A. (2006). Qualitative determination of proanthocyanidin cleavage products after acid-catalyzed degradation in the presence of excess phloroglucinol by cyclodextrin-modified micellar electrokinetic chromatography. *Pharmaceutical Biology* 44: 309–314.

Qa'dan, F., Nahrstedt, A., Schmidt, M., and Mansoor, K. (2010). Polyphenols from *Ginkgo biloba*. *Scientia Pharmaceutica* 78: 897–907.

Quijada, J., Drake, C., Gaudin, E., et al. (2018). Condensed tannin changes along the digestive tract in lambs fed with sainfoin pellets or hazelnut skins. *Journal of Agricultural and Food Chemistry* 66: 2136–2142.

Quijada, J., Fryganas, C., Ropiak, H.M., et al. (2015). Anthelmintic activities against *Haemonchus contortus* or *Trichostrongylus colubriformis* from small ruminants are influenced by structural features of condensed tannins. *Journal of Agricultural and Food Chemistry* 63: 6346–6354.

Ramsay, A., Brown, L., Luu, T., and Mueller-Harvey, I. (2014). Comparison between Hydrodynamic-CCC and Sephadex gel-chromatography to resolve highly polar, complex mixtures of polymeric tannin (polyphenol) molecules in sainfoin. *Proceedings of the 8th International Conference on Countercurrent Chromatography*, London, UK (23–25 July 2014).

Ramsay, A., Drake, C., Grosse Brinkhaus, A., et al. (2015). NaOH enhances extractability and analysis of proanthocyanidins in ensiled sainfoin (*Onobrychis viciifolia*). *Journal of Agricultural and Food Chemistry* 63: 9471–9479.

Ramsay, A., Williams, A.R., Thamsborg, S.M., and Mueller-Harvey, I. (2016). Galloylated proanthocyanidins from shea (*Vitellaria paradoxa*) meal have potent anthelmintic activity against *Ascaris suum*. *Phytochemistry* 122: 146–153.

Rebello, L.P.G., Lago-Vanzela, E.S., Barcia, M.T., et al. (2013). Phenolic composition of the berry parts of hybrid grape cultivar BRS Violeta (BRS Rubea × IAC 1398-21). using HPLC–DAD–ESI-MS/MS. *Food Research International* 54: 354–366.

Reed, J.D., Krueger, C.G., and Vestling, M.M. (2005). MALDI-TOF mass spectrometry of oligomeric food polyphenols. *Phytochemistry* 66: 2248–2263.

Reid, D.G., Bonnet, S.L., Kemp, G., and van der Westhuizen, J.H. (2013). Analysis of commercial proanthocyanidins. Part 4: Solid state 13C NMR as a tool for in situ analysis of proanthocyanidin tannins, in heartwood and bark of quebracho and acacia, and related species. *Phytochemistry* 94: 243–248.

Rhodes, D.H., Hoffmann, Jr., L., Rooney, W.L., et al. (2014). Genome-wide association study of grain polyphenol concentrations in global sorghum [*Sorghum bicolor* (L.) Moench] germplasm. *Journal of Agricultural and Food Chemistry* 62: 10916–10927.

Riaz, M.Q., Südekum, K.H., Clauss, M., and Jayanegara, A. (2014). Voluntary feed intake and digestibility of four domestic ruminant species as influenced by dietary constituents: A meta-analysis. *Livestock Science* 162: 76–85.

Rigaud, J., Perez-Ilzarbe, J., da Silva, J.M.R., and Cheynier, V. (1991). Micro method for the identification of proanthocyanidin using thiolysis monitored by high-performance liquid chromatography. *Journal of Chromatography A* 540: 401–405.

Robbins, R.J., Leonczak, J., Johnson, J.C., et al. (2009). Method performance and multi-laboratory assessment of a normal phase high pressure liquid chromatography–fluorescence detection method for the quantitation of flavanols and procyanidins in cocoa and chocolate containing samples. *Journal of Chromatography A* 1216: 4831–4840.

Romer, F.H., Underwood, A.P., Senekal, N.D., et al. (2011). Tannin fingerprinting in vegetable tanned leather by solid state NMR spectroscopy and comparison with leathers tanned by other processes. *Molecules* 16: 1240–1252.

Ropiak, H.M., Lachmann, P., Ramsay, A., et al. (2017). Identification of structural features of condensed tannins that affect protein aggregation. *PLoS ONE* 12 (1): e0170768.

Ropiak, H.M., Ramsay, A., and Mueller-Harvey, I. (2016). Condensed tannins in extracts from European medicinal plants and herbal products. *Journal of Pharmaceutical and Biomedical Analysis* 121: 225–231.

Rouméas, L., Billerach, G., Aouf, C., et al. (2018). Furylated flavonoids: fully biobased building blocks produced by condensed tannins depolymerization. *ACS Sustainable Chemistry and Engineering* 6, 1112–1120.

Salaritabar, A., Darvishi, B., Hadjiakhoondi, F., et al. (2017). Therapeutic potential of flavonoids in inflammatory bowel disease: A comprehensive review. *World Journal of Gastroenterology* 23: 5097–5114.

Salminen, J.-P. (2018). Two-dimensional tannin fingerprints by liquid chromatography tandem mass spectrometry offer a new dimension to plant tannin analyses and help to visualize the tannin diversity in plants. *Journal of Agricultural and Food Chemistry* 66: 9162–9171.

Saura-Calixto, F., Perez-Jimenez, J., Tourino, S., et al. (2010). Proanthocyanidin metabolites associated with dietary fibre from in vitro colonic fermentation and proanthocyanidin metabolites in human plasma. *Molecular Nutrition and Food Research* 54: 939–946.

Schofield, P., Mbugua, D.M., and Pell, A.N. (2001). Analysis of condensed tannins: a review. *Animal Feed Science and Technology* 91: 21–40.

Seddon, T.J., and Downey, M.O. (2008). Comparison of analytical methods for the determination of condensed tannins in grape skin. *Australian Journal of Grape and Wine Research* 14: 54–61.

Selmi, C., Mao, T.K., Keen, C.L., et al. (2006). The antiinflammatory properties of cocoa flavanols. *Journal of Cardiovascular Pharmacology*, 47 (2): S163–S171.

Shoji, T., Mutsuga, M., Nakamura, T., et al. (2003). Isolation and structural elucidation of some procyanidins from apple by low-temperature nuclear magnetic resonance, *Journal of Agricultural and Food Chemistry* 51: 3806–3813.

Spranger, I., Sun, B., Mateus, A.M., et al. (2008). Chemical characterization and antioxidant activities of oligomeric and polymeric procyanidin fractions from grape seeds. *Food Chemistry* 108: 519–532.

Stewart, J.L., Mould, F., and Mueller-Harvey, I. (2000). The effect of drying treatment on the fodder quality and tannin content of two provenances of *Calliandra calothyrsus* Meissner. *Journal of the Science of Food and Agriculture* 80: 1461–1468.

Steynberg, P.J., Bezuidenhoudt, B.C.B., Ferreira, D., and Steynberg, J.P. (1994). Cleavage of the interflavanyl bond in 5-deoxy (A-ring) proanthocyanidins. *Journal of the Chemical Society, Chemical Communications*, 31–32.

Stodt, U., and Engelhardt, U.H. (2013). Progress in the analysis of selected tea constituents over the past 20 years. *Food Research International* 53: 636–648.

Stringano, E., Cramer, R., Hayes, W., et al. (2011). Deciphering the complexity of sainfoin (*Onobrychis viciifolia*) proanthocyanidins by MALDI-TOF mass spectrometry with a judicious choice of isotope patterns and matrices. *Analytical Chemistry* 83: 4147–4153.

Sun, P., Li, K., Wang, T., et al. (2019). Procyanidin C1, a component of cinnamon extracts, is a potential insulin sensitizer that targets adipocytes. *Journal of Agricultural and Food Chemistry* 67: 8839–8846.

Sun, X., Jin, Z., Yang, L., et al. (2013). Ultrasonic-assisted extraction of procyanidins using ionic liquid solution from *Larix gmelinii* bark. *Journal of Chemistry* 2013. doi: 10.1155/2013/541037.

Tanaka, Y., Yanagida, A., Komeya, S., et al. (2014). Comprehensive separation and structural analyses of polyphenols and related compounds from bracts of hops (*Humulus lupulus* L.). *Journal of Agricultural and Food Chemistry* 62: 2198–2206.

Tarascou, I., Barathieu, K., Simon, C., et al. (2006). A 3D structural and conformational study of procyanidin dimers in water and hydro-alcoholic media as viewed by NMR and molecular modeling. *Magnetic Resonance in Chemistry* 44: 868–880.

Taylor, A.W., Barofsky, E., Kennedy, J.A., and Deinzer, M.L. (2003). Hop (*Humulus lupulus* L.) proanthocyanidins characterized by mass spectrometry, acid catalysis, and gel permeation chromatography. *Agricultural and Food Chemistry* 51: 4101–4110.

Terra, X., Valls, J., Vitrac, X., et al. (2007). Grape-seed procyanidins act as antiinflammatory agents in endotoxin-stimulated RAW 264.7 macrophages by inhibiting NFkB signaling pathway. *Journal of Agricultural and Food Chemistry* 55: 4357–4365.

Terrill, T.H., Miller, J.E., Burke, J.M., et al. (2012). Experiences with integrated concepts for the control of *Haemonchus contortus* in sheep and goats in the United States. *Veterinary Parasitology* 186: 28–37.

Terrill, T.H., Rowan, A.M., Douglas, G.B., and Barry, T.N. (1992). Determination of extractable and bound condensed tannin concentrations in forage plants, protein concentrate meals and cereal grains. *Journal of the Science of Food and Agriculture* 58: 321–329.

Terrill, T.H., Waghorn, G.C., Woolleg, D.J., et al. (1994). Assay and digestion of ^{14}C-labelled condensed tannin in the gastro intestinal tract of sheep. *British Journal of Nutrition* 72: 467–477.

Theodoridou, K., Aufrère, J., Andueza, D., et al. (2010). Effects of condensed tannins in fresh sainfoin (*Onobrychis viciifolia*) on *in vivo* and *in situ* digestion in sheep. *Animal Feed Science and Technology* 160: 23–38.

Toden, S., Ravindranathan, P., Gu, J., et al. (2018). Oligomeric proanthocyanidins (OPCs) target cancer stem-like cells and suppress tumor organoid formation in colorectal cancer. *Scientific Reports* 8: 3335. doi: 10.1038/s41598-018-21478-8.

Totlani, V.M., and Peterson, D.G. (2006). Epicatechin carbonyl-trapping reactions in aqueous Maillard systems: Identification and structural elucidation. *Journal of Agricultural and Food Chemistry* 54: 7311–7318.

van Beek, T.A., and Montoro, P. (2009). Chemical analysis and quality control of Ginkgo biloba leaves, extracts, and phytopharmaceuticals. *Journal of Chromatography A* 1216: 2002–2032.

van Dooren, I., Foubert, K., Theunis, M., et al. (2018). Advantages of a validated UPLC–MS/MS standard addition method for the quantification of A-type dimeric and trimeric proanthocyanidins in cranberry extracts in comparison with well-known quantification methods. *Journal of Pharmaceutical and Biomedical Analysis* 148: 32–41.

Van Huynh, A., and Bevington, J. (2014). MALDI-TOF MS analysis of proanthocyanidins in two lowland tropical forest species of *Cecropia*: a first look at their chemical structures. *Molecules* 19: 14484–14495.

Venter, P.B., Senekal, N.D., Kemp, G., et al. (2012a) Analysis of commercial proanthocyanidins. Part 3: The chemical composition of wattle (*Acacia mearnsii*) bark extract. *Phytochemistry* 83: 153–167.

Venter, P.B., Sisa, M., van der Merwe, M.J., et al. (2012b) Analysis of commercial proanthocyanidins. Part 1: The chemical composition of quebracho (*Schinopsis lorentzii* and *Schinopsis balansae*) heartwood extract. *Phytochemistry* 73: 95–105.

Verardo, V., Cevoli, C., Pasini, F., et al. (2015). Analysis of oligomer proanthocyanidins in different barley genotypes using high-performance liquid chromatography–fluorescence detection–mass spectrometry and near-infrared methodologies. *Journal of Agricultural and Food Chemistry* 63: 4130–4137.

Vernhet, A., Dubascoux, S., Cabane, B., et al. (2011). Characterization of oxidized tannins: comparison of depolymerization methods, asymmetric flow field-flow fractionation and small-angle X-ray scattering. *Analytical and Bioanalytical Chemistry* 401: 1559–1569.

Vidal, C.M.P., Zhu, W., Manohar, S., et al. (2016). Collagen-collagen interactions mediated by plant-derived proanthocyanidins: A spectroscopic and atomic force microscopy study. *Acta Biomaterialia* 41: 110–118.

Vu, H.T., Scarlett, C.J., and Vuong, Q.V. (2017). Optimization of ultrasound-assisted extraction conditions for recovery of phenolic compounds and antioxidant capacity from banana (*Musa cavendish*) peel. *Journal of Food Processing and Preservation*, 41: e13148. doi.org/10.1111/jfpp.13148

Waghorn, G. (2008). Beneficial and detrimental effects of dietary condensed tannins for sustainable sheep and goat production – progress and challenges. *Animal Feed Science and Technology* 147: 116–139.

Waghorn, G.C., Douglas, G.B., Niezen, J.H., et al. (1998). Forages with condensed tannins – their management and nutritive value for ruminants. *Proceedings of the New Zealand Grassland Association* 60: 89–98.

Wang, Y., Majak, W., and McAllister, T.A. (2012). Frothy bloat in ruminants: cause, occurrence, and mitigation strategies. *Animal Feed Science and Technology* 172: 103–114.

Wang, Y., Singh, A.P., Hurst, W.J., et al. (2016). Influence of degree-of-polymerization and linkage on the quantification of proanthocyanidins using 4-dimethylaminocinnamaldehyde (DMAC) assay. *Journal of Agricultural and Food Chemistry* 64: 2190–2199.

Wawer, I., Wolniak, M., and Paradowska, K. (2006). Solid state NMR study of dietary fiber powders from aronia, bilberry, black currant and apple. *Solid State Nuclear Magnetic Resonance* 30: 106–113.

Whitacre, E., Oliver, J., Van Den Broek, R., et al. (2003). Predictive analysis of cocoa procyanidins using near-infrared spectroscopy techniques. *Journal of Food Science* 68: 2618–2622.

Williams, A.R., Fryganas, C., Reichwald, K., et al. (2016). Polymerization-dependent activation of porcine γδ T-cells by proanthocyanidins. *Research in Veterinary Science* 105: 209–215.

Wojnicz, D., Sycz, Z., Walkowski, S., et al. (2012). Study on the influence of cranberry extract Żuravit S·O·S® on the properties of uropathogenic *Escherichia coli* strains, their ability to form biofilm and its antioxidant properties. *Phytomedicine* 19: 506–514.

Yu, P., Jonker, A., and Gruber, M. (2009). Molecular basis of protein structure in proanthocyanidin and anthocyanin-enhanced Lc-transgenic alfalfa in relation to nutritive value using synchrotron-radiation FTIR microspectroscopy: A novel approach. *Spectrochimica Acta Part A* 73, 846–853.

Zamora, R., and Hidalgo, F.J. (2018). Carbonyl–phenol adducts: an alternative sink for reactive and potentially toxic lipid oxidation products. *Journal of Agricultural and Food Chemistry* 66: 1320–1324.

Zanchi, D., Narayanan, T., Hagenmuller, D., et al. (2008). Tannin-assisted aggregation of natively unfolded proteins. *EPL* 82: 58001.

Zeller, W.E., Ramsay, A., Ropiak, H.M., et al. (2015) ^{1}H–^{13}C HSQC NMR spectroscopy for estimating procyanidin/prodelphinidin and *cis/trans* flavanol ratios of condensed tannin fractions: correlation with thiolysis. *Journal of Agricultural and Food Chemistry* 63: 1967–1973.

Zeller, W.E. and Schatz, P.F. (2017). The U.S. dairy forage research center (USDFRC) condensed tannin NMR database. *Journal of Agricultural and Food Chemistry* 65: 5104–5106.

Zeller, W.E., Sullivan, M.L., Mueller-Harvey, I., et al. (2015) Protein precipitation behavior of condensed tannins from *Lotus pedunculatus* and *Trifolium repens* with different mean degrees of polymerization. *Journal of Agricultural and Food Chemistry* 63: 1160–1168.

Zhang, A., Li, J., Zhang, S., et al. (2017). Characterization and acid-catalysed depolymerization of condensed tannins derived from larch bark. *RSC Advances* 7: 35135–35146.

Zhang, L.-L., Lin, Y.-M., Zhou, H.-C., et al. (2010). Condensed tannins from mangrove species *Kandelia candel* and *Rhizophora mangle* and their antioxidant activity. *Molecules* 15: 420–431.

Zhou, H.-C., Tam, N. F-y., Lin, Y.-M., et al. (2014). Relationships between degree of polymerization and antioxidant activities: a study on proanthocyanidins from the leaves of a medicinal mangrove plant *Ceriops tagal*. *PLoS ONE* 9 (10): e107606.

Zhou, F-Z., Yan, L., Yin, S.-W., et al. (2018). Development of pickering emulsions stabilized by gliadin/proanthocyanidins hybrid particles (GPHPS) and the fate of lipid oxidation and digestion. *Journal of Agricultural and Food Chemistry* 66: 1461–1471.

Zurita, J., Díaz-Rubio, M.E., and Saura-Calixto, F. (2012). Improved procedure to determine non-extractable polymeric proanthocyanidins in plant foods. *International Journal of Food Sciences and Nutrition* 63: 936–939.

7

Lignin Monomers Derived from the Flavonoid and Hydroxystilbene Biosynthetic Pathways

José C. del Río[1], Jorge Rencoret[1], Ana Gutiérrez[1], Wu Lan[2,3], Hoon Kim[3], and John Ralph[3,4]

[1] *Instituto de Recursos Naturales y Agrobiología de Sevilla (IRNAS), CSIC, Seville, Spain*
[2] *Department of Biological System Engineering, University of Wisconsin–Madison, Madison, USA*
[3] *Department of Energy Great Lakes Bioenergy Research Center, Wisconsin Energy Institute, University of Wisconsin–Madison, Madison, USA*
[4] *Department of Biochemistry, University of Wisconsin–Madison, Madison, USA*

7.1 Lignin Monomers Derived from the Monolignol Biosynthetic Pathway

7.1.1 'Canonical' Monolignols

Lignin has long been considered to be a complex phenylpropanoid polymer derived essentially from the oxidative radical coupling of three *p*-hydroxycinnamyl alcohols differing in their degree of methoxylation, the so-called monolignols: *p*-coumaryl, coniferyl, and sinapyl alcohols (Figure 7.1a). Extensive research has been devoted during the past decades to elucidating the monolignol biosynthetic pathway and the mechanisms of gene regulation, which are now basically clarified (Boerjan et al. 2003; Dixon and Reddy, 2003; Vanholme et al. 2008, 2010, 2019). The biosynthesis of the monolignols starts with the deamination of phenylalanine (and/or tyrosine), which is derived from the shikimate biosynthetic pathway, and involves successive hydroxylation reactions of the aromatic ring, followed by phenolic *O*-methylation and reduction of the side-chain carboxylic acid group to an alcohol (Dixon et al. 2001; Boerjan et al. 2003; Ralph et al. 2004; Liu, 2012; Vanholme et al. 2008, 2010, 2019; Barros et al. 2016). These reactions are catalyzed by different enzymes that are well documented, and include phenylalanine ammonia-lyase (PAL), tyrosine ammonia-lyase (TAL), cinnamate 4-hydroxylase (C4H), 4-coumarate:coenzyme A ligase (4CL), ferulate 5-hydroxylase (F5H), *p*-coumarate 3-hydroxylase (C3H), *p*-hydroxycinnamoyl-CoA:quinate/shikimate hydroxycinnamoyltransferase (HCT), caffeoyl shikimate esterase (CSE), caffeoyl-CoA *O*-methyltransferase (CCoAOMT), caffeoyl-CoA reductase (CCR), caffeic acid *O*-methyltransferase (COMT), and cinnamyl alcohol dehydrogenase (CAD).

Recent Advances in Polyphenol Research, Volume 7, First Edition. Edited by Jess Dreher Reed, Victor Armando Pereira de Freitas, and Stéphane Quideau.

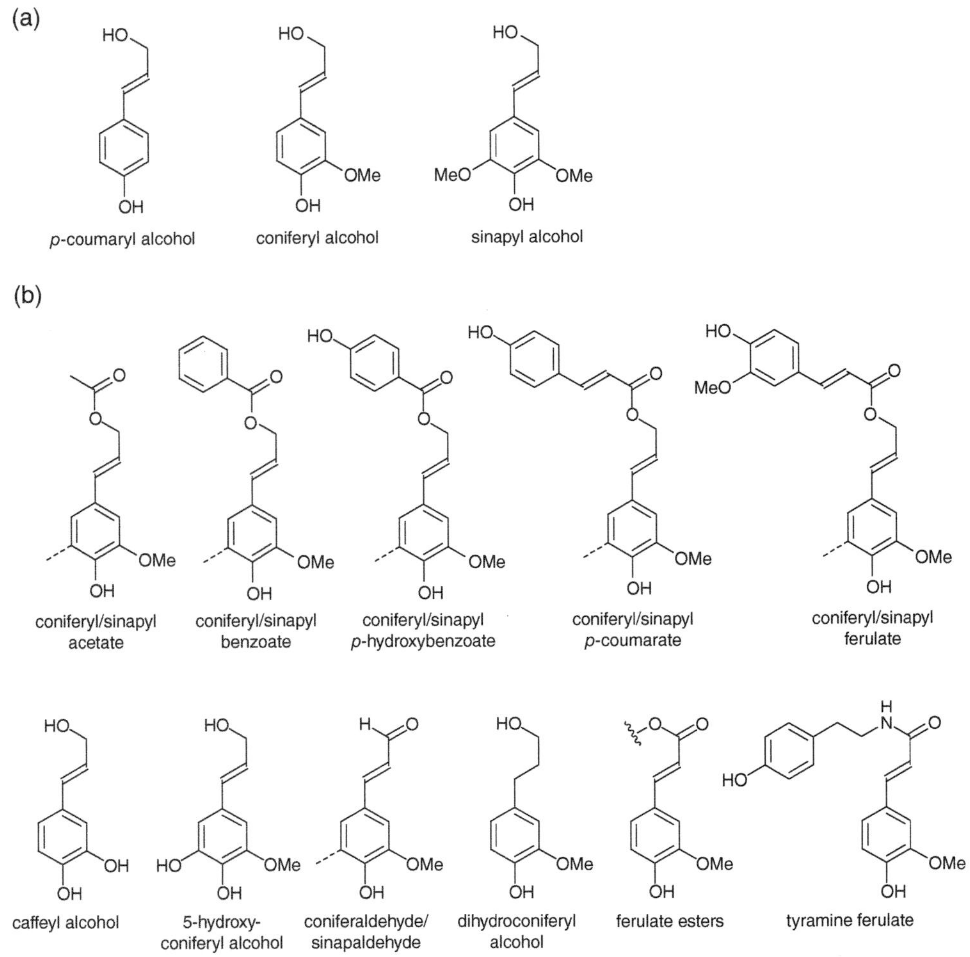

Figure 7.1 (a) Structures of the three canonical monolignols, *p*-coumaryl, coniferyl, and sinapyl alcohols. (b) 'Non-conventional' lignin monomers also derived from the monolignol biosynthetic pathway. Lignin monomers acylated at the γ-OH with acetate, benzoate, *p*-hydroxybenzoate, *p*-coumarate, and ferulate; caffeyl alcohol; 5-hydroxyconiferyl alcohol; hydroxycinnamaldehydes; dihydroconiferyl alcohol; ferulate esters; tyramine ferulate.

After their synthesis, the monolignols are delivered to the cell wall (by largely unknown mechanisms) where they are oxidized in a reaction mediated by peroxidases and/or laccases and then polymerized in a combinatorial fashion by free-radical coupling mechanisms, generating a variety of structures and linkages within the polymer (Boerjan et al. 2003; Ralph et al. 2004, 2019). The most frequent interunit linkage is the β–*O*–4 alkyl-aryl ether; other common linkages and structures in the lignin polymer include β–5 phenylcoumarans, β–β resinols, 5–5 dibenzodioxocins, 5–*O*–4 biphenyl ethers, and β–1 spirodienones. The relative abundance of each of these linkages largely depends on the supply of the individual monolignols to the lignifying zone, which greatly varies among taxa and cell type as well as environmental conditions or growth stage (Terashima et al. 1993; Campbell and Sederoff, 1996; Vermerris and Boon, 2001; Donaldson, 2001; Rencoret et al. 2011; Lourenço et al. 2016).

7.1.2 Other 'Nonconventional' Lignin Monomers

During the last few years, a number of papers have demonstrated that several other phenolic compounds, all deriving from the shikimate-derived monolignol biosynthetic pathway (Figure 7.1b), also behave as lignin monomers in many plants, participating in radical coupling reactions during lignification. This is the case for monolignol ester conjugates with acetates, *p*-hydroxybenzoates, *p*-coumarates, or the newly found ferulate and benzoate analogs, that are also used as lignin monomers in a variety of natural plants (del Río et al. 2007, 2008; Martínez et al. 2008; Ralph, 2010; Lu et al. 2015; Karlen et al. 2016, 2017). Monolignol acetates are ubiquitous in angiosperms and have been found in high amounts (sometimes reaching up to 80% acetylation degree) in the lignins of kenaf, sisal, and abaca, as well as in many hardwoods (Ralph, 1996; Lu and Ralph, 2002; del Río et al. 2004, 2007, 2008); *p*-hydroxybenzoates are widely found in palms, poplars, and willows (Smith, 1955; Landucci et al. 1992; Rencoret et al. 2013, 2018; Lu et al. 2004, 2015) and have also been recently found in high amounts in the lignin of the seagrass *Posidonia oceanica* (Kaal et al. 2018; Rencoret et al. 2020); *p*-coumarates are typical of grasses and are also found in significant levels in other monocots, such as curaua and abaca (Ralph et al. 1994; Lu and Ralph, 1999; Martínez et al. 2008; del Río et al. 2008; Ralph, 2010); monolignol ferulates have recently been described in a large number of plant species (Karlen et al. 2016); likewise, benzoates have also been recently reported in the lignins of some palms, such as date and macaúba palms (Karlen et al. 2017; Rencoret et al. 2018) and C3H and C4H downregulated transgenic *Populus trichocarpa* (Kim et al. 2020). Lignins derived exclusively from caffeyl alcohol were discovered in seed coats of vanilla orchid and in some members of the Cactaceae, Euphorbiaceae, and Cleomaceae families and are almost homopolymers linked through β–*O*–4-coupling and producing chains of benzodioxane structures (Chen et al. 2012, 2013; Tobimatsu et al. 2013), and unusual lignins composed of 5-hydroxyguaiacyl units derived from 5-hydroxyconiferyl alcohol were found in the seeds of three species of Escobaria (Chen et al. 2013). Hydroxycinnamaldehydes, the immediate precursors of monolignols, have been detected in various mutant and transgenic plants (Kim et al. 2003; Lapierre et al. 2004), but are also evidenced at low levels in essentially all lignins; in 'normal' wild-type plants, it remains unclear whether the aldehyde is delivered to the wall as such, i.e., as an intermediate of incomplete monolignol biosynthesis, or if it results from the oxidation of the monolignols in the oxidizing environment required for lignification. Dihydroconiferyl alcohol has also been found in 'all' gymnosperm lignins, and is an abundant component in a CAD-deficient pine mutant (Ralph, Kim et al. 1999; Lapierre et al. 2000). Polysaccharide hydroxycinnamate esters (particularly ferulates) are also incorporated into lignins, principally in grasses, where they appear to act as nucleation sites for lignin polymerization (Ralph et al. 1995; Hatfield et al. 1999). Finally, the amide tyramine ferulate has been found incorporated into the lignins of tobacco plants and presumably other Solanaceae (Ralph et al. 1998), and the related diferuloyl putrescine has recently been identified in the lignin of maize kernels (del Río et al. 2018).

The incorporation of these 'non-conventional' lignin monomers into the lignin polymer generates characteristic and diagnostic structures, such as the tetrahydrofuran structures produced from the β–β coupling of γ-acylated monolignols (Lu and Ralph, 2002, 2005; del Río et al. 2007, 2008; Ralph, 2010; Lu et al. 2004, 2015) or the benzodioxane structures produced from the incorporation of caffeyl and 5-hydroxyconiferyl alcohols (Chen et al. 2012, 2013; Tobimatsu et al. 2013), that provided compelling evidence for their participation in coupling and cross-coupling reactions during lignification. Therefore, these phenolic compounds should be considered to be authentic lignin monomers, in addition to the three canonical monolignols. Indeed, from examination of a large number of lignins, it appears unlikely that any lignin polymer is derived solely from the three monolignols!

In all of the cases above, however, the 'non-conventional' lignin monomers are still derived from the established monolignol biosynthetic pathway – they are either intermediates in the pathway or arise from derivatization reactions of these intermediates or, in the case of the variously acylated monolignols, of the monolignols themselves. Recent investigations have demonstrated that other phenolic compounds derived from outside the classical shikimate-derived monolignol biosynthetic pathway – namely, the flavonoid and hydroxystilbene biosynthetic pathways (i.e., the flavone tricin, the hydroxystilbenes piceatannol, resveratrol, and isorhapontigenin, and the hydroxystilbene glucosides piceid, isorhapontin and astringin; Figure 7.2) – can also participate in radical coupling reactions with monolignols and/or lignin oligomers and become integrally incorporated into the lignin polymer in several plants (del Río et al. 2012, 2017, 2020; Lan et al. 2015; Lan, Morreel et al. 2016; Lan, Rencoret et al. 2016; Lan et al. 2019; Rencoret et al. 2018, 2019; Neiva et al. 2020). These phenolic compounds should therefore be also considered as authentic lignin monomers participating in lignification.

7.2 Flavonoid and Hydroxystilbene Biosynthetic Pathways

Flavonoids and hydroxystilbenes are metabolic hybrids as they derive from a combination of the shikimate-derived phenylpropanoid and the acetate/malonate-derived polyketide pathways (Figure 7.3). Flavonoids and hydroxystilbenes are synthesized from a coenzyme A (CoA) activated hydroxycinnamic acid unit and three malonyl-CoA extender units. The first step is the deamination of L-phenylalanine or tyrosine catalyzed by phenylalanine ammonia-lyase (PAL) or tyrosine ammonia-lyase (TAL) to *trans*-cinnamic or *p*-coumaric acids, respectively. The cinnamic acid is hydroxylated by cinnamate-4-hydroxylase (C4H) to produce *p*-coumaric acid, which is then activated by *p*-coumarate:CoA ligase (4CL) to

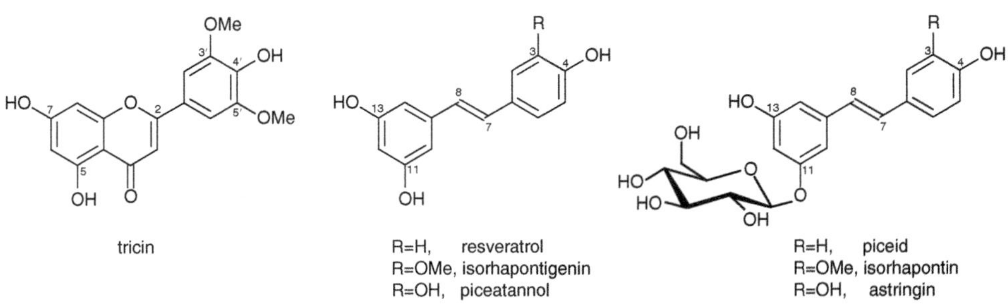

Figure 7.2 Phenolic compounds derived from the flavonoid (tricin) and hydroxystilbene (resveratrol, isorhapontigenin, and piceatannol, and their respective glucosides piceid, isorhapontin, and astringin) biosynthetic pathways that have been found to act as true lignin monomers in several plants. *Source:* del Río et al. 2012, 2017, 2020; Lan et al. 2015; Rencoret et al. 2019.

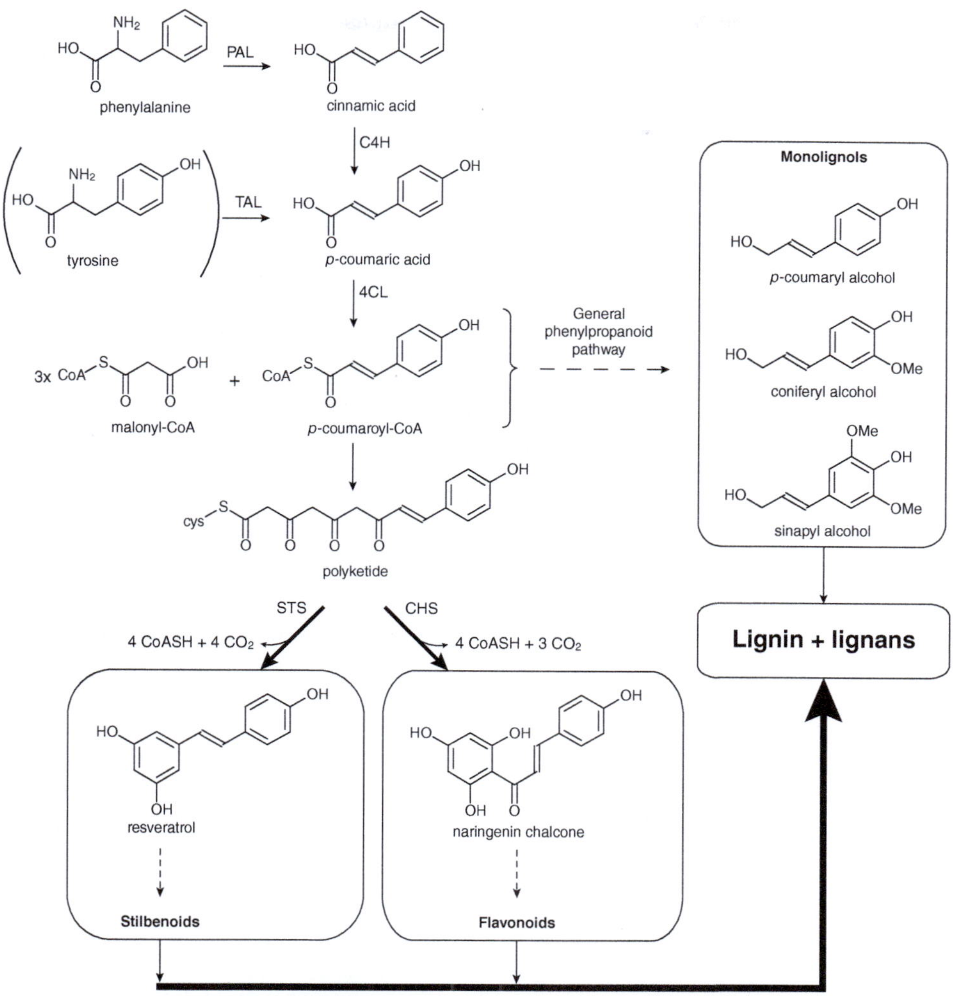

Figure 7.3 Simplified scheme of the general polyphenolics biosynthetic pathways leading to the biosynthesis of monolignols, flavonoids, and stilbenes. PAL, phenylalanine ammonia-lyase; TAL, tyrosine ammonia-lyase; C4H, cinnamate 4-hydroxylase; 4CL, 4-coumarate:CoA ligase; STS, stilbene synthase; CHS, chalcone synthase.

produce *p*-coumaroyl-CoA, which is also the entry point for the production of monolignols intended for lignin (and lignans). A polyketide synthase then sequentially adds three acetate extender units, derived from malonyl-CoA, to an activated *p*-coumaroyl-CoA unit. Depending on the polyketide synthase activity, chalcone synthase (CHS; EC 2.3.1.74) or stilbene synthase (STS; EC 2.3.1.95), cyclization of the generated tetraketide intermediate results in the production of either a chalcone or a stilbene (Watts et al. 2006). Whereas CHS cyclizes the polyketide intermediate by an intramolecular Claisen condensation, cyclization by STS involves an aldol condensation accompanied by an additional decarboxylation leading to the loss of one carbon as CO_2 (Austin and Noel, 2003). Flavonoids include different classes of compounds, such as flavones, flavanones, flavonols, isoflavones, anthocyanins, chalcones, aurones, xanthones, and flavanols (Andersen and Markham, 2006; Quideau et al. 2011). Simple hydroxystilbenes, however, have a more restricted number of compounds that consist of two phenolic moieties joined by an ethylene bridge and, similarly to monolignols, are particularly prone to participating in oxidative coupling reactions to produce dimers and higher oligomers (Lin and Yao, 2006; Quideau et al. 2011; Rivière et al. 2012; Keylor et al. 2015).

7.3 Radical Coupling of Flavonoids and Hydroxystilbenes with Monolignols – Flavonolignans and Stilbenolignans

Flavonoids and hydroxystilbenes are known to participate in oxidative radical cross-coupling reactions with monolignols to produce 'nonconventional' lignans (termed flavonolignans and stilbenolignans) that have two phenylpropanoid units linked together through a diversity of linkages (Begum et al. 2010; Chambers et al. 2015). Although lignification and lignan formation are distinct processes separated in time and space (Ralph, Peng et al. 1999; Umezawa, 2003), the wide natural occurrence in plants of these types of hybrid compounds arising from the cross-coupling with monolignols is an indication that at least some members of the flavonoids and hydroxystilbenes might also be compatible with lignification.

7.3.1 Flavonolignans

Flavonolignans have a broad structural diversity and include a wide variety of compounds that present different linkages and structure types (Figure 7.4), including simple 4′–*O*–β″ ethers (as in aegecin and salcolins A and B), benzodioxane structures (as in silybins A and B, dehydrosilybin, silandrins A and B, or hydnocarpin D), phenylcoumaran structures (as in silychristins A and B, dehydrosilychristin, silyhermin, and neosilyhermins A and B); cyclohexane bridges (as in neohydnocarpin), and tricyclic ketones (as in silydianin) (Begum et al. 2010). Different classes of flavonoids belonging to the families of flavanonols, flavonols, flavanones, and flavones have been found coupling with monolignols in several flavonolignans (Begum et al. 2010; Chang et al. 2010; Chambers et al. 2015). However, and despite the high number of flavonoids identified in plants, only a limited number of them (taxifolin, quercetin, eriodictyol, dihydrotricin, apigenin, luteolin, selgin, and tricin) have been reported to form flavonolignans (Figure 7.4a). The flavanonol taxifolin is the most important one, and was found in silybin, silychristin, isosilybin, isosilychristin, silydianin, and other flavonolignans isolated from milk thistle (Lee and Liu, 2003; Csupor et al. 2016; AbouZid et al. 2016, 2017; Martinelli et al. 2017). The flavonol quercetin was found in dehydrosilybin and dehydrosilychristin (Begum et al. 2010). The flavanone eriodictyol was found in neosilyhermin, silandrins A and B, silyhermin, and neosilyhermin, also isolated from milk thistle (Csupor et al. 2016), whereas the flavanone dihydrotricin was present in the calquiquelignans A and B from rattan palm (Chang et al. 2010). The flavone apigenin was found in lignoside and isolignoside (Begum et al. 2010); another flavone, luteolin, was present in hydnocarpin, hydnocarpin D, hydnowightin, isohydnocarpin, neohydnocarpin and sinaicitin (Begum et al. 2010), whereas selgin was found in palstatin, 5′-hydroxyhydnocarpin and 5′-hydroxyhydnocarpin D (Chambers et al. 2015), and tricin was found in aegecin and salcolins A and B as well as in their acetylated and *p*-coumaroylated γ-carbon

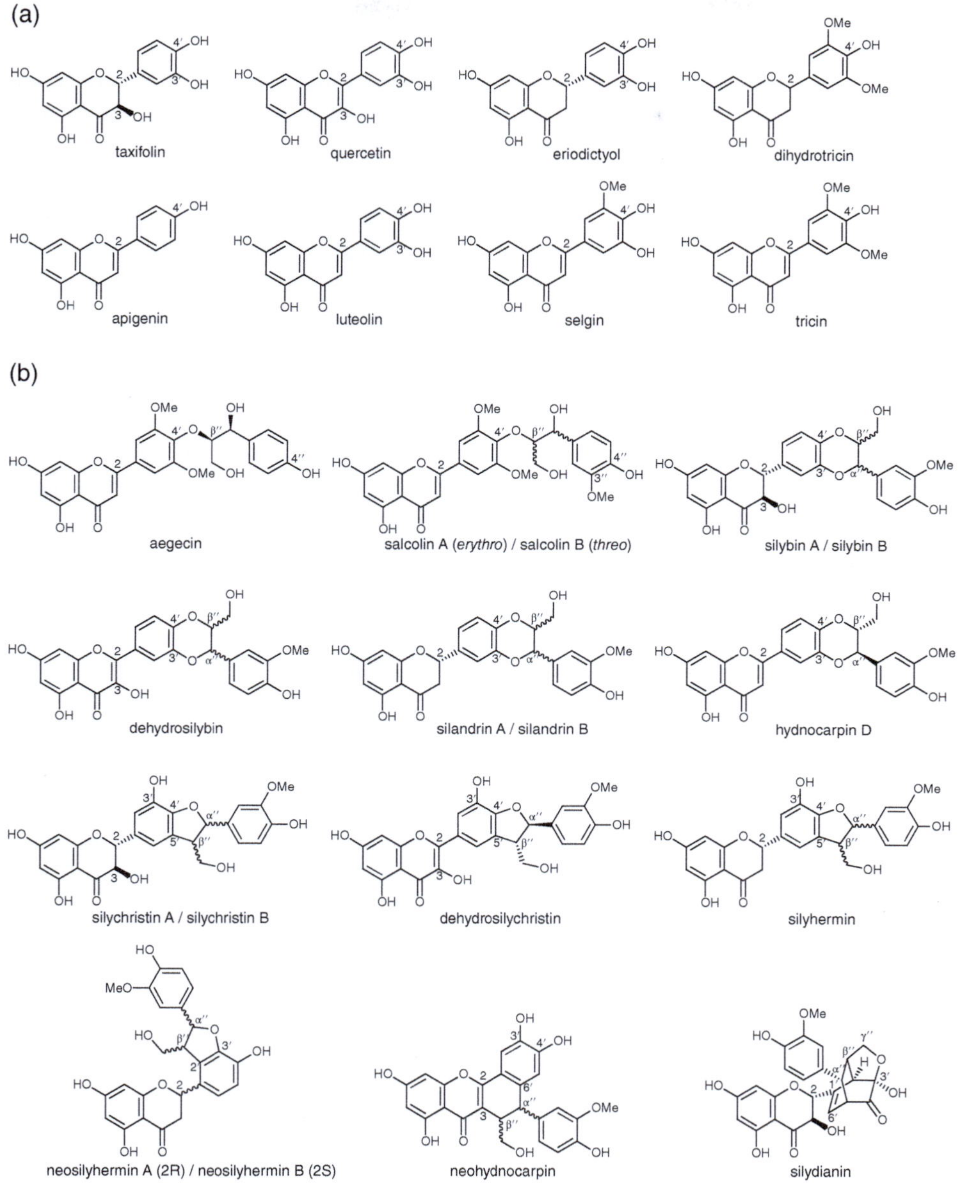

Figure 7.4 (a) Structures of the flavonoids that are known to form flavonolignans. (b) Examples of some flavonolignans identified in plants; all arise from putative radical coupling between the flavonoid and a monolignol.

monolignol conjugates in several kinds of grasses (Wenzig et al. 2005; Begum et al. 2010; Chang et al. 2010; Lee et al. 2015).

Flavonolignans are formed by oxidative radical coupling reactions generating a wide variety of regio- and stereoisomers and, like the lignans themselves, may involve proteins to direct stereospecific coupling (Davin et at., 1997; AbouZid et al. 2017). The coupling of flavonoids with monolignols may occur at different positions. Whereas tricin and dihydrotricin can only form 4–*O*–β linkages with monolignols, other flavonoids (eriodictyol, taxifolin, quercetin, apigenin, luteolin, or selgin) can form additional linkages. For example, silybin is formed by oxidative radical coupling between the β-position of coniferyl alcohol and the 4′-OH of ring-B of taxifolin, followed by internal trapping of the quinone methide intermediate by the 3′-OH group of taxifolin forming a benzodioxane ring. Silychristin, on the other hand, is formed by radical coupling of coniferyl alcohol (at its β-position) with the C-5′ position of taxifolin followed by a subsequent α–*O*–4′ bonding during rearomatization producing a phenylcoumaran structure. These radical coupling reactions produce different pairs of diastereomers, such as silybin A and B, and silychristin A and B, and the same occur with other flavonolignans (Lee and Liu, 2003). All of these examples of flavonolignans illustrate the compatibility of flavonoids with lignification.

7.3.2 Stilbenolignans

Stilbenolignans are similarly produced from cross-coupling of monolignols with the phenylpropanoid moiety of hydroxystilbenes through different linkage types, as shown in Figure 7.5. These structure types commonly include benzodioxanes (as in aiphanol and 5-hydroxyaiphanol), phenylcoumarans (as in gnetofuran A and gnetucleistol F), or

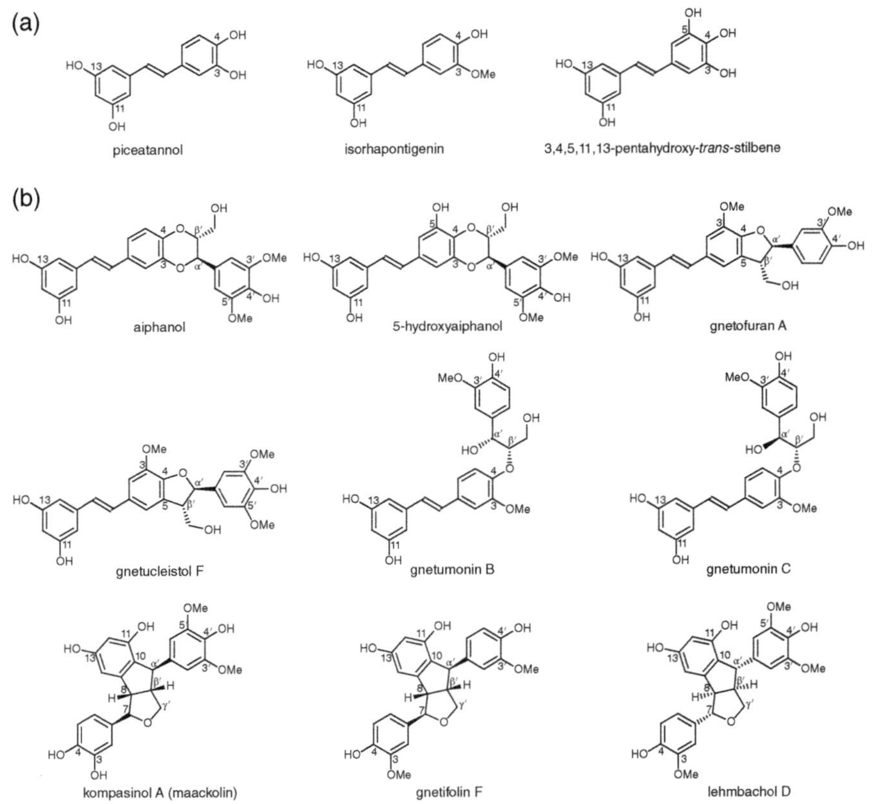

Figure 7.5 (a) Structures of the hydroxystilbenes that are known to form stilbenolignans. (b) Examples of some stilbenolignans identified in plants; all arise from putative radical coupling between the hydroxystilbene and a monolignol.

structures having a 3-oxabicyclo-[3.3.0]-octane ring system (as in kompasinol A, gnetifolin F, and lehmbachol D); stilbenolignans having simple 4–*O*–β′ ether structures (as in gnetumonins B and C) have recently been described in *Gnetum montanum* (Ma et al. 2017). Only a limited number of hydroxystilbenes (piceatannol, isorhapontigenin, and 3,4,5,11,13-pentahydroxy-*trans*-stilbene) have been found coupling with monolignols forming different stilbenolignans (Figure 7.5a). Piceatannol was found forming different stilbenolignans, such as aiphanol (Lee et al. 2001) and kompasinol A (Kobayahsi et al. 1996), from β–*O*–4′ and β–8′ couplings, respectively, of sinapyl alcohol with piceatannol. Isorhapontigenin was observed as part of the stilbenolignans gnetifolin F, gnetofuran A, lehmbachol D, gnetuclesitol F, and gnetumonins B and C, with different coupling types (Yao et al. 2006; Ma et al. 2017). Finally, the 3,4,5,11,13-pentahydroxy-*trans*-stilbene has also been found coupling with sinapyl alcohol in the stilbenolignan 5-hydroxyaiphanol (Lam and Lee, 2010) and 5-hydroxykompasinol A (Lam et al. 2008). As occurs with flavonolignans, all of these examples of stilbenolignans also illustrate the compatibility of some hydroxystilbenes with lignification because of the common radical coupling principles.

7.4 Lignin Monomers Derived from the Flavonoid and Hydroxystilbene Biosynthetic Pathways

Recent investigations have demonstrated that some members of the flavonoids (such as the flavone tricin) and hydroxystilbenes (particularly piceatannol), including also the respective hydroxystilbene glucosides (piceid, isorhapontin, and astringin), behave as authentic lignin monomers participating in radical coupling reactions to become integrally incorporated into the lignin polymer in some plants (del Río et al. 2012, 2017, 2020; Lan et al. 2015; Rencoret et al. 2018, 2019; Neiva et al. 2020), as is detailed in the following sections. These discoveries reveal the previously unknown interconnections between lignin biosynthesis and that of the flavonoids and hydroxystilbenes, and drastically expand, or even challenge, the traditional definition of lignin.

7.4.1 Lignin Monomers from the Flavonoid Biosynthetic Pathway – Tricin (and Naringenin and Apigenin in Rice Mutants)

The flavone tricin (5,7,4′-trihydroxy-3′,5′-dimethoxyflavone, Figure 7.2) was the first phenolic compound derived from outside the canonical monolignol biosynthetic pathway that was established as a true lignin monomer participating in cross-coupling reactions with monolignols (del Río et al. 2012). Tricin was first discovered in lignin preparations from wheat straw, and its structure was established by two-dimensional nuclear magnetic resonance (2D-NMR) (Figure 7.6). Previously unassigned signals in the heteronuclear single-quantum correlation (HSQC) spectra of grass lignins could be definitively assigned to tricin. The diagnostic signals corresponding to the C_8–H_8 and C_6–H_6 correlations appeared in the HSQC spectrum at δ_C/δ_H 94.1/6.56 and 98.8/6.20; the spectrum also showed the C_3–H_3 correlations at δ_C/δ_H 104.5/7.04, whereas the correlations for $C_{2'6'}$–$H_{2'6'}$ were observed at δ_C/δ_H 103.9/7.31 (del Río et al. 2012). The structure of tricin presents two phenolic hydroxyls at C-5 and C-7 of the chroman-4-one skeleton, with diagnostic phenolic

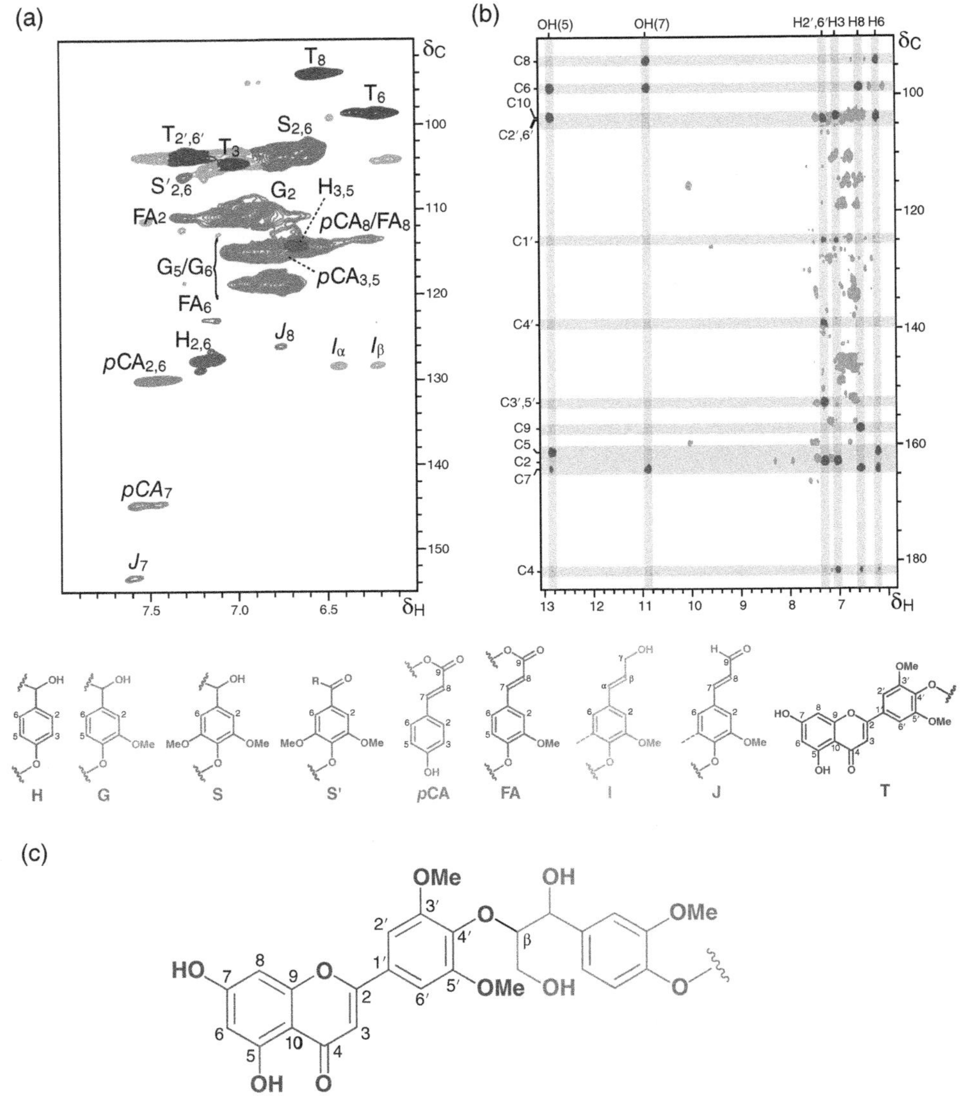

Figure 7.6 (a) Aromatic region of the 2D-HSQC-NMR spectrum of the lignin isolated from wheat straw showing the main correlation signals and their structural assignments to the color-coordinated structures below. (b) Partial HMBC spectrum (δ_C/δ_H 90–185/6.0–13.0) of wheat straw lignin that was crucial in establishing that the non-monolignol component of the lignin was the flavone tricin (5,7,4′-trihydroxy-3′,5′-dimethoxyflavone, **T**). (c) Mode of incorporation of tricin into the lignin structure through 4′–*O*–β ether linkages. *Source:* (a, b, c) Adapted from del Río et al. 2012. (*See insert for color representation of the figure.*)

proton chemical shifts that were readily apparent from their long-range correlations in the heteronuclear multiple-bond correlation (HMBC) spectrum at δ_H 12.86 and 10.88. More importantly, the HMBC spectrum revealed that the 4′-OH of ring-B was not free, suggesting that tricin was etherified to the lignin polymer through 4′–*O*–β ether linkages (as shown in Figure 7.6c), as also occurred in the flavonolignans *threo/erythro* tricin 4′–*O*–β-guaiacylglycerol ethers, also known as salcolins A and B, that have been described in several plants (Wenzig et al. 2005; Chang et al. 2010; Lee et al. 2015). In fact, the signal for the correlation of the tricin C4′ carbon (at 139.5 ppm) and the proton at the β-position of a G-unit at 4.28 ppm was observed in the HMBC spectrum, providing evidence for this incorporation (del Río et al. 2012), which was later confirmed in separate studies (Lan et al. 2015, 2018). Additional biomimetic radical coupling reactions confirmed that tricin can only form 4′–*O*–β cross-coupled products with monolignols; no other type of cross-coupled products could be produced, nor were homodimeric coupling products seen (Lan et al. 2015).

Tricin cannot undergo radical dehydrodimerization and its only possible mode of incorporation into the lignin polymer is via 4′–*O*–β-coupling with a monolignol. Tricin can therefore only appear at the starting end of the lignin chain and this suggests a potential role as a nucleation site for lignification in grasses (as well as in other monocots), a role that was previously assigned to ferulates (Ralph et al. 1995). A detailed study of gel-permeation chromatography (GPC) fractions of maize lignin with different molecular weights showed that the percentage of tricin was slightly higher in lower-molecular-weight fractions, whereas the content of β–β resinol units was lower, adding evidence to the hypothesis that tricin acts as a nucleation site for lignification (Lan, Morreel et al. 2016).

A detailed analysis of the phenolic metabolites present in the lignifying zone of maize indicated the occurrence of a wide array of tricin-containing flavonolignan or low-molecular-weight flavonolignin (flavono-oligolignol) metabolites (Lan, Morreel et al. 2016). The compounds included a wide variety of products resulting from the coupling of all three monolignols (*p*-coumaryl, coniferyl, and sinapyl alcohols), with tricin (at its 4′–*O*- position) forming the respective tricin–4′–*O*–(β-arylglyceryl) ethers, along with the products arising from the coupling of tricin with the γ-acylated monolignols (the coniferyl and sinapyl acetates, and the *p*-coumarate conjugates). Trimeric compounds arising from the coupling of tricin-(4′–*O*–β)-monolignol with a further 4–*O*–β or 5–β linkage to another monolignol (including acylated monolignols) were also identified. In all cases, these tricin-containing metabolites were found to be fully racemic, further evidencing the combinatorial nature of the cross-coupling reactions between tricin and monolignols. It was therefore evident that the dimeric compounds resulting from the coupling of tricin with monolignols (and with their acetate and *p*-coumarate conjugates) should not be termed flavonolignans (that should be optically active); hence, these compounds should be considered to be oligomers that are destined for the lignin polymer, and the more appropriate term 'flavonolignols' or specifically 'tricin-oligolignols' was suggested for them (Lan, Morreel et al. 2016).

Soon after the discovery of the flavone tricin in the lignin of wheat straw, several other studies reported its occurrence in the lignins from other grasses, including giant reed (You et al. 2013), bamboo (Wen et al. 2013), barley (Rencoret et al. 2015), rice (Wu et al. 2013), maize (Lan et al. 2015), and sugarcane (del Río et al. 2015), as well as in other monocots such as in coconut coir (Rencoret et al. 2013). All these studies indicated that tricin occurred widely among lignins, particularly in grasses. A further survey of more than 50 plants from different origins (that included angiosperm monocots and eudicots, as well as gymnosperms) expanded the range of species with tricin incorporated into their lignins and indicated that it was widely distributed in the lignins from species of the Poaceae family and that it also occurred in other monocots such as in the vanilla plant (*Vanilla planifolia* and *V. phalaenopsis*) from the Orchidaceae (Lan, Rencoret et al. 2016). Minor amounts of tricin were also found in the lignin from curaua (*Ananas erectifolius*) from the Bromeliaceae. In general terms, tricin was absent in the lignins from eudicotyledons, although some amounts could be detected in the lignins in alfalfa (*Medicago sativa*) and in *M. truncatula* from the Fabaceae (Bickoff et al. 1964; Lan, Rencoret et al. 2016; Lui et al. 2020; Ralph, 2020). The exact course of the evolution of tricin-lignins is therefore not yet clear but it appears to have occurred independently multiple times. No traces of tricin were observed in the lignins from gymnosperms, however. In the monocots, the content of tricin integrated into the lignin polymer was found to be higher than the content of extractable tricin, reaching up to 4841 and 5250 mg/kg in wheat and oat straws (in comparison to only 1014 and 1377 mg/kg of free tricin), which indicates that lignin, an abundant material that is usually regarded as a waste side-stream, can be an attractive and potential source for this valuable compound (Lan, Morreel et al. 2016).

As noted above, the entry into the biosynthesis of flavones is controlled by chalcone synthase (CHS), which converts *p*-coumaroyl-CoA into naringenin chalcone (Figure 7.7). The resulting naringenin chalcone is then isomerized by chalcone isomerase (CHI) to the flavanone naringenin, which is the common precursor for the biosynthesis of all major classes of flavonoids. The biosynthetic pathway leading from naringenin to tricin in grasses has only been recently elucidated (Lam et al. 2015, 2017). These investigations indicated that naringenin is converted into apigenin by a flavone synthase II (FNSII), and subsequent sequential hydroxylations and *O*-methylations at the flavone B-ring produce the respective luteolin, chrysoeriol, selgin, and finally the flavone tricin. It has also been recently demonstrated in several grasses (maize, rice, sorghum) that the caffeate *O*-methyltransferase (COMT) involved in the synthesis of sinapaldehyde or sinapyl alcohol in the monolignol biosynthetic pathway also participates in the biosynthesis of tricin *in planta* (Fornalé et al. 2017; Eudes et al. 2017; Lam et al. 2017; Lam, Tobimatsu et al. 2019). The interconnections of the monolignol and flavonoid biosynthetic pathways in the biosynthesis of grass lignins were clearly evidenced in a CHS-deficient mutant in maize, that lacked tricin and, obviously, its incorporation into the lignin polymer (Eloy et al. 2017). The study showed that the lignin for the mutant plant was enriched in β–β (resinol and tetrahydrofuran structures) and β–5 linkages (phenylcoumaran structures), further supporting the contention that tricin acts as an initiation site for lignin chains and that, in the absence of tricin, more monolignol dimerization reactions occur (Eloy et al. 2017).

Likewise, a recent study in rice showed that disruption of flavone synthase II (FNSII), the enzyme that catalyzes the direct conversion of the flavanone naringenin to the flavone apigenin, resulted in an altered cell wall lignin incorporating the intermediate naringenin, instead of tricin, into the lignin polymer (Lam et al. 2017). Naringenin, as tricin, can be incorporated into the lignin polymer via 4′–*O*–β cross-coupling with monolignols (as shown in Figure 7.8a). However, in contrast to tricin, naringenin has a nonsubstituted *p*-hydroxyphenyl B-ring, and thus can also couple with monolignols via 3′–β-type coupling, producing a phenylcoumaran structure, as depicted in Figure 7.8b. These phenylcoumaran structures were suggested to occur in the lignin of an *fnsII* rice mutant on the basis of the increase of signals for phenylcoumarans in the HSQC spectra (Lam et al. 2017). It is therefore evident that the flavanone naringenin can also be considered as an authentic lignin

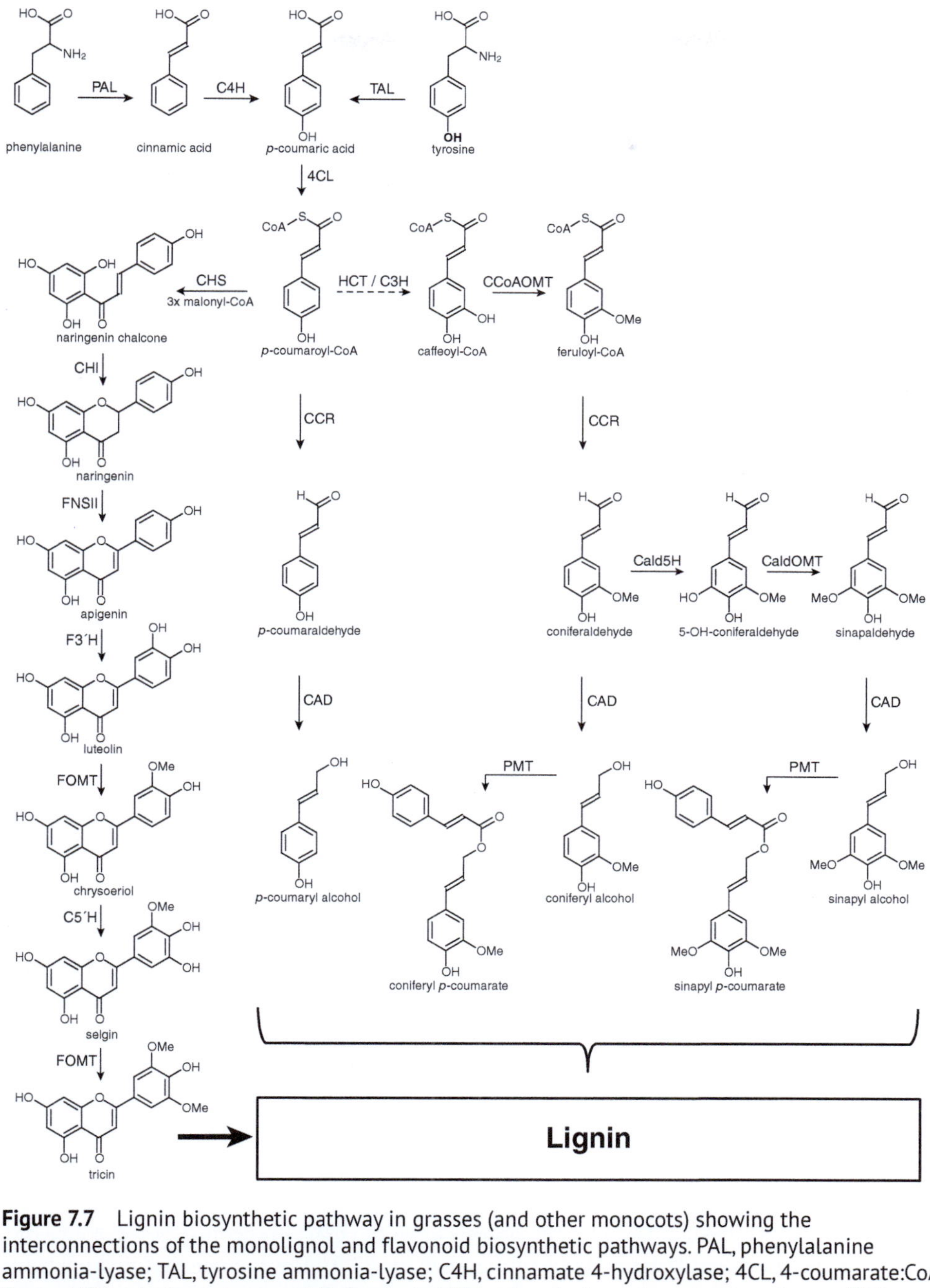

Figure 7.7 Lignin biosynthetic pathway in grasses (and other monocots) showing the interconnections of the monolignol and flavonoid biosynthetic pathways. PAL, phenylalanine ammonia-lyase; TAL, tyrosine ammonia-lyase; C4H, cinnamate 4-hydroxylase; 4CL, 4-coumarate:CoA ligase; HCT: *p*-hydroxycinnamoyl-CoA:quinate/shikimate *p*-hydroxycinnamoyltransferase; C3H: *p*-coumarate 3-hydroxylase; CCoAOMT: caffeoyl-CoA *O*-methyltransferase; CCR: cinnamoyl-CoA reductase; Cald5H (= F5H: ferulate 5-hydroxylase): coniferaldehyde 5-hydroxylase; CaldOMT (= COMT: caffeate *O*-methyltransferase): 5-hydroxyconiferaldehyde *O*-methyltransferase; CAD: cinnamyl alcohol dehydrogenase; PMT: *p*-coumaroyl-CoA:monolignol transferase (this step also requires *p*-coumaroyl-CoA from further up in the pathway as a co-substrate); CHS, chalcone synthase; CHI: chalcone isomerase; FNSII: flavone synthase II; F3′H: flavonoid 3′-hydroxylase; FOMT: flavonoid *O*-methyltransferase; C5′H: crysoeriol 5′-hydroxylase.

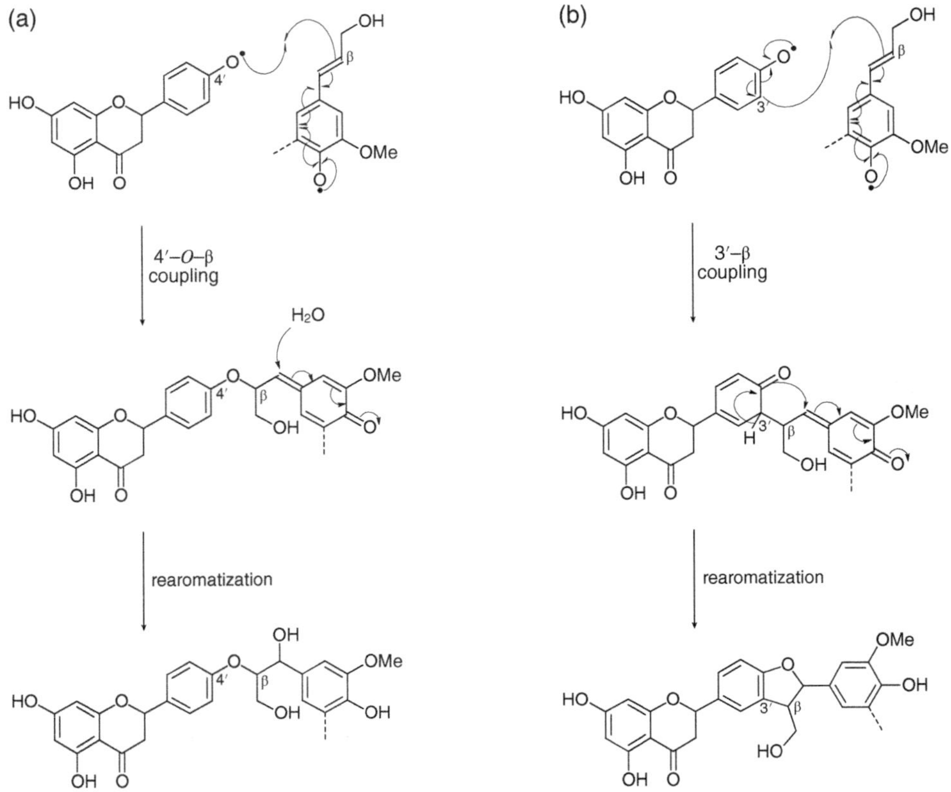

Figure 7.8 Naringenin cross-coupling modes with monolignols. (a) 4′–*O*–β coupling of naringenin with monolignols. (b) 3′–β coupling of naringenin with monolignols producing a phenylcoumaran structure. *Source:* (a, b) Adapted from Lam et al. 2017.

monomer in the *fnsII* rice mutant, participating in cross-coupling reactions with monolignols and being incorporated into the lignin polymer. Additional biomimetic radical coupling reactions of naringenin with monolignols further confirmed the compatibility of naringenin with lignification (Lam et al. 2017). Interestingly, the *fnsII* mutant plants presented normal growth performances, similar to the wild-type plants, suggesting that incorporating naringenin instead of tricin into the lignin structure did not affect normal growth and development; removing tricin altogether in the CHS mutant didn't appear to affect the stem biomass but did significantly increase leaf biomass (for largely unknown reasons) (Eloy et al. 2017).

More recently, it was claimed that the flavone apigenin can also be incorporated into the lignin polymer of culm tissues in a rice mutant lacking F3′H, the enzyme that hydroxylates the B-ring of apigenin to produce luteolin, an intermediate in the biosynthesis of tricin (Lam, Lui et al. 2019). This rice mutant cannot produce luteolin (and tricin in the last instance). Instead it will accumulate apigenin, which is the phenolic compound that is incorporated into the lignin polymer. Additional biomimetic radical coupling reactions of apigenin with monolignols further confirmed its compatibility with lignification. Thus, apigenin can also be considered as another noncanonical lignin monomer participating in radical coupling reactions during lignification in this particular rice mutant (Lam, Lui et al. 2019).

As shown in Section 7.3.1, the wide diversity of flavonolignans identified in plants illustrates the compatibility of flavonoids with lignification. Although only the flavone tricin has been found in significant amounts in the lignins of a large number of plants, and the flavanone naringenin in the lignin of *fnsII* rice mutants and apigenin in rice mutants lacking F3′H, it is likely that other flavonoids, such as the flavanonol taxifolin, the flavonol quercetin, the flavanones eriodictyol and dihydrotricin, or the flavones luteolin and selgin, that are known to cross-couple with monolignols to form flavonolignans, could also be incorporated into the lignins of several plants through radical coupling reactions. Continuing lignin structural research and a broad survey of lignins from different plant origins is needed to discover other flavonoids that might be incorporated into the lignin polymers.

7.4.2 Lignin Monomers from the Hydroxystilbene Biosynthetic Pathway – Resveratrol, Isorhapontigenin, and Piceatannol, and Their *O*-glucosides Piceid, Isorhapontin, and Astringin

Recently, we also reported the occurrence of a second class of polyphenolic compounds, hydroxystilbenes (piceatannol, resveratrol, and isorhapontigenin), that behave as authentic lignin monomers participating in coupling and cross-coupling reactions during lignification of palm fruit endocarps and that become integrally incorporated into the lignin structure (del Río et al. 2017; Rencoret et al. 2018). Likewise, hydroxystilbene glucosides (piceid, isorhapontin, and astringin) were also found incorporated into the lignin polymer of Norway spruce bark, where they appear to act as genuine lignin monomers participating in radical coupling reactions during lignification (Rencoret et al. 2019; Neiva et al. 2020). Hydroxystilbenes are a class of nonflavonoid polyphenolics that, like the flavonoids, are metabolic hybrids resulting from a combination of the shikimate-derived phenylpropanoid and the acetate/malonate-derived polyketide pathways (Figure 7.3). However, and contrary to flavonoids that exhibit an enormous chemical diversity, simple hydroxystilbenes constitute a rather restricted class of compounds, the most important being resveratrol (3,11,13-trihydroxy-*trans*-stilbene), widely present in Vitaceae and Fabaceae, or pinosylvin (3,5-dihydroxy-*trans*-stilbene), which is present in Pinaceae. Hydroxystilbenes occur in a number of plant families where they play a role in plant protection and present antiviral, antibacterial, and antioxidant properties that are known to contribute to plant disease resistance (Roupe et al. 2006; Jeandet et al. 2010; Piotrowska et al. 2012).

The biosynthesis of hydroxystilbenes is controlled by stilbene synthase (STS) (Figure 7.9). STS catalyzes the formation of the stilbene backbone from three malonyl-CoA units and one CoA-ester of a cinnamic derivative; hence resveratrol is formed from *p*-coumaroyl-CoA whereas pinosylvin is formed from cinnamoyl-CoA. STS is a member of the CHS superfamily of type III polyketide synthases (Austin and Noel, 2003). STS enzymes are classified according to their preferred substrate into a *p*-coumaroyl-CoA-specific type such as resveratrol synthase (EC 2.3.1.95), which occurs primarily in angiosperms, or a cinnamoyl-CoA-specific type such as pinosylvin synthase (EC 2.3.1.146), which is typical for gymnosperms (Chong et al. 2009). The introduction of an additional hydroxyl group into the B-ring of the

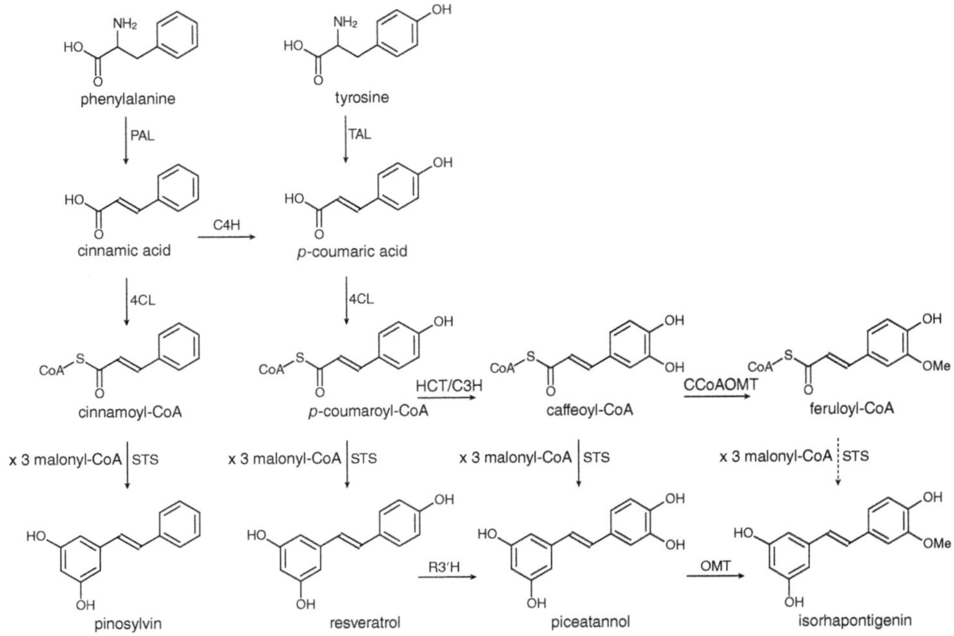

Figure 7.9 Biosynthetic pathway of simple hydroxystilbenes. PAL, phenylalanine ammonia-lyase; TAL, tyrosine ammonia-lyase; C4H, cinnamate 4-hydroxylase; 4CL, 4-coumarate:CoA ligase; HCT: *p*-hydroxycinnamoyl-CoA:quinate/shikimate *p*-hydroxycinnamoyltransferase; C3H: *p*-coumarate 3-hydroxylase; CCoAOMT: caffeoyl-CoA *O*-methyltransferase; R3′H: resveratrol 3′-hydroxylase; STS: stilbene synthase.

stilbene molecule, as in piceatannol (3,4,11,13-tetrahydroxy-*trans*-stilbene), seems to take place at the level of cinnamic acids. The specific STS responsible for the formation of piceatannol from caffeoyl-CoA has not been identified so far, although pinosylvin synthase was active with caffeoyl-CoA and was proposed to be responsible for the biosynthesis of piceatannol in *Picea* sp. (Raiber et al. 1995; Chong et al. 2009; Dubrovina and Kiselev, 2017); similarly, isorhapontigenin may presumably derive from feruloyl-CoA. However, it has also been proposed that resveratrol can undergo subsequent hydroxylation and methylation reactions by the corresponding stilbene 3-hydroxylase (S3H) and *O*-methyltransferase (OMT) to produce piceatannol and then isorhapontigenin (Hammerbacher et al. 2011).

Hydroxystilbenes, as the monolignols, can be oxidized to form radicals that are resonance-stabilized, as shown for piceatannol in Figure 7.10. Hydroxystilbene dimerization or cross-coupling with other hydroxystilbenes produces a wide variety of dimers (ε-viniferin, cassigarol E, scirpusin B, and others) and higher oligomers (Lin and Yao, 2006; Quideau et al. 2011; Rivière et al. 2012; Keylor et al. 2015). Hydroxystilbenes can also cross-couple with monolignols, and several stilbenolignans (see Figure 7.5) have been identified in a variety of plants from different families, as discussed in Section 7.3.2 (Kobayashi et al. 1996; Lee et al. 2001; Banwell et al. 2005; Yao et al. 2006; Lam and Lee, 2010; Begum et al. 2010; Ma et al. 2017), thus illustrating their chemical compatibility with lignification.

The first evidence for the occurrence of hydroxystilbenes incorporated into lignin polymers came from the release of significant amounts of piceatannol (together with minor

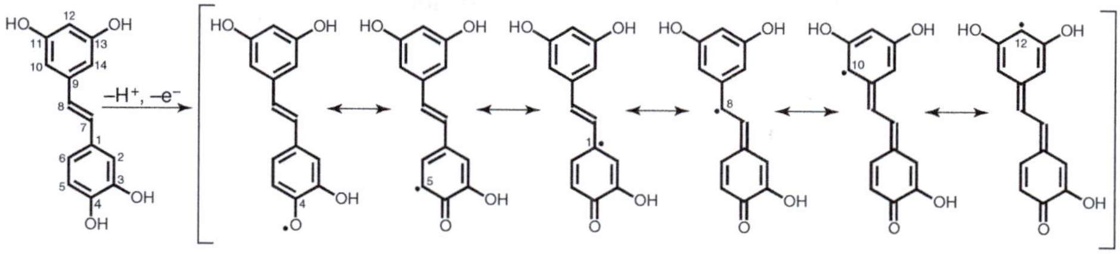

Figure 7.10 Piceatannol's phenolic radical and its different resonance forms. Although there are, in principle, different phenolic radicals that could be formed via the 3-, 11-, and 13-OH, all of the products formed are consistent with reactions only from the one arising from the 4-OH, presumably because the most stable phenolic radical is that from dehydrogenation of the 4-OH. Resonance forms show how coupling can occur at the 4–*O*-, 5-, 8- and 10-positions, among others.

amounts of resveratrol and isorhapontigenin) from lignin preparations isolated from several palm fruit (macaúba, carnauba, and coconut) endocarps by the Derivatization Followed by Reductive Cleavage (DFRC) degradation method (Figure 7.11) (del Río et al. 2017; Rencoret et al. 2018). DFRC is a chemical degradative method that selectively cleaves β–*O*–4 bonds in lignins, releasing the corresponding lignin monomers involved in those linkages (Lu and Ralph, 1997a, 1997b, 1998). This indicated that piceatannol (and to a lower extent resveratrol and isorhapontigenin) is an important component in the lignins of palm fruit endocarps, and that at least a part of them is present in the lignin polymer as β-ether linked structures (similar to the structures of gnetumonins B and C depicted in Figure 7.5), the linkages amenable to DFRC cleavage. However, and unlike tricin, which has only one possible mode of coupling with monolignols, piceatannol would be expected to couple and cross-couple with other piceatannols as well as with monolignols (and oligolignols) in a variety of ways, forming condensed structures that are not amenable to release by DFRC. It is therefore likely that the real amounts of piceatannol and other hydroxystilbene monomers incorporated into these lignins could be significantly higher than realized from the levels released by DFRC.

Useful information regarding the different linkages involving piceatannol in the lignin polymer was provided by 2D-NMR. Signals from piceatannol units were clearly observed in the 2D-HSQC-NMR spectra of the lignins isolated from several palm fruit endocarps, as shown for macaúba in Figure 7.12. The most striking feature in the aromatic region of the spectra was the presence of a previously unreported group of signals (labeled P_b and P_c) appearing at δ_C/δ_H 100-107/5.8-6.8 from piceatannol-derived units. Likewise, the oxygenated-aliphatic region of the spectra also showed signals other than those commonly observed from conventional lignin structures (β-aryl ethers A, phenylcoumarans B, resinols C, or cinnamyl alcohol end-groups I). These new correlation signals, labeled P_b, P_c and V, corresponded to different structures involving piceatannol units that are depicted in Figure 7.12. Definitive identification of these structures was achieved by detailed HSQC-TOCSY and HMBC experiments and by comparing with piceatannol polymerization products and *in vitro* biomimetic cross-coupling reactions (del Río et al. 2017). Thus, the structure labeled P_b was identified as the dehydrodimerization product of two piceatannol units involving 8–*O*–4′/3′–*O*–7 linkages and producing a benzodioxane structure, similar to the stilbene dimer cassigarol E (Baba et al. 1994; Li et al. 2005; Morikawa et al. 2010); the structure labeled P_c was identified as another dehydrodimerization product of two

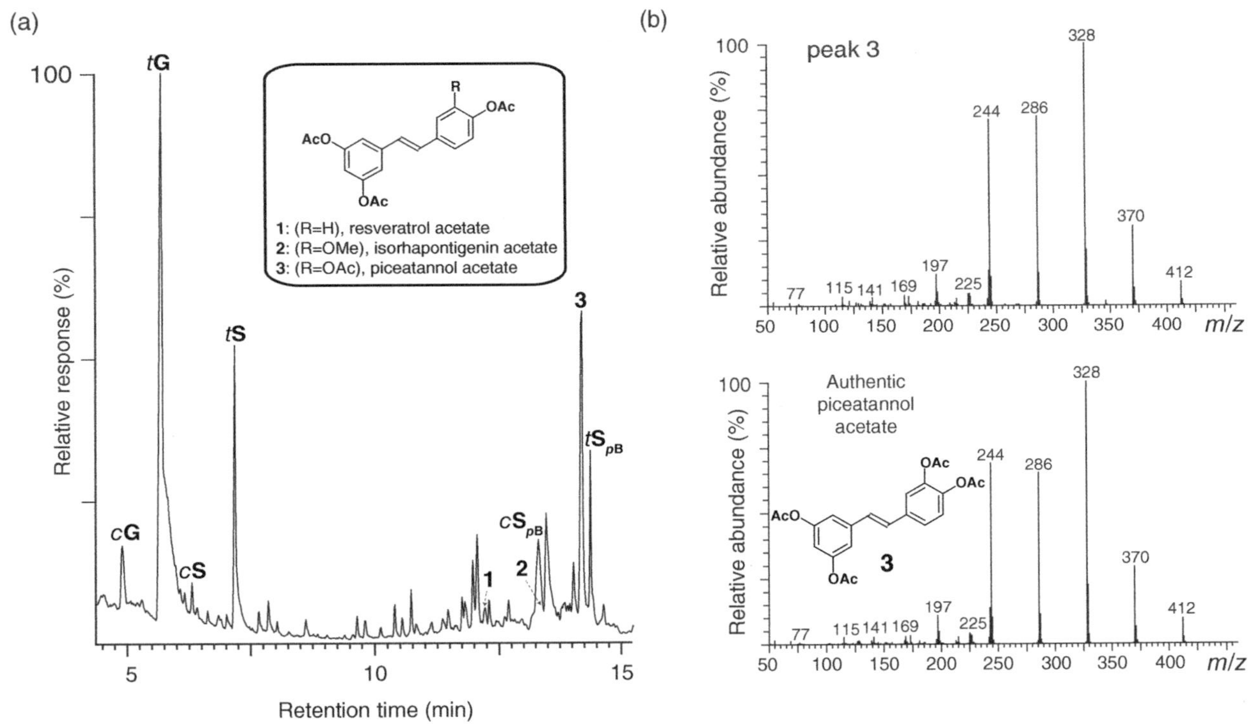

Figure 7.11 (a) Chromatogram of the DFRC degradation products released from the lignin from macaúba (*Acrocomia aculeata*) fruit endocarp showing the presence of hydroxystilbenes (**1**, resveratrol; **2**, isorhapontigenin; **3**, piceatannol, as their acetate derivatives). *c***G**, *t***G**, *c***S**, and *t***S** are the normal *cis*- and *trans*-coniferyl (**G**) and sinapyl (**S**) alcohol monomers (as their acetate derivatives). $c\mathbf{S}_{\mathbf{pB}}$ and $t\mathbf{S}_{\mathbf{pB}}$ are the *cis*- and *trans*-sinapyl *p*-hydroxybenzoates (as their acetate derivatives). (b) Comparison of the electron-impact (EI) mass spectra from peak **3** (acetylated) produced in the DFRC of the lignin from macaúba endocarp (top) and from an authentic standard of piceatannol **3** (acetylated) (bottom). *Source:* (a,b) Adapted from del Río et al. 2017.

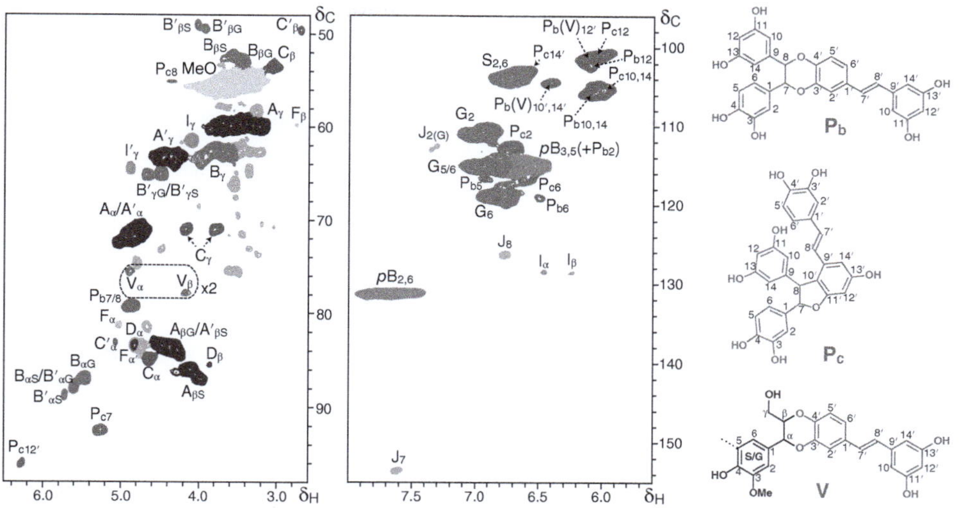

Figure 7.12 Side-chain (δ_C/δ_H 48-98/2.6-6.5) and aromatic (δ_C/δ_H 96-155/5.6-8.0 ppm) regions of the 2D-HSQC-NMR spectrum (in DMSO-d_6) of the lignin isolated from macaúba (*Acrocomia aculeata*) fruit endocarp showing the main signals from lignin and piceatannol (**P**) units. The structures of the main piceatannol homodimeric coupling (**P**$_b$, **P**$_c$) and piceatannol-lignin cross-coupling structures (**V**) are depicted. Signals from typical lignin structures are as follows: **A**: β–*O*–4′ alkyl-aryl ethers; **A′**: β-ether structures with *p*-hydroxybenzoates acylating the γ-OH; **B**: phenylcoumarans; **B′**: phenylcoumarans with *p*-hydroxybenzoates acylating the γ-OH; **C**: resinols; **C′**: tetrahydrofuran structures formed by β–β′-coupling of monolignols acylated at the γ-OH; **D**: dibenzodioxocins; **F**: spirodienones; **I**: cinnamyl alcohol end-groups; **I′**: cinnamyl alcohol end-groups acylated at the γ-OH; **J**: cinnamaldehyde end-groups; ***pB***: *p*-hydroxybenzoates; **G**: guaiacyl units; **S**: syringyl units. *Source:* Adapted from del Río et al. 2017. (*See insert for color representation of the figure.*)

piceatannol units involving 8–10′/11′-*O*–7 linkages and producing a phenylcoumaran structure, similar to the stilbene dimer scirpusin B (Nakajima et al. 1978; Powell et al. 1987; Wang et al. 2011); structure V, on the other hand, was identified as a cross-coupling product of a piceatannol unit and a monolignol producing a benzodioxane bridge, similar to the stilbenolignan aiphanol (Lee et al. 2001).

The biosynthetic mechanisms of the different structures involving piceatannol are shown in Figure 7.13. Structure P_b was formed via 8–*O*–4′-type radical coupling of a piceatannol unit at its 8-position with another piceatannol unit (at its *O*–4′ position) followed by internal trapping of the quinone methide intermediate by the 3′-hydroxyl group; structure P_c was formed by the radical coupling of a piceatannol unit (at its 8-position) with another piceatannol unit (at its 10′-position) followed by a subsequent 11′-*O*–7 bonding during rearomatization of the quinone methide intermediate, illustrating the importance of electron delocalization throughout both aromatic rings of the structure, as illustrated in Figure 7.10. Structure V was identified as arising from the cross-coupling of piceatannol and monolignols and are formed via radical coupling of a monolignol (at its β-position) and the catechol moiety of piceatannol (at its *O*–4′ position) followed by internal trapping of the quinone methide intermediate by the 3′-hydroxyl group in the piceatannol unit to form the benzodioxane structure V. Biomimetic cross-coupling reactions between piceatannol and monolignols successfully proved that benzodioxane structures V could be easily formed

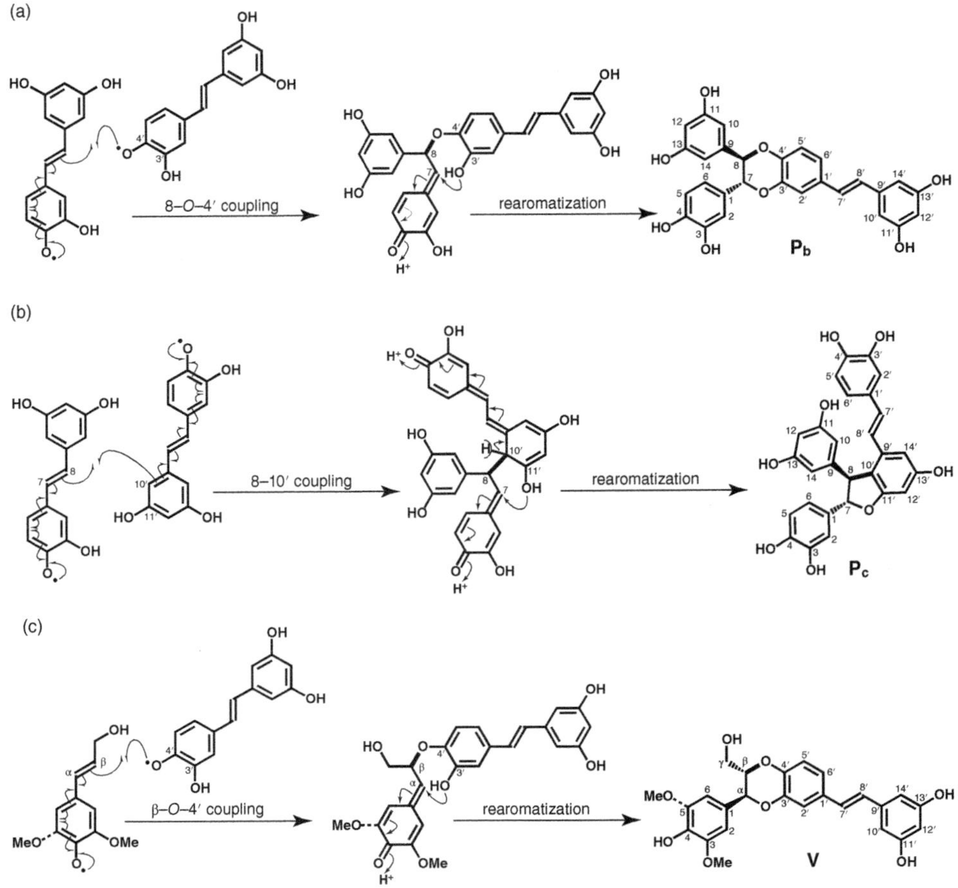

Figure 7.13 (a) Piceatannol dehydrodimerization by 8–*O*–4′ coupling to give the benzodioxane structure $\mathbf{P_b}$. (b) Piceatannol dehydrodimerization by 8–10′ coupling to give the phenylcoumaran structure $\mathbf{P_c}$. (c) Cross-coupling of monolignols M, sinapyl and coniferyl alcohol, with piceatannol 1 via β–*O*–4′ coupling to give the benzodioxane structure **V**. *Source:* (a, b, c) Adapted from del Río et al. 2017.

during the radical reaction (del Río et al. 2017). The occurrence of these benzodioxane structures in the lignins from palm fruit endocarps conclusively demonstrates that the hydroxystilbene piceatannol acts as an authentic lignin monomer, participating in coupling and cross-coupling reactions during lignification of these tissues and being integrally incorporated into the lignin polymer. Likewise, the study of the thermodynamics of these reactions using density functional theory (DFT) calculations indicated that, in general, the energetics of both homo-coupling and cross-coupling reactions are comparable with those of the monolignol coupling, demonstrating the compatibility of piceatannol with the lignification process (Elder et al. 2019, 2020).

As noted in Section 7.3.2, only a limited number of hydroxystilbenes (piceatannol, isorhapontigenin and 3,4,5,11,13-pentahydroxy-*trans*-stilbene) have been found coupling with monolignols and forming different stilbenolignans, indicating their compatibility with lignification. In principle, all of these hydroxystilbenes can potentially occur as lignin monomers in plants. However, as far as we know, resveratrol has only been found forming dehydrodimers and higher oligomers but has not been found coupling with monolignols forming stilbenolignans (Begum et al. 2010; Rivière et al. 2012; Keylor et al. 2015), perhaps due to some type of chemical incompatibility, and this might explain why it has only been detected in minor amounts in the lignins of palm fruit endocarps. A broader survey of plant lignins is required to expand the range of plants with hydroxystilbenes incorporated into their lignins.

The incorporation of hydroxystilbenes into the lignins of palm fruit endocarps has been suggested to have a potential role in seed protection. The incorporation of piceatannol (and other hydroxystilbenes) into the lignins may allow the production of higher amounts of lignin by incorporating other phenolic compounds present in the cell wall into the lignin polymer (which indeed will form more condensed structures, as the phenylcoumarans and benzodioxanes shown in Figure 7.13), contributing to endocarp hardening. Hydroxystilbenes can also provide additional antioxidant, antifungal, and antiviral properties (Roupe et al. 2006; Jeandet et al. 2010; Piotrowska et al. 2012), thus further contributing to seed protection.

More recently, it has also been reported that the hydroxystilbene glucosides isorhapontin (isorhapontigenin-*O*-glucoside) and, at lower levels, astringin (piceatannol-*O*-glucoside) and piceid (resveratrol-*O*-glucoside), are incorporated into the lignin polymer in Norway spruce bark (Rencoret et al. 2019; Neiva et al. 2020). The corresponding aglycones isorhapontigenin, piceatannol, and resveratrol, along with glucose (as their peracetylated derivatives), were released by DFRC, a chemical degradative method that cleaves β-ether

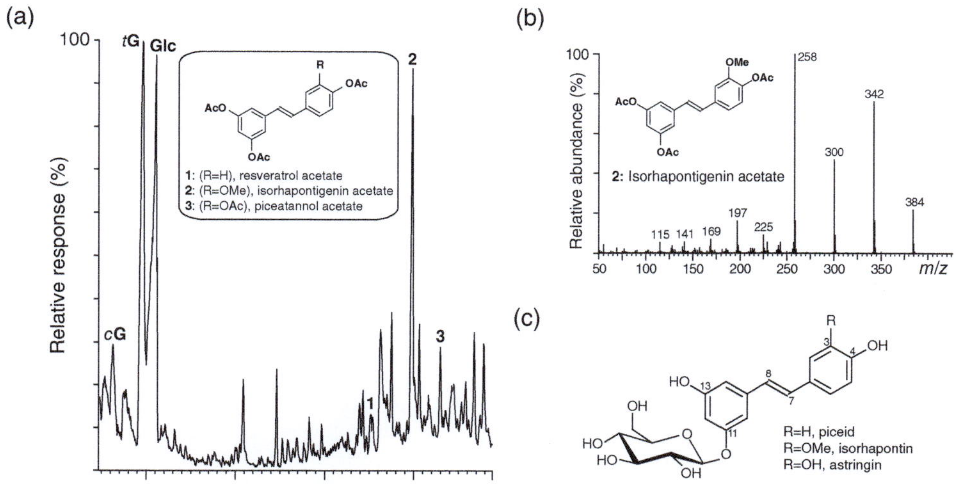

Figure 7.14 (a) Chromatogram of the DFRC degradation products released from the lignin of Norway spruce (*Picea abies*) bark showing the presence of hydroxystilbenes (**1**, resveratrol; **2**, isorhapontigenin; **3**, piceatannol, as their acetate derivatives). *c***G** and *t***G**, are the *cis*- and *trans*-coniferyl alcohol monomers (as their acetate derivatives). Note the occurrence of a peak from glucose **Glc** (as its peracetylated derivative). (b) Electron-impact mass spectrum of isorhapontigenin acetate (peak **2**), the major hydroxystilbene released upon DFRC. (c) Structures of the hydroxystilbene glucosides. *Source:* (a,b,c) Adapted from Rencoret et al. 2019.

bonds in lignin (Figure 7.14). The DFRC data, therefore, indicated that at least a part of the hydroxystilbene glucosides, particularly isorhapontin, was incorporated into the lignin of Norway spruce bark as β-ether linked structures, the ones cleaved by the DFRC degradation method. Additional information regarding the composition and structure of the lignin isolated from Norway spruce bark, including the mode of incorporation of the hydroxystilbene glucosides into the lignin polymer, was obtained from 2D-HSQC-NMR analysis (Figure 7.15). Signals from hydroxystilbenes were also clearly present in the aromatic

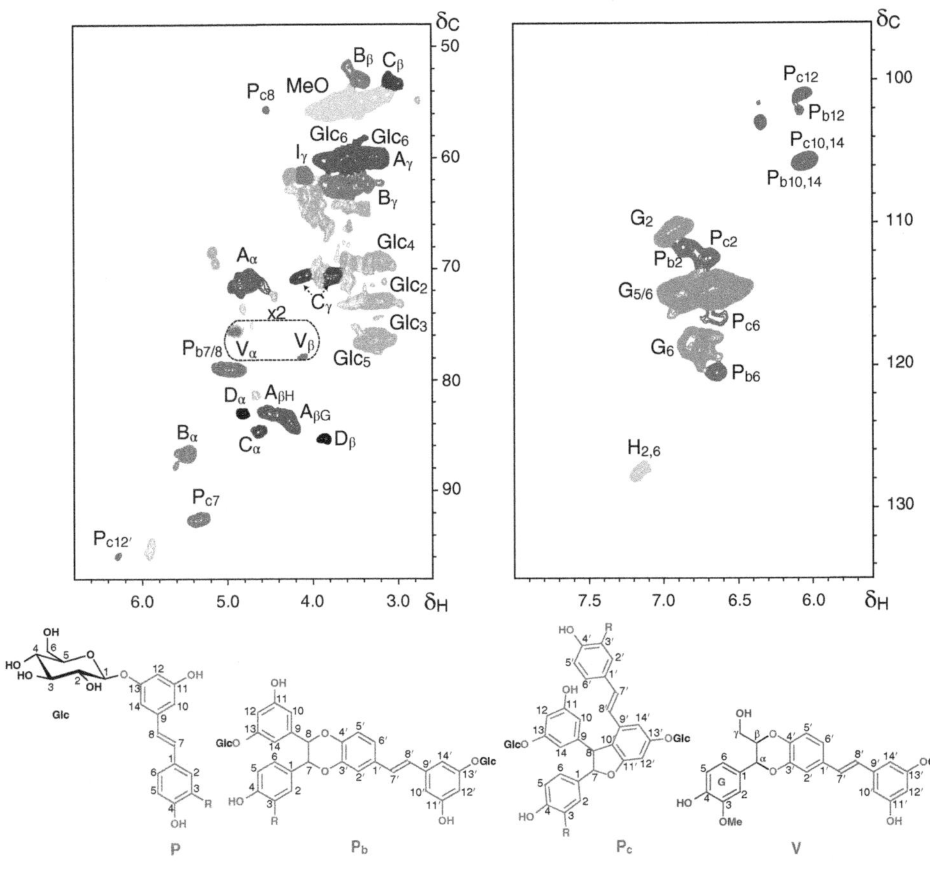

Figure 7.15 Aliphatic-oxygenated (δ_C/δ_H 48–98/2.6–6.8), and aromatic (δ_C/δ_H 96–135/5.6–8.0) regions of the 2D-HSQC-NMR spectrum (in DMSO-d_6) of the 'milled-wood' lignin preparation isolated from Norway spruce bark showing the main signals from lignin and hydroxystilbene glucoside units. The main signals from hydroxystilbene glucosides are: **P**: hydroxystilbene glucosides (isorhapontin, R= OMe; astringin, R = OH; piceid, R = H); $\mathbf{P_b}$: 8–*O*–4′/3′–*O*–7 benzodioxane structures involving isorhapontin (R = OMe), astringin (R = OH) or piceid (R = H) units; $\mathbf{P_c}$: 8–10′/11′–7 phenylcoumaran structures involving isorhapontin (R = OMe), astringin (R = OH) or piceid (R = H) units (please note that other phenylcoumaran structures arising from other linkages, such as 8–5′ and 8–12′, may also occur, but only the 8–10′ linkage has been authenticated due to the occurrence of signal $\mathbf{P_{c12'}}$); **V**: β–*O*–4′/3′–*O*–α benzodioxane structure formed by cross-coupling of astringin and coniferyl alcohol. Note that the signals from glucose (**Glc**) units appear in the aliphatic-oxygenated region. Signals from typical lignin structures are as follows: **A**: β–*O*–4′ alkyl-aryl ethers; **B**: phenylcoumarans; **C**: resinols; **D**: dibenzodioxocins; **I**: cinnamyl alcohol end-groups; **H**: *p*-hydroxyphenyl units; **G**: guaiacyl units. *Source:* Adapted from Rencoret et al. 2019. (*See insert for color representation of the figure.*)

region of the spectrum and were similar to those from the hydroxystilbenes observed in the 2D-HSQC-NMR spectra of lignins from palm fruit endocarps (del Río et al. 2017; Rencoret et al. 2018), including signals from isorhapontigenin and piceatannol, and confirming the results observed by DFRC. In addition, signals from the glucose moiety were clearly apparent in the aliphatic-oxygenated region of the spectrum, confirming the occurrence of the hydroxystilbene glucosides already advanced by DFRC. Most importantly, signals from structures involving the coupling of hydroxystilbene glucosides were clearly observed in the HSQC spectrum (Figure 7.15). Signals for benzodioxane (**P**$_\mathbf{b}$) and phenylcoumaran structures (**P**$_\mathbf{c}$) involving coupling of two hydroxystilbene glucosides, and signals for benzodioxane structures (**V**) formed by cross-coupling of astringin and coniferyl alcohol presented similar correlations to those observed for the incorporation of the hydroxystilbene piceatannol into the lignins of palm fruit endocarps (del Río et al. 2017). The occurrence of all of these coupled and cross-coupled structures involving hydroxystilbene glucosides indicates that these phenolic compounds, particularly isorhapontin and astringin, also behave as authentic lignin monomers that participate in radical coupling reactions during lignification in Norway spruce bark.

7.5 Conclusions and Future Prospects

Different phenolic compounds from outside the canonical monolignol biosynthetic pathway, namely the flavonoid and hydroxystilbenoid pathways, have been found to behave as authentic lignin monomers in several plants, particularly in monocots. The lignins from grasses and other monocots incorporate the flavone tricin (and naringenin and apigenin have been found incorporated into the lignins in some rice mutants) whereas the lignins from palm fruit endocarps incorporate the hydroxystilbene piceatannol (as well as resveratrol and isorhapontigenin at lower levels), and the lignin from Norway spruce bark incorporates hydroxystilbene glucosides, particularly isorhapontin. These discoveries broaden the definition of lignin monomers, and indicate that lignification is a flexible mechanism and that the plant is capable of using a variety of phenolic compounds from different biosynthetic pathways for the formation of the lignin polymers, further expanding the traditional definition of lignin. The discovery of 'nonconventional' phenolic precursors, different from the three canonical monolignols, illustrates yet again the high metabolic plasticity of lignification and reveals that any phenolic compound that is delivered to the cell wall may be oxidized and incorporated into the lignin polymer during lignification via radical coupling reactions, and subject exclusively to simple chemical compatibility (Ralph et al. 2008, 2019; Grabber et al. 2010, 2012, 2015; Vanholme et al. 2008, 2010, 2012, 2019; Mottiar et al. 2016; del Río et al. 2020). The incorporation of 'nonconventional' lignin monomers, not usually present in the lignins of other plants, as is the case for the tricin or the piceatannol described here (or any other flavonoid or hydroxystilbene that are compatible with lignification), can open up new ways to design and engineer the lignin structure to produce polymers and plant-based biomaterials with altered properties. On the other hand, valuable flavonoids and hydroxystilbenes could also be obtained from agricultural residues (cereal straws, palm fruit shells, or spruce bark) or from low-value lignin side-streams in lignocellulose processing mills, which will open new opportunities for the valorization of these residues that are considered as wastes (Rinaldi et al. 2016).

7.6 Acknowledgments

This study has been funded by the Spanish projects CTQ2014-60764-JIN and AGL2017-83036-R (financed by Agencia Estatal de Investigación, AEI, and Fondo Europeo de Desarrollo Regional, FEDER). John Ralph, Wu Lan, and Hoon Kim were funded by the DOE Great Lakes Bioenergy Research Center (DOE BER Office of Science DE-FC02-07ER64494 and DE-SC0018409).

References

AbouZid, S.F., Ahmed. H.S., Moawad, A.S., et al. (2017). Chemotaxonomic and biosynthetic relationships between flavonolignans produced by *Silybum marianum* populations. *Fitoterapia* 119: 175–184.

AbouZid, S.F., Chen, S.N., and Pauli, G.P. (2016). Silymarin content in *Silybum marianum* populations growing in Egypt. *Industrial Crops and Products* 83: 729–737.

Andersen, Ø.M., and Markham, K.R. (2006). *Flavonoids - Chemistry, Biochemistry and Applications*. Boca Raton: CRS Press/Taylor & Francis.

Austin, M.B., and Noel, J.P. (2003). The chalcone synthase superfamily of type III polyketide synthases. *Natural Product Reports* 20: 79–110.

Baba, K., Kido, T., Taniguchi, M., and Kozawaqa, M. (1994). Stilbenoids from *Cassia garrettiana*. *Phytochemistry* 36: 1509–1513.

Banwell, M.G., Chand, S., and Savage, G.P. (2005). An enantioselective total synthesis of the stilbenolignan (−)-aiphanol and the determination of its absolute stereochemistry. *Tetrahedron: Asymmetry* 16: 1645–1654.

Barros, J., Serrani-Yarce, J.C., Chen F., et al. (2016). Role of bifunctional ammonia-lyase in grass cell wall biosynthesis. *Nature Plants* 2: 16050.

Begum, S.A., Sahai, M., and Ray, A.B. (2010). Non-conventional lignans: coumarinolignans, flavonolignans, and stilbenolignans. *Fortschritte der Chemie Organischer Naturstoffe (Progress in the Chemistry of Organic Natural Products)* 93: 1–70.

Bickoff, E.M., Livingston, A.L., and Booth, A.N. (1964). Tricin from alfalfa: isolation and physiological activity. *Journal of Pharmaceutical Sciences* 53: 1411–1412.

Boerjan, W., Ralph, J., and Baucher, M. (2003). Lignin biosynthesis. *Annual Review of Plant Biology* 54: 519–546.

Campbell, M.M. and Sederoff, R.R. (1996). Variation in lignin content and composition. *Plant Physiology* 110: 3–13.

Chambers, S.C., Valentová, K., and Křen, V. (2015). Non-taxifolin derived flavonolignans: phytochemistry and biology. *Current Pharmaceutical Design* 21: 5489–5500.

Chang, C.L., Wang, G.J., Zhang, L.J. et al. (2010). Cardiovascular protective flavolignans and flavonoids from *Calamus quiquesetinervius*. *Phytochemistry* 71: 271–279.

Chen, F., Tobimatsu, Y., Havkin-Frenkel, D., et al. (2012). A polymer of caffeyl alcohol in plant seeds. *Proceedings of the National Academy of Sciences USA* 109: 1772–1777.

Chen, F., Tobimatsu, Y., Jackson, L., et al. (2013). Novel seed coat lignins in the Cactaceae: structure, distribution and implications for the evolution of lignin diversity. *Plant Journal* 73: 201–211.

Chong, J., Poutaraud, A., and Hugueney, P. (2009). Metabolism and roles of stilbenes in plants. *Plant Science*177: 143–155.

Csupor, D., Csorba, A., and Hohmann, J. (2016). Recent advances in the analysis of flavonolignans of *Silybum marianum*. *Journal of Pharmaceutical and Biomedical Analysis* 130: 301–317.

Davin, L.B., Wang, H.B., Crowell, A.L., et al. (1997). Stereoselective biomolecular phenoxy radical coupling by an auxiliary (dirigent) protein without an active center. *Science* 275: 362–366.

del Río, J.C., Gutiérrez, A., and Martínez, A.T. (2004). Identifying acetylated lignin units in non-wood fibers using pyrolysis-gas chromatography/mass spectrometry. *Rapid Communications in Mass Spectrometry* 18: 1181–1185.

del Río, J.C., Lino, A.G., Colodette, J.L., et al. (2015). Differences in the chemical structure of the lignins from sugarcane bagasse and straw. *Biomass & Bioenergy* 81: 322–338.

del Río, J.C., Marques, G., Rencoret, J., et al. (2007). Occurrence of naturally acetylated lignin units. *Journal of Agricultural and Food Chemistry* 55: 5461–5468.

del Río, J.C., Rencoret, J., Gutiérrez, A., et al. (2017). Hydroxystilbenes are monomers in palm fruit endocarp lignins. *Plant Physiology* 174: 2072–2082.

del Río, J.C., Rencoret, J., Gutiérrez, A., et al. (2018). Structural characterization of lignin from maize (*Zea mays* L.) fibers: Evidence for diferuloylputrescine incorporated into the lignin polymer in maize kernels. *Journal of Agricultural and Food Chemistry* 66: 4402–4413.

del Río, J.C., Rencoret, J., Gutiérrez, A., et al. (2020). Lignin monomers from beyond the canonical monolignol biosynthetic pathway: Another brick in the wall. *ACS Sustainable Chemistry & Engineering* 8: 4997—5012.

del Río, J.C., Rencoret, J., Marques, G. et al. (2008). Highly acylated (acetylated and/or *p*-coumaroylated) native lignins from diverse herbaceous plants. *Journal of Agricultural and Food Chemistry* 56: 9525–9534.

del Río, J.C., Rencoret, J., Prinsen, P., et al. (2012). Structural characterization of wheat straw lignin as revealed by analytical pyrolysis, 2D-NMR, and reductive cleavage methods. *Journal of Agricultural and Food Chemistry* 60: 5922–5935.

Dixon, R.A., Chen, F., Guo, D., and Parvathi, K. (2001). The biosynthesis of monolignols: a "metabolic grid" or independent pathways to guaiacyl and syringyl units? *Phytochemistry* 57: 1069–1084.

Dixon, R.A. and Reddy, M.S.S. (2003). Biosynthesis of monolignols. Genomic and reverse genetic approaches. *Phytochemical Reviews* 2: 289–306.

Donaldson, L.A. (2001). Lignification and lignin topochemistry – an ultrastructure view. *Phytochemistry* 57: 859–873.

Dubrovina, A.S., and Kiselev, K.V. (2017). Regulation of stilbene biosynthesis in plants. *Planta* 246: 597–623.

Elder, T., del Río, J.C., Ralph, J., et al. (2019). Radical coupling reactions of piceatannol and monolignols: A density functional theory study. *Phytochemistry* 164: 12–23.

Elder, T., del Río, J.C., Ralph, J., et al. (2020). Coupling and reactions of lignols and new lignin monomers: A density functional theory study. *ACS Sustainable Chemistry & Engineering* 8: 11033–11045.

Eloy, N.B., Voorend, W., Lan, W., et al. (2017). Silencing *CHALCONE SYNTHASE* in maize impedes the incorporation of tricin into lignin and increases lignin content. *Plant Physiology* 173: 998–106.

Eudes, A., Dutta, T., Deng, K., et al. (2017). SbCOMT (Bmr12) is involved in the biosynthesis of tricin-lignin in sorghum. *PLoS ONE* 12 (6): e0178160.

Fornalé, S., Rencoret, J., García-Calvo, L., et al. (2017). Changes in cell wall polymers and degradability in maize mutants lacking 3′- and 5′-*O*-methyl-transferases involved in lignin biosynthesis. *Plant and Cell Physiology* 58: 240–255.

Grabber, J.H., Ress, D., and Ralph, J. (2012). Identifying new lignin bioengineering targets: impact of epicatechin, quercetin glycoside, and gallate derivatives on the lignification and fermentation of maize cell walls. *Journal of Agricultural and Food Chemistry* 60: 5152–5160.

Grabber, J.H., Santoro, N., Foster, C.E., et al. (2015). Incorporation of flavonoid derivatives or pentagalloyl glucose into lignin enhances cell wall saccharification following mild alkaline or acidic pretreatments. *BioEnergy Research* 8: 1391–1400.

Grabber, J.H., Schatz, P.F., Kim, H., et al. (2010). Identifying new lignin bioengineering targets: 1. Monolignol substitute impacts on lignin formation and cell wall fermentability. *BMC Plant Biology* 10: 1–13.

Hammerbacher, A., Ralph, S.G., Bohlmann, J., et al. (2011). Biosynthesis of the major tetrahydroxystilbenes in spruce, astringin and isorhapontin, proceeds via resveratrol and is enhanced by fungal infection. *Plant Physiology* 157: 876–890.

Hatfield, R.D., Ralph, J., and Grabber, J.H. (1999). Cell wall cross-linking by ferulates and diferulates in grasses. *Journal of the Science of Food and Agriculture* 79: 403–407.

Jeandet, P., Delaunois, B., Conreux, A., et al. (2010). Biosynthesis, metabolism, molecular engineering, and biological functions of stilbene phytoalexins in plants. *BioFactors* 36: 331–341.

Kaal, J., Serrano, O., del Río, J.C., and Rencoret, J. (2018). Radically different lignin composition in *Posidonia* species may link to differences in organic carbon sequestration capacity. *Organic Geochemistry* 124: 247–256.

Karlen, S.D., Smith, R.A., Kim, H., et al. (2017). Highly decorated lignins in leaf tissues of the Canary Island date palm *Phoenix canariensis*. *Plant Physiology* 175: 1058–1067.

Karlen, S.D., Zhang, C., Peck, M.L., et al. (2016). Monolignol ferulate conjugates are naturally incorporated into plant lignins. *Science Advances* 2 (10): e1600393.

Keylor, M.H., Matsuura, B.S., and Stephenson, C.R.J. (2015). Chemistry and biology of resveratrol-derived natural products. *Chemical Reviews* 115: 8976–9027.

Kim, H., Ralph, J., Lu, F., et al. (2003). NMR analysis of lignins in CAD-deficient plants. Part 1. Incorporation of hydroxycinnamaldehydes and hydroxybenzaldehydes into lignins. *Organic & Biomolecular Chemistry* 1: 268–281.

Kim, H., Li, Q., Karlen, S.D., et al. (2020). Monolignol benzoates incorporate into the lignin of transgenic *Populus trichocarpa* depleted in C3H and C4H. *ACS Sustainable Chemistry & Engineering* 8: 3644–3654

Kobayashi, M., Mahmud, T., Yoshioka, N., et al. (1996). Indonesian medicinal plants. XVIII. Kompasinol A, a new stilbeno-phenylpropanoid from the bark of *Koompassia malaccensis* (Fabaceae). *Chemical and Pharmaceutical Bulletin* 44: 2249–2253.

Lam, P.Y., Lui, A.C.W., Yamamura, M., et al. (2019) Recruitment of specific flavonoid B-ring hydroxylases for two independent biosynthesis pathways of flavone-derived metabolites in grasses. *New Phytologist* 223: 204–219.

Lam, P.Y., Liu, H., and Lo, C. (2015). Completion of tricin biosynthesis pathway in rice: cytochrome P450 75B4 is a unique chrysoeriol 5′-hydroxylase. *Plant Physiology* 168: 1527–1536.

Lam, P.Y., Tobimatsu, Y., Matsumoto, N., et al. (2019) OsCAldOMT1 is a bifunctional *O*-methyltransferase involved in the biosynthesis of tricin-lignins in rice cell walls. *Scientific Reports* 9: 11597.

Lam, P.Y., Tobimatsu, Y., Takeda, Y., et al. (2017). Disrupting flavone synthase II alters lignin and improves biomass digestibility. *Plant Physiology* 174: 972–985.

Lam, S.H., Chen, J.M., Kang, C.J., et al. (2008). α-Glucosidase inhibitors from the seeds of *Syagrus romanzoffiana*. *Phytochemistry* 69: 1173–1178.

Lam, S.H., and Lee, S.S. (2010). Unusual stilbenoids and a stilbenolignan from seeds of *Syagrus romanzoffiana*. *Phytochemistry* 71: 792–797.

Lan, W., Lu, F., Regner, M., et al. (2015). Tricin, a flavonoid monomer in monocot lignification. *Plant Physiology* 167: 1284–1295.

Lan, W., Morreel, K., Lu, F., et al. (2016). Maize tricin-oligolignol metabolites and their implications for monocot lignification. *Plant Physiology* 171: 810–820.

Lan, W., Rencoret, J., del Río, J.C., and Ralph, J. (2019). Tricin in grass lignin: biosynthesis, characterization, and quantitation. In: *Lignin: Biosynthesis, Functions and Economic Significance* (eds F. Lu and F. Yue), 51–78. New York: Nova Science Publishers, Inc.

Lan, W., Rencoret, J., Lu, F., et al. (2016). Tricin-lignins: occurrence and quantitation of tricin in relation to phylogeny. *Plant Journal* 88: 1046–1057.

Lan, W., Yue, F., Rencoret, J., et al. (2018). Elucidating tricin-lignin structures: assigning correlations in HSQC spectra of monocot lignins. *Polymers* 10: 196.

Landucci, L.L., Deka, G.C., and Roy, D.N.A. (1992). ^{13}C NMR study of milled wood lignins from hybrid *Salix* clones. *Holzforschung* 46: 505–511.

Lapierre, C., Pilate, G., Pollet, B., et al. (2004). Signatures of cinnamyl alcohol dehydrogenase deficiency in poplar lignins. *Phytochemistry* 65: 313–321.

Lapierre, C., Pollet, B., MacKay, J.J., and Sederoff, R.R. (2000). Lignin structure in a mutant pine deficient in cinnamyl alcohol dehydrogenase. *Journal of Agricultural and Food Chemistry* 48: 2326–2331.

Lee, D., Cuendet, M., Vigo, J.S., et al. (2001). A novel cyclooxygenase-inhibitory stilbenolignan from the seeds of *Aiphanes aculeata*. *Organic Letters* 3: 2169–2171.

Lee, D.Y.-W., and Liu, Y. (2003). Molecular structure and stereochemistry of silybin A, silybin B, isosilybin A, and isosilybin B, isolated from *Silybum marianum* (milk thistle). *Journal of Natural Products* 66: 1171–1174.

Lee, S.S., Baek, Y.S., Eun, C.S., et al. (2015). Tricin derivatives as anti-inflammatory and anti-allergic constituents from the aerial parts of *Zizania latifolia*. *Bioscience, Biotechnology, and Biochemistry* 79: 700–706.

Li, W.-L., He, K.-K., Li, Y., and Hou, Z.-J. (2005). Total synthesis of (±)-shegansu B, gnetuhainin F, (±)-maackin A and (±)-cassigarol E. *Acta Chimica Sinica* 63: 1607–1612.

Lin, M., and Yao, C.-S. (2006). Natural oligostilbenes. In: *Studies in Natural Products Chemistry*, Vol. 33, (ed. A. Rahman), 601–644. Amsterdam: Elsevier B.V.

Liu, C.-J. (2012). Deciphering the enigma of lignification: precursor transport, oxidation and the topochemistry of lignin assembly. *Molecular Plant* 5: 304–317.

Lourenço, A., Rencoret, J., Chematova, C., et al. (2016). Lignin composition and structure differs between xylem, phloem and phellem in *Quercus suber* L. *Frontiers in Plant Science* 7: 1612.

Lu, F., Karlen, S., Regner, M., et al. (2015). Naturally *p*-hydroxybenzoylated lignins in palms. *BioEnergy Research* 8: 934–952.

Lu, F. and Ralph, J. (1997a) Derivatization followed by reductive cleavage (DFRC method), a new method for lignin analysis: protocol for analysis of DFRC monomers. *Journal of Agricultural and Food Chemistry* 45: 2590–2592.

Lu, F. and Ralph, J. (1997b) The DFRC method for lignin analysis. Part 1. A new method for β-aryl ether cleavage: lignin model studies. *Journal of Agricultural and Food Chemistry* 45: 4655–4660.

Lu, F. and Ralph, J. (1998). The DFRC method for lignin analysis. Part 2. Monomers from isolated lignins. *Journal of Agricultural and Food Chemistry* 46: 547–552.

Lu, F. and Ralph, J. (1999). Detection and determination of *p*-coumaroylated units in lignin. *Journal of Agricultural and Food Chemistry* 47: 1985–1992.

Lu, F. and Ralph, J. (2002). Preliminary evidence for sinapyl acetate as a lignin monomer in kenaf. *Chemical Communications*, 90–91.

Lu, F. and Ralph, J. (2005). Novel β–β structures in lignins incorporating acylated monolignols. *Appita*, 233–237.

Lu, F., Ralph, J., Morreel, K., Messens, E., and Boerjan, W. (2004). Preparation and relevance of a cross-coupling product between sinapyl alcohol and sinapyl *p*-hydroxybenzoate. *Organic & Biomolecular Chemistry* 2: 2888–2890.

Lui, A.C.W., Lam, P.Y., Chan, K.H., et al. (2020). Convergent recruitment of 5'-hydroxylase activities by CYP75B flavonoid B-ring hydroxylases for tricin biosynthesis in *Medicago* legumes. *New Phytologist* 228: 269–284.

Ma, Y.Q., Zhai, Y.M., Deng, Y., et al. (2017). Stilbeno-phenylpropanoids from *Gnetum montanum* Markgr. *Phytochemistry Letters* 21: 42–45.

Martinelli, T., Whittaker, A., Benedettelli, S., et al. (2017). The study of flavonolignan association patterns in fruits of diverging *Silybum marianum* (L.) Gaertn. chemotypes provides new insights into the silymarin biosynthetic pathway. *Phytochemistry* 144: 9–18.

Martínez, A.T., Rencoret, J., Marques, G. et al. (2008). Monolignol acylation and lignin structure in some non-woody plants: a 2D NMR study. *Phytochemistry* 69: 2831–2843.

Morikawa, T., Xu, F., Matsuda, H., and Yoshikawa, M. (2010). Structures of novel norstilbene dimer, longusone A, and three new stilbene dimers, longusols A, B, and C, with antiallergic and radical scavenging activities from Egyptian natural medicine *Cyperus longus*. *Chemical and Pharmaceutical Bulletin* 58: 1379–1385.

Mottiar, Y., Vanholme, R., Boerjan, W., et al. (2016). Designer lignins: harnessing the plasticity of lignification. *Current Opinion in Biotechnology* 37: 190–200.

Nakajima, K., Taguchi, H., Endo, T., and Yosioka, I. (1978). The constituents of *Scirpus fluviatilis* (Torr.) A. Gray. I. The structures of two new hydroxystilbene dimers, scirpusin A and B. *Chemical and Pharmaceutical Bulletin* 26: 3050–3057.

Neiva, D., Rencoret, J., Marques, G., et al. (2020). Lignin from tree barks: chemical structure and valorization. *ChemSusChem* 13: 4537–4547.

Piotrowska, H., Kucinska, M., and Murias, M. (2012). Biological activity of piceatannol: leaving the shadow of resveratrol. *Mutation Research* 750: 60–82.

Powell, R.G., Bajaj, R., and McLaughlin, J.L. (1987). Bioactive stilbenes of *Scirpus maritimus*. *Journal of Natural Products* 50: 293–296.

Quideau, S., Deffieux, D., Douat-Casassus, C., and Pouysegu, L. (2011). Plant polyphenols: chemical properties, biological activities, and synthesis. *Angewandte Chemie International Edition* 50: 586–621.

Raiber, S., Schroder, G., and Schroder, J. (1995). Molecular and enzymatic characterization of two stilbene synthases Eastern white pine (*Pinus strobus*). A single Arg/His difference determines the activity and the pH dependence of the enzymes. *FEBS Letters* 361: 299–302.

Ralph, J. (1996). An unusual lignin from kenaf. *Journal of Natural Products* 59: 341–342.

Ralph, J. (2010). Hydroxycinnamates in lignification. *Phytochemistry Reviews* 9: 65–83.

Ralph, J. (2020). Tricin and tricin-lignins in *Medicago* versus in monocots. *New Phytologist* 228: 11–14.

Ralph, J., Brunow, G., Harris, P.J., et al. (2008). Lignification: are lignins biosynthesized via simple combinatorial chemistry or via proteinaceous control and template replication? In: *Recent Advances in Polyphenol Research*, Vol. 1, (eds F. Daayf and V. Lattanzio), 36–66. Oxford: Wiley Blackwell.

Ralph, J., Grabber, J.H., and Hatfield, R.D. (1995). Lignin-ferulate cross-links in grasses: active incorporation of ferulate polysaccharide esters in ryegrass lignins. *Carbohydrate Research* 275: 167–178.

Ralph, J., Hatfield, R.D., Piquemal, J., et al. (1998). NMR characterization of altered lignins extracted from tobacco plants down-regulated for lignification enzymes cinnamyl-alcohol dehydrogenase and cinnamoyl-CoA reductase. *Proceedings of the National Academy of Science USA* 95: 12803–12808.

Ralph, J., Hatfield, R.D., Quideau, S., et al. (1994). Pathway of *p*-coumaric acid incorporation into maize lignin as revealed by NMR. *Journal of the American Chemical Society* 116: 9448–9456.

Ralph, J., Kim, H., Peng, J., and Lu, F. (1999) Arylpropane-1,3-diols in lignins from normal and CAD-deficient pines. *Organic Letters* 1: 323–326.

Ralph, J., Lapierre, C., and Boerjan, W. (2019). Lignin structure and its engineering. *Current Opinion in Biotechnology* 56: 240–249.

Ralph, J., Lundquist, K., Brunow, G., et al. (2004). Lignins: natural polymers from oxidative coupling of 4-hydroxyphenylpropanoids. *Phytochemistry Reviews* 3: 29–60.

Ralph, J., Peng, J., Lu, F., et al. (1999) Are lignins optically active? *Journal of Agricultural and Food Chemistry* 47: 2991–2996.

Rencoret, J., Gutiérrez, A., Nieto, L., et al. (2011). Lignin composition and structure in young versus adult *Eucalyptus globulus* plants. *Plant Physiology* 155: 667–682.

Rencoret, J., Kim, H., Evaristo, A.B., et al. (2018). Variability in lignin composition and structure in cell walls of different parts of macaúba (*Acrocomia aculeata*) palm fruit. *Journal of Agricultural and Food Chemistry* 66: 138–153.

Rencoret, J., Marques, G., Serrano, O., et al. (2020). Deciphering the unique structure and acylation pattern of *Posidonia oceanica* lignin. *ACS Sustainable Chemistry & Engineering* 8: 12521–12533.

Rencoret, J., Neiva, D., Marques, G., et al. (2019). Hydroxystilbene glucosides are incorporated into Norway spruce bark lignin. *Plant Physiology* 180: 1310–1321.

Rencoret, J., Prinsen, P., Gutiérrez, A., et al. (2015). Isolation and structural characterization of the milled wood lignin, dioxane lignin, and cellulolytic lignin preparations from brewer's spent grain. *Journal of Agricultural and Food Chemistry* 63: 603–613.

Rencoret, J., Ralph, J., Marques, G., et al. (2013). Structural characterization of lignin isolated from coconut (*Cocos nucifera*) coir fibers. *Journal of Agricultural and Food Chemistry* 61: 2434–2445.

Rinaldi, R., Jastrzebshi, R., Clough, M.T., et al. (2016). Paving the way for lignin valorisation: recent advances in bioengineering, biorefining and catalysis. *Angewandte Chemie (International Edition)* 55: 8164–8215.

Rivière, C., Pawlus, A.D., and Mérillon, J.M. (2012). Natural stilbenoids: distribution in the plant kingdom and chemotaxonomic interest in Vitaceae. *Natural Product Reports* 29: 1317–1333.

Roupe, K.A., Remsberg, C.M., Yáñez, J.A., and Davies, N.M. (2006). Pharmacometrics of stilbenes: seguing towards the clinic. *Current Clinical Pharmacology* 1: 81–101.

Smith, D.C.C. (1955). *p*-Hydroxybenzoate groups in the lignin of aspen (*Populus tremula*). *Journal of the Chemical Society*, 2347–2351.

Terashima, N., Fukushima, K., He, L.F., and Takabe, K. (1993). Comprehensive model of the lignified plant cell wall. In: *Forage Cell Wall Structure and Digestibility*, (eds H.G. Jung, D.R. Buxton, R.D. Hatfield, and J. Ralph), 247–270. Madison, WI: American Society of Agronomy, Crop Science Society of America, Soil Science Society of America.

Tobimatsu, Y., Chen, F., Nakashima, J., et al. (2013). Coexistence but independent biosynthesis of catechyl and guaiacyl/syringyl lignin polymers in seed coats. *Plant Cell* 25: 2587–2600.

Umezawa, T. (2003). Diversity in lignan biosynthesis. *Phytochemistry Reviews* 2: 371–390.

Vanholme, R., Demedts, B., Morreel, K., et al. (2010). Lignin biosynthesis and structure. *Plant Physiology* 152: 895–905.

Vanholme, R., De Meester, B., Ralph, J., and Boerjan, W. (2019). Lignin biosynthesis and its integration into metabolism. *Current Opinion in Biotechnology* 56: 230–239.

Vanholme, R., Morreel, K., Darrah, C., et al. (2012). Metabolic engineering of novel lignin in biomass crops. *New Phytologist* 196: 978–1000.

Vanholme, R., Morreel, K., Ralph, J., and Boerjan, W. (2008). Lignin engineering. *Current Opinion in Plant Biology* 11: 278–285.

Vermerris, W. and Boon, J.J. (2001). Tissue specific patterns of lignification are disturbed in the *brown midrib2* mutant of maize (*Zea mays* L.). *Journal of Agricultural and Food Chemistry* 49: 721–728.

Wang, C.Y., Lam, S.H., Tseng, L.H., and Lee, S.S. (2011). Rapid screening of lignans from *Phyllanthus myrtifolius* and stilbenoids from *Syagrus romanzoffiana* by HPLC-SPE-NMR. *Phytochemical Analysis* 22: 352–360.

Watts, K.T., Lee, P.C., and Schmidt-Dannert, C. (2006). Biosynthesis of plant-specific stilbene polyketides in metabolically engineered *Escherichia coli*. *BMC Biotechnology* 6: 22.

Wen, J.L., Sun, S.L., Xue, B.L., and Sun, R.C. (2013). Quantitative structural characterization of the lignins from the stem and pith of bamboo (*Phyllostachys pubescens*). *Holzforschung* 67: 613–627.

Wenzig, E., Kunert, O., Ferreira, D., et al. (2005). Flavonolignans from *Avena sativa*. *Journal of Natural Products* 68: 289–292.

Wu, M., Pang, J., Lu, F., et al. (2013). Application of new expansion treatment method on agricultural waste. Part 1: influence of pretreatment on the properties of lignin. *Industrial Crops & Products* 50: 887–895.

Yao, C.S., Lin, M., and Wang, L. (2006). Isolation and biomimetic synthesis of anti-inflammatory stilbenolignans from *Gnetum cleistostachyum*. *Chemical and Pharmaceutical Bulletin* 54: 1053–1057.

You, T.T., Mao, J.Z., Yuan, T.Q., et al. (2013). Structural elucidation of the lignins from stems and foliage of *Arundo donax* Linn. *Journal of Agricultural and Food Chemistry* 61: 5361–5370.

8

Complex Regulation of Proanthocyanidin Biosynthesis in Plants by R2R3 MYB Activators and Repressors

Dawei Ma and C. Peter Constabel

Centre for Forest Biology and Biology Department, University of Victoria, Victoria, Canada

8.1 Introduction to PAs and Flavan-3-ols

Proanthocyanidins (PAs), also called condensed tannins, are important plant secondary metabolites. They are widespread, but especially abundant in woody plants (Barbehenn and Constabel 2011). PAs are commonly present in the human diet in nuts, seeds, fruits, and berries, as well as in drinks such as cocoa, tea, and wine (Blade et al. 2016). The PAs are responsible for the astringency of fruits, tea, and wine, and in addition can have multiple beneficial effects on health. High PA intake is associated with reduced risk of cardiovascular diseases, metabolic syndrome, and some neurological conditions and cancers (Blade et al. 2016; Nunes et al. 2016). In the plant, PAs are important in protection of seed and fruit, and a role of leaf PAs in defense against vertebrate herbivores and pathogens is documented (Bailey et al. 2004; Wang et al. 2017; Ullah et al. 2017). When deposited in soil as leaf litter, PAs can also impact nutrient cycling and decomposition (reviewed in Constabel et al. 2014). Recent evidence suggests that PAs may help to scavenge reactive oxygen species (ROS) during abiotic stress (Gourlay and Constabel 2019).

PAs are oligomers or polymers of flavan-3-ols. PAs are thus a product of the phenylpropanoid and flavonoid pathways, and share most enzymes with anthocyanin biosynthesis. The elucidation of the PA pathway was facilitated by the Arabidopsis *transparent testa* (*tt*) mutants, which show a light-colored seed coat endothelium phenotype due to their reduced PA accumulation (Tohge et al. 2017; Constabel 2018). Like many herbaceous plants, Arabidopsis accumulates PAs only in the seed coat (Debeaujon et al. 2003). Flavonoid biosynthesis begins with the condensation of *p*-coumaroyl-CoA and three malonyl-CoA units to naringenin chalcone by chalcone synthase (CHS), an enzyme belonging to the polyketide synthase superfamily. Chalcone isomerase (CHI) converts chalcone to flavanone, which is then oxidized at the 3-position by flavanone 3-hydroxylase (F3H) to produce dihydrokaempferol. Flavonoid 3'-hydroxylase (F3'H) and flavonoid 3'5'-hydroxylase (F3'5'H) catalyze the hydroxylation of dihydrokaempferol on the B-ring at 3'- and 5'-positions to dihydroquercetin and dihydromyricetin, respectively; hydroxylation can also occur on the

Recent Advances in Polyphenol Research, Volume 7, First Edition. Edited by Jess Dreher Reed, Victor Armando Pereira de Freitas, and Stéphane Quideau.

flavanone stage. Dihydroflavonols are reduced by dihydroflavonol reductase (DFR) to leucoanthocyanidins, and then acted on by anthocyanidin synthase (ANS) to produce the anthocyanidins, intermediates for both anthocyanin and PA. Interestingly, ANS and a homolog of leucoanthocyanidin dioxygenase (LDOX) were recently shown to have roles in generating extension and starter units, respectively, for PA dimers (Jun et al. 2018). Dihydroflavonols can also be reduced by flavonol synthase, producing flavonols in a competing branch of flavonoid biosynthesis.

Leucoanthocyanidin reductase (LAR) and anthocyanidin reductase (ANR) are PA-specific enzymes which produce catechin (2,3-*trans*-flavan-3-ol) and epicatechin (2,3-*cis*-flavan-3-ol), respectively. LAR was first cloned and characterized in the legume *Desmodium uncinatum* (Tanner et al. 2003), although its activity was described much earlier (Stafford and Lester 1984, 1985). The first characterized ANR gene was *BANYULS* (*BAN*) from Arabidopsis. The *ban* mutant has a reddish testa phenotype and was ultimately shown to encode ANR, the enzyme which reduces anthocyanidins to epicatechin (Xie et al. 2003). Interestingly, Arabidopsis does not contain an LAR gene and thus does not synthesize catechin. Most plant species, however, utilize both catechin and epicatechin subunits for PA synthesis.

Despite extensive knowledge of flavan-3-ol synthesis, new enzymes are still being discovered. For example, an Arabidopsis chalcone isomerase-like (CHIL) gene specifically involved in PA and flavonol biosynthesis was recently described (Jiang et al. 2015). CHIL mutants showed reduced PA and flavonol accumulation in seeds but without a transparent testa phenotype. Overexpression of CHIL in *chs* or *chi* mutants did not rescue the transparent testa phenotype, which suggests a distinct function of CHIL from CHS or CHI. Interestingly, CHIL physically interacts with CHI in yeast two-hybrid and protein florescence complementation assays, leading the authors to propose a mechanism whereby CHIL interacts with CHI to enhance either PA or anthocyanin biosynthesis (Jiang et al. 2015). This gives rise to the possibility that PA and anthocyanin biosynthesis may separate earlier in the flavonoid pathway than previously believed.

The final steps in PA biosynthesis, comprised of glycosylation, transport, and polymerization of the monomers, are long-standing and mostly unresolved questions (Dixon et al. 2005; Zhao and Dixon 2010; Zhao et al. 2010; Zhao 2015; Constabel 2018). UDP-glycosyltransferases (UGTs) are proposed to catalyze catechin or epicatechin glycosylation prior to transport to the vacuole. The UGT72L1 enzyme from *Medicago truncatula* specifically converts epicatechin to epicatechin 3'-*O*-glucoside (Pang et al. 2008), and UGT84A22 from tea (*Camellia sinensis*) glycosylates precursors of galloylated catechin (Cui et al. 2016). Arabidopsis UGT80B1, encoded by the TT15 gene, was found to act upstream of PA biosynthesis although its substrates remain to be defined (Xu et al. 2017). Catechin or epicatechin glycosides are thought to be translocated to the vacuole by specific transporters (Zhao 2015). Medicago MATE1 and Arabidopsis TT12 genes encode multidrug and toxic compound extrusion (MATE) transporters that can move epicatechin glycoside across membranes (Marinova et al. 2007; Zhao and Dixon 2009). In addition, a glutathione S-transferase (GST) encoded by Arabidopsis TT19 is implicated in PA, anthocyanin, and flavonol transport via a glutathione pump (Li et al. 2011). Substrate specificity of GSTs is seen in grapevine (*Vitis vinifera*), where VvGST1 and VvGST4 transport both anthocyanin and PA precursors, but VvGST3 appears to specifically transport the latter (Pérez-Díaz et al. 2016).

PA polymerization is presumed to occur in the vacuole but remains enigmatic. Both enzymatic and nonenzymatic radical-based mechanisms have been proposed (reviewed in Constabel 2018). In Arabidopsis, a laccase-type phenol oxidase (*tt10*) mutant shows delayed browning of the seed coat and an increase in soluble PAs, which suggests a role of this enzyme in a possible oxidative polymerization mechanism (Pourcel et al. 2005). Recently, the LAR enzyme in Medicago was shown to have the ability to cleave S-cysteinyl-epicatechin, a newly proposed intermediate of the PA polymerization reaction (Liu, Wang et al. 2016). Loss of this enzyme leads to an increase in PA polymer length, and greater proportion of insoluble PA. Despite these recent advances, however, the biochemistry of PA polymer assembly remains very much a mystery.

8.2 Regulation of PA and Flavonoid Biosynthesis by MYB Transcription Factors

PA and flavonoid biosynthesis is regulated primarily by MYB transcription factors. These are encoded by large gene families in plants, and they participate in many aspects of plant development, stress responses, and metabolism. Their central role in control of phenylpropanoid biosynthesis, including flavonoids, has been extensively documented (Liu et al. 2015; Xu et al. 2014). MYB factors are classified by the number of N-terminal DNA binding repeats; those containing two repeats are the most common and referred to as R2R3 MYBs (Kranz et al. 1998; Dubos et al. 2010). The R2R3 MYBs are by far the most important for phenylpropanoid and PA metabolism; for the sake of convenience, here we will simply refer them as MYB factors.

In addition, in anthocyanin and PA biosynthesis the MYB factor requires two additional transcription factors, a basic helix-loop-helix (bHLH) and a WD40-repeat (WDR) protein, which together form the so-called MBW complex. MYB and bHLH regulators of anthocyanin biosynthesis were first identified in studies of anthocyanin biosynthesis in maize (*Zea mays*) (Paz-Ares et al. 1987; Goff et al. 1992; Ludwig et al. 1989). WDR genes were later found to also be required (de Vetten et al. 1997; Walker 1999). Other MBW complexes participate in determining cell differentiation, including trichome and root hair development as well as other epidermal cell fates (Ramsay and Glover 2005). The same bHLH and WDR cofactors can interact with several different MYB factors, and these combinatorial interactions allow them to participate in multiple regulatory functions. However, the MYB acts as the key regulator that determines the specificity of the MBW complex. While the TT8 bHLH factor is involved in regulation of both PAs and anthocyanins, it partners with different MYBs for each of these functions. Likewise, bHLH protein GL3 participates in MBW complexes that regulate anthocyanin accumulation as well as trichome and root hair formation (Ramsay and Glover 2005).

The first characterized PA-specific MYB factor to be discovered was Arabidopsis TT2 (MYB123) (Nesi et al. 2001). *tt2* mutants showed reduced PA accumulation in the endothelium of Arabidopsis seeds, while anthocyanin accumulation in vegetative tissues was unaffected. This suggested that distinct MYB transcription factors are responsible for these pathways. Arabidopsis TT2 does not affect the expression of 'early' flavonoid pathway genes such as CHS, CHI, F3H, and F3'H or FLS, but is specific to 'late' flavonoid pathway genes DFR, LDOX, and ANR (Nesi et al. 2001). Many other TT2-type PA-specific MYB transcription factors have now been characterized (Figure 8.1). For example, grape R2R3 MYB VvMYBPA2 and Medicago MtMYB14 and MtPAR are the TT2-type MYBs that promote PA accumulation in these systems (Terrier et al. 2008; Verdier et al. 2012; Liu et al. 2014). TT2 orthologs from lotus (*Lotus japonicus*) activate the ANR promoter when transiently expressed in Arabidopsis leaves (Yoshida et al. 2008). In poplar, PtMYB134 is the TT2 ortholog. When the PtMYB134 regulator is overexpressed in transgenic poplar, PA accumulation in leaves and many other tissues is strongly enhanced. In these transgenic poplars, both early and late flavonoid pathway genes are upregulated, and gel shift assays confirm that MYB134 can interact with both late and early flavonoid gene promoters (Mellway et al. 2009; James et al. 2017). However, the synthesis of other phenolic secondary metabolites is reduced, suggesting additional complexity in the transcriptional regulation of phenylpropanoids in poplars (Mellway et al. 2009).

A second type of PA-specific MYB was first described from grapevine berries and seeds (Bogs et al. 2007). The VvMYBPA1 gene falls into a different phylogenetic clade from the TT2 orthologs (Figure 8.1), and induces ectopic PA accumulation in vegetative organs when overexpressed in Arabidopsis (Bogs et al. 2007). MYBPA1-type MYBs have been identified in other plant species, including persimmon (*Diospyros kaki*), cacao (*Theobroma cacao*), and poplar (*Populus* spp.). In poplar, the MYBPA1-type PtMYB115 regulates PA accumulation in leaves via activation of biosynthesis genes, and like the TT2-like PtMYB134 gene, its overexpression in transgenic poplar leads to a high-PA phenotype. Interestingly, overexpression of these MYBs also enhances hydroxylation of PA by promoting flavonoid 3'5'-hydroxylase (F3'5'H) expression (James et al. 2017). Whether MYBPA1-type MYBs are widely distributed in other species, and if this type of PA MYB has additional functions, still requires further study.

A third class of R2R3 MYB activator, the MYB5-type, can regulate both anthocyanin and PA accumulation and was first identified in grape as VvMYB5a and VvMYB5b. Overexpression of either gene in tobacco induces both anthocyanins and PAs by stimulating expression of phenylpropanoid biosynthesis genes (Deluc et al. 2006, 2008). Arabidopsis *myb5* mutants show abnormal trichome production and mucilage in the outer seed coat, but are not affected in PA accumulation. However, *myb5* and *tt2* double Arabidopsis mutants showed an enhanced transparent testa phenotype compared to the single mutants (Gonzalez et al. 2009), and overexpression of AtMYB5 in *tt2* mutants also partially rescues the transparent testa phenotype (Xu et al. 2014). This suggests that MYB5 is also indirectly involved in PA biosynthesis. Likewise, the VvMYB5a ortholog in Medicago, MtMYB5, also induces PA accumulation (Liu et al. 2014). In poplar, VvMYB5 orthologs PtMYB006 and PtMYB126 are downregulated in PA and anthocyanin MYB repressor-overexpressing poplars (Ma et al. 2018). Overexpression of PtMYB006 in *Populus tomentosa* promotes anthocyanin and PA accumulation but reduces secondary cell wall formation (Wang et al. 2019). MYB5-type MYBs may show different functions in different systems; it is likely they indirectly regulate PA biosynthesis by affecting other branches of the phenylpropanoid pathway or other transcription factors.

Other classes of MYB transcription factors may also be involved in PA biosynthesis. Recently a distinct type of PA MYB regulator, PpMYB7, was identified in peach (*Prunus persica*). PpMYB7 does not belong to one of the three PA MYB clades described earlier.

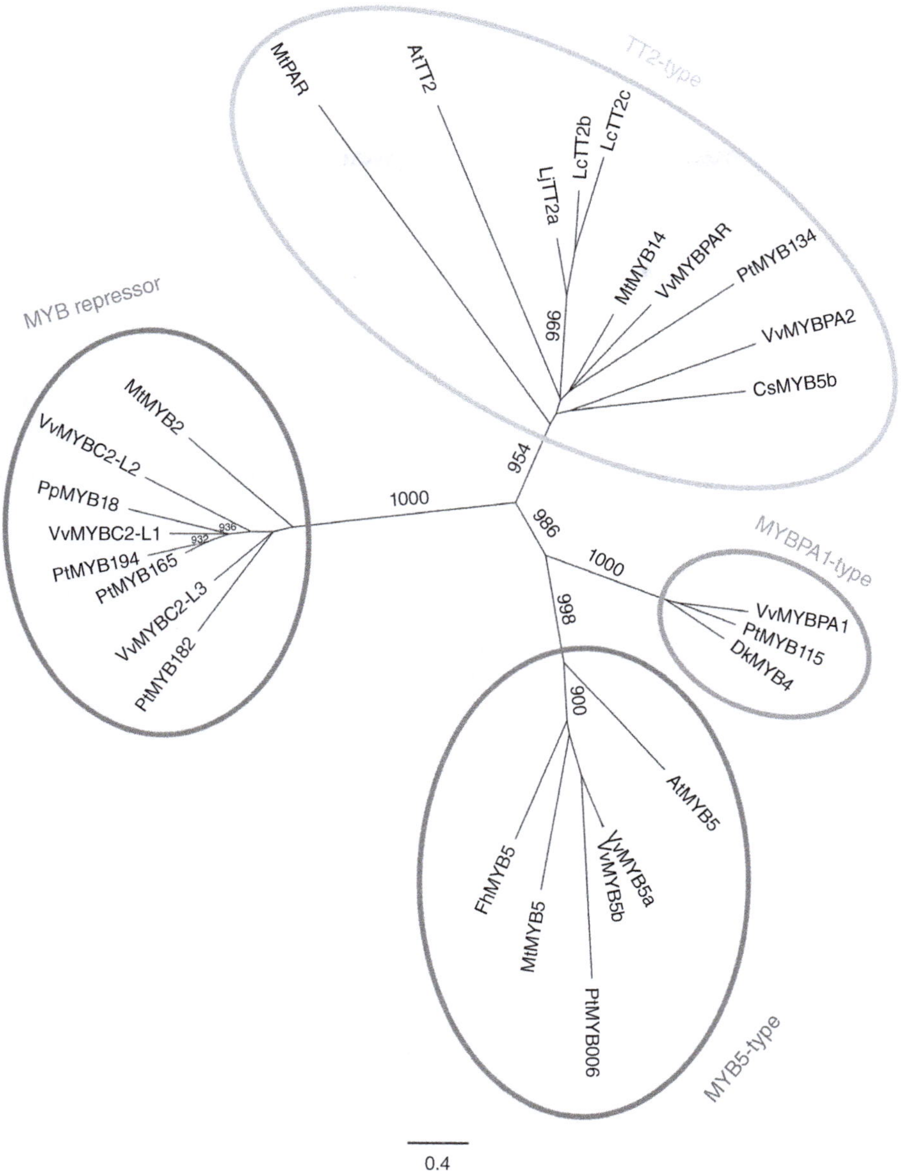

Figure 8.1 Phylogenetic analysis of characterized PA-regulating MYBs. Phylogenetic analysis of three major groups of PA MYB activators and one group of PA MYB repressors. The protein sequences were taken from NCBI (www.ncbi.nlm.nih.gov/). The multiple sequence alignment was performed using the Multiple Sequence Comparison by Log-Expectation (MUSCLE) method in Mega-X (version 10.1 www.megasoftware.net/). The best-fit model was determined by Prottest3 (Darriba et al. 2011) to be Jones-Taylor-Thornton with gamma distribution and invariant sites (JTT+I+G). The phylogenetic tree was generated using the maximum likelihood method in Mega-X. Scale bar represents mean substitutions per site. Protein sequences were obtained from NCBI (www.ncbi.nlm.nih.gov/). AtMYB5 (At3G13540, *Arabidopsis thaliana*); AtTT2 (NM_122946.3, *A. thaliana*); CsMYB5b (KY827397.1, *Camellia sinensis*); DkMYB4 (BAI49721.1, *Diospyros kaki*), FhMYB5 (MK168337.1, *Freesia hybrida*); LjTT2a (BAG12893.1, *Lotus japonicus*); LcTT2b (KY196475.1, *L. corniculatus*); LcTT2c (KY196476.1, *L. corniculatus*); MtMYB2 (EF468475.1, *Medicago truncatula*); MtMYB5 (XM_003601561.3, *M. truncatula*); MtMYB14 (JN225514.1, *M. truncatula*); MtPAR (HQ337434.1, *M. truncatula*); PpMYB18 (KT159234.1, *Prunus persica*); PtMYB006 (KF811025.1, *Populus tomentosa*); PtMYB115 (XP_002302644.1, *P. tremula* x *P. tremuloides*); PtMYB134 (ACR83705.1, *P. tremuloides*); PtMYB165 (XP_002315890.2, *P. tremula* x *P. tremuloides*); PtMYB182 (XP_002305872.1, *P. tremula* x *P. tremuloides*); PtMYB194 (XP_002311495.2, *P. tremula* x *P. tremuloides*); VvMYB5a (AAS68190.1, *Vitis vinifera*); VvMYB5b (AAX51291.1, *V. vinifera*); VvMYBC2-L1 (JJX050227.1, *V. vinifera*); VvMYBC2-L2 (NM_001281251.1, *V. vinifera*); VvMYBC2-L3 (KM046932.1, *V. vinifera*); VvMYBPA1 (CAJ90831.1, *V. vinifera*); VvMYBPA2 (ACK56131.1, *V. vinifera*); VvMYBPAR (AB911341.1, *V. vinifera*). (*See insert for color representation of the figure.*)

Interestingly, PpMYB7 activates DFR and LAR but not ANR in transient promoter activation assays (Zhou et al. 2015), and thus has the potential to alter PA subunit composition. Orthologs of this MYB in other plant species, and their potential targets, await more investigation. Another MYB, the apple (*Malus domestica*) MdMYB23, was found to promote PA accumulation when overexpressed in apple callus and Arabidopsis, and can directly bind to MYB recognition sites in ANR promoters (An et al. 2018). However, its ortholog in Arabidopsis, AtMYB23, is involved in trichome and root hair development (Kirik et al. 2005; Kang et al. 2009). Again this could suggest different roles of this subgroup of MYBs in different species. As mentioned, MYBs controlling anthocyanin and PA biosynthesis and epidermis cell differentiation share similar regulatory mechanisms, as in both cases the R2R3 MYBs are only active within a MBW complex. Thus these two groups of MYBs may also share other functions.

8.3 The Importance of Repressor MYBs in PA and Flavonoid Metabolism

Repressor MYBs that affect the transcriptional regulation of phenylpropanoid biosynthesis were first reported 20 years ago (Tamagnone et al. 1998; Jin et al. 2000), generally in the context of monolignol biosynthesis. Only recently have repressor MYBs become recognized as important components in the regulation of flavonoid synthesis as well (Ma and Constabel 2019). These are best characterized in flavonoid-rich model systems, including poplar leaf Medicago seed coat, and grape berry skin and seed. These MYB repressors usually fall into subgroup four of the R2R3 MYB clade, and are characterized by the conserved C-terminal repression motifs called the ethylene-responsive element binding factor-associated amphiphilic repression (EAR) motif. Compared to the MYB activators, flavonoid MYB repressors have broader effects; rather than being specific for one type of flavonoid end product, MYB repressors can regulate both anthocyanin and PA biosynthesis.

In poplar, where PA biosynthesis has been studied extensively and two PA MYB activators, a suite of repressor-type MYBs were found to be upregulated in PtMYB134- and PtMYB115-overexpressing poplars, suggesting they also participate in PA regulation. The roles of these MYB repressors, PtMYB182, PtMYB165, and PtMYB194 – all R2R3 MYBs – have now been studied in detail (Yoshida et al. 2015; Ma et al. 2018) and found to suppress both PA and anthocyanin biosynthesis when overexpressed in hybrid poplar. Transcriptome analysis of these repressor-overexpressing plants indicates that expression of many genes encoding flavonoid enzymes as well as the MBW components is reduced. In transient expression assays, the repressor MYBs all directly inhibit the activation of PA biosynthesis genes DFR, LAR, ANR, and ANS by PtMYB134 and PtMYB115 (Ma et al. 2018). They therefore directly interfere with activation of relevant promoters by the MBW complex. Based on yeast two-hybrid analyses, this is at least in part due to direct binding of the MYB repressors with the activating bHLH cofactor, in effect preventing the bHLH from participating in the MBW complex (James et al. 2017; Ma et al. 2018). The R3-MYB repressor PtMYB179 is also able to suppress activation of PA genes in transient promoter activation assays, and therefore may influence PA biosynthesis (Yoshida et al. 2015). Other mechanisms of repression may also be involved (see below).

Similar R2R3 MYB repressors have been discovered in grapevine berries, Medicago seeds, and peach fruits. In grapevine, three R2R3 MYB repressors, VvMYBC2-L1, VvMYBC2-L2, and VvMYBC2-L3, have been characterized. VvMYBC2-L1 and VvMYBC2-L3 are highly expressed in berries, and appear to impact both anthocyanin and PA biosynthesis (Cavallini et al. 2015). Similarly, the Medicago MYB repressor MtMYB2 is observed to suppress PA biosynthesis when overexpressed in hairy roots, while *MtMYB2* mutants are impacted in both anthocyanin accumulation in leaves and PA accumulation in seeds (Jun et al. 2015). Mutation of the EAR motif affects its repression function, but the mechanism of action is not known (Jun et al. 2015). In peach, the PpMYB18 gene was recently characterized as a flavonoid repressor expressed at the ripening and juvenile stage of peach fruits. PbMYB18 inhibits DFR, LAR, and UFGT promoter activation in transient expression in tobacco, and suppresses PAs in seeds when overexpressed in Arabidopsis. PpMYB18 may thus serve to prevent cells over accumulating anthocyanin or PA during fruit development (Zhou et al. 2019).

The large number of MYB repressors with impact on the flavonoid pathways in poplar and other plants suggests they are important components in the transcriptional regulatory network of flavonoid biosynthesis. Much work remains to be done to elucidate their mechanisms of action, in particular on the role of the conserved repressor (EAR) motif (reviewed in Ma and Constabel 2019). Whereas some studies have shown it to be essential (Jun et al. 2015), in other cases its deletion did not affect repressor activity. Likewise, unlike their lignin-specific counterparts, the flavonoid R2R3 repressor MYBs have not yet been shown to directly bind DNA or promoter elements. However, the ability of both R2R3 and R3 repressor MYBs to bind the bHLH cofactor is well established, as well as on their *in vivo* interactions with other components of the regulatory network in yeast two-hybrid or bimolecular fluorescence complementation assays (Ma and Constabel 2019).

8.4 The Complex Interaction of PA MYB Activators, MYB Repressors, and bHLH Transcription Factors

Transcriptional regulation of PA biosynthesis involves multiple PA MYB activators working with bHLH and WDR factors in the MBW complex, as well a several MYB repressors. Together these transcription factors constitute the basis of a regulatory network for PA biosynthesis, the breadth of which is beginning to be described in poplar. As outlined in Section 8.2, the PA pathway in poplar is activated by at least two PA-specific MYBs, PtMYB134 and PtMYB115 (James et al. 2017). These activator MYBs act within the MBW complex to directly stimulate transcription of the PA pathway genes, with the repressor MYBs suppressing this activation. In addition to their direct regulation of PA enzyme promoters, these activator MYBs also regulate each other and cooperate in a regulatory network. This was demonstrated by assaying PtMYB134 and PtMYB115 promoters with both MYB proteins in reciprocal transient promoter activation assays in poplar suspension cells. As predicted, PtMYB134 induces PtMYB115 expression; surprisingly, in the reciprocal experiment PtMYB134 expression was found to be regulated by PtMY115. Furthermore, both MYBs regulate their own promoters, leading to autoregulation and additional positive feedback (James et al. 2017). In addition, both MYB115 and MYB134 can induce expression of the required cofactor, PtbHLH131 (Yoshida et al. 2015; James et al. 2017). The observed interactions are summarized in a model of MYB and bHLH interactions in poplar, and highlight the feedback loops (Figure 8.2). An induction of bHLH cofactors by the corresponding R2R3 MYB activator has been observed in other species: Arabidopsis bHLH TT8 is regulated by TT2 and the anthocyanin regulator MYB75 (Baudry et al. 2004), and in Medicago, both PA- and anthocyanin-specific MYBs stimulate expression of their required cofactor, MtTT8 (Li et al. 2016).

As mentioned in Section 8.2, in PtMYB134 and PtMYB115 overexpressor plants, the R2R3 MYB repressors PtMYB182, PtMYB165, and PtMYB194, as well as the CPC-type R3-MYB repressor MYB179, are upregulated and thus predicted to be controlled by the activator MYBs (James et al. 2017). To test this directly, we cloned PtMYB165 and PtMYB179 promoters into reporter gene constructs for promoter activation assays in poplar suspension cells. Both MYB repressor promoters were clearly activated by PtMYB134 in the presence of the appropriate bHLH cofactor (Figure 8.3), indicating that repressor MYBs can be controlled by the activating MYBs. Similar regulation was shown previously in the forage legume (*Trifolium repens*). The R2R3 MYB repressor TrMYB134 promoter could be

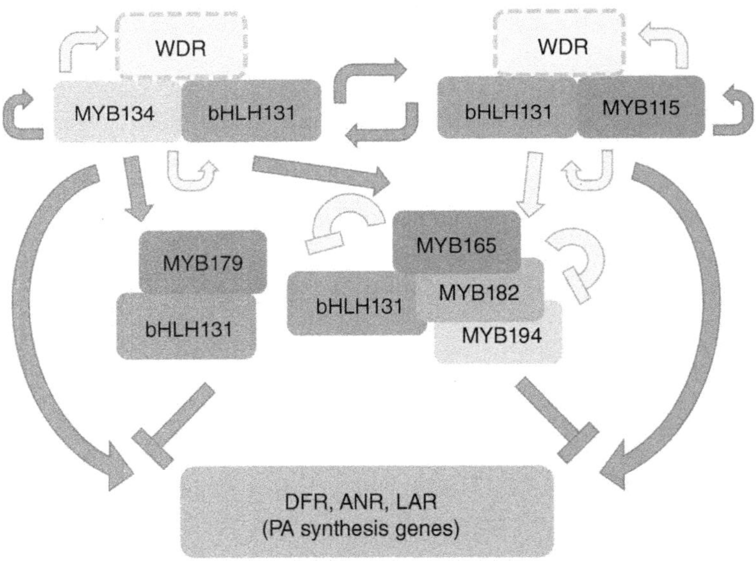

Figure 8.2 Model depicting the interaction of MYB activators, MYB repressors, and bHLH cofactors in poplar. PtMYB134- and PtMYB115-containing MBW complexes (top) activate genes encoding PA enzymes (bottom) as well as numerous other regulators. Solid arrows represent activation or repression demonstrated directly by promoter activation assays in transiently transformed poplar cells. Hollow arrows represent activation or repression inferred solely from transcriptomic analysis of MYB-overexpressing transgenic poplars. Microarray analysis suggests that PtMYB134 and PtMYB115 upregulate PtWDR, PtbHLH131, PtMYB165, PtMYB194, PtMYB182, and PtMYB179 expression (James et al. 2017). Transient promoter activation assays further confirm PtMYB165 and PtMYB194 induction by MYB134 (see Figure 8.3), and that MYB134 and MYB115-containing MBW complexes activate each other's and their own promoters (James et al. 2017). RNA-seq analysis of MYB165-overexpressor plants indicates that MYB182, MYB194, and bHLH131 expression is downregulated by MYB165 (Ma et al. 2018). (*See insert for color representation of the figure.*)

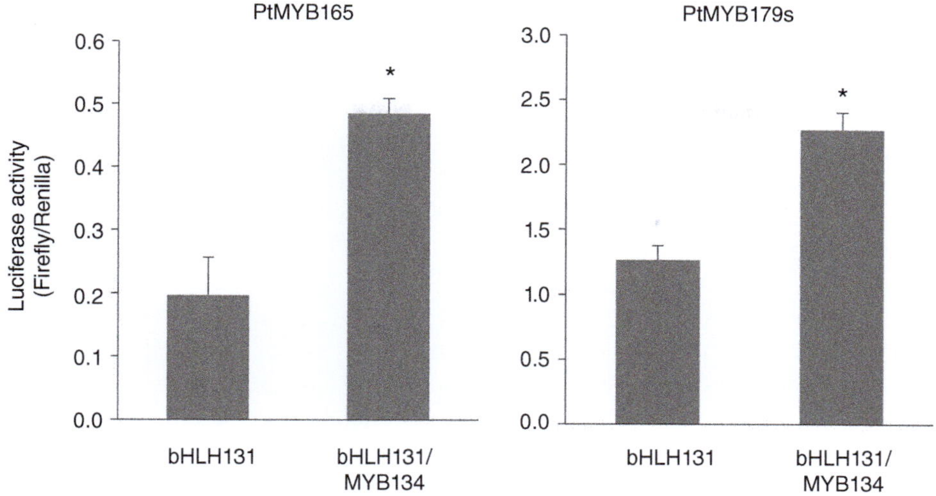

Figure 8.3 Transactivation of poplar PtMYB165 and PtMYB179 promoters by PtMYB134 and PtbHLH131 in transiently transformed poplar cells. The PtMYB165 and PtMYB179 promoters were fused to the firefly luciferase reporter gene in the dual luciferase reporter vector pGREEN800-LUC. This vector also contains constitutively expressed Renilla luciferase used to normalize for transformation efficiency. Plasmid were introduced into poplar suspension cells by biolistic bombardment, together with overexpression constructs encoding the PtbHLH131 and PtMYB134 activators (see Ma et al. 2018 for details). The activation of PtMYB165 (left) and PtMYB179 (right) promoters by PtMYB134 in the presence of PtbHLH131, normalized by Renilla activity, is shown. Asterisks represent significant differences from the bHLH-only controls ($P < 0.05$ in Student's t-test).

activated by PA activator TrMYB14 together with a bHLH factor, TrAN1. Moreover, this activation could be suppressed by TrMYB134 itself (Albert et al. 2015). Although it first appears counterintuitive that PA activators can also induce repressor MYBs, one of their roles may be to prevent overaccumulation of PAs after induction of the PA pathway stress, as well as to repress competing pathways. Additional information of the *in vivo* interactions of activator and repressor MYBs will come with more detailed transcriptome analysis at different time intervals after stress.

An additional layer of complexity is added to the network of transcription factors by varying degrees of specificity of the MYB-bHLH protein interactions. Earlier work had shown that a given MYB can use one of several bHLH cofactors to generate a functional MBW complex. For example, Arabidopsis TT2 can interact with the bHLHs TT8, GL3, or EGL3 to activate BAN promoter (Xu et al. 2014). Similarly, in poplar PtMYB134 can have productive interactions with either bHLH079 (GL3-type) or bHLH131 (TT8-type) (Yoshida et al. 2015). The bHLH transcription factors form a very large family, and these poplar bHLHs belong to subgroup IIIf (Feller et al. 2011). Group III bHLH proteins contain an N-terminal MYB interaction region (MIR), and subgroup IIIf includes bHLHs that participate in MBW complexes that regulate epidermal cell fate in Arabidopsis, the accumulation of anthocyanins in leaves, and PAs in seed coat (Feller et al. 2011). Subgroup IIIf can be further divided into two subclades, the GL3 clade and TT8 clade (Figure 8.4). GL3-type bHLHs are associated with anthocyanin accumulation and epidermal cell differentiation (Payne et al. 2000; Bernhardt et al. 2003; Feyissa et al. 2009), and it has been suggested that

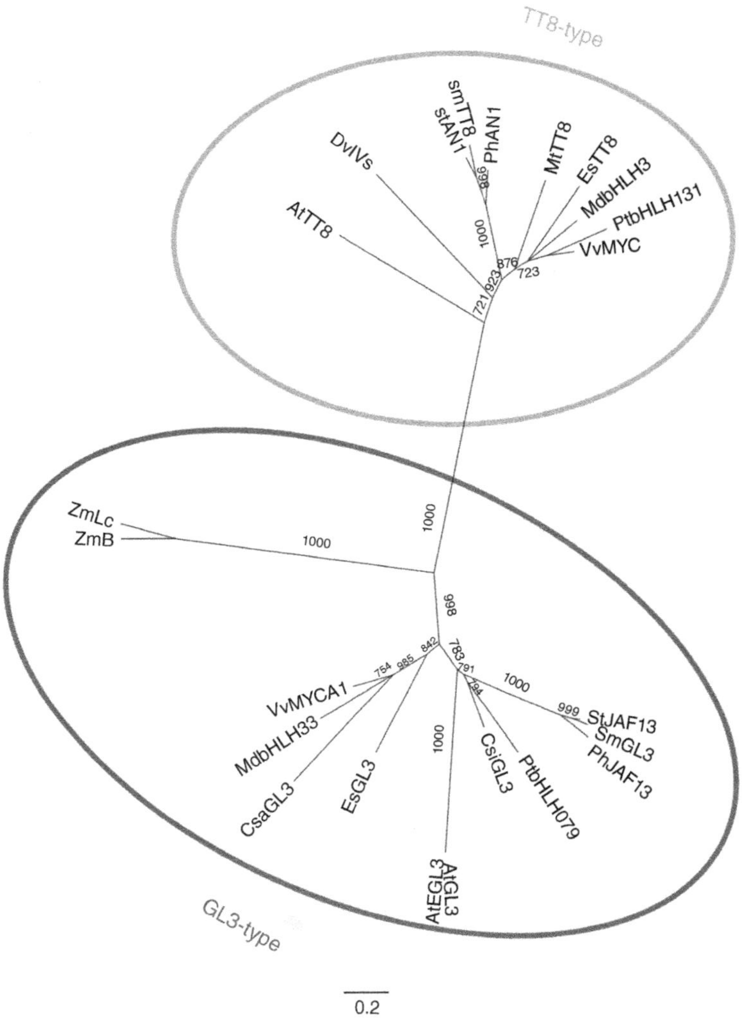

Figure 8.4 Phylogenetic analysis of two major groups of bHLHs active in MBW complexes for PA regulation. The tree was generated using the maximum likelihood method in Mega-X as outlined for Figure 8.1. Scale bar represents mean substitutions per site. AtEGL3 (NM_001198373.2, *Arabidopsis thaliana*); AtGL3 (NM_148067.4, *A. thaliana*); AtTT8 (NM_117050.3, *A. thaliana*); CsaGL3 (JN190488.1, *Cucumis sativus*); CsiGL3 (KU886538.1, *Camellia sinensis*); DvIVS (AB787560.1, *Dahlia pinnata*); EsGL3 (KJ010529.1, *Epimedium sagittatum*); EsTT8 (KC686401.1, *E. sagittatum*); MdbHLH3 (HM122458.1, *Malus domestica*); MdbHLH33 (DQ266451.1, *M. domestica*); MtTT8 (KM892777.1, *Medicago truncatula*); PhAN1 (AF260919.1, *Petunia hybrida*); PhJAF13 (AF020545.1, *P. hybrida*); PtbHLH079 (KP723395.1, *Populus tremula* x *P. tremuloides*); PtbHLH131 (KJ909757.1, *P. tremula* x *P. tremuloides*); SmGL3 (KP006502.1, *Solanum melongena*); SmTT8 (KT591486.1, *S. melongena*); StAN1 (NP_001275098.1, *S. tuberosum*); StJAF13 (KP317176.1, *S. tuberosum*); VvMYC (NM_001281253.1, *Vitis. vinifera*); VvMYCA1 (NM_001281025.1, *V. vinifera*); ZmB (X57276.1, *Zea mays*); ZmLc (DQ414252.1, *Z. mays*). (*See insert for color representation of the figure.*)

different bHLH cofactors work partially redundantly in a tissue-specific manner (Xu et al. 2015). This flexibility of MBW components may explain the broader repressive function that is typical for MYB repressors; binding bHLH cofactors would disrupt both a PA and anthocyanin-specific MBW complex and thus impact both types of end products.

The transcriptional regulation network of PA biosynthesis is complex and includes many potential positive and negative feedbacks and regulatory interactions, including autoregulation of several MYBs. How these regulatory loops act *in vivo* still needs to be explored.

8.5 Developmental and Plant Hormone-Mediated Regulation of the PA Pathway via MYBs

The regulation of PA biosynthesis by the interplay of MYB activators, repressors, and the MBW complex described in Section 8.4 is ultimately controlled by developmental and hormonal signals. This aspect of developmental regulation of PAs has mainly been studied in *Arabidopsis thaliana* and *Medicago truncatula,* where PAs accumulate exclusively within the developing seed. In Arabidopsis, PA accumulates in three different seed cell types: micropylar, endothelial, and chalazal cells (Debeaujon et al. 2003). PA promoter ß-glucuronidase fusion constructs expressed in wild type and *tt2* (MYB), *tt8* (bHLH), and *ttg1* (WDR) mutant backgrounds have provided detailed insight into PA pathway regulation (Xu et al. 2014). In the *ttg1* mutant background, no promoter activity can be seen; however, in *tt2* and *tt8* mutants, weak but detectable promoter activity indicates that additional MYB and bHLH genes (i.e. AtMYB5 and AtGL3) also contribute to the transcriptional control of PA accumulation (Xu et al. 2014). In addition, a MADS box transcription factor (TT16) acts upstream of the MBW complex and regulates cell differentiation in seeds, and affects PA accumulation in the endothelial cells (Nesi et al. 2002; Xu et al. 2017). Likewise, a zinc finger protein involved in cell differentiation, TT1, interacts directly with TT2, and thus functions in PA accumulation in endothelial cells (Sagasser et al. 2002). In Medicago seeds, a model of the MBW complex as the major transcriptional regulator of PAs has also been established, with the additional involvement of the MtMYB2 repressor (Liu et al. 2014; Li et al. 2016; Pang et al. 2009; Jun et al. 2015). Here, PAs mainly accumulate in the epidermal cell layer on the hilar side of the seed coat, and are spatially restricted to these cells by the MtMYB2 repressor; *Mtmyb2* mutant seeds show PAs diffusion from the hilar side of the seed to the middle of the seed (Jun et al. 2015). The repressor MYBs thus help to fine-tune PA accumulation and distribution during development.

Seed coat development and PA accumulation are also affected by the plant hormones, as shown by experiments on Arabidopsis seeds with exogenously applied auxin and gibberellin (GA) (Figueiredo et al. 2016). Furthermore, the bHLH transcription factor, TCP3, can negatively modulate the auxin response, but also induces anthocyanin and PA accumulation by interacting with the R2R3-MYB and TT8 protein complex (Li and Zachgo 2013). TCP3 may thus act as a bridge between auxin signaling and PA accumulation in Arabidopsis seeds. In fruit, hormones such as ethylene, jasmonic acid (JA), and abscisic acid (ABA) are known to enhance PA accumulation. For example, in apple, a recent study showed that ethylene induces anthocyanin and PA accumulation via ERF1B, an ethylene response factor (ERF). ERF1B is a transcriptional regulator, which can directly interact with two TT2-type PA-MYB transcription factors from apple, MdMYB9 and MdMYB11, and binds to their promoters (Zhang et al. 2018). Consistent with this, overexpression of ERF1B in apple calli results in increased accumulation of both anthocyanin and PA. Likewise, in grape berry skins, the ethylene-producing compound ethephon as well as light treatment stimulate expression of VvMYBPA1 and increase PA content (Liu, Song et al. 2016). A grape ERF,

VviERF045, was also shown to induce PA accumulation in grapevine leaves when overexpressed (Leida et al. 2016). Together, these results suggest that in grapevine and in apple fruit, ethylene may regulate PA accumulation through ERF and MYB interactions.

The regulation of the PA pathway by jasmonates, including methyl jasmonate (MeJA), has also been demonstrated in fruit. MeJA-treated apple showed increased accumulation of anthocyanin and PA, together with upregulation of MdMYB9 and MdMYB11. Further investigation showed that the Jasmonate ZIM-domain (JAZ) protein MdJAZ2 can also bind to MdbHLH3 (An et al. 2015). Since JAZ proteins are negative regulators of JA signalling, removal of JAZ2 by jasmonate signaling can increase MBW activity and PA accumulation. Similarly, in strawberry (*Fragaria* × *ananassa*) fruits, methyl jasmonate (MeJA) treatment can also induce PA accumulation (Delgado et al. 2018). A role for the hormone abscisic acid (ABA) in developmental regulation PA synthesis has been demonstrated in persimmon fruit (Akagi et al. 2012). Here, a basic leucine-zipper transcription factor, DkZIP5, was shown to bind to promoter elements of the PA-MYB DkMYB4 and thus induce PA accumulation (Akagi et al. 2012). Similarly, in peach, PA MYB PpMYB7 and PpMYBPA1 promoters can also be activated by PpbZIP5 in response to ABA (Zhou et al. 2015). Thus, a variety of plant hormones are implicated in the developmental regulation of PAs. Several of these hormones are also important in plant stress responses.

8.6 Stress Activation of PA Synthesis by MYBs in Poplar and Other Woody Plants

In trees and woody plants, as well as some herbaceous species, the PA pathway is very active in vegetative tissues and PAs accumulate in bark, roots, and leaves. Furthermore, in some plants, such as poplar and aspen trees, PA biosynthesis pathways can be induced in leaves by both biotic and abiotic stresses. The first demonstration that trees can respond to herbivory via transcriptional activation of the PA pathway enzymes was reported for *Populus tremuloides* (Peters and Constabel 2002). Interestingly, MeJA was effective in PA induction in this system. The TT2-type PtMYB134 and MYBPA1-type PtMYB115 were subsequently identified as key activators of PAs in poplar, and shown to be inducible as well. This revealed the critical role of the MYBs in stress activation of PAs (Mellway et al. 2009; James et al. 2017). Subsequently, other biotic stresses were shown to stimulate PA biosynthesis in poplar, in particular pathogens (Miranda et al. 2007). Experiments with both MYB134- and MYB115-overexpressing poplars demonstrated that high PA content increases resistance to fungal *Melampsora* leaf rust as well as *Dothiorella gregaria*, confirming the functional significance of PAs and these MYBs in pathogen defense (Ullah et al. 2017; Wang et al. 2017). PA induction in *P. nigra* involves plant stress hormones, as *Melampsora* leaf rust infection stimulates increases in salicylic acid (SA), JA, and ABA content. The SA analog benzothiadiazole also upregulates MBW complex genes including PtMYB134, PtMYB115, PtbHLH131, and PtWDR-1, thus connecting plant hormones to PA biosynthesis (Ullah et al. 2019). In bilberry (*Vaccinium myrtillus*), PA accumulation can be promoted by fungal endophyte and pathogen infection (Koskimäki et al. 2009). Whether PA induction in this system also involves multiple MYBs needs further investigation.

While biotic stress induction of PAs is consistent with their proposed roles as antinutritive and antimicrobial metabolites, the adaptive value of induction of the PA pathway by abiotic stress has been more difficult to explain. Both UV and high light stress strongly stimulate PA biosynthesis, again preceded by a rapid induction of PtMYB134 expression (Mellway et al. 2009). Nitrogen deficiency likewise enhances leaf PA content in poplar (Harding et al. 2005; Gourlay and Constabel 2019). In tea (*Camellia sinensis*), PA accumulation is induced under water deficit, together with enhanced hydroxylation of the flavan-3-ol PA precursors (Hernández et al. 2006). In apple, the induction of the PA activator MdMYB23 by cold stress was recently reported, and its binding to the MdANR promoter as well as promoters of two C-repeat binding factors (CBFs), MdCBF1 and MdCBF2 (An et al. 2018). CBFs play important role in cold stress response by activating cold response genes. Similarly, sugar supply induces flavonoid accumulation, and sucrose can promote anthocyanin and PA accumulation in apple leaves (Liu et al. 2017). This response was mediated by the protein kinase MdSnRK1.1 and again implicates jasmonate signaling. Specifically, under high sucrose, MdSnRK1.1 phosphorylates MdJAZ18, which leads to its degradation. MdJAZ18 is a negative regulator of JA signalling that interacts with the flavonoid MBW complex protein MdbHLH3. Degradation of MdJAZ18 thus releases MdbHLH3 from a JAZ-bHLH complex and promotes anthocyanin and PA accumulation (Liu et al. 2017). Altogether, a broad set of independent abiotic stresses can induce PA biosynthesis in trees or woody plants.

The importance of PAs in biotic stress resistance can be understood on the basis of the antimicrobial nature and protein-binding capacity of PAs; by contrast, the significance of abiotic stress induction of PAs and its adaptive role in abiotic stress resistance has been unclear. Recently, the demonstration that PAs can act as *in planta* antioxidants and protect against oxidative stress has provided a physiological rationale for these observations (Gourlay and Constabel 2019). Many abiotic stresses cause a build-up of reactive oxygen species and ultimately oxidative stress, in particular in leaves where disruption of photosynthetic electron transport leads to superoxide formation. PAs are known to possess very high *in vitro* antioxidant capacity (Quideau et al. 2011), and an *in vivo* antioxidant function has been hypothesized (Hernández et al. 2009). Our recent work using the PtMYB134-overexpressing and PtMYB115-overexpressing poplar directly demonstrate that PAs can in fact protect against oxidative stress. High-PA leaves sustain significantly less damage from ROS produced by the herbicide methyl viologen than do control leaves (Gourlay and Constabel 2019). These effects are also observed following UV and drought stress (G. Gourlay and C.P. Constabel, work in preparation), suggesting that PA induction could be an adaptive response to the ROS buildup during abiotic stress. Together, these data can explain the broad inducibility of PA biosynthesis in poplar laves by a wide array of stressors.

8.7 Summary and Conclusions

The transcriptional control of PA biosynthesis by multiple MYB activators and MBW complexes is well established. Recent research is suggesting that MYB repressors also contribute to PA pathway regulation, and provides the outlines of a complex network of positive and negative regulators. Significant progress in poplar and apple, where the PA pathway is strongly stress-induced, has provided direct evidence of these interactions, which could act to limit or fine-tune a stress response. In a developmental context, similar interactions of MYB activators, repressors, and bHLH cofactors have been demonstrated and suggest that MYB repressors could act to limit PA accumulation to certain tissues. How activation of the MYBs and the MBW complex are externally linked to signaling cascades, hormones, and the environment continues to be an active area of research.

8.8 Acknowledgments

Research in the Constabel lab is funded by the Natural Sciences and Engineering Research Council (NSERC) of Canada. The support of the University of Victoria's Centre for Forest Biology is gratefully acknowledged.

References

Akagi, T., Katayama-Ikegami, A., Kobayashi, S., et al. (2012). Seasonal abscisic acid signal and a basic leucine zipper transcription factor, DkbZIP5, regulate proanthocyanidin biosynthesis in persimmon fruit. *Plant Physiology* 158: 1089–1102.

Albert, N.W. (2015). Subspecialization of R2R3-MYB repressors for anthocyanin and proanthocyanidin regulation in forage legumes. *Frontiers in Plant Science* 6: 1–13.

An, J.P., Li, R., Qu, F.J., et al. (2018). R2R3-MYB transcription factor MdMYB23 is involved in the cold tolerance and proanthocyanidin accumulation in apple. *Plant Journal* 96: 562–577.

An, X.H., Tian, Y., Chen, K.Q., et al. (2015). MdMYB9 and MdMYB11 are involved in the regulation of the JA-induced biosynthesis of anthocyanin and proanthocyanidin in apples. *Plant and Cell Physiology* 56: 650–662.

Bailey, J.K., Schweitzer, J.A., Rehill, B.J., et al. (2004). Beavers as molecular geneticists: A genetic basis to the foraging of an ecosystem engineer. *Ecology* 85: 603–608.

Barbehenn, R. V. and Constabel, P.C. (2011). Tannins in plant-herbivore interactions. *Phytochemistry* 72: 1551–1565.

Baudry, A., Heim, M.A., Dubreucq, B., et al. (2004). TT2, TT8, and TTG1 synergistically specify the expression of BANYULS and proanthocyanidin biosynthesis in *Arabidopsis thaliana*. *Plant Cell* 39: 366–380.

Bernhardt, C., Lee, M.M., Gonzalez, A., et al. (2003). The bHLH genes *GLABRA3* (*GL3*). and *ENHANCER OF GLABRA3* (*EGL3*). specify epidermal cell fate in the Arabidopsis root. *Development* 130: 6431–6439.

Blade, C., Aragones, G., Arola-Arnal, A., et al. (2016). Proanthocyanidins in health and disease. *Biofactors* 42: 5–12.

Bogs, J., Jaffe, F.W., Takos, A.M., et al. (2007). The grapevine transcription factor VvMYBPA1 regulates proanthocyanidin synthesis during fruit development. *Plant Physiology* 143: 1347–1361.

Cavallini, E., Matus, J.T., Finezzo, L., et al. (2015). The phenylpropanoid pathway is controlled at different branches by a set of R2R3-MYB C2 repressors in grapevine. *Plant Physiology* 167: 1448–1470.

Constabel, C.P. (2018). Molecular controls of proanthocyanidin synthesis and structure: prospects for genetic engineering in crop plants. *Journal of Agricultural and Food Chemistry* 66: 9882–9888.

Constabel, C.P., Yoshida, K. and Walker V. (2014). Diverse ecological roles of plant tannins: plant defense and beyond. In: *Recent Advances in Polyphenol Research, Vol. 4* (eds. A. Romani, V. Lattanzio, and S. Quideau) 115–142. Wiley.

Cui, L., Yao, S., Dai, X., et al. (2016). Identification of UDP-glycosyltransferases involved in the biosynthesis of astringent taste compounds in tea (*Camellia sinensis*). *Journal of Experimental Botany* 67: 2285–2297.

Darriba, D., Taboada, G.L., and Posada, D. (2011). ProtTest 3 : fast selection of best-fit models of protein evolution. *Bioinformatics* 27: 1–4.

Debeaujon, I., Nesi, N., Perez, P., et al. (2003). Proanthocyanidin-accumulating cells in Arabidopsis testa: regulation of differentiation and role in seed development. *Plant Cell* 15: 2514–2531.

Delgado, L.D., Zúñiga, P.E., Figueroa, N.E., et al. (2018). Application of a JA-Ile biosynthesis inhibitor to methyl jasmonate-treated strawberry fruit induces upregulation of specific MBW complex-related genes and accumulation of proanthocyanidins. *Molecules* 23: 1433

Deluc, L., Barrieu, F., Marchive, C.C., et al. (2006). Characterization of a grapevine R2R3-MYB transcription factor that regulates the phenylpropanoid pathway. *Plant Physiology* 140: 499–511.

Deluc, L., Bogs, J., Walker, A.R., et al. (2008). The transcription factor VvMYB5b contributes to the regulation of anthocyanin and proanthocyanidin biosynthesis in developing grape berries. *Plant Physiology* 147: 2041–2053.

de Vetten, N., Quattrocchio, F., Mol, J., and Koes, R. (1997). The an11 locus controlling flower pigmentation in petunia encodes a novel WD-repeat protein conserved in yeast, plants, and animals. *Genes and Development* 11: 1422–1434.

Dixon, R. a, Xie, D.-Y., and Sharma, S.B. (2005). Proanthocyanidins – a final frontier in flavonoid research? *New Phytologist* 165: 9–28.

Dubos, C., Stracke, R., Grotewold, E., et al. (2010). MYB transcription factors in Arabidopsis. *Trends in Plant Science* 15: 573–581.

Feller, A., MacHemer, K., Braun, E.L., and Grotewold, E. (2011). Evolutionary and comparative analysis of MYB and bHLH plant transcription factors. *Plant Journal* 66: 94–116.

Feyissa, D.N., Løvdal, T., Olsen, K.M., et al. (2009). The endogenous GL3, but not EGL3, gene is necessary for anthocyanin accumulation as induced by nitrogen depletion in arabidopsis rosette stage leaves. *Planta* 230: 747–754.

Figueiredo, D.D., Batista, R.A., Roszak, P.J., et al. (2016). Auxin production in the endosperm drives seed coat development in Arabidopsis. *Elife* 5: 1–23.

Goff, S.A., Cone, K.C., and Chandler, V.L. (1992). Functional analysis of the transcriptional activator encoded by the maize B gene: Evidence for a direct functional interaction between two classes of regulatory proteins. *Genes and Development* 6: 864–875.

Gonzalez, A., Mendenhall, J., Huo, Y., and Lloyd, A. (2009). TTG1 complex MYBs, MYB5 and TT2, control outer seed coat differentiation. *Developmental Biology*, 325: 412–421.

Gourlay, G. and Constabel, C.P. (2019). Condensed tannins are inducible antioxidants and protect hybrid poplar against oxidative stress. *Tree Physiology* 39: 345–355.

Harding, S.A., Jiang, H., Jeong, M.L., et al. (2005). Functional genomics analysis of foliar condensed tannin and phenolic glycoside regulation in natural cottonwood hybrids. *Tree Physiology* 25: 1475–1486.

Hernández, I., Alegre, L., and Munné-Bosch, S. (2006). Enhanced oxidation of flavan-3-ols and proanthocyanidin accumulation in water-stressed tea plants. *Phytochemistry* 67: 1120–1126.

Hernández, I., Alegre, L., Van Breusegem, F., and Munné-Bosch, S. (2009). How relevant are flavonoids as antioxidants in plants? *Trends in Plant Science* 14: 125–132.

James, A.M., Ma, D., Mellway, R.D., et al. (2017). Poplar MYB115 and MYB134 transcription factors regulate proanthocyanidin synthesis and structure. *Plant Physiology* 174: 154–171.

Jiang, W., Yin, Q., Wu, R., et al. (2015). Role of a chalcone isomerase-like protein in flavonoid biosynthesis in *Arabidopsis thaliana*. *Journal of Experimental Botany* 66: 7165–7179.

Jin, H., Cominelli, E., Bailey, P., et al. (2000). Transcriptional repression by AtMYB4 controls production of UV-protecting sunscreens in *Arabidopsis*. *EMBO Journal* 19: 6150–6161.

Jun, J.H., Liu, C., Xiao, X., and Dixon, R.A. (2015). The transcriptional repressor MYB2 regulates both spatial and temporal patterns of proanthocyandin and anthocyanin pigmentation in *Medicago truncatula*. *Plant Cell* 27: 2860–2879.

Jun, J.H., Xiao, X., Rao, X., and Dixon, R.A. (2018). Proanthocyanidin subunit composition determined by functionally diverged dioxygenases. *Nature Plants* 4: 1034–1043.

Kang, Y.H., Kirik, V., Hulskamp, M., et al. (2009). The MYB23 gene provides a positive feedback loop for cell fate specification in the Arabidopsis root epidermis. *Plant Cell* 21: 1080–1094.

Kirik, V., Lee, M.M., Wester, K., et al. (2005). Functional diversification of MYB23 and GL1 genes in trichome morphogenesis and initiation. *Development* 132: 1477–1485.

Koskimäki, J.J., Hokkanen, J., Jaakola, L., et al. (2009). Flavonoid biosynthesis and degradation play a role in early defence responses of bilberry (*Vaccinium myrtillus*) against biotic stress. *European Journal of Plant Pathology*, 125: 629–640.

Kranz, H.D., Denekamp, M., Greco, R., et al. (1998). Towards functional characterisation of the members of the R2R3-MYB gene family from Arabidopsis thaliana. *Plant Journal* 16: 263–276.

Leida, C., Dal Rì, A., Dalla Costa, L., et al. (2016). Insights into the role of the berry-specific ethylene responsive factor VviERF045. *Frontiers in Plant Science* 7: 1–17.

Li, P., Chen, B., Zhang, G., et al. (2016). Regulation of anthocyanin and proanthocyanidin biosynthesis by *Medicago truncatula* bHLH transcription factor MtTT8. *New Phytologist* 210: 905–921.

Li, S. and Zachgo, S. (2013). TCP3 interacts with R2R3-MYB proteins, promotes flavonoid biosynthesis and negatively regulates the auxin response in *Arabidopsis thaliana*. *Plant Journal* 76: 901–913.

Li, X., Gao, P., Cui, D., et al. (2011). The Arabidopsis tt19-4 mutant differentially accumulates proanthocyanidin and anthocyanin through a 3' amino acid substitution in glutathione S-transferase. *Plant, Cell and Environment* 34: 374–388.

Liu, C., Jun, J.H., and Dixon, R.A. (2014). MYB5 and MYB14 play pivotal roles in seed coat polymer biosynthesis in *Medicago truncatula*. *Plant Physiology* 165: 1424–1439.

Liu, C., Wang, X., Shulaev, V., and Dixon, R.A. (2016). A role for leucoanthocyanidin reductase in the extension of proanthocyanidins. *Nature Plants* 2: 1–7.

Liu, J., Osbourn, A., and Ma, P. (2015). MYB transcription factors as regulators of phenylpropanoid metabolism in plants. *Molecular Plant* 8: 689–708.

Liu, M., Song, C., Chi, M., et al. (2016). The effects of light and ethylene and their interaction on the regulation of proanthocyanidin and anthocyanin synthesis in the skins of *Vitis vinifera* berries. *Plant Growth Regulation* 79: 377–390.

Liu, X.-J., An, X.-H., Liu, X., et al. (2017). MdSnRK1.1 interacts with MdJAZ18 to regulate sucrose-induced anthocyanin and proanthocyanidin accumulation in apple. *Journal of Experimental Botany* 68: 2977–2990.

Ludwig, S.R., Habera, L.F., Dellaportat, S.L., and Wessler, S.R. (1989). Lc, a member of the maize R gene family responsible for tissue-specific anthocyanin production, encodes a protein similar to transcriptional activators and contains the myc-homology region. *Proceedings of the National Academy of Sciences of the USA* 86: 7092–7096.

Ma, D., and Constabel, C.P. (2019). MYB repressors as regulators of phenylpropanoid metabolism in plants. *Trends in Plant Science* 24: 275–289.

Ma, D., Reichelt, M., Yoshida, K., et al. (2018). Two R2R3-MYB proteins are broad repressors of flavonoid and phenylpropanoid metabolism in poplar. *Plant Journal* 96: 949–965.

Marinova, K., Pourcel, L., Weder, B., et al. (2007). The Arabidopsis MATE transporter TT12 acts as a vacuolar flavonoid/H^+-antiporter active in proanthocyanidin-accumulating cells of the seed coat. *Plant Cell* 19: 2023–2038.

Mellway, R.D., Tran, L.T., Prouse, M.B., et al. (2009). The wound-, pathogen-, and ultraviolet B-responsive MYB134 gene encodes an R2R3 MYB transcription factor that regulates proanthocyanidin synthesis in poplar. *Plant Physiology* 150: 924–941.

Miranda, M., Ralph, S.G., Mellway, R., et al. (2007). The transcriptional response of hybrid poplar (*Populus trichocarpa x P. deltoides*) to infection by *Melampsora medusae* leaf rust involves induction of flavonoid pathway genes leading to the accumulation of proanthocyanidins. *Molecular Plant-Microbe Interactions* 20: 816–831.

Nesi, N., Debeaujon, I., Jond, C., et al. (2002). The *TRANSPARENT TESTA16* Locus encodes the ARABIDOPSIS BSISTER MADS domain protein and is required for proper development and pigmentation of the seed coat. *Plant Cell* 14: 2463–2479.

Nesi, N., Jond, C., Debeaujon, I., et al. (2001). The Arabidopsis TT2 gene encodes an R2R3 MYB domain protein that acts as a key determinant for proanthocyanidin accumulation in developing seed. *Plant Cell* 13: 2099–2114.

Nunes, M.A., Pimentel, F., Costa, A.S.G., et al. (2016). Cardioprotective properties of grape seed proanthocyanidins: An update. *Trends in Food Science and Technology* 57: 31–39.

Pang, Y., Peel, G.J., Sharma, S.B., et al. (2008). A transcript profiling approach reveals an epicatechin-specific glucosyltransferase expressed in the seed coat of Medicago truncatula. *Proceedings of the National Academy of Sciences of the USA* 105: 14210–14215.

Pang, Y., Wenger, J.P., Saathoff, K., et al. (2009). A WD40 repeat protein from Medicago truncatula Is necessary for tissue-specific anthocyanin and proanthocyanidin biosynthesis but not for trichome development. *Plant Physiology* 151: 1114–1129.

Payne, C.T., Zhang, F. and Lloyd, A.M. (2000). GL3 encodes a bHLH protein that regulates trichome development in Arabidopsis through interaction with GL1 and TTG1. *Genetics* 156: 1349–1362.

Paz-Ares, J., Ghosal, D., Wienand, U., et al. (1987). The regulatory c1 locus of Zea mays encodes a protein with homology to myb proto-oncogene products and with structural similarities to transcriptional activators. *EMBO Journal* 6: 3553–3558.

Pérez-Díaz, R., Madrid-Espinoza, J., Salinas-Cornejo, J., et al. (2016). Differential roles for VviGST1, VviGST3, and VviGST4 in proanthocyanidin and anthocyanin transport in *Vitis vinífera*. *Frontiers in Plant Science* 7: 1–13.

Peters, D.J. and Constabel, C.P. (2002). Molecular analysis of herbivore-induced condensed tannin synthesis: Cloning and expression of dihydroflavonol reductase from trembling aspen (*Populus tremuloides*). *Plant Journal* 32: 701–712.

Pourcel, L., Routaboul, J.-M., Kerhoas, L., et al. (2005). *TRANSPARENT TESTA10* encodes a laccase-like enzyme involved in oxidative polymerization of flavonoids in Arabidopsis seed coat. *Plant Cell* 17: 2966–2980.

Quideau, S., Deffieux, D., Douat-Casassus, C., and Pouysegu, L. (2011). Plant polyphenols: chemical properties, biological activities, and synthesis. *Angewandte Chemie* 50: 586–621.

Ramsay, N.A. and Glover, B.J. (2005). MYB-bHLH-WD40 protein complex and the evolution of cellular diversity. *Trends in Plant Science* 10: 63–70.

Sagasser, M., Lu, G., Hahlbrock, K. and Weisshaar, B. (2002). A. thaliana *TRANSPARENT TESTA1* is involved in seed coat development and defines the WIP subfamily of plant zinc finger proteins. *Genes and Development* 16: 138–149.

Stafford, H.A. and Lester, H.H. (1985). Flavan-3-ol biosynthesis. The conversion of (+)-dihydromyricetin to its flavan-3,4-diol (leucodelphinidin) and to (+)-gallocatechin by reductases extracted from tissue cultures of *Ginkgo biloba* and *Psudotsuga menziesii*. *Plant Physiology* 78: 791–794.

Stafford, H.A. and Lester, H.H. (1984). Flavan-3-ol biosynthesis. The conversion of (+)-dihydroquercetin and flavan-3,4-cis-diols (leucocyanidin) to (+)-catechin by reductases extracted from cell suspension cultures of Douglas fir. *Plant Physiology* 76: 184–186.

Tamagnone, L., Merida, A., Parr, A., et al. (1998). The AmMYB308 and AmMYB330 transcription factors from *Antirrhinum* regulate phenylpropanoid and lignin biosynthesis in transgenic tobacco. *Plant Cell* 10: 135–154.

Tanner, G.J., Francki, K.T., Abrahams, S., et al. (2003). Proanthocyanidin biosynthesis in plants. Purification of legume leucoanthocyanidin reductase and molecular cloning of its cDNA. *Journal of Biological Chemistry* 278: 31647–31656.

Terrier, N., Torregrosa, L., Ageorges, A., et al. (2008). Ectopic expression of VvMybPA2 promotes proanthocyanidin biosynthesis in grapevine and suggests additional targets in the pathway. *Plant Physiology* 149: 1028–1041.

Tohge, T., de Souza, L.P., and Fernie, A.R. (2017). Current understanding of the pathways of flavonoid biosynthesis in model and crop plants. *Journal of Experimental Botany* 68: 4013–4028.

Ullah, C., Tsai, C.-J., Unsicker, S.B., et al. (2019). Salicylic acid activates poplar defense against the biotrophic rust fungus *Melampsora larici-populina* via increased biosynthesis of catechin and proanthocyanidins. *New Phytologist* 221: 960–975.

Ullah, C., Unsicker, S.B., Fellenberg, C., et al. (2017). Flavan-3-ols are an effective chemical defense against rust infection. *Plant Physiology* 175: 1560–1578.

Verdier, J., Zhao, J., Torres-Jerez, I., et al. (2012). MtPAR MYB transcription factor acts as an on switch for proanthocyanidin biosynthesis in *Medicago truncatula*. *Proceedings of the National Academy of Sciences of the USA* 109: 1766–1771.

Walker, A.R. (1999). The *TRANSPARENT TESTA GLABRA1* Locus, which regulates trichome differentiation and anthocyanin biosynthesis in Arabidopsis, encodes a WD40 repeat protein. *Plant Cell* 11: 1337–1350.

Wang, L., Lu, W., Ran, L., et al. (2019). R2R3-MYB transcription factor MYB6 promotes anthocyanin and proanthocyanidin biosynthesis but inhibits secondary cell wall formation in *Populus tomentosa*. *Plant Journal* 99: 733–751.

Wang, L., Ran, L., Hou, Y., et al. (2017). The transcription factor MYB115 contributes to the regulation of proanthocyanidin biosynthesis and enhances fungal resistance in poplar. *New Phytologist* 215: 351–367.

Xie, D.-Y., Sharma, S.B., Paiva, N.L., et al. (2003). Role of anthocyanidin reductase, encoded by *BANYULS* in plant flavonoid biosynthesis. *Science* 299: 396–399.

Xu, W., Bobet, S., Gourrierec, J. Le, et al. (2017). *TRANSPARENT TESTA 16* and *15* act through different mechanisms to control proanthocyanidin accumulation in Arabidopsis testa. *Journal of Experimental Botany* 68: 2859–2870.

Xu, W., Dubos, C., and Lepiniec, L. (2015). Transcriptional control of flavonoid biosynthesis by MYB-bHLH-WDR complexes. *Trends in Plant Science* 20: 176–185.

Xu, W., Grain, D., Bobet, S., et al. (2014). Complexity and robustness of the flavonoid transcriptional regulatory network revealed by comprehensive analyses of MYB-bHLH-WDR complexes and their targets in Arabidopsis seed. *New Phytologist* 202: 132–144.

Yoshida, K., Iwasaka, R., Kaneko, T., et al. (2008). Functional differentiation of *Lotus japonicus* TT2s, R2R3-MYB transcription factors comprising a multigene family. *Plant and Cell Physiology* 49: 157–169.

Yoshida, K., Ma, D. and Constabel, C.P. (2015). The MYB182 protein down-regulates proanthocyanidin and anthocyanin biosynthesis in poplar by repressing both structural and regulatory flavonoid genes. *Plant Physiology* 167: 693–710.

Zhang, J., Xu, H., Wang, N., et al. (2018). The ethylene response factor MdERF1B regulates anthocyanin and proanthocyanidin biosynthesis in apple. *Plant Molecular Biology* 98: 205–218.

Zhao, J. (2015). Flavonoid transport mechanisms: How to go, and with whom. *Trends in Plant Science* 20: 576–585.

Zhao, J. and Dixon, R.A. (2009). MATE transporters facilitate vacuolar uptake of epicatechin 3'-O-glucoside for proanthocyanidin biosynthesis in *Medicago truncatula* and Arabidopsis. *Plant Cell* 21: 2323–2340.

Zhao, J. and Dixon, R.A. (2010). The "ins" and "outs" of flavonoid transport. *Trends in Plant Science* 15: 72–80.

Zhao, J., Pang, Y., and Dixon, R.A. (2010). The mysteries of proanthocyanidin transport and polymerization. *Plant Physiology* 153: 437–443.

Zhou, H., Lin-Wang, K., Liao, L., et al. (2015). Peach MYB7 activates transcription of the proanthocyanidin pathway gene encoding leucoanthocyanidin reductase, but not anthocyanidin reductase. *Frontiers in Plant Science* 6: 1–13.

Zhou, H., Lin-Wang, K., Wang, F., et al. (2019). Activator-type R2R3-MYB genes induce a repressor-type R2R3-MYB gene to balance anthocyanin and proanthocyanidin accumulation. *New Phytologist*: 221: 1919–1934.

9

Conservation and Divergence Between Bryophytes and Angiosperms in the Biosynthesis and Regulation of Flavonoid Production

Kevin M. Davies, Rubina Jibran, Nick W. Albert, Yanfei Zhou, and Kathy E. Schwinn

Plant and Food Research, Palmerston North, New Zealand

9.1 Introduction

9.1.1 The Bryophytes

We are often most familiar with the vascular plants, particularly the seed-bearing gymnosperms and angiosperms, as these contain the species most important to agriculture and forestry. Not surprisingly, they are also the plant groups for which the majority of genetic and molecular biology studies have been published. However, the second largest group of land plants after angiosperms is the bryophytes, with more than 20000 species. Bryophytes is the collective name for the nonvascular land plant group comprised of the liverworts (Marchantiophyta, approximately 9000 species), hornworts (Anthocerotophyta, approximately 300 species), and mosses (Bryophyta, approximately 12000 species). They are the major contributors to the green carpet that covers the floor of many forests, but mosses and liverworts actually occupy a great variety of environments and are found in some of the most diverse and challenging ecological niches. They grow from the Arctic to the Antarctic, and from lakes to deserts, and are the dominant flora in some of the most extreme environments – for example, there are only two flowering plant species native to Antarctic but around 130 moss and liverwort species.

Traditionally, evidence such as morphological comparison with the fossil record placed liverworts at the base of the land plant evolutionary tree – thus having a 'sister' relationship to all extant land plants. However, although it is generally accepted that land plants evolved from an ancestral charophycean algae, with the last common ancestor being bryophytelike, the placement of liverworts as basal has been challenged by recent data from gene sequencing studies (Wickett et al. 2014; Puttick et al. 2018). Such studies have developed two principal hypotheses alternative to liverworts as basal: hornworts as sister to all other land plants or a single bryophyte clade that is sister to the vascular/seed plants. The extant algal sister group is thought to be within the Chlorophyta green algae, probably the order

Recent Advances in Polyphenol Research, Volume 7, First Edition. Edited by Jess Dreher Reed, Victor Armando Pereira de Freitas, and Stéphane Quideau.

Zygnematales (a charophyte) or a clade of the Zygnematales and Coleochaetales together (Zhong et al. 2014; Delwiche and Cooper 2015; de Vries and Archibald 2018).

The bryophytes are becoming an increasing focus of genomic and molecular biology research. Principally this is because they can inform us about characters that may have an ancient evolutionary origin and, by comparison with the other major plant groups, how genetic mechanisms and genomes have changed during land plant evolution. However, there is also interest in their ability to survive extreme conditions, and because bryophytes are key flora in some of the most climate change–vulnerable ecosystems.

9.1.2 Evolution of the Phenylpropanoid Pathway

While there are remarkably well preserved fossils that are informative on fine anatomical detail, such as those found in the Rhynie Chert in Scotland (Early Devonian; about 410 million years ago, MYA) (Ligrone et al. 2012), the fossil record provides little information on the possible secondary metabolism of the first land plants. One approach to gain understanding of what metabolite pathways might have been present in the first land plants, and how they have since evolved, is to examine the pathways present in species representative of the diversity of extant land plant taxa. In particular, examining those taxa with a large evolutionary separation, such as the bryophytes and angiosperms, may provide indications of which pathways are in common and so might have originated with the last common ancestor.

The flavonoid pathway is a major branch of the larger phenylpropanoid pathway, a central component of land plant metabolism. Although there is debate on which aspects of the larger phenylpropanoid pathway may be present in extant algal relatives, based on current evidence flavonoid biosynthesis is restricted to land plants. It is hypothesised to have arisen during the period when plants were colonizing the land as a key adaptation for coping with the additional abiotic stresses faced by a terrestrial lifestyle. In particular, the 'colourless' flavonoid groups of flavonols and flavones are proposed to have arisen as UV 'sunscreens'. The production of flavonols/flavones is certainly near-ubiquitous in land plants, and their role in providing increased tolerance to UVB is conserved between liverworts and angiosperms (Clayton et al. 2018). During evolution, the diversity of phenylpropanoid biosynthesis increased greatly, with more than 8000 different flavonoid structures having been reported (Andersen and Markham 2006). This is thought to have facilitated colonization of the wide range of environments which plants now occupy (Banks et al. 2011).

The most common coloured flavonoids are the anthocyanins, which form when sugar residues are added to the core chromophore, the anthocyanidin, making the pigment water-soluble. Anthocyanins are located to the cell vacuole. Anthocyanins provide most of the red to blue colours of flowers and fruits. However, anthocyanins are also found in all types of vegetative organs, including underground organs such as roots and tubers. Moreover, some divisions of land plants that lack flowers or fruit, such as the ferns and gymnosperms, still produce anthocyanins. Thus, anthocyanins first arose as compounds with benefits to vegetative tissue function, with additional roles arising later in evolution. Those first functions were probably in abiotic stress tolerance, as there is evidence from many studies showing anthocyanin production improves the general abiotic stress tolerance of plants (Agati et al. 2012; Landi et al. 2015).

Anthocyanins are induced by a wide range of stresses, including cold, drought, excess white light, and senescence. Anthocyanins can directly absorb green light energy. They can therefore reduce the stress imposed on chloroplasts from excess sunlight, which is exacerbated when other stressors such as cold and drought are also present. However, this cannot explain why anthocyanins are produced in some other situations, such as in roots. One prominent hypothesis is that anthocyanins are antioxidants that scavenge the reactive oxygen species (ROS) generated during normal plant processes, production of which is promoted by stress. However, flavonols are also very effective antioxidants, so it may be that anthocyanins combine direct light screening with antioxidant properties (Neill and Gould 2003; Agati et al. 2012; Cheynier et al. 2013; Landi et al. 2015). Regardless of the mechanism(s), it has become clear that anthocyanins are a major stress tolerance character of at least flowering plants.

The extant relatives of the proposed algal ancestors of plants are generally accepted not to have flavonoid production. The most common photoprotective compounds produced are the mycosporine-like amino acids (MAAs), colourless, water-soluble, polar zwitterionic amino acid derivatives. However, while prevalent in red algae (Rhodophyta), MAAs are not found in all algal taxa examined, and there are few reports for the Chlorophyta. Perhaps as MAA replacements, some green algal species show strong phenolic accumulation in response to stresses such as cold or increased light. Although some articles report flavonoid production in algae, more complete analysis has typically found gallic acid derivatives (based on galloylglucoside), which are presumed products from steps in the shikimate pathway prior to the start of the phenylpropanoid branch. The biosynthetic pathway to similar compounds in brown algae (Heterokonts), green algae, and red algae has been established (Richards et al. 2006), and one of the key genes required, shikimate dehydrogenase, has been identified in green algae (Carrington et al. 2018). The phenolics produced range from glycosylated monomers through to red/purple pigments that resemble the hydrolysable tannins/gallotannins of land plants (Han et al. 2012; Aigner et al. 2013; Holzinger and Pichrtová 2016). These may perform similar functions to flavonoids in land plants. For example, *Zygogonium ericetorum*, a filamentous charophyte green alga that can live in challenging environments such as open soils, high altitudes, or acidic ponds, generates strong purple vacuolar pigmentation under higher photosynthetic stress conditions (Aigner et al. 2013). Phenolic production may be advantageous over MAAs under nutrient stress conditions because phenolics have a lower metabolic cost and do not contain nitrogen.

In the next sections we examine the knowledge of the flavonoid pathways for basal plants in context of the angiosperm data, with an emphasis on liverworts.

9.2 Flavonoid Biosynthesis in Basal Plants

In angiosperms, the phenylpropanoid pathway has been the subject of hundreds of research articles, and is one of the best-understood plant metabolite pathways. There is knowledge in depth on the biosynthesis, regulation, and function of not only the 'core' compounds, such as flavonols, anthocyanins, and proanthocyanidins, but also many of the more taxonomically restricted flavonoid branches, such as the isoflavonoids, stilbenes, and 5-deoxyflavonoids. In contrast, while structural diversity has been examined for some of the nonvascular plants (liverworts, hornworts, and mosses) and the basal vascular plants (lycophytes and ferns), we lack knowledge on the biosynthetic pathways involved. Not only are the pathways to bryophyte-specific compounds such as the *bis*-bibenzyls, auronidins, and sphagnorubins generally not understood, but we also do not know whether the biosynthetic pathways producing the compounds in common between bryophytes and angiosperms, such as flavones and flavonols, are conserved across these distantly related plant groups.

Until recently, molecular research on basal plants has been hampered by the lack of sequence resources – of the more than 100 published land plant genome sequences completed by 2017, all but four were of angiosperms (Rensing 2017). However, the knowledge gap on the phenylpropanoid pathway of basal plants may be rapidly filled with the advent of the new bryophyte and lycophyte molecular genetic models, specifically *Marchantia polymorpha* (marchantia, the liverwort model), *Anthoceros agrestis* (a hornwort), *Physcomitrella patens* (a moss), *Selaginella moellendorffi* (a lycophyte), and *Azolla filiculoides* (a fern) (Rensing et al. 2008; Banks et al. 2011; Szövényi et al. 2015; Bowman et al. 2017; Albert et al. 2018; F.-W. Li et al. 2018).

There is a general lack of knowledge on phenylpropanoid biosynthesis in hornworts. What information is available is limited to analysis of compounds present in the thallids of a few species and preliminary studies on the response of one species to exposure to additional UV light (Markham 1988; Asakawa et al. 2013; Monforte et al. 2018). Remarkably, flavonoids have yet to be identified in hornworts, with the main phenolics detected being cinnamic acid derivatives such as lignans and rosmarinic acid.

At present, most is known for liverworts. The phenylpropanoid chemistry of liverworts has been extensively characterised, partly because of the biological and medicinal activities identified for liverwort extracts (Asakawa et al. 2013; Asakawa 2016). Additionally, the biosynthetic pathways and the regulatory mechanisms for some of these compounds are now starting to emerge. Progress has been greatly aided by the advantages of marchantia as a model for molecular and physiology studies. Marchantia is well suited as a laboratory species, as it is small, has a rapid growth rate, and can reproduce asexually in large numbers through single-cell-derived clonal gemmae. The genome is small (approximately 230 Mb), with a high-quality genome sequence published (Bowman et al. 2017), and significantly fewer genes than Arabidopsis (around 19000 gene models compared with around 28000 protein-coding genes in Arabidopsis). Moreover, it offers efficient CRISPR/Cas9 mutagenesis in a dominant haploid gametophytic generation.

For mosses, there is a wealth of data on the structural diversity but little on the underlying biosynthetic pathways. A review on flavonoids of 300 moss species representing 59 families found that, although some of the flavone and flavonol glycosides typical of angiosperms do occur, the most commonly occurring compounds were aglycone biflavonoids and cinnamic acid derivatives (Asakawa et al. 2013). The main phenolic pigment characterised from mosses is sphagnorubin, a reddish-violet cell wall–located compound with five rings that is thought to be anthocyanidin related (Figure 9.1). Other compounds have been found only rarely; for example, aurones have been reported only from one moss and three liverwort species.

The situation for lycophytes is similar to that of mosses; there are many data on structural diversity but not biosynthetic pathways. *S. moellendorffi* and related lycophytes produce a rich variety of phenylpropanoids, including variously substituted flavones, 4′7-dihydroxyflavanones, coumarins, biflavonoids, stilbene-like compounds, and lignans (Weng and Noel 2013). *S. moellendorffi* also produces the red selaginellin pigments, which have yet to be reported from other plant groups. It is suggested that selaginellins are polymers derived from orsellinic acid, which could be formed by a Type III POLYKETIDE SYNTHASE (PKS) using the

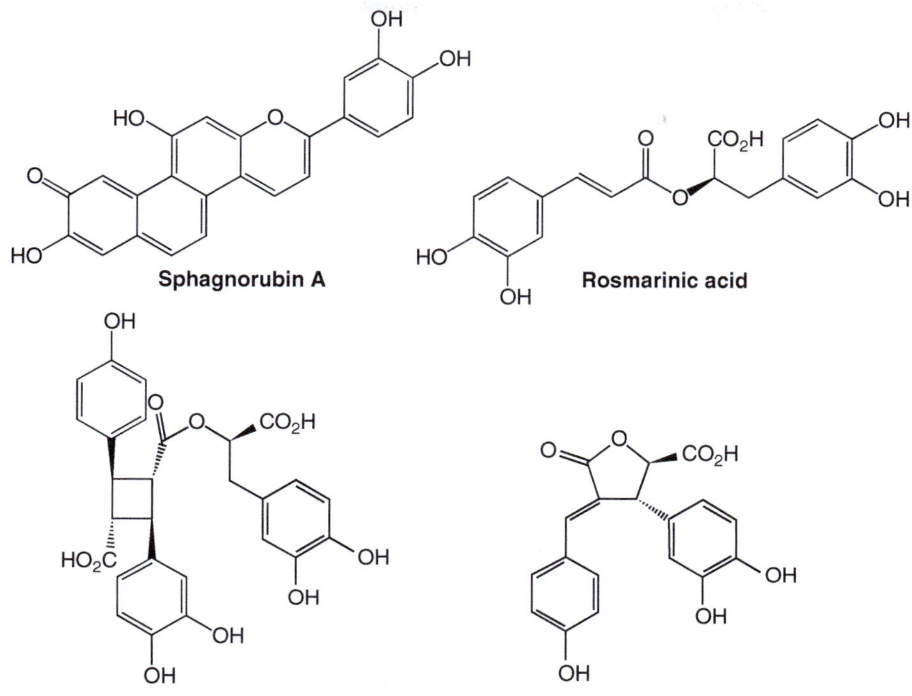

Figure 9.1 Notable phenolic compounds of mosses and hornworts. Sphagnorubin A is a red pigment found in moss cell walls. Rosmarinic acid and the lignans anthocerotonic acid and megacerotonic acid are characteristic compounds of hornworts.

same starting substrates as for flavonoids – malonyl-CoA and an acetyl-CoA starter unit (Shi et al. 2012).

In the next part of this review we discuss the flavonoid and related pathway data for bryophytes in more detail. We focus most on the liverworts, as these are the proposed basal group and there have been significant recent advances in understanding the molecular genetics of the pathway in marchantia. We begin with the early biosynthetic steps of the phenylpropanoid pathway, and then discuss some of the specific compound groups that are of particular interest or for which there have been significant recent research advances.

9.3 Origins of the Phenylpropanoid Biosynthetic Pathway and Conservation Across the Embryophytes

9.3.1 Phenylalanine Ammonia Lyase

As the first dedicated enzyme of the phenylpropanoid pathway, PHENYLALANINE AMMONIA LYASE (PAL) directs substrate from primary amino acid metabolism into the biosynthesis of specialised metabolites that are essential for both plant growth and development and the response to the biotic and abiotic environment, such as flavonoids, phenolic glycosides, and lignin. In dicots, PAL predominantly uses L-phenylalanine (L-Phe) as its substrate to produce *trans*-cinnamic acid. However, in monocots and some fungi and photosynthetic bacteria the enzyme also has TYROSINE AMMONIA LYASE (TAL) activity, using L-tyrosine to produce coumaric acid and bypassing the requirement for CINNIMATE 4-HYDROXYLASE (C4H) (Ferrer et al. 2008; Weise et al. 2017). Phylogenetically, PAL is related to the bacterial and fungal TAL and HISTIDINE AMMONIA LYASE (HAL) gene families (Xiang and Moore 2005; Ferrer et al. 2008; Wu et al. 2014; Weise et al. 2017). Although the genetics and biochemistry of PAL have been studied extensively for angiosperms and gymnosperms, until recently there were few studies on basal land plants and algae. However, with the completion of genome sequences for moss, liverwort, and algal species attention has been given to the evolutionary origin of PAL and how this links to the development of the phenylpropanoid pathway in land plants. Whether PAL arose during land plant evolution or is also present in algae has yet to be resolved (de Vries et al. 2017). Authentic PAL genes are present in basal plants, as we discuss in more detail later. The uncertainty concerns whether PAL or its ancestral gene occur in algae, and at what point in evolution they may have been acquired by horizontal gene transfer (HGT). Initially, Emiliani et al. (2009) proposed that the land plant ancestor acquired PAL via a HGT event during symbioses with soil bacteria and fungi during the early period of land colonization, not from their algal ancestor. Similarly, Labeeuw et al. (2015) suggested that PAL is absent from algae. However, PAL orthologues were identified by de Vries et al. (2017) in the streptophyte algae species *Klebsormidium flaccidum*, and the authors suggested the possibility of endosymbiotic gene transfer from cyanobacteria prior to the divergence of algae and land plants. Kf*PAL* has some conserved PAL motifs, and structural modelling suggests phenylalanine as a substrate. However, Kf*PAL* is much shorter than land plant PALs and lacks their distinctive 'fanlike' pattern of residues, having more structural similarities to PALs of cyanobacteria and HALs of soil-dwelling methylobacteria. Nevertheless, this could be an evolutionary intermediate state of the protein from which the land plant PALs evolved, and so represent the ancestral PAL state (de Vries et al. 2017). Functional analysis of the encoded activity, along with additional sequence analysis of other algal species, would help to resolve this. In Figure 9.2 we present a phylogeny for PAL and related sequences incorporating streptophyte algae, a fungus, and various groups of land plants. A clear clade is formed of PAL sequences from land plants, including from all the bryophyte groups. For liverworts, genes encoding PAL have been cloned and

Ceratodon purpureus CHIL (isotig15641), FAPa1 (isotig05846), FAPb1 (isotig15077), FAPb2 (isotig14426); *Ginkgo biloba* 4CL1 (AFJ73426), CHS (AY647263); *Huperzia serrata* ACS (ABI94386); *Humulus lupulus* CHIL1 (AVR53896), CHIL2 (AVR53897), CHS (BAB47195), VPS (BAB12102); *Lilium hybrida* CHS (AAD49355); *Lolium perenne* 4CL1 (AAF37732); *Oryza sativa* 4CL1 (CAA36850); *Phaeodactylum tricornutum* FAPd2 (XP_002181351); *Physcomitrella patens* CHILa (XP_001773128), CHILb (XP_001769093), FAPa1 (DC930902), FAPa2 (XP_001783129), FAPb (DC956762), 4CL1 (ABV60447), 4CL2 (Pp019G044600), 4CL3 (Pp022G056000), 4CL4 (Pp021G054800), ACoASyn (XP_024389050); *Picea sitchensis* FAP1 (ABK25337); *Pinus strobus* STS (CAA87013); *Pinus sylvestris* CHS (CAA43166); *Pinus taeda* 4CL1 (P41636); *Plagiochasma appendiculatum* 4CL (AJT43268), STCS (AUG98244), STCS1 (AHY39237), SCTS2 (AIV42295); *Psilotum nudum* CHS (BAA87922), STS (BAA87924), VPS (BAA87923); *Ricciocarpos natans* CHI (WJLO 9530), CHIL (WJLO 8822), FAPb (WJLO 406); *Selaginella moellendorffii* CHI (XP_002980623), CHIL (FE452267), FAPa1 (FE510754), FAPa2 (XP_002987458), FAPb (FE511008), 4CL1 (XP_002969881), 4CL2 (Smoe_177393), MCL (XP_024540059); *Sphaerocarpos texanus* CHI (HERT 7773), CHIL (HERT 829), FAPb (HERT 8792); *Sphagnum fallax* CHIL (Sphfalx0068s0053), FAPa1 (Sphfalx0013s0179), FAPa2 (Sphfalx0032s0188), FAPb (Sphfalx0071s0061); *Syntrichia ruralis* CHIL (CN201507); *Vitis vinifera* CHS (BAB84112), STS (ABD64685); *Zea mays* CHS (P24824). *Source:* (a, b, c, d) Kearse et al. 2012. (*See insert for color representation of the figure.*)

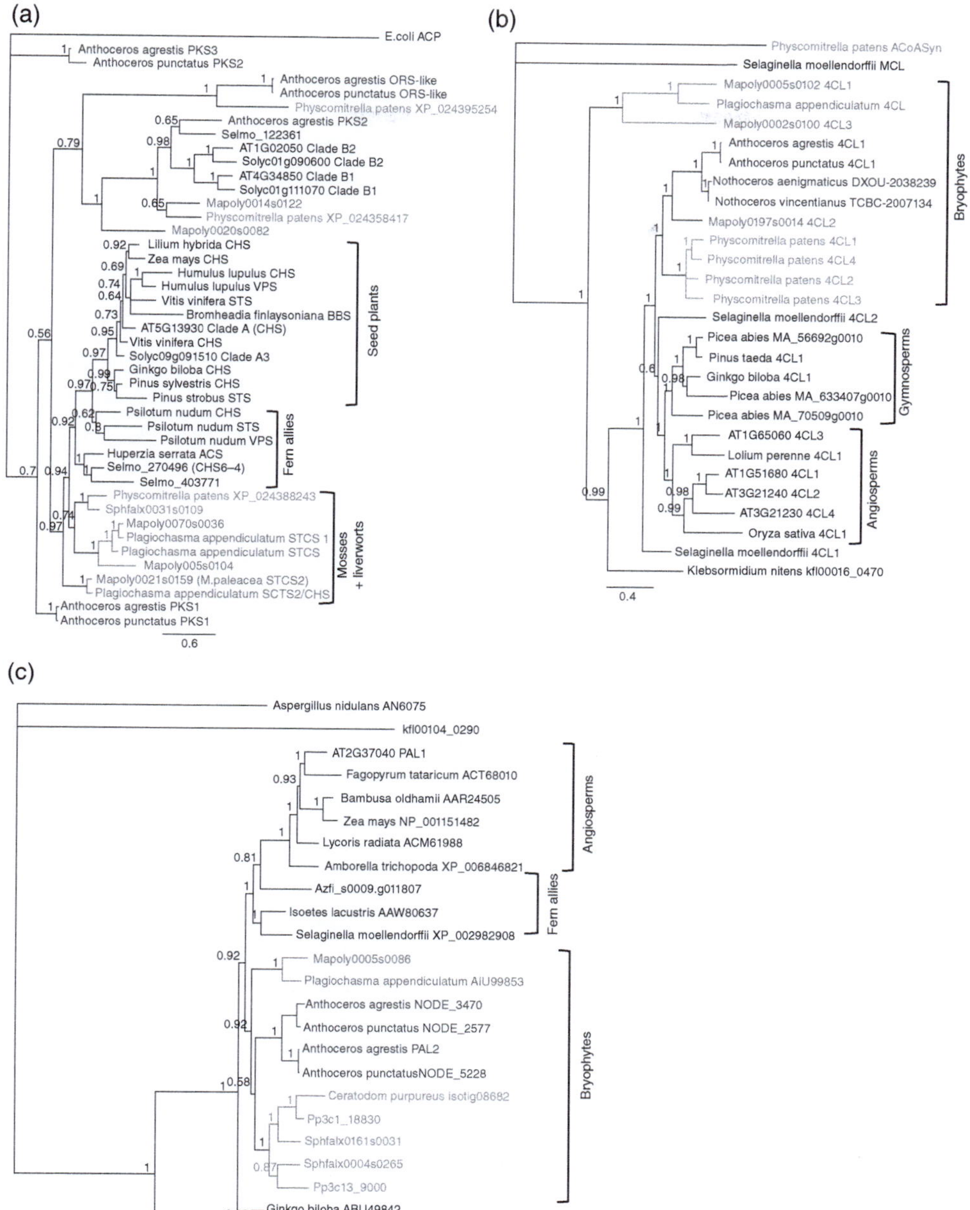

Figure 9.2 Phylogenetic analysis of sequences related to phenylpropanoid biosynthesis. Deduced amino acid sequences were aligned using ClustalW and trees formed using MrBayes within the Geneious (v 10.0.9) sequence analysis software package using default parameters, with the posterior probabilities of the branches shown. Trees are for **(a)** Polyketide Synthases, **(b)** 4-Coumarate-CoA ligase, **(c)** Phenylalanine ammonia-lyase, and **(d)** Chalcone isomerase. Sequence accession details not given directly on the tree or in the chapter text are as follows: *Amborella trichopoda* CHI (XP_006850622), CHIL (XP_011622657), FAPa1 (AmTr_v1.0_scaffold00103.21), FAPa2 (AmTr_v1.0_scaffold00024.257), FAPa3 (AmTr_v1.0_scaffold00061.192); *Anthoceros agrestis* PAL2 (SPO49995); *Arabidopsis thaliana* CHI (At5g05270), CHIL (At3g55120), FAPa1 (AT3G63170), FAPa2 (At2g26310), FAPa3 (XP_020245856), FAPb (At1g53520); *Aspergillus nidulans* AN6075 (XP_663679); *Bazzania trilobata* CHIL (WZYK 6680, from the data of Wickett et al. 2014), FAPb1 (WZYK 8096), FAPb2 (WZYK 7662); *Bromheadia finlaysoniana* BBS (CAA10514); *Cannabis sativa* CHIL (AFN42529); *Carica papaya* FAP1 (XP_021898058), FAPa3 (XP_021911943);

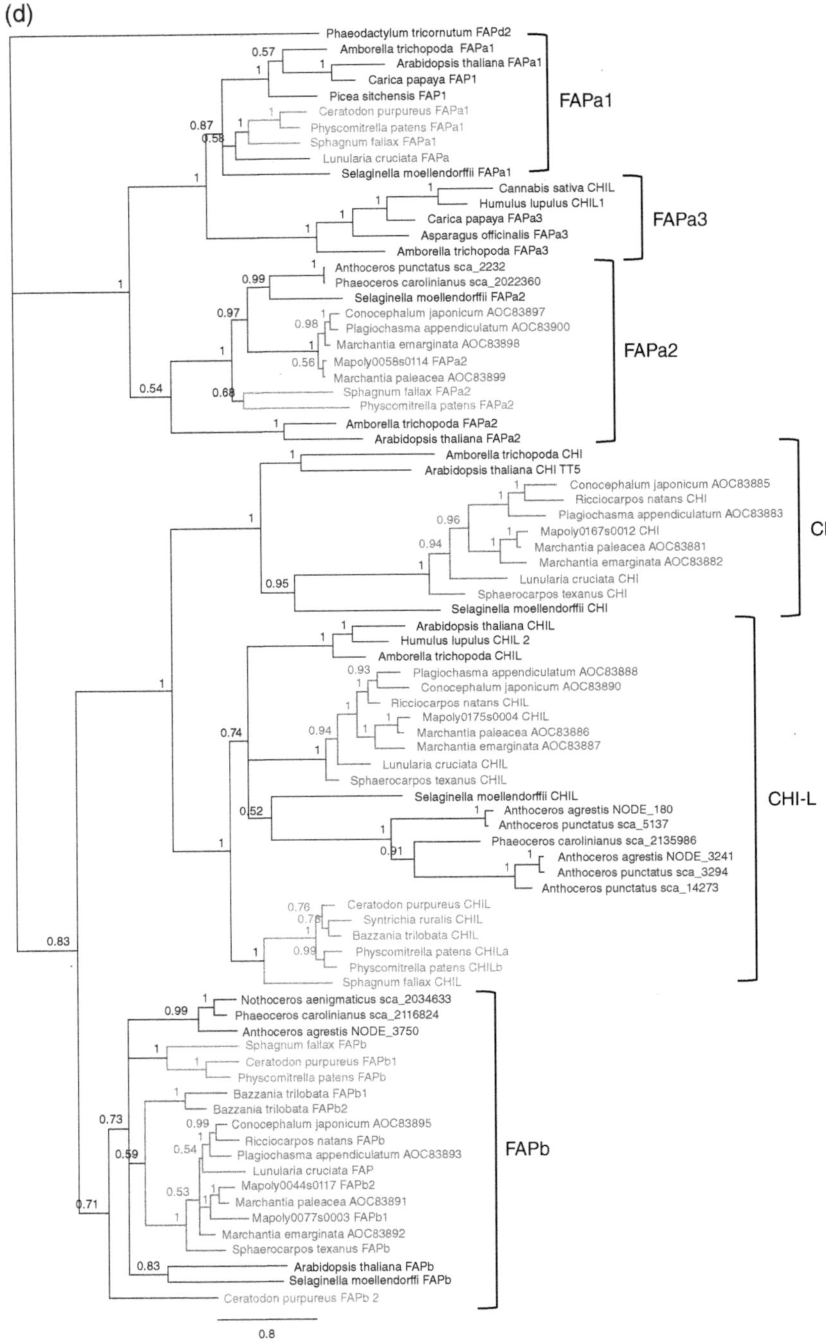

Figure 9.2 (Continued)

characterized from marchantia (Bowman et al. 2017; Clayton et al. 2018) and *Plagiochasma appendiculatum* (Yu et al. 2014). The major land plant groups form distinct clades, including for monocots and dicots, rather than distinct PAL subclades forming across the plant groups. This supports a monophyletic origin of PAL in the major plant groups, with extensive gene duplication within the groups. However, there is evidence for PAL genes with separate origins within the gymnosperms and also the basal dicot *Nelumbo nucifera* (sacred lotus) (Wu et al. 2014).

PAL is commonly encoded by multigene families of three to six genes, with exceptions such as the *Solanum* species potato (*S. tuberosum*) and tomato (*S. lycopersicon*) having 40 and 26 genes per haploid genome, respectively (Joos and Hahlbrock 1992; Chang et al. 2008). Our annotation of the marchantia genome found 10 candidate PAL genes, including an apparently recent multiple duplication of a pairing of a *PAL* and *CHALCONE SYNTHASE* (*CHS*) gene (Bowman et al. 2017). Unlike in tomato (Chang et al. 2008), most of the Mp*PAL* genes are actively expressed (Albert et al. 2018; Clayton et al. 2018), and at least six encode proteins with PAL activity (Yoshikawa et al. 2018).

In seed plants, *PAL* gene expression is regulated by both developmental cues, such as for lignification or flower and fruit development, and environmental stimuli, including pathogen infection, wounding, UVB irradiation, drought stress, and nutrient depletion (Huang et al. 2010; Pawlak-Sprada et al. 2011; Kim and Hwang 2014; Zhang and Liu 2015). The regulatory systems established for seed plants probably have many conserved elements in the bryophytes, as wounding, pathogen infection, and UVB irradiation induce *PAL* transcript accumulation in either marchantia or *P. patens* (Wolf et al. 2010; Ponce De Leon et al. 2012; Clayton et al. 2018; Yoshikawa et al. 2018). Expression studies suggest both distinct and overlapping expression patterns of *PAL* gene family members during development and biotic and abiotic stresses.

9.3.2 Cinnamate 4-hydroxylase

C4H belongs to the CYP73A subfamily of cytochrome P450 monooxygenases. It catalyses the first hydroxylation step of the phenylpropanoid pathway, converting *trans*-cinnamic acid into *para*-coumaric acid (Koopmann et al. 1999; Blount et al. 2000; Blee et al. 2001; Ehlting et al. 2006; Ferrer et al. 2008). Although the principal physiological substrate of C4H is *trans*-cinnamic acid, *in vitro* it can use alternative compounds that fit the substrate recognition pocket, including naphthalenes, quinolines, and indoles (Urban et al. 1994; Schalk et al. 1997). C4H is a key regulatory point for the production of a range of phenylpropanoids, including flavonoids, coumarins, artemisinin, phenolic esters, amides, and monolignols (Bassard et al. 2012; Kumar et al. 2016).

Genes encoding C4H have been characterized from many plant species. Some species, including marchantia, have only a single copy whereas others contain small gene families (Betz et al. 2001; Raes et al. 2003; Lu et al. 2006; Kawai et al. 2014; Bowman et al. 2017; Xia et al. 2017). Phylogenetic analysis suggests a relatively ancient duplication of C4H prior to the separation of gymnosperms and angiosperms (about 350 MYA), as distinct clades (Class I and II) of *C4H* sequences occur containing both monocot and dicot sequences, with gymnosperm sequences falling closer to Class I (Ro et al. 2001; Xia et al. 2017). However, some dicots have lost Class II, including Arabidopsis (Raes et al. 2003). C4H gene activity is induced by both biotic and abiotic stresses, including light, drought stress, salinity, cold, elicitors, and wounding (J. Kim et al. 2013; Phimchan et al. 2014; Wang et al. 2017; Xia et al. 2017). In many species, it has a significant role in the UVB light stress response, promoting flavonoid and/or hydroxycinnamic acid production. In Arabidopsis, the *C4H* gene is the key target of the repressor transcription factor *At*MYB4, with UVB exposure repressing At*MYB4* gene expression and thus promoting *C4H* activity and sinapate ester production (Jin et al. 2000). Besides flavonoid biosynthesis, the other central biosynthetic pathway of vascular plants for which C4H is key is that for lignin: reducing C4H activity can reduce lignin content (Sykes et al. 2015). In *Eucalyptus* the Class II *C4H* is mainly associated with stress responses and wood lignin biosynthesis, and Class I *C4H* is constitutively expressed during general synthesis of phenylpropanoid metabolites (Harakava 2005).

Recently *C4H* genes have been characterized from basal land plants, including from the liverwort species *P. appendiculatum* (Pa*C4H*), marchantia (Mp*C4H*), and *Marchantia paleacea* (Mpal*C4H1* and Mpal*C4H2*) (X.-Y. Liu et al. 2017). Additionally, we find a conserved *C4H* in the draft genome sequences of the hornworts *A. agrestis* and *A. punctatus* (Figure 9.2). Phylogenetic analysis reveals strong conservation of C4H across land plants, including characteristic motifs and residues, but has failed to detect authentic sequences in algal genomes. Functional assays reflect the high sequence conservation of C4H sequences in bryophytes, as the three liverwort C4H gene products examined for function *in vitro* (*Pa*C4H1, *Mpal*C4H1, and *Mpal*C4H2) were all able to catalyse the *p*-hydroxylation of both *trans*-cinnamic and 3-hydroxycinnamic acids, but were unable to use *p*-coumaric acid, 2-hydroxycinnamic acid, or benzoic acid, a substrate specificity which mirrors that of Arabidopsis C4H.

9.3.3 4-Coumarate-CoA Ligase (4CL)

4-COUMARATE-CoA LIGASE (4CL) catalyses the ligation of CoA with cinnamic acid and its methoxy and hydroxy derivatives, forming the various monomers that feed into the different branches of the phenylpropanoid pathway (Lavhale et al. 2018). It has been extensively characterised in vascular plants for its role in lignin and flavonoid biosynthesis (Sun et al. 2013; Oliveira et al. 2015; Mandoulakani et al. 2017). 4CL is part of small family of adenylate-forming enzymes. Multiple isoforms of 4CL are common, with individual genes dedicated to different pathway branches. For example, in Arabidopsis, At*4CL3* contributes to flavonoid biosynthesis, whereas At*CL1*, At*CL2*, and At*CL4* are part of lignin production (Li et al. 2015). Similarly, in *Pueraria lobata* (Kudzu, part of the Fabaceae) Pl*4CL1* transcript amounts positively correlated with isoflavone production, while Pl*4CL2* was suggested to be involved in lignification (Li et al. 2014). The different isoforms generally display similar substrate utilization profiles, although there are some reports of variable catalytic efficiency towards specific substrates. Thus, it is plausible that there is diversity in *4CL* gene family function as well as expression pattern (Li et al. 2015; Lavhale et al. 2018). In rice, 4CL sequences separate into four groups, which is more than have been characterised for dicots (Sun et al. 2013). Although these additional groups may not have variant functions, it does suggest that monocot and dicot 4CL gene families evolved independently after the separation of the two plant lineages (Lavhale et al. 2018). The data for *Pinus taeda* suggest another evolutionary divergence of 4CL sequences prior to the angiosperm and gymnosperm split (Chen et al. 2014). In the basal vascular plant *S. moellendorffi*, *Sm*4CL1 and *Sm*4CL2 can utilize coumaric acid, cinnamic acid, caffeic acid and ferulic acid, but not sinapic acid as substrates (Liu et al. 2018). Sm*4CL1* was strongly expressed in stems while Sm*4CL2* was higher in leaves, and both were highly expressed in roots. 4CL sequences are conserved in the bryophytes, including hornworts (Figure 9.2). Sequences with similarity to 4CL have also been found in red algae and green algae, suggesting the existence of the enzyme in a shared ancestor of embryophytes and green algae, even before the ancestral divergence of the red algae (Labeeuw et al. 2015a).

Some specialisations in 4CL function have come to light recently. In *Scutellaria baicalensis* a variant 4CL (4CL-like) with a strong preference for cinnamic acid forms cinnamoyl-CoA as the starting molecules for a pathway to the formation of 4′-deoxyflavones. The subsequent enzymes, such as CHS and CHALCONE ISOMERASE (CHI) have probably coevolved to use these variant substrates (Zhao et al. 2016). These compounds occur in other taxonomically distant species, such as *Cephalocereus senilis* ('old man' cactus) (Jiang et al. 2016), and so 4CL enzymes with these activities may be widespread.

9.3.4 Chalcone Synthase

The entry point to the flavonoid-specific section of the phenylpropanoid pathway is CHS, a polyketide synthase (PKS) that catalyses a stepwise condensation reaction of three acetate residues from malonyl-CoA with the substrate 4-coumaroyl-CoA to produce naringenin-chalcone. CHS is ubiquitous in all plant groups examined to date, including liverworts and mosses. Xie et al. (2016) examined PKS gene occurrence across angiosperms, *S. moellendorffii*, *P. patens*, and diverse algal species, and found convincing PKS genes across all the land plants but none within the algae. Subclades with possible functional significance were identified within angiosperm species, but the moss and lycophyte sequences formed basal outliers on their own.

There are no publications concerning hornworts, so we examined the *Anthoceros* draft genome sequences (NCBI accessions ERR771108 and SRR1278954 for *A. agrestis* and *A. punctatus*, respectively). We identified PKS sequences from *A. punctatus* and *A. agrestis* that claded with the subgroup A of characterised land plant CHS sequences (Figure 9.2). Additionally, on a reciprocal Basic Local Alignment Search Tool (BLAST) screen the top hits for these were characterised CHS sequences. *Anthoceros* sequences were also identified that claded with the sequences encoding variant 'basal' PKS activities, such as the 2-oxoalkylresorcinol synthase of *P. patens* (S.Y. Kim et al. 2013; L. Li et al. 2018). Within subgroup A the pattern followed predicted evolutionary lines, with bryophytes basal, then lycophytes and ferns, then gymnosperms and angiosperms. However, functional analysis is required before an unequivocal CHS function can be assigned to these sequences. The Type III PKS gene family, of which CHS is part, is large and contains enzymes with very high amino-acid sequence identity but distinct activities. For example, STILBENE SYNTHASE (STS) can vary from CHS by only a few amino acid residues and uses the same substrates as CHS (4-coumaroyl/cinnamoyl-CoA and three molecules of malonyl-CoA), but instead of producing chalcones yields products with two rings and a reduced 'bridge' because of the loss of a carbon.

9.3.5 Chalcone Isomerase and Chalcone Isomerase-like

CHI and CHALCONE ISOMERASE-LIKE (CHI-L) have emerged as two of the most interesting flavonoid biosynthetic enzymes with regard to their occurrence and function in basal plants. CHI catalyses the enantioselective formation of (*S*)-flavanones (naringenin, eriodictyol, and pinocembrin) from chalcone precursors. However, the absolute requirement of CHI for flavonoid biosynthesis has been the subject of debate. The same closure to form the C-ring can occur spontaneously *in vitro* for chalcones with an available hydroxyl (the C4′/6′ position), but this yields a racemic mix of the isomers. However, this spontaneous closure is prevented if the hydroxyl is glycosylated or absent (the 6′-deoxychalcones). Mutants for *chi* in Arabidopsis (*tt5*) and rice (*gh1*) (Hong et al. 2012; Jiang et al. 2015) prevent most, but not all, flavonoid biosynthesis. However, *chi* mutants of marchantia completely lose production of flavones (Clayton et al. 2018), showing CHI is an essential *in planta* activity in this species. Remarkably, in the case of the *anthocyanin-free* mutant of tomato, corresponding to a *chi* gene, not only is flavonoid biosynthesis lost in affected tissues (leaves and trichomes) but there are pleotropic effects that include loss of monoterpene and sesquiterpene biosynthesis (Kang et al. 2014).

The role of CHI-L is not resolved, although recent studies have provided putative functions. In hop (*Humulus lupulus*), *Hl*CHIL2 enhances the activities of CHS and an aromatic prenyltransferase (*Hl*PT1L) through direct protein–protein interaction (Ban et al. 2018). The single *CHI-L* gene of marchantia may have a similar activity, as the production of both flavones and auronidins was reduced to around 10% of wild-type amounts in a *chi-l* mutant, suggesting a general action in assisting biosynthetic steps prior to CHI (Clayton et al. 2018). Arabidopsis *chi-l* mutants also have reduced proanthocyanidin and flavonol concentrations in seeds, but the proposed action is through direct interaction with CHI (Jiang et al. 2015). The *chi-l* mutants of Japanese morning glory (*Ipomoea nil*) also have reduced anthocyanin biosynthesis (Morita et al. 2014). Thus, a principal function of the ancestral CHI-L of land plants may be to promote CHI and/or CHS activity. However, the second hop gene, *Hl*CHIL1, is suggested to have a divergent function, binding and stabilizing the ring-open configuration of chalcones during formation of the methylated prenylchalcones, such as xanthohumol, preventing the spontaneous formation of the C-ring.

CHI and *CHI-L* are thought to have evolved from a nonenzymatic ancestor related to extant Fatty-Acid binding Protein genes, which are found in algae and bacteria in addition to land plants (Ngaki et al. 2012). To date, no good candidate *CHI* or *CHI-L* sequences have been found outside land plants. Additionally, although sequences with demonstrated CHI activity (Cheng et al. 2018; Clayton et al. 2018) occur in liverworts, no *CHI* sequences have been found in genome or transcriptomes of moss species (Ngaki et al. 2012). Nor are *CHI* sequences present in the limited transcriptomes available for hornworts (Ngaki et al. 2012) or the draft genome sequence of *A. agrestis* and *A. punctatus* (Figure 9.2). Given the phenotypes of *chi* mutants in other plant groups, how mosses produce significant amounts of flavonoids in the absence of CHI is not clear.

The principal chalcones for flavonoid production have a 6′-hydroxyl group. However, in the Fabaceae (legumes) the coaction of CHALCONE REDUCTASE (CHR) with CHS causes the production of the 6′-deoxychalcones, cyclisation of which to the corresponding 4′7-dihydroxyflavanone requires a specialised CHI (Davies et al. 1998). There have been few studies of similar activities outside of angiosperms, but *S. moellendorffii* produces 6′-deoxychalcones and 4′7-dihydroxyflavanones (Weng and Noel 2013).

9.3.6 Hydroxylation Activities

Hydroxylation activities conducted by members of the CypP450 or 2-oxoglutarate-dependent dioxygenase (2OGD) superfamilies occur at many points of the phenylpropanoid pathway. These range from steps of the core pathway (Figure 9.3), such as C4H,

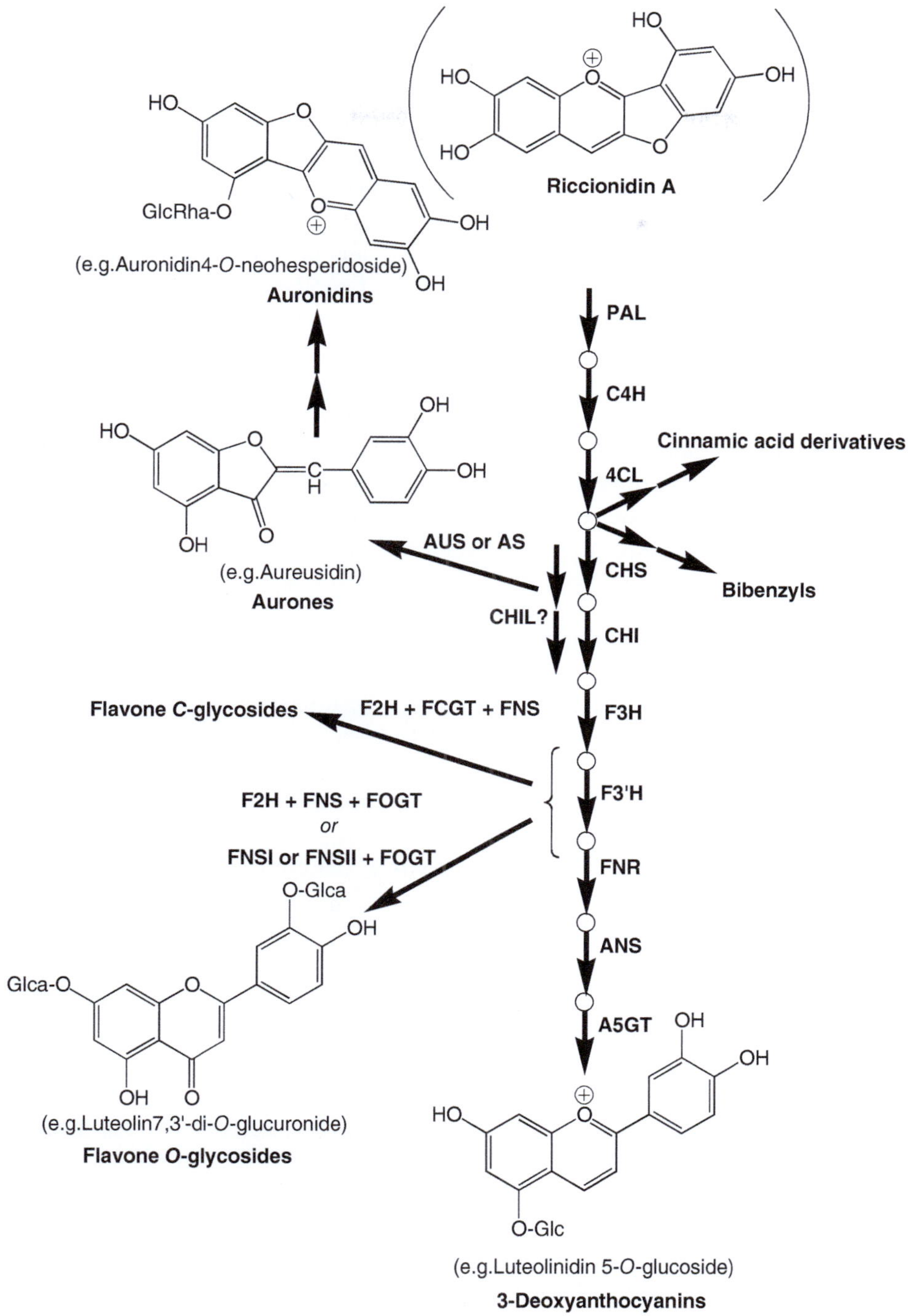

Figure 9.3 A section of the core phenylpropanoid/flavonoid pathway leading to production of flavones, anthocyanins, and auronidins. Flavones are prominent compounds of liverworts, and have also been reported from mosses. Multiple biosynthetic routes to flavones have been reported from angiosperms; which of these occur in bryophytes is not known. The pigments 3-deoxyanthocyanidins and auronidins have been reported from mosses and liverworts, respectively. The original structure published for riccionidinA (as an anthocyanidin) is also shown. Enzyme abbreviations are: PAL, phenylalanine ammonia-lyase; C4H, cinnamate 4-hydroxylase; 4CL, 4-coumarate:CoA ligase; CHS, chalcone synthase; CHI, chalcone isomerase; CHIL, chalcone isomerase-like; F3H, flavanone 3-hydroxylase; F3′H, flavonoid 3′-hydroxylase; FNR, flavanone reductase; ANS, anthocyanidin synthase; A5GT, anthocyanidin 5-*O*-glucosyltransferase; AS, aurone synthase; AUS, aureusidin synthase; FNS, flavone synthase (2OGD = I, Cyp450 = II); F2H, flavanone 2-hydroxylase; FOGT, flavone *O*-glucosyltransferase; FCGT, flavone *C*-glucosyltransferase.

CINNAMATE 3-HYDROXYLASE (C3H, CYP98A), FLAVANONE 3-HYDROXYLASE (F3H, a 2OGD), ANTHOCYANIDIN SYNTHASE (ANS, a 2OGD), FLAVONE SYNTHASE (either the CypP450 FNSII or the 2OGD FNSI), FLAVONOL SYNTHASE (FLS, a 2OGD), FLAVONOID 3′-HYDROXYLASE (F3′H) and FLAVONOID 3′5′-HYDROXYLASE (F3′5′H), through to activities acting on compounds of limited taxonomic occurrence. In the following sections we discuss specific CypP450 and 2OGD activities in the sections on production of different phenylpropanoid groups. General reviews of these enzyme groups are already available (Nelson and Werck-Reichhart 2011; Kawai et al. 2014), as well as the phylogenies in Bowman et al. (2017) that align all the CypP450 and 2OGD sequences of marchantia against representatives of the characterised dicot clades. It is worth noting the difficulties of assigning function based on sequence similarity and phylogenetics alone, and some proposed identifications of genes such as the *F3′H* and *F3′5′H* in bryophytes should be treated with caution until functional assays or genetic mutants are available.

9.4 Notable Phenylpropanoids of Bryophytes

In the following sections we examine some of the notable phenylpropanoid compound groups of bryophytes. Rosmarinic acid, lignans, bibenzyls, and coumarins are phenolics derived from the pathway intermediates from cinnamic acid onwards, while the aurones, auronidins, flavones, flavonols, and 3-deoxyanthocyanins are flavonoids, as they all originate from chalcones. One of the most intriguing aspects of phenylpropanoid biosynthesis in the bryophytes is the apparent lack of flavonoids in the hornworts (Markham 1988; Asakawa et al. 2013; Monforte et al. 2018). Flavones and flavonols are central to the UVB tolerance mechanism of liverworts and angiosperms, yet hornworts lack these compounds. Possible alternatives for assisting with UV tolerance are cell wall–bound or soluble phenolics derived from branches of the phenylpropanoid pathway prior to the formation of flavanones or the use of nonphenolic compounds. As mentioned earlier, some algae show significant tolerance to terrestrial conditions, possibly based on compounds such as MAAs, gallic acid–derived red pigments (Holzinger and Pichrtová 2016). There is also evidence of compounds similar to lignin and related biosynthetic genes in *K. flaccidum* that could assist with UVB tolerance (Harholt et al. 2016). The only study of UVB treatments of hornworts published to date suggests that both soluble and insoluble phenolics may be important, in particular the cinnamic acid derivative rosmarinic acid and related compounds (Monforte et al. 2018). However, rosmarinic acid has yet to be characterised for the liverworts or mosses (Asakawa et al. 2013). In cell suspension cultures of the hornwort *A. agrestis* it was found that rosmarinic acid and rosmarinic acid 3′-*O*-glucoside could accumulate to more than 6% total dry weight (Vogelsang et al. 2006).

9.4.1 Hydroxycinnamic Acid Derivatives: Rosmarinic Acid and Lignans

Cinnamic acid derivatives are prominent in hornworts: not only rosmarinic acid but lignans such as megacerotonic acid and anthocerotonic acid (Figure 9.1) (Asakawa et al. 2013). Lignans are structurally diverse compounds found in a wide range of plant groups, including liverworts but not yet mosses. They are dimers of monolignols such as *p*-coumaroyl alcohol and coniferyl alcohol. The biosynthetic steps to monolignols proceed through formation of *p*-coumaroyl-CoA by 4CL in the core phenylpropanoid pathway, and then successively *p*-coumaraldehyde and *p*-coumaroyl alcohol by the enzymes CINNAMOYL-CoA REDUCTASE (CCR) and CINNAMOYL ALCOHOL DEHYDROGENASE (CAD). For formation of coniferyl alcohol there are additional steps: C3H forms caffeoyl-CoA from *p*-coumaroyl-CoA, and then CAFFEOYL-CoA *O*-METHYLTRANSFERASE (CCoAOMT) yields feruloyl-CoA. CCR and CAD then produce coniferaldehyde and coniferyl alcohol. The biosynthetic pathways to lignans from monolignols are diverse, but all start with the coupling of two monolignols units to form pinoresinol (Satake et al. 2017). Pinoresinol may be glycosylated and accumulate in itself, but is also the basis for at least two distinct lignan pathways, which start with the action of PINORESINOL-LARICIRESINOL REDUCTASE (PLR) or PIPERITOL/(+)-SESAMIN SYNTHASE (CYP81Q). From these activities the pathways can branch further, contributing to the diversity of lignan structures identified to date. Dirigent proteins are important in conferring the stereochemistry of the lignans formed.

Cinnamic acid derivatives are abundant in mosses, and may contribute to cell wall structure; for example, sphagnum acid (*p*-hydroxy-*beta*-[carboxymethyl]-cinnamic acid) is abundant in the *Sphagnum* species that form peat bogs. These are some of the most significant terrestrial plant communities – it is estimated that between 10 and 15% of the terrestrial carbon stock is held in dead and living *Sphagnum* peat bogs (Bonnett et al. 2010). It's uncertain what components of the various lignan and other cell wall cinnamic acid biosynthetic pathways characterised for angiosperms may also be present in bryophytes. However, bryophytes certainly produce a great variety of lignan structures, suggesting diverse biosynthetic activities across the species. The marchantia genome contains strong candidate sequences for CAD, CCR, C3H, and methyltransferases (Bowman et al. 2017), as well as several sequences for dirigent proteins.

9.4.2 Hydroxycinnamic Acid Derivatives: Coumarins

Coumarins and furanocoumarins are plant defence compounds that occur sporadically throughout land plants, including liverworts and mosses. There is a great variety of structures described based on the central coumarin structure. There are also a range of proposed *in planta* functions, especially in plant defence, as well as pharmacological properties. Major subgroups within the overall coumarin classification are simple coumarins, furanocoumarins, pyranocoumarins, and coumarins with modifications of the pyrone ring. Both aglycones and glycosides of these structural types may accumulate. The furanocoumarins and pyranocoumarins contain an additional five or six member rings, respectively, compared with the simple coumarins, and may be produced through prenylation of umbelliferone.

The taxonomic pattern of occurrence of coumarins suggests their biosynthesis may have arisen on independent occasions, therefore the enzymology of the pathway could differ between plant groups. Experimental evidence from different dicot species supports this. In Arabidopsis the initial simple coumarin, scopoletin, is formed by hydroxylation of feruloyl-CoA at the 6′ by a 2OGD (F6′H) and subsequent spontaneous formation of the

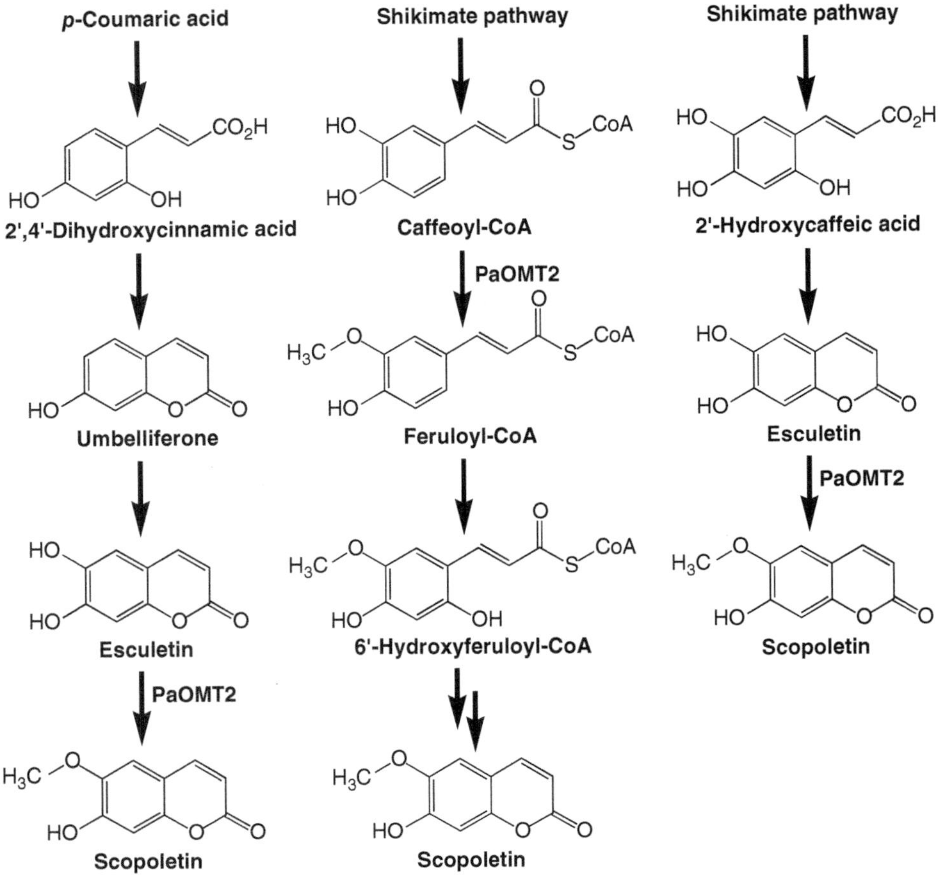

Figure 9.4 Examples of varied biosynthetic routes to the simple coumarin scopoletin. Different biosynthetic routes for the formation of coumarins are thought to have evolved across the major plant groups. Three possible routes to scopoletin are shown. In Arabidopsis, the route via feruloyl-CoA has been reported. The route(s) present in bryophytes have yet to be determined, but two of the different activities reported for OMT2 from the liverwort *Plagiochasma appendiculatum* are indicated.

lactone ring (Figure 9.4). Similarly, umbelliferone is formed by *p*-coumaroyl CoA 2′-hydroxylation, and there are examples of 2OGD enzymes that have C2′H activity, F6′H activity, or both activities (the 2′ and 6′ positions are equivalent). These enzymes thus represent the branching activity that directs phenylpropanoid substrate to coumarins rather than lignans or flavonoids. In contrast, the 5-methyl coumarins of species such as *Gerbera hybrida* are formed from acetate units via a pentaketide intermediate (via the acetate-malonate pathway rather than phenylpropanoid pathway), with PKSs as the key initial enzymes (Pietiäinen et al. 2016). Several other genes for later reactions in coumarin biosynthesis have also been characterised. These include a coumarin-specific prenyltransferase that catalyses formation of the precursor for both linear (psoralen derived) and angular (angelicin derived) furanocoumarins from umbelliferone (Karamat et al. 2013); the CypP450 enzymes psoralen synthase and angelicin synthase (Cyp71A clade), which conduct subsequent reactions; a scopoletin 8-hydroxylase, and bergaptol *O*–methyltransferase (Hehmann et al. 2004; Larbat et al. 2007, 2009; Xu et al. 2016; Krieger et al. 2018; Siwinska et al. 2018). There are few data yet on the biosynthetic pathway(s) in liverworts, but the widespread occurrence across diverse liverwort taxa suggests that the core phenylpropanoid pathway may have arisen relatively early in liverwort evolution. A Mg^{2+}-dependent *O*-methyltransferase has been characterised from *P. appendiculatum*, which methylates esculetin (6,7-dihydroxycoumarin) to scopoletin (Figure 9.4) and isoscopoletin (Xu et al. 2016).

9.4.3 *Bis*-bibenzyls

More than 70 different bibenzyls have been identified from liverworts (Asakawa and Ludwiczuk 2012; Asakawa 2016), including prenyl-bibenzyls that are structural analogues of the psychopharmacological tetrahydrocannabinol compound of *Cannabis sativa* (Hussain et al. 2018). The cyclic *bis*-bibenzyls characteristic of marchantia, the marchantins, are thought to be produced from dimerization of dihydrostilbenes, which are initially formed by a PKS using similar substrates as CHS; dihydro-*p*-coumaroyl-CoA (rather than *p*-coumaroyl-CoA) and 3x malonyl-CoA, followed by various hydroxylations (Figure 9.5) (Friederich, Maier, et al. 1999; Friederich, Rueffer, et al. 1999; Takahashi and Asakawa 2017). Candidate PKS genes encoding STILBENECARBOXYLATE SYNTHASE activity have been identified from *M. paleacea* (Yu et al. 2018), along with candidate sequences of *Radula marginata*, a species that makes cannabinol analogues (Hussain et al. 2018). The diversity and abundance of bibenzyls found in liverworts is not reported for any other plant group. There are only a few reports of any bibenzyls elsewhere in the plant kingdom: specifically, riccardin H and riccardin C in the fern *Hymenophyllum barbatum* and angiosperm *Primula macrocalyx*, respectively (Asakawa 2016). They are assumed to be important for plant defence, and at least one bibenzyl, lunularic acid, can mimic the action of the plant hormone abscisic acid (Yoshikawa et al. 2002).

A prominent feature of liverworts are the oil bodies, which may correspond to the first type of secretory structures to arise in terrestrial plants (Aigner et al. 2013; Lange 2015). Current understanding is that they are intracellular organelles bound in a single membrane, and contain lipophilic globules surrounded by a proteinaceous matrix in which various metabolites can accumulate to high concentrations. In particular, liverworts produce a great variety of terpenoids, with more than 700 unique structures reported, and these are major components in the oil bodies. Additionally, some bibenzyls, which are typically non-glycosylated and hydrophobic, have also been shown to be localised to the oil bodies of marchantia (Tanaka et al. 2016). Mosses, hornworts, and lycophytes appear to lack an equivalent structure that facilitates accumulation of high secondary metabolite concentrations, with glandular trichomes not appearing in the land plant lineage until the ferns (Lange 2015).

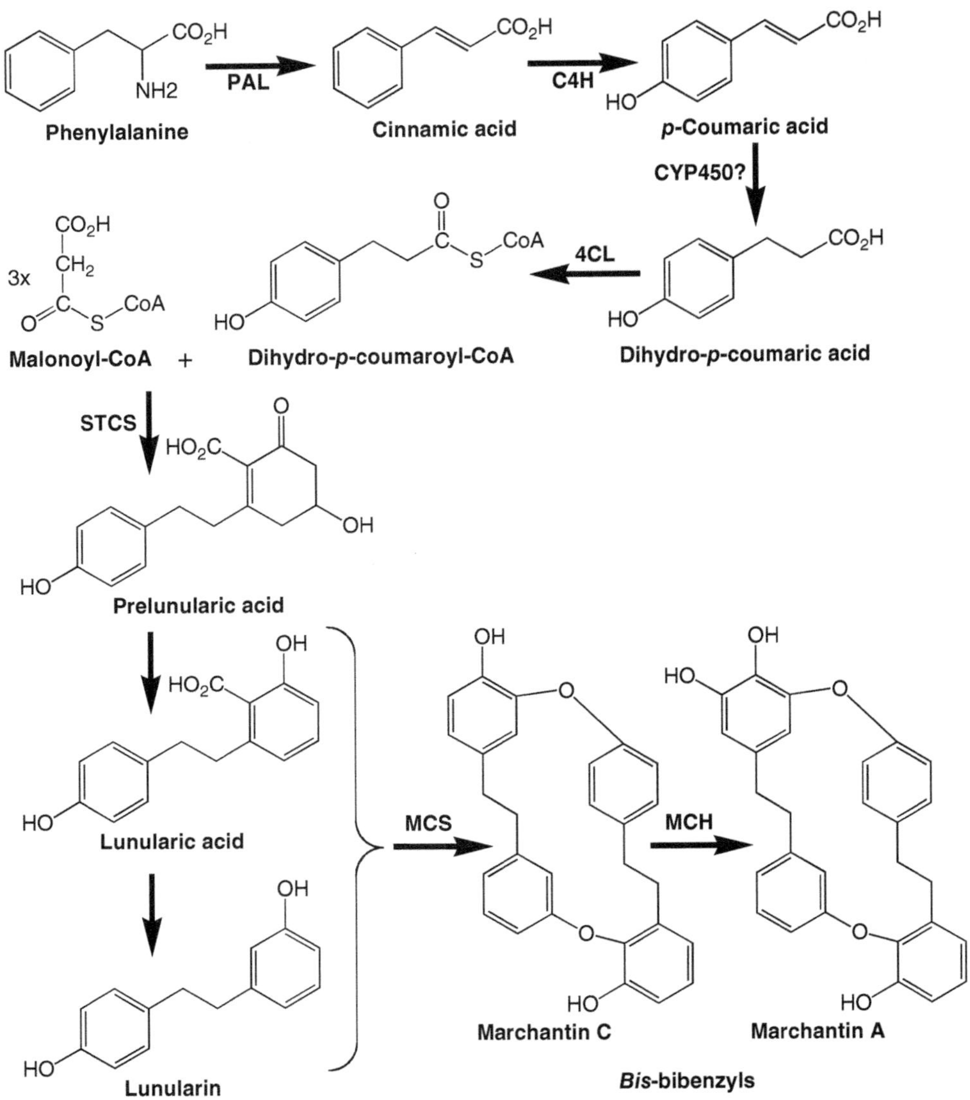

Figure 9.5 A section of the proposed biosynthetic route to the *bis*-benzyls marchantin A and C. Enzyme abbreviations are as given in Figure 9.4 and STCS, stilbenecarboxylate synthase; MCS, Marchantin C synthase; MCH, Marchantin C hydroxylase.

9.4.4 Flavones and Flavonols

Flavones and/or flavonols are found across the liverworts and mosses, but are absent from the hornworts. They are key compounds for UVB tolerance of angiosperms, and analysis of genetic mutants lacking components of the flavonoid or the UVB signalling pathways has confirmed the importance of flavone *O*-glycosides for UVB acclimation of marchantia (Clayton et al. 2018). Flavones are ubiquitous in mosses, and flavonols have also been reported for some species. Whether they have such a key function in UVB tolerance for mosses has yet to be confirmed, with studies on *P. patens* finding a high tolerance of UVB but only a small concurrent induction of quercetin glycoside production (Wolf et al. 2010). Flavones and flavonols can directly screen UVB radiation. However, there are other phenolics, such as hydroxycinnamic acids, that have broader absorption spectra for UV, require a small carbon investment, and involve fewer biosynthetic steps. Thus, flavones and flavonols may have other beneficial properties that led to the evolution of their biosynthesis and its retention across the major land plant groups, such as strong activity against ROS (Pollastri and Tattini 2011; Brunetti et al. 2018; Davies et al. 2018). As with some angiosperms, liverworts also produce a range of flavone *C*-glycosides. In angiosperms, such flavones can have anti-insect, antinematode, and antifungal activities (Jiang et al. 2016), but the functions in liverworts have yet to be established. Flavones can also be involved in interactions with beneficial microorganisms, such as for root mycorrhiza and nodulation.

The occurrence of flavones and flavonols in bryophytes suggests they probably have the same section of the phenylpropanoid pathway for their production as angiosperms, specifically the F3H, FLS, FNS, and appropriate UDP-glycosyltransferases. However, this has yet to be established. The analysis is complicated by the close sequence similarity and evolutionary relationship of the 2OGD enzymes FLS, FNS, F3H, and ANS and the parallel evolution of multiple routes to the compounds in angiosperms (Figure 9.3). For flavones there are both 2OGD (FNSI of the DOXC28 clade) and CytP450 (FNSII, CYP93B or CYP93G1) enzymes. Initially these showed taxonomic separation, with FNSI only in the Apiaceae, but species with both enzyme types have now been found (Martens and Mithofer 2005; Falcone Ferreyra et al. 2015; Jiang et al. 2016). Both enzymes form flavones directly from flavanones by introducing a double bond between the C2 and C3, and the flavones are then glycosylated to produce flavone *O*-glycosides. However, flavone *C*-glycosides are biosynthesised by alternative routes. A CytP450 (CYP93B1 or CYP93G2), the FLAVANONE 2-HYDROXYLASE (F2H), performs 2-hydroxylation of the flavanone to form 2-hydroxyflavanone (Du et al. 2010). This is then the substrate for a *C*-glycosyltransferase, with the product undergoing dehydration to convert to the flavone *C*-glycoside (Brazier-Hicks et al. 2009). In maize, the UFGT (UGT708A6) is bifunctional, being able to make both *C*- and *O*-glycosylated flavonoids. Additionally, when CYP93G1 was expressed in Arabidopsis the intermediate was stable and 2-hydroxyflavanone *O*-glycosides accumulated (Lam et al. 2014). In cereals, there is a third route to flavones, for the methoxylated tricin *O*-glycosides. For tricin, the 3′5′-dimethoxyflavone nucleus is formed first and *O*-glycosylated, then CYP75B4 conducts 5′-hydroxylation to convert chrysoeriol to selgin, the immediate precursor of tricin. Expression of CYP93G2 and CYP75B4 in Arabidopsis resulted in accumulation of different tricin *O*-glycosides. CYP93G1, CYP93G2, and CYP75B4 have so far only been characterised for Poaceae species, representing a potentially unique evolutionary branch of the flavonoid pathway in this order. There is also the alternative biosynthetic route mentioned earlier for *S. baicalensis* leading to 4′-deoxyflavones, in which a differing 4CL activity provides cinnamoyl-CoA as an alternative starter molecule. The FNSII, F2H, ISOFLAVONE SYNTHASE, and, potentially F3H, could all have evolved from an ancestral gene of the CYP93 type (Jiang et al. 2016).

The assignment of function to candidate bryophyte F3H/FNS/F2H sequences is further complicated by the observation that the only sequence characterised to date as having FNS activity (*in vitro*) is from a liverwort, *P. appendiculatum*, and is a 2OGD with both FNSI and F2H activity (Han et al. 2014). The residue Tyr240 is the critical residue for F2H activity, since a Tyr240Pro mutant could convert naringenin to apigenin, but could not produce 2-hydroxynaringenin. Two marchantia 2OGD genes have strongly induced expression accompanying flavone *O*-glycoside production in response to UVB exposure, including the only marchantia gene in the DOXC28 clade (Clayton et al. 2018). However, function cannot be assigned on the basis of sequence alone, as the top BLAST matches include a mix of F3H, F2H, FNSI, and FLS sequences. Liverworts and mosses produce flavones and/or flavonols that are highly substituted with glycosyl and methyl moieties. Based on phylogenetics (e.g. Bowman et al. 2017), it is probable that at least some of these reactions are conducted by enzymes similar to those characterised for angiosperms, and an apigenin 4′-*O*-methyltransferase has been characterised from *P. appendiculatum* that phylogenetically clades with flavone OMTs of angiosperms (H. Liu et al. 2017).

9.4.5 Aurones and Auronidins

Until recently, anthocyanins were thought to be present in all extant lineages of land plants, with liverworts producing 'primitive', nonglycosylated, anthocyanidins. Next, in evolutionary terms, the glycosylated 3-deoxyanthocyanins that are prominent in mosses and ferns were thought to have arisen. Finally, with the seed plants, the 3-hydroxyanthocyanins themselves appeared. At the base of this proposed evolutionary pathway was the red pigment of liverworts, riccionidinA. RiccionidinA was initially thought to be an anthocyanidin bound into the cell wall (Kunz et al. 1993). However, from genetic analysis of marchantia it was discovered that those proposals were based on a misinterpretation of the riccionidinA data (Berland et al. 2019). Using marchantia transgenics overproducing riccionidinA, the chemical structure of a soluble glycoside of riccionidinA was obtained, and it was realised that it was not a 'primitive' anthocyanin but rather a new class of flavonoids, for which the name 'auronidins' was proposed (Figure 9.3). CRISPR/Cas9-mediated mutagenesis of the marchantia *CHI*, *CHIL*, and *aurone synthase* (a polyphenol oxidase) genes was used to show that auronidin 4-*O*-neohesperidoside is biosynthesised from the aurone branch of the flavonoid pathway rather than that leading to anthocyanins (Figure 9.3) (Berland et al. 2019). There is a single report of riccionidinA for angiosperms, from root cultures of *Rhus chinensis* (syn. *R. javanica*) (Taniguchi et al. 2000).

The role(s) of auronidins have yet to be fully explored, but auronidins and anthocyanins both provide red colouration suitable for screening in the visible part of the light spectrum. Indeed, auronidins seem to provide similar protective properties to anthocyanins with regard to reducing damage from excess white light. Studies on the leafy liverwort species *Jamesoniella colorata* and *Isotachis lyallii*, which are commonly green in shade and red in exposed situations, found that under higher irradiance red leaves maintained higher quantum efficiencies for photosynthesis, had larger photochemical and nonphotochemical quenching values, and had improved desiccation tolerance (Hooijmaijers and Gould 2007; Hooijmaijers 2008). Interestingly, auronidins are also induced by UVB exposure, at least in some Antarctic liverwort species (Newsham 2009). Auronidins respond differently to pH than anthocyanins (having strong fluorescence that is absent from anthocyanins) and, as they are cell wall–bound, they may not be as effective against ROS. As mentioned earlier, some terrestrial algae also have red phenolic pigments. This suggests that achieving red colouration in itself could be a strong evolutionary driver, rather than the structural details *per se*. The presence of auronidins rather than anthocyanin in liverworts also raises questions regarding when anthocyanin biosynthesis arose during evolution, and what the pigmentation status of the last common ancestor of land plants may have been. Current understanding is that the red pigments of mosses are derivatives of anthocyanins (see the next sections) but there is no information on any red pigmentation for hornworts.

The aurones, from which auronidins are probably directly made, are yellow pigments found sporadically across the land plants. They are present in at least liverworts, mosses, and some angiosperms. Notably, they provide the yellow colouration of the flowers of species such as *Antirrhinum majus*. The key biosynthetic enzyme, converting chalcones to aurones, has been found for several species to be a variant form of POLYPHENOL OXIDASE (PPO), and termed the AURUESIDIN SYNTHASE (AUS), or AURONE SYNTHASE (AUS1). The PPO family contains both tyrosinases and catechol oxidases and is widespread in plants, fungi, and bacteria. Until the discovery of the AUS of *A. majus* (Nakayama et al. 2000; Davies et al. 2006), PPOs in plants were considered to be plastid-located enzymes that came in contact with substrates only following cell damage, to form brown phenolic polymers as part of wounding and plant defence responses. However, while the *A. majus* AUS closely resembles known PPO sequences, it lacks an N-terminal plastid transit peptide and is vacuolar-located (Ono et al. 2006). The *A. majus* AUS performs 4-hydroxylation and subsequent formation of aurones, but the AUS1 of the Asteraceae species *Coreopsis grandiflora* retains the chloroplast transit peptide and lacks the hydroxylation step towards tyrosine, probably contributing to the formation of 4-deoxyaurones (Kaintz et al. 2014). Tyrosinases are capable of both hydroxylation and subsequent oxidation of monophenols (e.g. tyrosine), and so AUS was classified as a tyrosinases. Catechol oxidases were thought to perform only oxidation of *o*-diphenolic compounds, and AUS1 was classified as such. However, Molitor et al. (2016) found that while AUS1 does indeed lack activity towards tyrosine, it is capable of both hydroxylation and oxidation of its target chalcone monophenol substrate. Albert et al. (2018) and Kubo et al. (2018) identified a PPO sequence (*Mapoly0021s0041*) that was strongly upregulated coincident with auronidin biosynthesis that probably encodes an AUS as CRISPR/Cas9-generated mutant lines (Berland et al. 2019) had auronidin amounts only 10% that of wild-type plants. Marchantia contains many candidate PPO sequences, including several lacking N-terminal and C-terminal domains typical of chloroplast-located PPOs. Thus, the PPOs of marchantia may contribute to a variety of secondary metabolite biosynthetic pathways.

9.4.6 3-Deoxyanthocyanins

The water-soluble red pigments of mosses are reported to be 3-deoxyanthocyanins, presumably produced by the same biosynthetic pathway as those of angiosperms. In species such as the grasses maize (*Zea mays*) and *Sorghum bicolor* and the Gesneriad *Sinningia cardinalis*, the 3-deoxyanthocyanins are probably formed by the action of a variant dihydroflavonol 4-reductase, termed the flavanone 4-reductase, using flavanones rather than the usual dihydroflavonol substrates (Winefield et al. 2005; Shih et al. 2006; Ibraheem et al. 2015). As the C3 hydroxyl group that is the usual site for initial glycosylation is absent, a specific UDP-GLUCOSE:FLAVONOID 5-*O*-GLUCOSYLTRANSFERASE may be required (Nakatsuka and Nishihara 2010; Kawahigashi et al. 2016). The 3-deoxyanthocyanins are particularly notable compounds of ferns. In the water fern *Azolla* (*A. filiculoides* or *A. pinnata*), they are responsible for the formation of 'red carpets' across the surface of lakes and canals. The leaves, which are initially green, produce large amounts of 3-deoxyanthocyanin (principally apigeninidin and luteolinidin 5-*O*-glucoside) in response to stresses such as strong sunlight, temperature fluctuations, or cadmium exposure. It might be assumed that, given the induction profile, they provide a similar photoprotective benefit as anthocyanins in angiosperms. However, they may have additional roles. The 3-deoxyanthocyanins have also been shown to promote the formation of *Azolla* symbiosis with the nitrogen-fixing cyanobacterium *Nostoc punctiforme*, and to contribute feeding deterrence against snails and tadpoles (Cohen, Meziane et al. 2002; Cohen, Sakihama et al. 2002).

9.4.7 Sphagnorubins

Several *Sphagnum* species, along with many other moss species such as *Ceratodon purpureus*, are characterised by their ability to become intensely red pigmented when under abiotic stress, such as high-light, dehydration, temperature below 5°C or nutrient imbalance (Gerdol et al. 1998; Dunn and Robinson 2006; Bonnett et al. 2010; Manninen et al. 2011). The principal pigment produced is sphagnorubin, a compound with five rings that may be the result of conjugation of a 3-deoxyanthocyanidin with a caffeic acid moiety. The pattern of sphagnorubin production in *Sphagnum*, when nights are cold but light amounts still high (such as on clear autumn days) matches the conditions that induce anthocyanin biosynthesis in angiosperms. There is a general lack of information on the possible function(s) of sphagnorubins. As yet, 3-hydroxyanthocyanins or proanthocyanidins have not been shown conclusively to be produced in liverworts, hornworts, or mosses.

9.5 Regulation of Flavonoid Production

Identifying which components of the flavonoid regulatory mechanisms identified in angiosperms are conserved in basal plants, such as bryophytes and lycophytes, is key for formulating proposals on how such mechanisms may have first arisen and how they have been subsequently modified in response to evolution of new plant structures (flowers, fruit, seeds) or to movement into additional ecological niches with differing biotic or abiotic stresses.

9.5.1 Activation of Flavonoid Biosynthesis in Response to UVB Light Exposure

One of the best characterised induction pathways for flavonoids is the UVB light response of Arabidopsis that leads to increased flavonol biosynthesis. The UV RESISTANCE LOCUS8 photoreceptor (UVR8) is present constitutively as an inactive dimer (Wargent and

Jordan 2013; Cao et al. 2017), with the active UVR8 monomer generated upon UVB exposure. UVR8 then moves to the nucleus and triggers response pathways by binding to the E3 ubiquitin ligase CONSTITUTIVELY PHOTOMORPHOGENIC1 (COP1), removing COP1 from its role in tagging target proteins for degradation (Cao et al. 2017). The key COP1 target in respect to flavonoid induction is the bZIP transcription factor ELONGATED HYPOCOTYL5 (HY5) and its homologue ELONGATED HYPOCOTYL5 HOMOLOG (HYH) (Huang et al. 2013; Binkert et al. 2014). HY5 upregulates the transcription of numerous genes required for the UVB response, including the R2R3MYB subgroup (SG) 7 gene *MYB12*, which is a key direct activator of flavonol biosynthetic genes. HY5 also activates the REPRESSOR OF UV-B PHOTOMORPHOGENESIS WD40-repeat genes (*RUP1* and *RUP2*) that modulate the UVR8 response through a negative feedback loop, perhaps by promoting redimerisation of UVR8 (Gruber et al. 2010; Cao et al. 2017).

As the central components of the system, such as UVR8, COP1, HY5/HYH, and SG7 MYBs, are conserved across all dicots examined, it is probable that the system is ubiquitous in dicots. Furthermore, studies on the basal Rosid grape (*Vitis vinifera*) have found strong conservation with the response system of Arabidopsis (Loyola et al. 2016; Czemmel et al. 2017). Recent studies have now shown that much of the UVR8 pathway for flavonoid induction in response to UVB is also conserved in marchantia, and indeed, some components are also found in algae. The UVR8 protein sequence is strongly conserved across the plant lineage (Fernández et al. 2016; Monforte et al. 2018), and so was present prior to the evolution of land plants. The UVR8 sequence of the Chlorophyceae green alga *Chlamydomonas reinhardtii* shows conservation of action with Arabidopsis UVR8 (Tilbrook et al. 2013; Allorent et al. 2016), although the downstream target components such as HY5 and the flavonoid pathway are lacking in green algae. The UVR8 sequence is conserved in marchantia and *P. patens* and can functionally complement the Arabidopsis *uvr8* mutant. However, the marchantia UVR8 has notable differences from those of Arabidopsis and *P. patens*, as dimeric UVR8 was barely detectable in marchantia thallids and the monomer was present in both the cytoplasm and nucleus with or without UVB exposure (Soriano et al. 2018). However, despite this apparent difference in marchantia UVR8 properties, the response pathway is remarkably conserved between marchantia and Arabidopsis (Clayton et al. 2018). Increased UVB exposure of marchantia triggers *HY5* gene upregulation, the activation of the phenylpropanoid biosynthetic pathway (*PAL*, *CHS*, and *CHIL* gene upregulation and increased flavone glycoside content), production of photoprotective proteins such as ELIP1 and protein chaperones, and induction of genes for RUP1 and components of the COP1 complex (COP1, DET1, SPA1, SUB1). Moreover, marchantia *hy5* mutants are hypersensitive to UVB exposure, and *rup1* mutants or UVR8 overexpression transgenics are hyposensitive. Thus, the UVR8 induction pathway for flavonoid production in response to UVB is probably a system liverworts and angiosperms inherited from a common early land plant ancestor.

Notable differences between marchantia and Arabidopsis seem to be the lack of genes in marchantia equivalent to HYH and MYB12. Only one of the 14 marchantia *bZIP* genes (Mp*HY5*) clades in 'Group H' with characterised *HY5* sequences (Bowman et al. 2017). The single R2R3MYB candidate gene significantly induced by UVB exposure of marchantia is Mp*MYB14* (Clayton et al. 2018), which is indeed an activator of flavonoid biosynthesis (Albert et al. 2018). However, UVB induction of flavone production is not altered in *myb14* plants (Clayton et al. 2018). We discuss the flavonoid-related MYB genes of basal plants in more detail in the next section. One possibility for the UVB response in marchantia is that *Mp*HY5 directly activates the flavonoid biosynthetic genes, as *At*HY5 directly activates the flavonoid biosynthetic genes as well as activating At*MYB12* and At*MYB75* (PAP1) (Hwang et al. 2016; Bhatia et al. 2018). The apparent lack of a dimer/monomer UVR8 system in marchantia is also a notable difference from Arabidopsis and *P. patens*, although this aspect requires further genetic and molecular analysis. For example, it questions the proposal that RUP1 acts through promoting accumulation of an inactive dimer.

9.5.2 The Role of MYB Transcription Factors in Regulating Flavonoid Biosynthesis in Land Plants

Studies across many different angiosperm plant species have established that R2R3MYB transcription factors are central to the regulation of flavonoid biosynthesis in response to developmental and environmental signals (Schwinn et al. 2006; Albert et al. 2011; Feller et al. 2011; Cheynier et al. 2013; Xu et al. 2015; Lloyd et al. 2017; Allan and Espley 2018; Chaves-Silva et al. 2018). For activation of the anthocyanin or proanthocyanidin pathways, SG5 or 6 R2R3MYBs form part of an MYB-bHLH-WDR (MBW) transcriptional activation complex. For activation of flavonol and flavone biosynthesis, SG7 MYBs act independently of the MBW complex. The transcriptional activity of the R2R3MYBs appears to be the principal determinant of the patterns of anthocyanin biosynthesis, although the bHLHs also contribute through their requirement for the MBW complex (Schwinn et al. 2006; Shang et al. 2011; Montefiori et al. 2015). Additionally, a range of R3 and R2R3MYBs have now been identified that have a repressive action. The R3MYBs act through competitively titering out the bHLH, preventing MBW complex formation, while the repressive R2R3MYBs form an MBW complex that actively represses transcription through the presence of repressive domains in the R2R3MYB (such as the ERF-associated amphiphilic repression (EAR) domain present in many SG4 sequences). For vegetative pigmentation at least, the core role of the R2R3MYB repressor may be to limit flavonoid biosynthesis under 'steady-state' conditions, while production of the R3MYB repressor is activated by the MBW activation complex to provide negative feedback modulation of the MBW pathway induction process (Albert et al. 2011; Cavallini et al. 2015). The repressors may also contribute to establishment of anthocyanin pigment patterning (Davies et al. 2012; Jiang and Rausher 2018). The importance of these repressive MYBs has become increasingly apparent with their ongoing identification as the basis of genetic loci controlling pigmentation (Albert et al. 2014; Cao et al. 2017; Colanero et al. 2018; Song et al. 2018).

It is not yet clear when the MBW activator complex for regulation of flavonoid production arose during plant evolution. However, an MBW complex able to activate flavonoid biosynthesis has been characterised for the gymnosperm Norway spruce (*Picea abies*) (Nemesio-Gorriz et al. 2017), indicating it preceded the development of colourful flowers and so arose for vegetative pigmentation. It also supports an ancient origin for the MBW complex in the plant lineage, as the last common ancestor of gymnosperms and angiosperms is thought to have existed around 300 million years ago. Basal plants have much smaller transcription factor families than angiosperms, with 22 R2R3MYB genes identified in the marchantia genome, 49 in *S. moellendorffii* and 62 in *P. patens*, as compared for example with the estimated 138 in Arabidopsis and 157 to 188 in maize (*Zea mays*) (Feller et al. 2011; Bowman et al. 2017; Albert et al. 2018; Kubo et al. 2018). Rather than there being clear orthologues in marchantia to the angiosperm phenylpropanoid regulators, two genes (Mp*MYB02* and Mp*MYB14*) are basal to a clade containing 19 Arabidopsis R2R3MYBs comprising SGs 4, 5, 6, 7, 15, and 44 (Albert et al. 2018; Kubo et al. 2018). Both Mp*MYB02* and Mp*MYB14* do indeed activate phenylpropanoid biosynthesis. MYB14 is the key activator of auronidin pigmentation, as *myb14* mutants lose all auronidin production and *MYB14* overexpression induces activation of the biosynthetic pathway and strong accumulation of auronidins (Albert et al. 2018; Kubo et al. 2018). MYB02 activates the production of *bis*-bibenzyls, with *MYB02* overexpression inducing activation of phenylpropanoid biosynthetic genes and accumulation of marchantins (Kubo et al. 2018). In *myb02* mutants marchantin production is lost, with the proposed precursor dihydrostilbenes lunularic acid and prelunularic acid accumulating (Kubo et al. 2018).

Whether these flavonoid-related R2R3MYBs identified for marchantia act within a MBW complex has yet to be determined. The amino acid motif ([D/E]Lx_2[R/K]x_3Lx_6Lx_3R) necessary for R2R3-MYB proteins to bind bHLH partners has been determined for angiosperms (Zimmerman et al. 2004). A complete version of this motif is present in SmXP002978781 of the lycophyte *S. moellendorffi*, but has not been found within bryophyte MYB proteins to date; MpMYB14 and MpMYB02 lack two residues, and PpXP001752936 from *P. patens* lacks one (Albert et al. 2018). WDR sequences are present in marchantia that can functionally complement the petunia flavonoid-related WDR mutant *anthocyanin11* (the authors' unpublished data), although data supporting a role in flavonoid biosynthesis are lacking. Also, a bHLH gene has been identified that may activate *bis*-bibenzyl biosynthesis in the liverwort *P. appendiculatum* (Wu et al. 2018). Phylogenetic analysis places *Pa*bHLH as being most closely related to the flavonoid MBW bHLH sequences of angiosperms, and overexpression in *P. appendiculatum* callus upregulated *PAL* transcript abundance and increased *bis*-bibenzyl amounts. While this supports the presence of the three necessary components, R2R3MYBs, bHLH, and WDR, there is as yet no evidence that an MBW complex is formed.

The R2R3-MYB genes with repressive activity are increasingly recognised as having important roles in regulating phenylpropanoid and flavonoid biosynthesis in angiosperms (Jin et al. 2000; Aharoni et al. 2001; Dubos et al. 2008; Albert et al. 2014, Albert 2015). These MYB genes belong to SG4 and are characterised by the presence of an EAR motif in the C terminus (LxLxL or DLNxxP) (Stracke et al. 2001), which mediates transcriptional repression. This occurs through the recruitment of the corepressor TOPLESS (TPL; a Gro-TLE protein), which may alter the chromatin environment of the promoter to inhibit access for the basal transcription machinery (Kagale et al. 2010; Kaul et al. 2014). The first characterised MYB repressors in plants were *Am*MYB308 from *A. majus* and its orthologue *At*MYB4 in Arabidopsis, which repress *C4H* to redirect flux towards cinnamic acid biosynthesis (sinapate esters) for UVB protection (Tamagnone et al. 1998; Jin et al. 2000). Similarly, *Ph*MYB4 in petunia acts upon *C4H*, which helps redirect flux towards floral volatile production for pollinator attraction (Colquhoun et al. 2011). This suggests that these repressive MYB proteins have been recruited to serve as key break points within the phenylpropanoid pathway, enabling synthesis of distinct metabolites in response to environmental cues or at different stages of development. While MYB repressors have well-established roles in angiosperms, it remains unknown whether they are ubiquitous in land plants. For example, we were unable to identify (known) repression motifs in any of the 21 marchantia R2R3-MYB sequences, yet *P. patens* and *S. moellendorffi* both have R2R3-MYB genes with putative EAR motif sequences (LxLxL). Marchantia does contain an orthologue of TOPLESS, which is involved in auxin signalling (Flores-Sandoval et al. 2015).

It seems likely that the ancestral R2R3MYB regulators of phenylpropanoid metabolism were activators, with repressive transcription factors evolving subsequently, as secondary metabolic pathways expanded and required additional degrees of control. Evidence from marchantia supports this: it lacks (recognisable) repressive R2R3-MYB genes and the two known phenylpropanoid regulators, *Mp*MYB14 and *Mp*MYB02, both activate their target genes (Albert et al. 2018; Kubo et al. 2018). Furthermore, the SG7 R2R3MYB genes in angiosperms that regulate flavonol/flavone biosynthesis also act on primary metabolism to deliver precursor metabolites into the phenylpropanoid pathway (Luo et al. 2008). This more general role of activating a much larger set of genes could suggest the ancestral phenylpropanoid/flavonoid R2R3MYB regulator acted to boost primary metabolism first, and was recruited to the phenylpropanoid pathway genes as the pathway evolved.

9.6 Concluding Remarks

Our understanding of the biosynthesis and function of phenylpropanoids in nonseed plants is lacking compared with the wealth of knowledge for flowering plant species. Recent studies on species such as marchantia, *P. appendiculatum*, and *P. patens* have started to elucidate both the phenylpropanoid pathways in common across land plants, such as for flavones, as well as those specific to different nonseed plant species, such as the bibenzyls. Such studies have also contributed knowledge on which pathways may have a common evolutionary origin across major land plant groups. However, molecular studies to date have been on only a few thallid liverwort species and one moss; and more genetic models are needed from across the bryophytes, lycophytes, and ferns.

9.7 Acknowledgements

The authors' research on bryophytes is supported by The Marsden Fund of New Zealand Grant PAF1701.

References

Agati, G., Azzarello, E., Pollastri, S., and Tattini, M. (2012). Flavonoids as antioxidants in plants: Location and functional significance. *Plant Science* 196: 67–76.

Aharoni, A., De Vos, C.H.R., Wein, M., et al. (2001). The strawberry *FaMYB1* transcription factor suppresses anthocyanin and flavonol accumulation in transgenic tobacco. *The Plant Journal* 28: 319–332.

Aigner, S., Remias, D., Karsten, U., and Holzinger, A. (2013). Unusual phenolic compounds contribute to ecophysiological performance in the purple-colored green alga *Zygogonium ericetorum* (zygnematophyceae, streptophyta) from a high-alpine habitat. *Journal of Phycology* 49: 648–660.

Albert, N.W. (2015). Subspecialization of R2R3-MYB repressors for anthocyanin and proanthocyanidin regulation in forage legumes. *Frontiers in Plant Science* 6: 1165.

Albert, N.W., Davies, K.M., Lewis, D.H., et al. (2014). A conserved network of transcriptional activators and repressors regulates anthocyanin pigmentation in eudicots. *Plant Cell* 26: 962–980.

Albert, N.W., Lewis, D.H., Zhang, H., et al. (2011). Members of an R2R3-MYB transcription factor family in *Petunia* are developmentally and environmentally regulated to control complex floral and vegetative pigmentation patterning. *The Plant Journal* 65: 771–784.

Albert, N.W., Thirmawithana, A.H., McGhie, T.K., et al. (2018). Genetic analysis of the liverwort *Marchantia polymorpha* reveals that R2R3MYB activation of flavonoid production in response to abiotic stress is an ancient character in land plants. *New Phytologist* 218: 554–566.

Allan, A.C., and Espley, R.V. (2018). MYBs drive novel consumer traits in fruits and vegetables. *Trends in Plant Science* 23: 693–705.

Allorent, G., Lefebvre-Legendre, L., Chappuis, R., et al. (2016). UV-B photoreceptor-mediated protection of the photosynthetic machinery in *Chlamydomonas reinhardtii*. *Proceedings of the National Academy of Sciences USA* 113: 14864–14869.

Andersen, O.M., and Markham, K.R., eds. (2006). *The Flavonoids: Chemistry, Biochemistry and Applications*. Boca Raton: CRC Press.

Asakawa, Y. (2016). Polyphenols in Bryophytes. In, *Recent Advances in Polyphenol Research Volume* 5, (eds. K. Yoshida, V. Cheynier, and S. Quideau), 36–66. U.K.: Wiley-Blackwell.

Asakawa, Y., and Ludwiczuk, A. (2012). Distribution of cyclic and acyclic *bis*-bibenzyls in the marchantiophyta (liverworts), ferns and higher plants and their biological activities, biosynthesis, and total synthesis. *Heterocycles* 86: 891–917.

Asakawa, Y., Ludwiczuk, A., and Nagashima, F. (2013). *Chemical Constituents of Bryophytes: Bio- and Chemical Diversity, Biological Activity, and Chemosystematics*. Vienna, Austria: Springer-Verlag.

Ban, Z., Qin, H., Mitchell, A.J., et al. (2018). Noncatalytic chalcone isomerase-fold proteins in *Humulus lupulus* are auxiliary components in prenylated flavonoid biosynthesis. *Proceedings of the National Academy of Sciences USA* 115: E5223–E5232.

Banks, J.A., Nishiyama, T., Hasebe, M., et al. (2011). The *Selaginella* genome identifies genetic changes associated with the evolution of vascular plants. *Science* 332: 960–963.

Bassard, J.-E., Richert, L., Geerinck, J., et al. (2012). Protein–protein and protein–membrane associations in the lignin pathway. *The Plant Cell* 24: 4465–4482.

Berland, H., Albert, N.W., Stavland, A., et al. (2019). Auronidins: A new flavonoid pigment class challenging when anthocyanin biosynthesis evolved. *Proceedings of the National Academy of Sciences USA* 116: 20232–20239.

Betz, C., McCollum, T.G., and Mayer, R.T. (2001). Differential expression of two cinnamate 4-hydroxylase genes in 'Valencia' orange (*Citrus sinensis* Osbeck). *Plant Molecular Biology* 46: 741–748.

Bhatia, C., Pandey, A., Gaddam, S.R., et al. (2018). Low temperature-enhanced flavonol synthesis requires light-associated regulatory components in *Arabidopsis thaliana*. *Plant and Cell Physiology* 59: 2099–2112.

Binkert, M., Kozma-Bognár, L., Terecskei, K., et al. (2014). UV-B-responsive association of the *Arabidopsis* bZIP transcription factor ELONGATED HYPOCOTYL5 with target genes, including its own promoter. *The Plant Cell* 26: 4200–4213.

Blee, K., Choi, J.W., O'Connell, A.P., et al. (2001). Antisense and sense expression of cDNA coding for CYP73A15, a class II cinnamate 4-hydroxylase, leads to a delayed and reduced production of lignin in tobacco. *Phytochemistry* 57: 1159–1166.

Blount, J.W., Korth, K.L., Masoud, S.A., et al. (2000). Altering expression of cinnamic acid 4-hydroxylase in transgenic plants provides evidence for a feedback loop at the entry point into the phenylpropanoid pathway. *Plant Physiology* 122: 107–116.

Bonnett, S.A.F., Ostle, N., and Freeman, C. (2010). Short-term effect of deep shade and enhanced nitrogen supply on *Sphagnum capillifolium* morphophysiology. *Plant Ecology* 207: 347–358.

Bowman, J.L., Kohchi, T., Yamato, K.T., et al. (2017). Insights into land plant evolution garnered from the *Marchantia polymorpha* genome. *Cell* 171: 287–304.

Brazier-Hicks, M., Evans, K.M., Gershater, M.C., et al. (2009). The *C*-glycosylation of flavonoids in cereals. *The Journal of Biological Chemistry* 284: 17926–17934.

Brunetti, C., Fini, A., Sebastiani, F., Gori, A., and Tattini, M. (2018). Modulation of phytohormone signaling: A primary function of flavonoids in plant–environment interactions. *Frontiers in Plant Science* 9: 1042.

Cao, X., Qiu, Z., Wang, X., et al. (2017). A putative R3 MYB repressor is the candidate gene underlying *atroviolacium*, a locus for anthocyanin pigmentation in tomato fruit. *Journal of Experimental Botany* 68: 5745–5758.

Carrington, Y., Guo, J., Le, C.H., et al. (2018). Evolution of a secondary metabolic pathway from primary metabolism: shikimate and quinate biosynthesis in plants. *The Plant Journal* 95: 823–833.

Cavallini, E., Tomas Matus, J., Finezzo, L., et al. (2015). The phenylpropanoid pathway is controlled at different branches by a set of R2R3-MYB C2 repressors in grapevine. *Plant Physiology* 167: 1448–1552.

Chang, A., Lim, M.H., Lee, S.W., et al. (2008). Tomato phenylalanine ammonia-lyase gene family, highly redundant but strongly underutilized. *The Journal of Biological Chemistry* 283: 33591–33601.

Chaves-Silva, S., Santos, A.L., Chalfun-Júnior, A., et al. (2018). Understanding the genetic regulation of anthocyanin biosynthesis in plants – tools for breeding purple varieties of fruits and vegetables. *Phytochemistry* 153: 11–27.

Chen, H.Y., Babst, B.A., Nyamdari, B., et al. (2014). Ectopic expression of a loblolly pine class II 4-coumarate:CoA ligase alters soluble phenylpropanoid metabolism but not lignin biosynthesis in *Populus*. *Plant and Cell Physiology* 55: 1669–1678.

Cheng, A.X., Zhang, X., Han, X.J., et al. (2018). Identification of chalcone isomerase in the basal land plants reveals an ancient evolution of enzymatic cyclization activity for synthesis of flavonoids. *New Phytologist* 217: 909–924.

Cheynier, V., Comte, G., Davies, K.M., et al. (2013). Plant phenolics: Recent advances on their biosynthesis, genetics, and ecophysiology. *Plant Physiology and Biochemistry* 71: 1–20.

Clayton, W.A., Albert, N.W., Thrimawithana, A.H., et al. (2018). UVR8-mediated induction of flavonoid biosynthesis for UVB tolerance is conserved between the liverwort *Marchantia polymorpha* and flowering plants. *The Plant Journal* 96: 503–517.

Cohen, M.F., Meziane, T., Tsuchiya, M., and Yamasaki, H. (2002). Feeding deterrence of *Azolla* in relation to deoxyanthocyanin and fatty acid composition. *Aquatic Botany* 74: 181–187.

Cohen, M.F., Sakihama, Y., Takagi, Y.C., et al. (2002). Synergistic effect of deoxyanthocyanins from symbiotic fern *Azolla* spp. on *hrmA* gene induction in the cyanobacterium *Nostoc punctiforme*. *Molecular Plant-Microbe Interactions* 15: 875–882.

Colanero, S., Perata, P., and Gonzali, S. (2018). The *atroviolacea* gene encodes an R3-MYB protein repressing anthocyanin synthesis in tomato plants. *Frontiers in Plant Science* 9: 830.

Colquhoun, T.A., Kim, J.Y., Wedde, A.E., et al. (2011). *PhMYB4* fine-tunes the floral volatile signature of *Petunia×hybrida* through *PhC4H*. *Journal of Experimental Botany* 62: 1133–1143.

Czemmel, S., Holl, J., Loyola, R., et al. (2017). Transcriptome-wide identification of novel UV-B- and light modulated flavonol pathway genes controlled by *VviMYBF1*. *Frontiers in Plant Science* 8: 1084.

Davies, K., Albert, N., Zhou, Y., and Schwinn, K. (2018). Functions of flavonoid and betalain pigments in abiotic stress tolerance in plants. *Annual Plant Reviews Online* 1: 1–41.

Davies, K.M., Albert, N.W., and Schwinn, K.E. (2012). From landing lights to mimicry: the molecular regulation of flower colouration and mechanisms for pigmentation patterning. *Functional Plant Biology* 39: 619–638.

Davies, K.M., Bloor, S.J., Spiller, G.B., and Deroles, S.C. (1998). Production of yellow colour in flowers: redirection of flavonoid biosynthesis in *Petunia*. *Plant Journal* 13: 259–266.

Davies, K.M., Marshall, G.B., Bradley, J.M., et al. (2006). Characterisation of aurone biosynthesis in *Antirrhinum majus*. *Physiologia Plantarum* 128: 593–603.

de Vries, J. and Archibald, J.M. (2018). Plant evolution: landmarks on the path to terrestrial life. *New Phytologist* 217: 1428–1434.

de Vries, J., de Vries, S., Slamovits, C.H., et al. (2017). How embryophytic is the biosynthesis of phenylpropanoids and their derivatives in streptophyte algae? *Plant and Cell Physiology* 58: 934–945.

Delwiche, Charles F., and Cooper, Endymion D. (2015). The evolutionary origin of a terrestrial flora. *Current Biology* 25: 899–910.

Du, Y., Chu, H., Chu, I.K., and Lo, C. (2010). CYP93G2 is a flavanone 2-hydroxylase required for *C*-glycosylflavone biosynthesis in rice. *Plant Physiology* 154: 324–333.

Dubos, C., Gourrierec, J.L., Baudry, A., et al. (2008). MYBL2 is a new regulator of flavonoid biosynthesis in *Arabidopsis thaliana*. *Plant Journal* 55: 940–953.

Dunn, J.L., and Robinson, S.A. (2006). Ultraviolet B screening potential is higher in two cosmopolitan moss species than in a co-occurring Antarctic endemic moss: implications of continuing ozone depletion. *Global Change Biology* 12: 2282–2296.

Ehlting, J., Hamberger, B., Million-Rousseau, R., and Werck-Reichhart, D. (2006). Cytochromes P450 in phenolic metabolism. *Phytochemistry Reviews* 5: 239–270.

Emiliani, G., Fondi, M., Fani, R., and Gribaldo, S. (2009). A horizontal gene transfer at the origin of phenylpropanoid metabolism: a key adaptation of plants to land. *Biology Direct* 4: 7.

Falcone Ferreyra, M.L., Emiliani, J., Rodriguez, E.J., et al. (2015). The identification of maize and arabidopsis type I FLAVONE SYNTHASEs links flavones with hormones and biotic interactions. *Plant Physiology* 169: 1090–1107.

Feller, A., Machemer, K., Braun, E.L., and Grotewold, E. (2011). Evolutionary and comparative analysis of MYB and bHLH plant transcription factors. *The Plant Journal* 66: 94–116.

Fernández, M.B., Tossi, V., Lamattina, L., and Cassia, R. (2016). A comprehensive phylogeny reveals functional conservation of the UV-B photoreceptor UVR8 from green algae to higher plants. *Frontiers in Plant Science* 7: 1698.

Ferrer, J.L., Austin, M.B., Stewart, C., and Noel, J.P. (2008). Structure and function of enzymes involved in the biosynthesis of phenylpropanoids. *Plant Physiology and Biochemistry* 46: 356–370.

Flores-Sandoval, E., Eklund, D.M., and Bowman, J.L. (2015). A simple auxin transcriptional response system regulates multiple morphogenetic processes in the liverwort *Marchantia polymorpha*. *PLoS genetics* 11: e1005207.

Friederich, S., Maier, U.H., deus-Neumann, B., et al. (1999). Biosynthesis of cyclic *bis*(bibenzyls) in *Marchantia polymorpha*. *Phytochemistry* 50: 589–598.

Friederich, S., Rueffer, M., Asakawa, Y., and Zenk, M.H. (1999). Cytochromes P-450 catalyze the formation of marchantins A and C in *Marchantia polymorpha*. *Phytochemistry* 52: 1195–1202.

Gerdol, R., Bonora, A., Marchesini, R., et al. (1998). Growth response of *Sphagnum capillifolium* to nighttime temperature and nutrient level: Mechanisms and implications for global change. *Arctic and Alpine Research* 30: 388–395.

Gruber, H., Heijde, M., Heller, W., et al. (2010). Negative feedback regulation of UV-B–induced photomorphogenesis and stress acclimation in Arabidopsis. *Proceedings of the National Academy of Sciences USA* 107: 20132–20137.

Han, J.W., Yoon, M., Kupper, F.D., et al. (2012). Accumulation of galloyl derivatives in a green alga, *Spirogyra varians*, in response to cold stress. *Journal of Applied Phycology* 24: 1279–1286.

Han, X.-J., Wu, Y.-F., Gao, S., et al. (2014). Functional characterization of a *Plagiochasma appendiculatum* flavone synthase I showing flavanone 2-hydroxylase activity. *FEBS Letters* 588: 2307–2314.

Harakava, R. (2005). Genes encoding enzymes of the lignin biosynthesis pathway in *Eucalyptus*. *Genetics and Molecular Biology* 28: 601–607.

Harholt, J., Moestrup, O., and Ulvskov, P. (2016). Why plants were terrestrial from the beginning. *Trends in Plant Science* 21: 96–101.

Hehmann, M., Lukacin, R., Ekiert, H., and Matern, U. (2004). Furanocoumarin biosynthesis in *Ammi majus* L.: cloning of bergaptol *O*–methyltransferase. *European Journal of Biochemistry* 271: 932–940.

Holzinger, A., and Pichrtová, M. (2016). Abiotic stress tolerance of Charophyte green algae: New challenges for Omics techniques. *Frontiers in Plant Science* 7: 678.

Hong, L., Qian, Q., Tang, D., et al. (2012). A mutation in the rice chalcone isomerase gene causes the *golden hull and internode 1* phenotype. *Planta* 236: 141–151.

Hooijmaijers, C.A. (2008). Desiccation tolerance in red and green gametophytes of *Jamesoniella colorata* in relation to photoprotection. *Planta* 227: 1301–1310.

Hooijmaijers, C.A.M., and Gould, K.S. (2007). Photoprotective pigments in red and green gametophytes of two New Zealand liverworts. *New Zealand Journal of Botany* 45: 451–461.

Huang, J., Gu, M., Lai, Z., et al. (2010). Functional analysis of the Arabidopsis PAL gene family in plant growth, development, and response to environmental stress. *Plant Physiology* 153: 1526–1538.

Huang, X., Ouyang, X., Yang, P., et al. (2013). Conversion from CUL4-based COP1-SPA E3 apparatus to UVR8-COP1-SPA complexes underlies a distinct biochemical function of COP1 under UV-B. *Proceedings of the National Academy of Sciences USA* 110: 16669–16674.

Hussain, T., Plunkett, B., Ejaz, M., et al. (2018). Identification of putative precursor genes for the biosynthesis of cannabinoid-like compound in *Radula marginata*. *Frontiers in Plant Science* 9: 537.

Hwang, J.-U., Song, W.-Y., Hong, D., et al. (2016). Plant ABC transporters enable many unique aspects of a terrestrial plant's lifestyle. *Molecular Plant* 9: 338–355.

Ibraheem, F., Gaffoor, I., Qixian, T., et al. (2015). A *Sorghum* MYB transcription factor induces 3-deoxyanthocyanidins and enhances resistance against leaf blights in maize. *Molecules* 20: 2388–2404.

Jiang, N., Doseff, A.I., and Grotewold, E. (2016). Flavones: From biosynthesis to health benefits. *Plants* 5: 27.

Jiang, P., and Rausher, M. (2018). Two genetic changes in cis-regulatory elements caused evolution of petal spot position in *Clarkia*. *Nature Plants* 4: 14–22.

Jiang, W., Yin, Q., Wu, R., et al. (2015). Role of a chalcone isomerase-like protein in flavonoid biosynthesis in *Arabidopsis thaliana*. *Journal of Experimental Botany* 66: 7165–7179.

Jin, H.L., Cominelli, E., Bailey, P., et al. (2000). Transcriptional repression by AtMYB4 controls production of UV-protecting sunscreens in Arabidopsis. *EMBO Journal* 19: 6150–6161.

Joos, H.J., and Hahlbrock, K. (1992). Phenylalanine ammonia-lyase in potato (*Solanum tuberosum* L.). Genomic complexity, structural comparison of two selected genes and modes of expression. *European Journal of Biochemistry* 204: 621–629.

Kagale, S., Links, M.G., and Rozwadowski, K. (2010). Genome-wide analysis of ethylene-responsive element binding factor-associated amphiphilic repression motif-containing transcriptional regulators in Arabidopsis. *Plant Physiology* 152: 1109–1134.

Kaintz, C., Molitor, C., Thill, J., et al. (2014). Cloning and functional expression in *E. coli* of a polyphenol oxidase transcript from *Coreopsis grandiflora* involved in aurone formation. *FEBS Letters* 588: 3417–3426.

Kang, J.-H., McRoberts, J., Shi, F., et al. (2014). The flavonoid biosynthetic enzyme chalcone isomerase modulates terpenoid production in glandular trichomes of tomato. *Plant Physiology* 164: 1161.

Karamat, F., Olry, A., Munakata, R., et al. (2013). A coumarin-specific prenyltransferase catalyzes the crucial biosynthetic reaction for furanocoumarin formation in parsley. *The Plant Journal* 77: 627–638.

Kaul, A., Schuster, E., and Jennings, B.H. (2014). The Groucho co-repressor is primarily recruited to local target sites in active chromatin to attenuate transcription. *PLoS Genetics* 10: e1004595.

Kawahigashi, H., Kasuga, S., Sawada, Y., et al. (2016). The sorghum gene for leaf color changes upon wounding (*P*) encodes a flavanone 4-reductase in the 3-deoxyanthocyanidin biosynthesis pathway. *G3*: Genes|*Genomes*|*Genetics* 6: 1439–1447.

Kawai, Y., Ono, E., and Mizutani, M. (2014). Evolution and diversity of the 2-oxoglutarate-dependent dioxygenase superfamily in plants. *The Plant Journal* 78: 328–343.

Kearse, M., Moir, R., Wilson, A., et al. (2012). Geneious Basic: an integrated and extendable desktop software platform for the organization and analysis of sequence data. *Bioinformatics* 28: 1647–1649.

Kim, D.S., and Hwang, B.K. (2014) An important role of the pepper phenylalanine ammonia-lyase gene (PAL1) in salicylic acid-dependent signalling of the defence response to microbial pathogens. *Journal of Experimental Botany* 65: 2295–2306.

Kim, J., Choi, B., Natarajan, S., and Bae, H. (2013). Expression analysis of kenaf cinnamate 4-hydroxylase (C4H) ortholog during developmental and stress responses. *Plant Omics* 6: 65.

Kim, S.Y., Colpitts, C.C., Wiedemann, G., et al. (2013). *Physcomitrella* PpORS, basal to plant type III polyketide synthases in phylogenetic trees, is a very long chain 2′-Oxoalkylresorcinol Synthase. *The Journal of Biological Chemistry* 288: 2767–2777.

Koopman, E., Logemann, E., and Hahlbrock, K. (1999). Regulation and functional expression of cinnamate 4-hydroxylase from parsley. *Plant Physiology* 119: 49–56.

Krieger, C., Roselli, S., Kellner-Thielmann, S., et al. (2018). The CYP71AZ P450 Subfamily: A driving factor for the diversification of coumarin biosynthesis in Apiaceous plants. *Frontiers in Plant Science* 9: 820.

Kubo, H., Nozawa, S., Hiwatashi, T., et al. (2018). Biosynthesis of riccionidins and marchantins is regulated by R2R3-MYB transcription factors in *Marchantia polymorpha*. *Journal of Plant Research* 131: 849–864.

Kumar, R., Vashisth, D., Misra, A., et al. (2016). RNAi down-regulation of cinnamate-4-hydroxylase increases artemisinin biosynthesis in *Artemisia annua*. *Scientific Reports* 6: 26458.

Kunz, S., Burkhardt, G., and Becker, H. (1993). Riccionidins A and B, anthocyanidins from the cell walls of the liverwort *Ricciocarpos natans*. *Phytochemistry* 35: 233–235.

Labeeuw, L., Martone, P.T., Boucher, Y., and Case, R.J. (2015). Ancient origin of the biosynthesis of lignin precursors. *Biology Direct* 10: 23.

Lam, P.Y., Zhu, F.-Y., Chan, W.L., et al. (2014). Cytochrome P450 93G1 is a flavone synthase II that channels flavanones to the biosynthesis of tricin *O*-linked conjugates in rice. *Plant Physiology* 165: 1315–1327.

Landi, M., Tattini, M., and Gould, K. (2015). Multiple functional roles of anthocyanins in plant-environment interactions. *Environmental and Experimental Botany* 119: 4–17.

Lange, B.M. (2015). The evolution of plant secretory structures and emergence of terpenoid chemical diversity. *Annual Review of Plant Biology* 66: 139–159.

Larbat, R., Hehn, A., Hans, J., et al. (2009). Isolation and functional characterization of CYP71AJ4 encoding for the first P450 monooxygenase of angular furanocoumarin biosynthesis. *The Journal of Biological Chemistry* 284: 4776–4785.

Larbat, R., Kellner, S., Specker, S., et al. (2007). Molecular cloning and functional characterization of psoralen synthase, the first committed monooxygenase of furanocoumarin biosynthesis. *The Journal of Biological Chemistry* 282: 542–554.

Lavhale, S.G., Kalunke, R.M., and Giri, A.P. (2018). Structural, functional and evolutionary diversity of 4-coumarate-CoA ligase in plants. *Planta* 248: 1063–1078.

Li, F.-W., Brouwer, P., Carretero-Paulet, L., et al. (2018). Fern genomes elucidate land plant evolution and cyanobacterial symbioses. *Nature Plants* 4: 460–472.

Li, L., Aslam, M., Rabbi, F., Vanderwel, M.C., et al. (2018). PpORS, an ancient type III polyketide synthase, is required for integrity of leaf cuticle and resistance to dehydration in the moss, *Physcomitrella patens*. *Planta* 247: 527–541.

Li, Y., Im Kim, J., Pysh, L., and Chapple, C. (2015). Four isoforms of *Arabidopsis thaliana* 4-coumarate: CoA ligase (4CL) have overlapping yet distinct roles in phenylpropanoid metabolism. *Plant Physiology* 169: 2409–2421.

Li, Z.B., Li, C.F., Li, J., and Zhang, Y.S. (2014). Molecular cloning and functional characterization of two divergent 4-coumarate: coenzyme A ligases from Kudzu (*Pueraria lobata*). *Biology and Pharmaceutical Bulletin* 37: 113–122.

Ligrone, R., Duckett, J.G., and Renzaglia, K.S. (2012). Major transitions in the evolution of early land plants: a bryological perspective. *Annals of Botany* 109: 851–871.

Liu, H., Xu, R.X., Gao, S., and Cheng, A.X. (2017). The functional characterization of a site-specific apigenin 4′-*O*-methyltransferase synthesized by the liverwort species *Plagiochasma appendiculatum*. *Molecules* 22: E759.

Liu, X.-Y., Wang, P.-P., Wu, Y.-F., et al. (2018). Cloning and functional characterization of two 4-Coumarate: CoA Ligase genes from *Selaginella moellendorffii*. *Molecules* 23: E595.

Liu, X.-Y., Yu, H.-N., Gao, S., et al. (2017). The isolation and functional characterization of three liverwort genes encoding cinnamate 4-hydroxylase. *Plant Physiology and Biochemistry* 117: 42–50.

Lloyd, A., Brockman, A., Aguirre, L., et al. (2017). Advances in the MYB–bHLH–WD Repeat (MBW) pigment regulatory model: Addition of a WRKY factor and co-option of an anthocyanin MYB for betalain regulation. *Plant and Cell Physiology* 58: 1431–1441.

Loyola, R., Herrera, D., Mas, A., et al. (2016). The photomorphogenic factors UV-B RECEPTOR 1, ELONGATED HYPOCOTYL 5, and HY5 HOMOLOGUE are part of the UV-B signalling pathway in grapevine and mediate flavonol accumulation in response to the environment. *Journal of Experimental Botany* 67: 5429–5445.

Lu, S., Zhou, Y., Li, L., and Chiang, V.L. (2006). Distinct roles of cinnamate 4-hydroxylase genes in *Populus*. *Plant and Cell Physiology* 47: 905–914.

Luo, J., Butelli, E., Hill, L., et al. (2008). AtMYB12 regulates caffeoyl quinic acid and flavonol synthesis in tomato: expression in fruit results in very high levels of both types of polyphenol. *The Plant Journal* 56: 316–326.

Mandoulakani, B.A., Eyvazpour, E., and Ghadimzadeh, M. (2017). The effect of drought stress on the expression of key genes involved in the biosynthesis of phenylpropanoids and essential oil components in basil (*Ocimum basilicum* L.). *Phytochemistry* 139: 1–7.

Manninen, S., Woods, C., Leith, I.D., and Sheppard, L.J. (2011). Physiological and morphological effects of long-term ammonium or nitrate deposition on the green and red (shade and open grown) Sphagnum capillifolium. *Environmental and Experimental Botany* 72: 140–148.

Markham, K.R. (1988). Distribution of flavonoids in the lower plants and its evolutionary significance. In, *The Flavonoids*, (ed. J.B. Harborne) 427–468. U.K: Chapman and Hall.

Martens, S., and Mithofer, A. (2005). Flavones and flavone synthases. *Phytochemistry* 66: 2399–2407.

Molitor, C., Mauracher, S.G., and Rompel, A. (2016). Aurone synthase is a catechol oxidase with hydroxylase activity and provides insights into the mechanism of plant polyphenol oxidases. *Proceedings of the National Academy of Sciences USA* 113: E1806–E1815.

Monforte, L., Soriano, G., Núñez-Olivera, E., and Martínez-Abaigar, J. (2018). Cell compartmentation of ultraviolet-absorbing compounds: An underexplored tool related to bryophyte ecology, phylogeny and evolution. *Functional Ecology* 32: 882–893.

Montefiori, M., Brendolise, C., Lin-Wang, K., et al. (2015). In the Solanaceae, a hierarchy of bHLHs confer distinct target specificity to the anthocyanin regulatory complex. *Journal of Experimental Botany* 66: 1427–1436.

Morita, Y., Takagi, K., Fukuchi-Mizutani, et al. (2014). A chalcone isomerase-like protein enhances flavonoid production and flower pigmentation. *The Plant Journal* 78: 294–304.

Nakatsuka, T., and Nishihara, M. (2010). UDP-glucose:3-deoxyanthocyanidin 5-*O*-glucosyltransferase from *Sinningia cardinalis*. *Planta* 232: 383–392.

Nakayama, T., Yonekura-Sakakibara, K., Sato, T., et al. (2000). Aureusidin synthase: a polyphenol oxidase homolog responsible for flower coloration. *Science* 290: 1163–1166.

Neill, S.O., and Gould, K.S. (2003). Anthocyanins in leaves: light attenuators or antioxidants? *Functional Plant Biology* 30: 865–873.

Nelson, D., and Werck-Reichhart, D. (2011). A P450-centric view of plant evolution. *The Plant Journal* 66: 194–211.

Nemesio-Gorriz, M., Blair, P.B., Dalman, K., et al. (2017). Identification of Norway Spruce MYB-bHLH-WDR transcription factor complex members linked to regulation of the flavonoid pathway. *Frontiers in Plant Science* 8: 305.

Newsham, K.K. (2009). The biology and ecology of the liverwort *Cephaloziella varians* in Antarctica. *Antarctic Science* 22: 131–143.

Ngaki, M.N., Louie, G.V., Philippe, R.N., et al. (2012). Evolution of the chalcone-isomerase fold from fatty-acid binding to stereospecific catalysis. *Nature* 485: 530–533.

Oliveira, M.B., de Andrade, R.V., Grossi-de-Sa, M.F., and Petrofeza, S. (2015). Analysis of genes that are differentially expressed during the *Sclerotinia sclerotiorum-Phaseolus vulgaris* interaction. *Frontiers in Microbiology* 6: 1162.

Ono, E., Hatayama, M., Isono, Y., et al. (2006). Localization of a flavonoid biosynthetic polyphenol oxidase in vacuoles. *Plant Journal* 45: 133–143.

Pawlak-Sprada, S., Arasimowicz-Jelonek, M., Podgórska, M., and Deckert, J. (2011). Activation of phenylpropanoid pathway in legume plants exposed to heavy metals. Part I. Effects of cadmium and lead on phenylalanine ammonia-lyase gene expression, enzyme activity and lignin content. *Acta Biochimica Polonica* 58: 211–216.

Phimchan, P., Chanthai, S., Bosland, P.W., and Techawongstien, S. (2014). Enzymatic changes in phenylalanine ammonia-lyase, cinnamic-4-hydroxylase, capsaicin synthase, and peroxidase activities in *Capsicum* under drought stress. *Journal of Agricultural and Food Chemistry* 62: 7057–7062.

Pietiäinen, M., Kontturi, J., Paasela, T., et al. (2016). Two polyketide synthases are necessary for 4-hydroxy-5-methylcoumarin biosynthesis in *Gerbera hybrida*. *The Plant Journal* 87: 548–558.

Pollastri, S., and Tattini, M. (2011). Flavonols: old compounds for old roles. *Annals of Botany* 108: 1225–1233.

Ponce De Leon, I., Schmelz, E.A., Gaggero, C., Castro, A., et al. (2012). *Physcomitrella patens* activates reinforcement of the cell wall, programmed cell death and accumulation of evolutionary conserved defence signals, such as salicylic acid and 12-oxo-phytodienoic acid, but not jasmonic acid, upon *Botrytis cinerea* infection. *Molecular Plant Pathology* 13: 960–974.

Puttick, M.N., Morris, J.L., Williams, T.A., et al. (2018). The interrelationships of land plants and the nature of the ancestral embryophyte. *Current Biology* 28: 733–745.

Raes, J., Rohde, A., Christensen, J.H., Van de Peer, Y., and Boerjan, W. (2003). Genome-wide characterization of the lignification toolbox in Arabidopsis. *Plant Physiology* 133: 1051–1071.

Rensing, S.A. (2017). Why we need more non-seed plant models. *New Phytologist* 216: 355–360.

Rensing, S.A., Lang, D., Zimmer, et al. (2008). The *Physcomitrella* genome reveals evolutionary insights into the conquest of land by plants. *Science* 319: 64–69.

Richards, T.A., Dacks, J.B., Campbell, S.A., et al. (2006). Evolutionary origins of the eukaryotic shikimate pathway: gene fusions, horizontal gene transfer, and endosymbiotic replacements. *Eukaryotic Cell* 5: 1517–1531.

Ro, D.K., Mah, N., Ellis, B.E., and Douglas, C.J. (2001). Functional characterization and subcellular localization of poplar (*Populus trichocarpa*× *Populus deltoides*) cinnamate 4-hydroxylase. *Plant Physiology* 126: 317–329.

Satake, H., Koyama, T., Matsumoto, E., et al. (2017). Metabolic engineering of lignan biosynthesis pathways for the production of transgenic plant-based foods. In, *Transgenesis and Secondary Metabolism, Reference Series in Phytochemistry* (ed. S. Jha), Springer International Publishing, Switzerland.

Schalk, M., Batard, Y., Seyer, A., et al. (1997). Design of fluorescent substrates and potent inhibitors of CYP73As, P450s that catalyze 4-hydroxylation of cinnamic acid in higher plants. *Biochemistry* 36: 15253–15261.

Schwinn, K., Venail, J., Shang, Y. J., et al. (2006). A small family of MYB-regulatory genes controls floral pigmentation intensity and patterning in the genus *Antirrhinum*. *The Plant Cell* 18: 831–851.

Shang, Y., Venail, J., Mackay, S., et al. (2011). The molecular basis for venation patterning of pigmentation and its effect on pollinator attraction in flowers of *Antirrhinum*. *New Phytologist* 189: 602–615.

Shi, S., Wang, Y.-Z., Zheng, X.-K., et al. (2012). Chemotaxonomic significance and biosynthesis of selaginellin from *Selaginella tamariscina*. *Biochemical Systematics and Ecology* 45: 151–154.

Shih, C.H., Chu, I.K., Yip, W.K., and Lo, C. (2006). Differential expression of two flavonoid 3′-hydroxylase cDNAs involved in biosynthesis of anthocyanin pigments and 3-deoxyanthocyanidin phytoalexins in *Sorghum*. *Plant and Cell Physiology* 47: 1412–1419.

Siwinska, J., Siatkowska, K., Olry, A., et al. (2018). Scopoletin 8-hydroxylase: a novel enzyme involved in coumarin biosynthesis and iron-deficiency responses in Arabidopsis. *Journal of Experimental Botany* 69: 1735–1748.

Song, H., Yi, H., Lee, M., et al. (2018). Purple *Brassica oleracea* var. *capitata F. rubra* is due to the loss of *BoMYBL2-1* expression. *BMC Plant Biology* 18: 82.

Soriano, G., Cloix, C., Heilmann, M., et al. (2018). Evolutionary conservation of structure and function of the UVR8 photoreceptor from the liverwort *Marchantia polymorpha* and the moss *Physcomitrella patens*. *New Phytologist* 217: 151–162.

Stracke, R., Werber, M., and Weisshaar, B. (2001). The *R2R3-MYB* gene family in *Arabidopsis thaliana*. *Current Opinion in Plant Biology* 4: 447–456.

Sun, H., Li, Y., Feng, S., et al. (2013). Analysis of five rice 4-coumarate: coenzyme A ligase enzyme activity and stress response for potential roles in lignin and flavonoid biosynthesis in rice. *Biochemical and Biophysical Research Communications* 430: 1151–1156.

Sykes, R.W., Gjersing, E.L., Foutz, K., et al. (2015). Down-regulation of *p*-coumaroyl quinate/shikimate 3′-hydroxylase (C3′H) and cinnamate 4-hydroxylase (C4H) genes in the lignin biosynthetic pathway of *Eucalyptus urophylla*× *E. grandis* leads to improved sugar release. *Biotechnology for Biofuels* 8: 128.

Szövényi, P., Frangedakis, E., Ricca, M., et al. (2015). Establishment of *Anthoceros agrestis* as a model species for studying the biology of hornworts. *BMC Plant Biology* 15: 1–7.

Takahashi, H., and Asakawa, Y. (2017). Transcriptome analysis of marchantin biosynthesis from the liverwort *Marchantia polymorpha*. *Natural Products Communications* 12: 1269–1272.

Tamagnone, L., Merida, A., Parr, A. et al. (1998). The AmMYB308 and AmMYB330 transcription factors from *Antirrhinum* regulate phenylpropanoid and lignin biosynthesis in transgenic tobacco. *The Plant Cell* 10: 135–154.

Tanaka, M., Esaki, T., Kenmoku, H., et al. (2016). Direct evidence of specific localization of sesquiterpenes and marchantin A in oil body cells of *Marchantia polymorpha* L. *Phytochemistry* 130: 77–84.

Taniguchi, S., Yazaki, K., Yabu-uchi, R., et al. (2000). Galloylglucoses and riccionidin A in *Rhus javanica* adventitious root cultures. *Phytochemistry* 53: 357–363.

Tilbrook, K., Arongaus, A.B., Binkert, M., et al. (2013). The UVR8 UV-B photoreceptor: perception, signaling and response. *The Arabidopsis Book* 11: e0164.

Urban, P., Werck-Reichhart, D., Teutsch, H.G., et al. (1994). Characterization of recombinant plant cinnamate 4-hydroxylase produced in yeast: Kinetic and spectral properties of the major plant P450 of the phenylpropanoid pathway. *European Journal of Biochemistry* 222: 843–850.

Vogelsang, K., Schneider, B., and Petersen, M. (2006). Production of rosmarinic acid and a new rosmarinic acid 3′-*O*-beta-D-glucoside in suspension cultures of the hornwort *Anthoceros agrestis* Paton. *Planta* 223: 369–373.

Wang, A., Zhu, M., Luo, Y., et al. (2017). A sweet potato cinnamate 4-hydroxylase gene, *IbC4H*, increases phenolics content and enhances drought tolerance in tobacco. *Acta Physiologiae Plantarum* 39: 276.

Wargent, J.J., and Jordan, B.R. (2013). From ozone depletion to agriculture: understanding the role of UV radiation in sustainable crop production. *New Phytologist* 197: 1058–1076.

Weise, N.J., Ahmed, S.T., Parmeggiani, F., et al. (2017). Zymophore identification enables the discovery of novel phenylalanine ammonia lyase enzymes. *Scientific Reports* 7: 13691.

Weng, J.K., and Noel, J.P. (2013). Chemodiversity in *Selaginella*: a reference system for parallel and convergent metabolic evolution in terrestrial plants. *Frontiers in Plant Science* 4: 119.

Wickett, N.J., Mirarab, S., Nguyen, N., et al. (2014). Phylotranscriptomic analysis of the origin and early diversification of land plants. *Proceedings of the National Academy of Sciences USA* 111: E4859–E4868.

Winefield, C.S., Lewis, D.H., Swinny, E.E., et al. (2005). Investigation of the biosynthesis of 3-deoxyanthocyanins in *Sinningia cardinalis*. *Physiologia Plantarum* 124: 419–430.

Wolf, L., Rizzini, L., Stracke, R., et al. (2010). The molecular and physiological responses of *Physcomitrella patens* to Ultraviolet-B radiation. *Plant Physiology* 153: 1123–1134.

Wu, Y., Zhao, Y., Liu, X. et al. (2018). A bHLH transcription factor regulates bisbibenzyl biosynthesis in the liverwort *Plagiochasma appendiculatum*. *Plant and Cell Physiology* 59: 1187–1199.

Wu, Z., Gui, S., Wang, S., and Ding, Y. (2014). Molecular evolution and functional characterisation of an ancient phenylalanine ammonia-lyase gene (*NnPAL1*) from *Nelumbo nucifera*: novel insight into the evolution of the PAL family in angiosperms. *BMC Evolutionary Biology* 14: 100.

Xia, J., Liu, Y., Yao, S., et al. (2017). Characterization and expression profiling of *Camellia sinensis* cinnamate 4-hydroxylase genes in phenylpropanoid pathways. *Genes* 8: 193.

Xiang, L., and Moore, B.S. (2005). Biochemical characterization of a prokaryotic phenylalalnine ammonia lyase. *Journal of Bacteriology* 187: 4286–4289.

Xie, L., Liu, P., Zhu, Z., et al. (2016). Phylogeny and expression analyses reveal important roles for plant PKS III family during the conquest of land by plants and angiosperm diversification. *Frontiers in Plant Science* 7: 1312.

Xu, R.-X., Gao, S., Zhao, Y., et al. (2016). Functional characterization of a $Mg2^{+}$-dependent *O*-methyltransferase with coumarin as preferred substrate from the liverwort *Plagiochasma appendiculatum*. *Plant Physiology and Biochemistry* 106: 269–277.

Xu, W., Dubos, C., and Lepiniec, L. (2015). Transcriptional control of flavonoid biosynthesis by MYB–bHLH–WDR complexes. *Trends in Plant Science* 20: 176–185.

Yoshikawa, H., Ichiki, Y., Sakakibara, K.D., et al. (2002). The biological and structural similarity between lunularic acid and abscisic acid. *Bioscience Biotechnology and Biochemistry* 66: 840–846.

Yoshikawa, M., Luo, W., Tanaka, G., et al. (2018). Wounding stress induces phenylalanine ammonia lyases, leading to the accumulation of phenylpropanoids in the model liverwort *Marchantia polymorpha*. *Phytochemistry* 155: 30–36.

Yu, H.-N., Liu, X.-Y., Gao, S., et al. (2014). Molecular cloning and functional characterization of a phenylalanine ammonia-lyase from liverwort *Plagiochasma appendiculatum*. *Plant Cell, Tissue and Organ Culture* 117: 265–277.

Yu, H.-N., Liu, X.-Y., Gao, S., et al. (2018). Structural and biochemical characterization of the plant type III polyketide synthases of the liverwort *Marchantia paleacea*. *Plant Physiology and Biochemistry* 125: 95–105.

Zhang, X., and Liu, C.-J. (2015). Multifaceted regulations of gateway enzyme phenylalanine ammonia-lyase in the biosynthesis of phenylpropanoids. *Molecular Plant* 8: 17–27.

Zhao, Q., Zhang, Y., Wang, G., et al. (2016). A specialized flavone biosynthetic pathway has evolved in the medicinal plant, *Scutellaria baicalensis*. *Science Advances* 2: e1501780.

Zhong, B., Xi, Z., Goremykin, V.V., et al. (2014). Streptophyte algae and the origin of land plants revisited using heterogeneous models with three new algal chloroplast genomes. *Molecular Biology and Evolution* 31: 177–183.

Zimmermann, I.M., Heim, M.A., Weisshaar, B. and Uhrig, J.F. (2004). Comprehensive identification of *Arabidopsis thaliana* MYB transcription factors interacting with R/B-like BHLH proteins. *Plant Journal* 40, 22–34.

10

Matching Proanthocyanidin Use with Appropriate Analytical Method

James A. Kennedy

Functional Phenolics, LLC, Corvallis, USA

10.1 Introduction

Proanthocyanidins are a flavonoid-based group of molecules that in their most simplified form, are oligomers and polymers of the flavan-3-ol base unit. With consideration given to flavan-3-ol base unit structure, interflavonoid bond location and molecular mass variance, this class of flavonoids is known for its complexity. In addition to this basic structure variation, additional structure variants have been identified in plant tissue extracts as well as in isolates from various applications as a result of the general reactivity of proanthocyanidins (Quideau et al. 2011).

Prior to the second World War, proanthocyanidins were used primarily in the tanning industry. With the development of alternative tanning methods, the use of proanthocyanidin isolates declined significantly after the war. This initial decline in use led to the exploration of alternative applications and today, proanthocyanidins have found utility in a variety of areas, from mining to the food and medical industries (various authors within Hemingway and Karchesy 1989; Hemingway and Laks 1992). Given the varied utility of proanthocyanidins and their diverse structure, it would seem evident that analytical chemistry endpoints would vary depending on intended use. As new uses emerge, therefore, an evaluation of appropriate analytical methodology that is based upon accuracy and precision of structure-function relationships should take place. Unfortunately, there is the tendency to maintain the use of analytical methodologies because of their historical value. These methodologies are almost exclusively based upon the method's ability to measure overall concentration; as a result, opportunities to improve upon analytical accuracy and precision as it relates to proanthocyanidin structure-function are slowed.

Many of the early proanthocyanidin analytical approaches predate our understanding of structure and generally emphasize a correlation with overall mass contribution (Broadhurst and Jones 1978; Hagerman and Butler 1981; Huemmer and Schreier 2008; Mole and Waterman 1987; Naczk and Shahidi 2004; Schofield et al. 2001; Singleton et al. 1999). The

Recent Advances in Polyphenol Research, Volume 7, First Edition. Edited by Jess Dreher Reed, Victor Armando Pereira de Freitas, and Stéphane Quideau.

analytical information provided by these methods continues to be useful although there are gaps in full structure-function understanding.

The understanding of structure that has occurred since the realization that flavan-3-ol subunits were the base unit of proanthocyanidins has made plain that there are structural features that specifically contribute to activity (e.g. molecular mass, secondary modification of structure). Although our understanding of proanthocyanidin structure is far from complete, our understanding is enough to conclude that proanthocyanidin structure can influence activity. Since its original structure elucidation in the 1960s, associations between specific proanthocyanidin structure and activity have steadily increased. Despite this, there is a persistent assumption that proanthocyanidin activity is determined by concentration determination alone.

This chapter discusses recent advances in understanding how specific proanthocyanidin structural features are associated with activity. Rather than review the published literature comprehensively, this chapter will focus on two areas where proanthocyanidin structure has become important: red wine astringency and biological activity. This chapter focuses on these two areas because in the first example an overall understanding and measurement of proanthocyanidin intermolecular interaction is desired, whereas in the second example an understanding and measurement of specific structure elements is desired.

10.2 General Proanthocyanidin Structure and Analysis

The general structure of proanthocyanidins is based upon the assembly of flavan-3-ol subunits via interflavonoid bonds (Figures 10.1 and 10.2) (Aron and Kennedy 2008). The size of proanthocyanidins can vary significantly from 2 subunits to over 80 subunits.

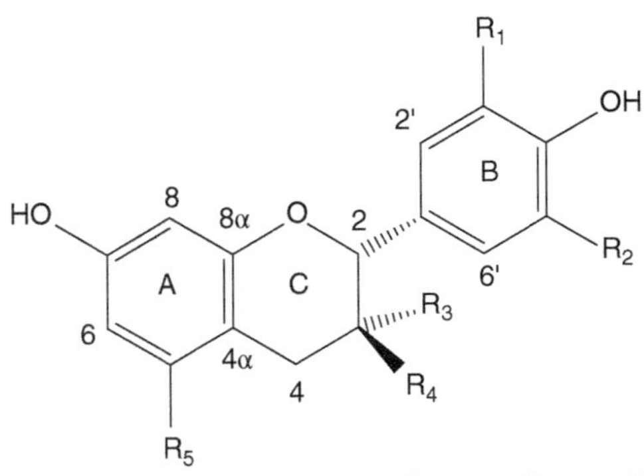

2R flavan-3-ol monomers	R_1	R_2	R_3	R_4	R_5
(+)-afzelechin	H	H	H	OH	OH
(−)-epiafzelechin	H	H	OH	H	OH
(+)-catechin	H	OH	H	OH	OH
(−)-epicatechin	H	OH	OH	H	OH
(+)-gallocatechin	OH	OH	H	OH	OH
(−)-epigallocatechin	OH	OH	OH	H	OH
(+)-fisetinidol	H	OH	H	OH	H
(−)-epifisetinidol	H	OH	OH	H	H
(+)-robinetinidol	OH	OH	H	OH	H

Figure 10.1 Flavan-3-ol monomer base units which could make assembly of proanthocyanidins.

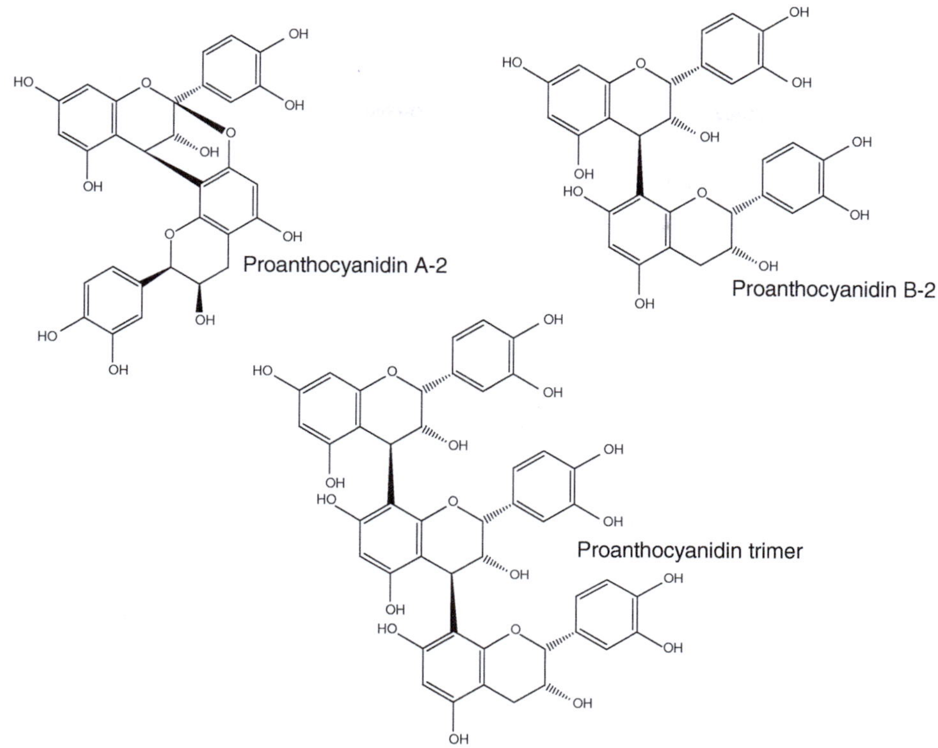

Figure 10.2 Examples of proanthocyanidin structures showing the most commonly found interflavonoid bond positions.

Interflavonoid bond positions are most commonly observed as linking the C-4 of the upper unit to C-8 of the lower unit (C-4→C-8) or C-4→C-6. Doubly linked subunits such as C-2→O-7; C-4→C-8 are observed, albeit less commonly so. Based upon this variation in subunit structure, the number and type of subunits within a given molecule, and the way that subunits are linked together, a complex number of possibilities in structure is possible.

In addition to the complexity derived from the assembly of flavan-3-ol subunits themselves, additional complexity results from structure modification within the plant or following extraction and as a result of general proanthocyanidin reactivity (Figure 10.3). The presence of glycosylation, anthocyanins, ethylidene bridges, and oxidative coupling, for example, can lead to extraordinarily complex proanthocyanidin structure (Bae et al. 1994; Cheynier 2005; De Taeye et al. 2017; Haslam 1980; Hayashi et al. 2018; Ma et al. 2018; Tanaka et al. 2018; Wollmann and Hofmann 2013; Zerbib, Cazals et al. 2018; Zerbib Mazauric et al. 2018). Our understanding of the role that this structure variance has on function has steadily increased as advances both in structural understanding as well as application have progressed. An obvious outcome of the structure-function relationship has been the development and appropriate use of analytical methodologies that fully explain and predict activity.

Historically, proanthocyanidin analysis has been based upon analytical methodologies that correlated with gravimetric yields and have been extensively reviewed elsewhere

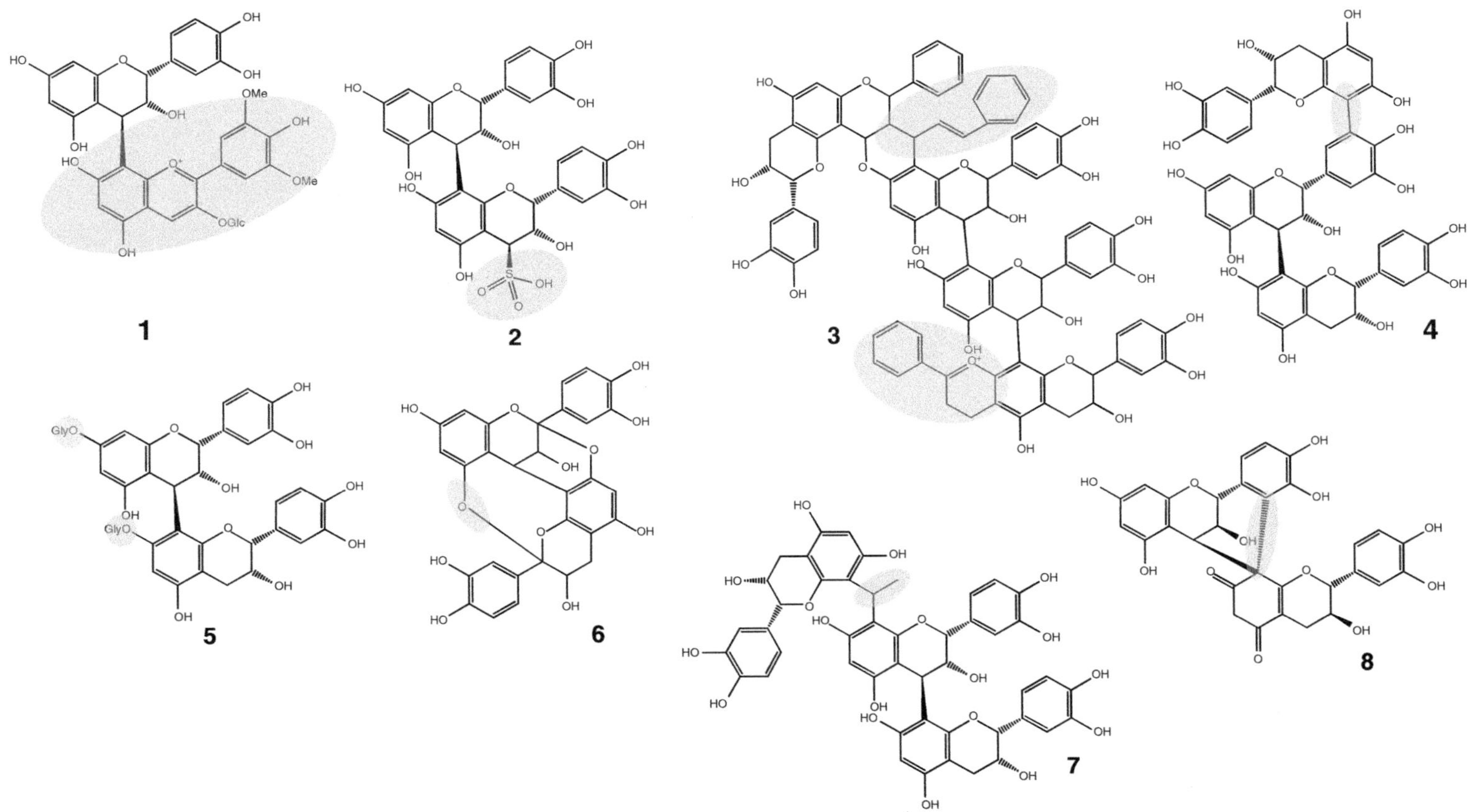

Figure 10.3 Examples of natural and synthetic modifications that could occur in proanthocyanidin structure, with modifications highlighted. In this case, the proanthocyanidin structure of compounds **1**, **2**, **3**, and **5** have been modified by the addition of nonproanthocyanidin moieties. Compounds **4**, **6**, **7** (addition of ethylidene bridge following ethanol oxidation to ethanal in wine), and **8** have been modified through direct or indirect oxidation.

(Huemmer and Schreier 2008; Naczk and Shahidi 2004; Schofield et al. 2001). These methods in large part have provided valuable information on the relationship between concentration and activity. There have been various analytical approaches taken to quantifying proanthocyanidins and correlative information with activity is generally available if proanthocyanidins are extracted and purified effectively. With recent developments in structure-activity relationships, however, there is an increasing realization that concentration-based methods alone do not provide the accuracy and precision required to fully assess utility.

The utility of proanthocyanidins in several applications relies on the ability of proanthocyanidins to noncovalently interact with other macromolecules (e.g. wine astringency, vegetable tanning). This interaction is considered to be based upon hydrophobic and hydrogen bonding interactions (Baxter et al. 1997; Frazier et al. 2010; Le Bourvellec and Renard 2012; McManus et al. 1985; Murray et al. 1994; Papadopoulou et al. 2005). Based upon this, an accurate assessment of proanthocyanidin activity would involve the measurement of noncovalent interaction. Historically, the prevailing assumption was that measured concentration was a reasonably good predictor of noncovalent interaction. The utility of proanthocyanidins may also depend on the concentration of specific proanthocyanidin molecules or structural components as opposed to the entire mass distribution (e.g. bioavailability, microbial accessibility). The following sections will provide an example of these two approaches.

10.3 Red Wine Mouthfeel

Red wine mouthfeel quality serves as a good example for how the selection of appropriate analytical methodology that is able to provide information on the ability of proanthocyanidins to noncovalently interact with macromolecules can assist in the improved understanding of and potential opportunity to better manage situations where proanthocyanidins have utility.

In addition to appearance, aroma and taste, mouthfeel is an important aspect of overall red wine quality (Gawel et al. 2000). For red wines, proanthocyanidins can be thought of as the focal point of mouthfeel, with many referring to mouthfeel quality as astringency quality. Proanthocyanidins contribute astringency to red wine. Previous research has shown that the magnitude of astringency perceived is correlated with the concentration of proanthocyanidins in the wine (Bajec and Pickering 2008). With that said, the quality of red wine mouthfeel is more complex than the concentration of proanthocyanidins per se. Additional factors which are known or expected to influence the quality of mouthfeel include wine matrix chemistry such as residual sugar, acidity, and the presence of other macromolecules such as polysaccharides (Fontoin et al. 2008; Watrelot et al. 2016; Figure 10.4). Historically the management of mouthfeel quality was seen as an exercise in the balance of proanthocyanidin concentration in the context of other wine matrix chemistry.

Over time, studies have shown that the beyond concentration, the chemistry of proanthocyanidins specifically can also have an impact on red wine astringency (Cheynier et al. 1992; Sáenz-Navajas et al. 2010; Sarni-Manchado et al. 1999; Vidal et al. 2003, 2004). Compositional elements such as size distribution, flavan-3-ol subunit composition, color incorporation, and direct and indirect oxidation have all been implicated in modifying how proanthocyanidins are perceived.

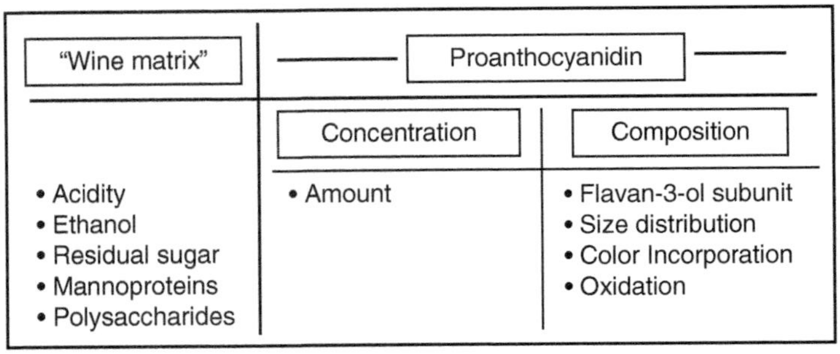

Figure 10.4 Red wine mouthfeel can be categorized by matrix composition (e.g. residual sugar, ethanol concentration) as well as proanthocyanidin concentration and composition.

During grape and wine production operations, there are many opportunities for the wine proanthocyanidin structure to become modified. The complexity of red wine proanthocyanidin structure and its determination limits our ability to fully understand how structure relates to perception/quality in practice. As a result, our ability to manage mouthfeel quality from a proanthocyanidin structure perspective is subject to conjecture. While the longer-term goal would be to have the ability to define structure completely, perhaps a shorter-term goal would be to provide analytical information on the ability of proanthocyanidins and modified proanthocyanidins to interact with macromolecular structures related to astringency perception.

Presumably, if proanthocyanidin structure has an impact on overall mouthfeel quality, it does so by modifying the ability of proanthocyanidins to interact with wine matrix chemistry, salivary proteins, and the oral mucosa, thereby modifying the sensory time-intensity response to proanthocyanidins (Bennick 2002; Payne et al. 2009; Poncet-LeGrand et al. 2007). Early work using NMR demonstrated that the energy of interaction between proanthocyanidin and protein is variable and depends upon structure (Baxter et al. 1997; Murray et al. 1994). Additional studies using isothermal titration calorimetry have also shown that the interaction between wine proanthocyanidins specifically and a model for salivary protein can be measured (Frazier et al. 2010; Poncet-LeGrand et al. 2007). Moreover, variations in grape and wine proanthocyanidin structure lead to variations in the strength of interaction between proanthocyanidins and a model salivary protein when measured by isothermal titration calorimetry (McRae et al. 2010).

There has been interest in developing additional methodologies for measuring intermolecular interaction so that the utility of this approach can be fully assessed. Fluorescence spectroscopy to measure fluorescence quenching of proteins, by proanthocyanidins, has been used with some success (Brandão et al. 2017; Soares et al. 2007). Alternative methods based upon NMR have also been investigated (Brandão et al. 2017; Delius et al. 2017).

Another approach has been the use of high-performance liquid chromatography (HPLC) to measure the energy of interaction between proanthocyanidins and a solid support (proanthocyanidin activity) and to use this information to predict potential interaction from an astringency perspective. HPLC-based interaction investigations indicate that structural features that are known to impact astringency perception also show differences when their proanthocyanidin activity is measured (Barak and Kennedy 2013; Revelette et al. 2014). Subsequent studies indicate that this approach does have reasonable throughput and, given parallel structure investigations, the approach also suggests that complex variations in proanthocyanidin composition correlate with changes in proanthocyanidin activity in a direction that is analogous to controlled sensory studies (Yacco et al. 2016). Specifically, proanthocyanidin size and extension unit galloylation are associated with corresponding proanthocyanidin activity. This approach to proanthocyanidin activity measurement has been used to better understand the nature of red wine mouthfeel quality under conditions where variation in mouthfeel quality have been observed and concentration and matrix do not adequately explain observations (Ma et al. 2018; Saenz-Navajas, Arias et al. 2018; Saenz-Navajas, Henschen et al. 2018; Watrelot et al. 2018). Initial sensory studies where this information has been collected supports the unique nature and potential value of this information with respect to understanding astringency subqualities (Saenz-Navajas et al. 2019, 2020).

The measurement of proanthocyanidin activity should have value whenever proanthocyanidins are being utilized for their ability to noncovalently interact with other macromolecules and for which the composition of proanthocyanidins is poorly understood. Under these applications, where concentration alone has been shown to incompletely explain observed activity, proanthocyanidin activity may provide complimentary and insightful analytical information.

10.4 Biological Activity

For some proanthocyanidin applications, specific proanthocyanidin structural features are being targeted for activity. There has been interest in the biological activity of proanthocyanidins for some time (Aron and Kennedy 2008; Zhang et al 2016). Recent evidence indicates that specific proanthocyanidin structural elements give rise to associated activity.

Proanthocyanidins have been shown to have biological activity as a result of their phenolic antioxidant properties and for their structural specificity (Bitzer et al. 2015; Kajiya et al. 2001; Santos-Buelga and Scalbert 2000). It has been shown, for example, that some proanthocyanidins upregulate endothelium nitric oxide synthase, thereby increasing nitric oxide production and subsequent relaxing of the endothelium (Antónia-Nunes et al. 2016; Pourová et al. 2018). This process is considered to be an important reason for the cardiovascular health–promoting properties of proanthocyanidins. Additionally, it has been shown that specific proanthocyanidin-derived gut microbial metabolites are bioavailable and have been identified in urine and plasma (Appeldoorn et al. 2009; Cueva et al. 2013; de Llano et al. 2019; M. Mena et al. 2019; Monagas et al. 2010; Stoupi et al. 2010; Wiese et al. 2015; Williamson et al. 2017). Furthermore, these metabolites have been found to cross the blood-brain barrier (Wang et al. 2015), where they have been implicated in cognition.

The challenge is that in most biological studies, proanthocyanidin concentration alone is measured and this concentration is being used to determine significance of biological activity. Most meta-analytical studies on the health effects of proanthocyanidins rely solely on proanthocyanidin concentration when assessing health effects, and in some studies the benefit of proanthocyanidin intake is in question (Hooper et al. 2008; Li et al. 2015). If the activity of proanthocyanidins can be ascribed to specific structural features, whether it be a specific molecule or a structural motif, then it is imperative that these elements be used to fully assess biological significance.

Recent studies support the conclusion that proanthocyanidin structure plays a role in the concentration of circulating microbial metabolites and biological activity (Marquez-Campos et al. 2019; Ottaviani et al. 2018, 2012; Wiese et al. 2015). In light of these studies showing that specific structural aspects play a more significant role than previously thought, it is important to provide compositional information to fully understand the role that proanthocyanidins play in human health.

With specific regard to biological activity, studies have shown that oligomers beyond trimers have limited absorption into the bloodstream (Ottaviani et al. 2012). In addition, there is evidence that proanthocyanidin molecular size influences general susceptibility to microbial degradation (Wiese et al. 2015) as well as activity in the gut (Lizarraga et al. 2007; M. Mena et al. 2019; Zhu et al. 2016). Therefore, in order to provide meaningful data, it would be necessary to quantify concentration data specific to oligomers.

In the case of the gut microbiome and known microbial metabolites (Figure 10.5), it would appear that specific proanthocyanidin features such as the possession of an unmodified proanthocyanidin B-ring would be critical. In other microbial communities, organic bond structure/energy has been shown to be related to susceptibility/resistance to microbial breakdown (Bonanomi et al. 2017; Salminen et al. 2011). Given the structure variation in proanthocyanidins (e.g. Figure 10.3) in comparison to known microbial metabolites, an analytical approach which provides information on overall size (i.e. susceptibility to microbial breakdown) and resemblance to base flavan-3-ol subunit (e.g. Kennedy and Jones 2001) would be of seeming importance as this information could be used to quantify flavan-3-ol material of a specific size which could be easily converted to known subunits. This approach would presumably be analogous to the ease with which gut microbiota can metabolize proanthocyanidins.

Based upon current evidence, the appropriate analytical approach to fully understanding the role of proanthocyanidins in human health, analytical methods that can quantify the conversion yield, as well as molecular mass distribution – and, more specifically, the concentration of individual oligomers – would be of most value. At a minimum, this analytical information should be used to characterize proanthocyanidins for clinical investigations so that recent research results could be accounted for with analytics.

There is recent awareness that in addition to variation in proanthocyanidin structure, variations in food composition and gut microflora (and the awareness of metabotypes) also play a role in the biological activity of proanthocyanidins (Le Bourvellec et al. 2019; P. Mena et al. 2019; Tomás-Barberán et al. 2019). It is reasonable to conclude that these collective observations will be increasingly considered when developing investigations in this area. Recent examples provide promising indications that proanthocyanidin structure-function associations will become more frequent (Rubert-Nason et al. 2019; Salminen 2018).

10.5 Summary

Proanthocyanidins have been utilized for many years and much of proanthocyanidin analysis has historically been targeted towards quantifying total mass contribution. Recent advances in the understanding of proanthocyanidin structure and the role that structure plays in activity has made it clear that proanthocyanidin function has moved beyond the measurement of mass contribution alone.

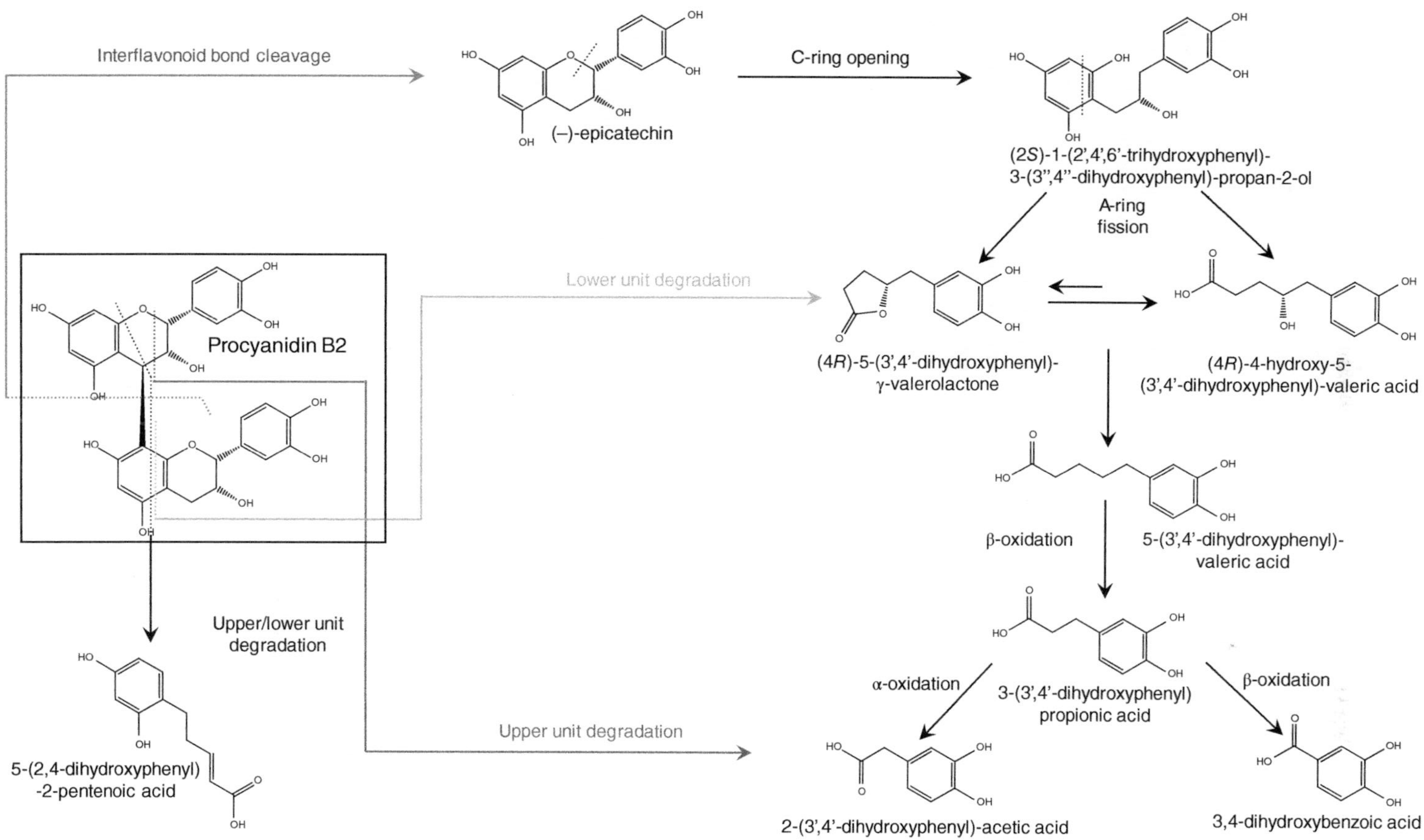

Figure 10.5 Microbial metabolites that have been identified and their plausible relationship to proanthocyanidins.

Two examples have been presented where the reporting of this information is becoming increasingly important. In the first example, studies have shown that the mouthfeel of red wine is dependent upon matrix composition such as residual sugar, ethanol, titratable acidity, and soluble polysaccharide content. Beyond this, it has also been shown that proanthocyanidin structural features such as molecular mass, galloylation, and color incorporation can influence the perception of astringency. In order to fully assess the role that proanthocyanidins contribute to mouthfeel, it would therefore be necessary to fully understand proanthocyanidin structure. Currently our understanding of red wine proanthocyanidin structure is far from complete. In the absence of full structure-function understanding, and alternatively, it would be advantageous to determine the activity of proanthocyanidins as a complimentary analytical method to concentration.

In the second example, current research supports that the biological activity of proanthocyanidins is in part due to very specific structural features that allow for the microbial breakdown of proanthocyanidins into specific metabolites that are then absorbed via the colon. Recent studies suggest that this route of biological activity has been underestimated and that emerging knowledge related to structure-activity promises to provide new directions for developing analytical methods that can appropriately be used to measure proanthocyanidin utility.

These examples support the idea that if we are to fully determine the utility of proanthocyanidins across all applications then the knowledge gained to date on structure-function relationships needs to be extended to the development and routine use of function-related analytics.

References

Antónia-Nunes, M., Pimentel, F., Costa, A.S.G., et al. (2016). Cardioprotective properties of grape seed proanthocyanidins: An update. *Trends in Food Science and Technology* 57 (Part A): 31–39.

Appeldoorn, M.M., Vincken, J.-P., Aura, A.-M., et al. (2009). Procyanidin dimers are metabolized by human microbiota with 2-(3,4-dihydroxyphenyl)acetic acid and 5-(3,4-dihydroxyphenyl)-γ-valerolactone as the major metabolites. *Journal of Agricultural and Food Chemistry* 57 (3): 1084–1092.

Aron, P.M. and Kennedy J.A. (2008). Flavan-3-ols: Nature, occurrence and biological activity. *Molecular Nutrition and Food Research* 52 (1): 79–104.

Bae, Y.S., Burger, J.F.W., Steynberg, J.P., et al. (1994). Flavan and procyanidin glycosides from the bark of blackjack oak. *Phytochemistry* 35 (2): 473–478.

Bajec, M.R. and Pickering, G.J. (2008). Astringency: Mechanisms and perception. *Critical Reviews in Food Science and Nutrition* 48 (9): 858–875.

Barak, J.A. and Kennedy, J.A. (2013). HPLC retention thermodynamics of grape and wine tannins. *Journal of Agricultural and Food Chemistry* 61 (18): 4270–4277.

Baxter, N.J., Lilley, T.H., Haslam, E., and Williamson, M.P. (1997). Multiple interactions between polyphenols and a salivary proline-rich protein repeat result in complexation and precipitation. *Biochemistry* 36 (18): 5566–5577.

Bennick, A. (2002). Interaction of plant polyphenols with salivary proteins. *Critical Reviews in Oral Biology and Medicine* 13 (2): 184–196.

Bitzer, Z.T., Glisan, S.L., Dorenkott, M.R., et al. (2015). Cocoa procyanidins with different degrees of polymerization possess distinct activities in models of colonic inflammation. *Journal of Nutritional Biochemistry* 26 (8): 827–831.

Bonanomi, G., Cesarano, G., Lombardi, N., et al. (2017). Litter chemistry explains contrasting feeding preferences of bacteria, fungi, and higher plants. *Scientific Reports* 7: 9208.

Brandão, E., Santos-Silva, M., García-Estévez, I., et al. (2017). Molecular study of mucin-procyanidin interaction by fluorescence quenching and saturation transfer difference (STD)-NMR. *Food Chemistry* 228: 427–434.

Broadhurst, R.B. and Jones, W.T. (1978). Analysis of condensed tannins using acidified vanillin. *Journal of the Science of Food and Agriculture* 29 (9): 788–794.

Cheynier, V. (2005). Polyphenols in foods are more complex than often thought. *American Journal of Clinical Nutrition* 81 (1): 223S–229S.

Cheynier, V., Rigaud, J., and Ricardo da Silva, J.M. (1992). Structure of procyanidin oligomers isolated from grape seeds in relation to some of their chemical properties. In: *Plant Polyphenols*, (eds. R. W. Hemingway and P. E. Laks), 281–294. New York: Plenum Press.

Cueva, C., Sánchez-Patán, F., Monagas, M., Walton, G.E., et al. (2013). In vitro fermentation of grape seed flavan-3-ol fractions by human faecal microbiota: changes in microbial groups and phenolic metabolites. *FEMS Microbiology Ecology* 83 (3): 792–805.

Delius, J., Oliver, F., and Hofmann, T. (2017). Label-free quantitative H-1 NMR spectroscopy to study low-affinity ligand-protein interactions in solution: A contribution to the mechanism of polyphenol-mediated astringency. *PLOS One* 12. https://journals.plos.org/plosone/article?id=10.1371/journal.pone.0184487.

de Llano, D.G., Liu, H.Y., Khoo, C., et al. (2019). Some new findings regarding the antiadhesive activity of cranberry phenolic compounds and their microbial-derived metabolites against uropathogenic bacteria. *Journal of Agricultural and Food Chemistry* 67 (8): 2166–2174.

De Taeye, C., Caullet, G., Evina, V.J.E., and Collin, S. (2017). Proacyanidin A2 and its degradation products in raw, fermented, and roasted cocoa. *Journal of Agricultural and Food Chemistry* 65 (8): 1715–1723.

Fontoin, H., Saucier, C., Teissedre, P.-L., and Glories, Y. (2008). Effect of pH, ethanol and acidity on astringency and bitterness of grape seed tannin oligomers in model wine solution. *Food Quality and Preference* 19 (3): 286–291.

Frazier, R.A., Deaville, E.R., Green, R.J., et al. (2010). Interactions of tea tannins and condensed tannins with proteins. *Journal of Pharmaceutical Sciences* 51 (2): 490–495.

Gawel, R., Oberholster, A., Francis, I.L. (2000). A "mouth-feel wheel": terminology for communicating the mouth-feel characteristics of red wine. *Australian Journal of Grape and Wine Research* 6 (3): 203–207.

Hagerman, A.E. and Butler, L.G. (1981). The specificity of proanthocyanidin-protein interactions. *Journal of Biological Chemistry* 256 (9): 4494–4497.

Haslam, E. (1980). In vino veritas: Oligomeric procyanidins and the ageing of red wines. *Phytochemistry* 19 (12): 2577–2582.

Hayashi, S., Nakano, K., Yanase, E. (2018). Investigation of color-deepening phenomenon in catechin (4→8)-dimer as a proanthocyanidin model and structural determination of its derivatives by oxidation. *Food Chemistry* 239: 1126–1133.

Hemingway, R.W. and Karchesy, J.J. (1989). *Chemistry and the Significance of Condensed Tannins*. New York: Plenum Press.

Hemingway, R.W. and Laks, P.E. (1992). *Plant Polyphenols: Synthesis, Properties, Significance.* New York: Plenum Press.

Hooper, L., Kroon, P.A., Rimm, E.B., et al. (2008). Flavonoids, flavonoid-rich foods, and cardiovascular risk: a meta-analysis of randomized controlled trials. *American Journal of Clinical Nutrition* 88 (1): 38–50.

Huemmer, W. and Schreier, P. (2008). Analysis of proanthocyanidins. *Molecular Nutrition and Food Research* 52 (12): 1381–1398.

Kajiya, K., Kumazawa, S., and Nakayama, T. (2001). Steric effects on interaction of tea catechins with lipid bilayers. *Bioscience, Biotechnology and Biochemistry* 65 (12): 2638–2643.

Kennedy, J.A. and Jones, G.P. (2001). Analysis of proanthocyanidin cleavage products following acid-catalysis in the presence of excess phloroglucinol. *Journal of Agricultural and Food Chemistry* 49 (4): 1740–1746.

Le Bourvellec, C., Boas, P.B.V., Lepercq, P., et al. (2019). Procyanidin-cell wall interactions within apple matrices decrease the metabolization of procyanidins by the human gut microbiota and the anti-inflammatory effect of the resulting microbial metabolome in vitro. *Nutrients* 11 (3): 664.

Le Bourvellec, C. and Renard, C. (2012). Interactions between polyphenols and macromolecules: Quantification methods and mechanisms. *Critical reviews in Food Science and Nutrition* 52 (3): 213–248.

Li, S.-H., Zhao, P., Tian, H.-B., et al. (2015). Effect of grape polyphenols on blood pressure: A meta analysis of randomized controlled trials. *PLOS One.* https://journals.plos.org/plosone/article?id=10.1371/journal.pone.0137665.

Lizarraga, D., Lozano, C., Briede, J.J., et al. (2007). The importance of polymerization and galloylation for the antiproliferative properties of procyanidin-rich natural extracts. *FEBS Journal* 274 (18): 4802–4811.

Ma, L.J., Watrelot, A.A., Addison, B., and Waterhouse, A.L. (2018). Condensed tannin reacts with SO2 during wine aging, yielding flavan-3-ol sulfonates. *Journal of Agricultural and Food Chemistry* 66 (35): 9259–9268.

Marquez-Campos, E., Stehle, P., Simon, M.-C. (2019). Microbial metabolites of flavan-3-ols and their biological activity. *Nutrients* 11 (10): 2260.

McManus, J.P., Davis, K.G., Beart, J.E., et al. (1985). Polyphenol interactions. part 1. introduction; some observations on the reversible complexation of polyphenols with proteins and polysaccharides. *Journal of the Chemical Society, Perkin Transactions* 2: 1429–1438.

McRae, J.M., Falconer, R.J., Kennedy, J.A. (2010). Thermodynamics of grape and wine tannin interaction with polyproline: Implications for red wine astringency. *Journal of Agricultural and Food Chemistry* 58 (23): 12510–12518.

Mena, M., Bresciani, L., Brindani, N., Ludwig, I.A., et al. (2019). Phenyl-γ-valerolactones and phenylvaleric acids, the main colonic metabolites of flavan-3-ols: synthesis, analysis, bioavailability, and bioactivity. *Natural Product Reports* 36 (5): 714–752.

Mena, P., Ludwig, I.A., Tomatis, V.B., et al. (2019). Inter-individual variability in the production of flavan-3-ol colonic metabolites: Preliminary elucidation of urinary metabotypes. *European Journal of Nutrition* 58: 1529–1543.

Mole, S. and Waterman, P.G. (1987). A critical analysis of techniques for measuring tannins in ecological studies. 1. Techniques for chemically defining tannins. *Oecologia* 72: 137–147.

Monagas, M., Urpi-Sarda, M., Sanchez-Patan, F., et al. (2010). Insights into the metabolism and microbial biotransformation of dietary flavan-3-ols and the bioactivity of their metabolites. *Food and Function* 1: 233–253.

Murray, N.J., Williamson, M.P., Lilley, T.H., and Haslam, E. (1994). Study of the interaction between salivary proline-rich proteins and a polyphenol by ^{1}H-NMR spectroscopy. *European Journal of Biochemistry* 219 (3): 923–935.

Naczk, M. and Shahidi, F. (2004). Extraction and analysis of phenolics in food. *Journal of Chromatography A* 1054 (1-2): 95–111.

Ottaviani, J.I., Heiss, C., Spencer, J.P.E., et al. (2018). Recommending flavanols and procyanidins for cardiovascular health: Revisited. *Molecular Aspects of Medicine* 61: 63–75.

Ottaviani, J.I., Kwik-Uribe, C., Keen, C.L., and Schroeter, H. (2012). Intake of dietary procyanidins does not contribute to the pool of circulating flavanols in humans. *American Journal of Clinical Nutrition* 95 (4): 851–858.

Papadopoulou, A., Green, R.J., and Frazier, R.A. (2005). Interactions of flavonoids with bovine serum albumin: A fluorescence quenching study. *Journal of Agricultural and Food Chemistry* 53 (1): 158–163.

Payne, C., Bowyer, P.K., Herderich, M., and Bastian, S.E.P. (2009). Interaction of astringent grape seed procyanidins with oral epithelial cells. *Food Chemistry* 115: 551–557.

Poncet-Legrand, C., Gautier, C., Cheynier, V.,and Imberty, A. (2007). Interactions between flavan-3-ols and poly(L-proline). studied by isothermal titration calorimetry: Effect of the tannin structure. *Journal of Agricultural and Food Chemistry* 55 (22): 9235–9240.

Pourová, J., Najmanová, M., Migkos, T., et al. (2018). Two flavonoid metabolites, 3,4-dihydroxyphenylacetic acid and 4-methylcatechol, relax arteries *ex vivo* and decrease blood pressure *in vivo*. *Vascular Pharmacology* 111: 36–43.

Quideau, S., Deffieux, D., Douat-Casassus, C., and Pouysegu, L. (2011). Plant Polyphenols: Chemical properties, biological activities, and synthesis. *Angewandte Chemie International Edition* 50 (3): 586–621.

Revelette, M.R., Barak, J.A., and Kennedy, J.A. (2014). High performance liquid chromatography determination of red wine tannin stickiness. *Journal of Agricultural and Food Chemistry* 62 (28): 6626–6631.

Rubert-Nason, K.F. and Lindroth, R.L. (2019). Analysis of condensed tannins in Populus spp. using reversed phase UPLC-PDA-(-)esi-MS following thiolytic depolymerization. *Phytochemical Analysis* 30 (3): 257–267.

Saenz-Navajas, M.P., Arias, I., Ferrero-Del-Teso, S., et al. (2018). Chemo-sensory approach for the identification of chemical compounds driving green character in red wines. *Food Research International* 109: 138–148.

Sáenz-Navajas, M.P., Ferrero-del-Teso, S., Jeffery, D.W., et al. (2020). Effect of aroma perception on taste and mouthfeel dimensions of red wines: Correlation of sensory and chemical measurements. *Food Research International* 131: 108945.

Sáenz-Navajas, M.P., Ferrero-del-Teso, S., Romero, M., et al. (2019). Modelling wine astringency from its chemical composition using machine learning algorithms. *Oeno One* 53 (3): 499–509.

Saenz-Navajas, M.P., Henschen, C., Cantu, A., et al. (2018). Understanding microoxygenation: Effect of viable yeasts and sulfur dioxide levels on the sensory properties of a Merlot red wine. *Food Research International* 108: 505–515.

Sáenz-Navajas, M.P., Tao, Y.S., Dizy, M., et al. (2010). Relationship between nonvolatile composition and sensory properties of premium Spanish red wines and their correlation to quality perception. *Journal of Agricultural and Food Chemistry* 58 (23): 12407–12416.

Salminen, J.-P. (2018). Two-dimensional tannin fingerprints by liquid chromatography tandem mass spectrometry offer a new dimension to plant tannin analyses and help to visualize the tannin diversity in plants. *Journal of Agricultural and Food Chemistry* 66 (35): 9162–9171.

Salminen, J.-P., Karonen, M., and Sinkkonen, J. (2011). Chemical ecology of tannins: Recent developments in tannin chemistry reveal new structures and structure-activity patterns. *Chemistry – A European Journal* 17 (10): 2806–2816.

Santos-Buelga, C. and Scalbert, A. (2000). Proanthocyanidins and tannin-like compounds – Nature, occurrence, dietary intake and effects on nutrition and health. *Journal of the Science of Food and Agriculture* 80 (7): 1094–1117.

Sarni-Manchado, P., Cheynier, V., and Moutounet, M. (1999). Interactions of grape seed tannins with salivary proteins. *Journal of Agricultural and Food Chemistry* 47 (1): 42–47.

Schofield, P., Mbugua, D.M., Pell, A.N. (2001). Analysis of condensed tannins: A review. *Animal Feed Science and Technology* 91 (1–2): 21–40.

Singleton, V.L., Orthofer, R., and Lamuela-Raventos, R.M. (1999). Analysis of total phenols and other oxidation substrates and antioxidants by means of Folin-Ciocalteu reagent. *Methods in Enzymology, Oxidants and Antioxidants, Part A* 299: 152–178.

Soares, S., Mateus, N., and De Freitas, V. (2007). Interaction of different polyphenols with bovine serum albumin (BSA). and human salivary alpha-amylase (HAS) by fluorescence quenching. *Journal of Agricultural and Food Chemistry* 55 (16): 6727–6735.

Stoupi, S., Williamson, G., Drynan, J.W., et al. (2010). A comparison of the *in vitro* biotransformation of procyanidin B2 by human faecal microbiota. *Molecular Nutrition and Food Research* 54 (6): 747–759.

Tanaka, T., Matsuo, Y., and Saito, Y. (2018). Solubility of tannins and preparation of oil-soluble derivatives. *Journal of Oleo Science* 67 (10): 1179–1187.

Tomás-Barberán, F.A. and Espín, J.C. (2019). Effect of food structure and processing on (poly) phenol–gut microbiota interactions and the effects on human health. *Annual Review of Food Science and Technology* 10: 221–238.

Vidal, S., Francis, L., Guyot, S., et al. (2003). The mouth-feel properties of grape and apple proanthocyanidins in a wine-like medium. *Journal of the Science of Food and Agriculture* 83 (6): 564–573.

Vidal, S., Francis, L., Noble, A., et al. (2004). Taste and mouth-feel properties of different types of tannin-like polyphenolic compounds and anthocyanins in wine. *Analytica Chimica Acta* 513 (1): 57–65.

Wang, D., Ho, L., Faith, J., et al. (2015). Role of intestinal microbiota in the generation of polyphenol-derived phenolic acid mediated attenuation of Alzheimer's disease β-amyloid oligomerization. *Molecular Nutrition and Food Research* 59 (6): 1025–1040.

Watrelot, A.A., Badet-Murat, M.-L., and Waterhouse, A.L. (2018). Oak barrel tannin and toasting temperature: Effects on red wine condensed tannin chemistry. *Lebensmittel-Wissenschaft Und-Technologie-Food Science and Technology* 91: 330–338.

Watrelot, A.A., Schulz, D.L., and Kennedy, J.A. (2016). Wine polysaccharides influence tannin-protein interactions. *Food Hydrocolloids* 63: 571–579.

Wiese, S., Esatbeyoglu, T., Winterhalter, P., et al. (2015). Comparative biokinetics and metabolism of pure monomeric, dimeric and polymeric flavan-3-ols: A randomized cross-over study in humans. *Molecular Nutrition and Food Research* 59 (4): 610–621.

Williamson, G. and Clifford, M.N. (2017). Role of the small intestine, colon and microbiota in determining the metabolic fate of polyphenols. *Biochemical Pharmacology* 139 (S1): 24–39.

Wollmann, N. and Hofmann, T. (2013). Compositional and sensory characterization of red wine polymers. *Journal of Agricultural and Food Chemistry* 61 (9): 2045–2061.

Yacco, R.S., Watrelot, A.A., and Kennedy, J.A. (2016). Red wine tannin structure-activity relationships during fermentation and maceration. *Journal of Agricultural and Food Chemistry* 64 (4): 860–869.

Zerbib, M., Cazals, G., Ducasse, M.-A., et al. (2018). Evolution of flavanol glycosides during red wine fermentation. *Molecules* 23 (12): 3300.

Zerbib, M., Mazauric, J.-P., Meudec, E., et al. (2018). New flavanol *O*-glycosides in grape and wine. *Food Chemistry* 266: 441–448.

Zhang, L., Wang, Y., Li, D., et al. (2016). The absorption, distribution, metabolism and excretion of procyanidins. *Food and Function* 7 (3): 1273–1281.

Zhu, W., Xiong, L., Peng, J.M., et al. (2016). Structure-dependent membrane-perturbing potency of four proanthocyanidin dimers on 3T3-L1 preadipocytes. *Journal of Agricultural and Food Chemistry* 64 (37): 7022–7032.

11

Imaging Polyphenolic Compounds in Plant Tissues

Marisa S. Otegui

Department of Botany and Center for Quantitative Cell Imaging, University of Wisconsin–Madison, Madison, USA

11.1 Introduction

Plants synthesize thousands of compounds derived from primary and specialized metabolic pathways (Hill et al. 2015). Although there are different analytical tools to determine the chemical nature and content of metabolites in total plant extracts (Kim et al. 2011; Dettmer et al. 2007), the analysis of the subcellular localization of small molecules in plant tissues remains a significant challenge.

(Poly)phenolic compounds are a large group of specialized metabolites in plants with diverse functions, including structural and functional roles in cell walls, plant resistance against pathogens and insects, oxidative stress, protection against UV-B radiation, attraction of pollinators, and fruit dispersal (Dixon and Paiva 1995; Agati and Tattini 2010; Gunnaiah et al. 2012; Tanaka et al. 2008). These compounds accumulate in different subcellular compartments, such as vacuoles (e.g. anthocyanins), nucleus (e.g. flavonoids), plastids (e.g. flavonoids), and cell walls (e.g. lignin and other phenolic compounds impregnating cutinized and suberized cell walls) (Agati et al. 2007; Polster et al. 2006; Agati et al. 2012). In some cases, modified organelles store massive amounts of phenolic compounds. For example, two types of plastid-derived organelles are the phenyloplasts, which accumulate phenol glucosides (4-*O*-[3-methoxybenzaldehyde] β-D-glucoside) that upon hydrolysis release vanillin in vanilla fruits (Brillouet et al. 2014) and the tannosomes, which contain condensed tannins (proanthocyanidins) together with thylakoid and plastid envelope components and are exported from plastids into vacuoles (Brillouet et al. 2013). Another remarkable example is tapesomes, endoplasmic reticulum- derived organelles that accumulate flavonoids and alkanes and form inside the tapetum (a secretory layer inside anthers) during the formation of pollen grains (Hsieh and Huang 2007).

Although the localization of many but not all the enzymes involved in their biosynthetic pathways has been studied, how the intermediate metabolites and precursors are trafficked within the cell and transported and deposited at their final destinations remains unclear for most polyphenols (Zhao and Dixon 2010; Zhao et al. 2010).

Recent Advances in Polyphenol Research, Volume 7, First Edition. Edited by Jess Dreher Reed, Victor Armando Pereira de Freitas, and Stéphane Quideau.

Analytical methods such as high-performance liquid chromatography (HPLC) followed by mass spectrometry (MS) of whole tissue extracts or isolated organelles have been critical to revealing the chemical identity of specialized metabolites in plants, but unfortunately they do not provide either temporal or spatial resolution to understand their trafficking and accumulation dynamics in living cells. A powerful approach to analyzing subcellular distribution of plant specialized metabolites is mass spectrometry imaging (MSI), which enables the determination of ionizable biological molecules within a tissue section. Among the MSI approaches, MALDI-MSI (matrix-assisted laser desorption/ionization mass spectrometry imaging) uses either whole tissues or 10–35 μm thick sections of frozen or chemically fixed samples. Intact plant organs can also be directly lyophilized and mounted on a sample plate to detect metabolites on their surfaces. A chemical matrix is then applied on the sample to promote desorption and ionization. Once inside the instrument, a UV/IR laser ionizes molecules (lipid, proteins, specialized metabolites) in the sample that then enter into a mass spectrometer for chemical determination and mapping (Sturtevant et al. 2016; Qin et al. 2018). The resolution of this method is affected by a number of parameters such as laser spot size, mass spectrometer sensitivity, matrix of choice, and abundance of metabolite under analysis, but in general it is in the range of 5–10 μm. MALDI-MSI has been used to detect flavonoids in *Arabidopsis* flowers (Cha et al. 2008), the flavonoids rutin and maysin in maize leaves (Korte et al. 2015), and the flavonoid hesperidin in *Citrus* leaves during pathogen infection (Soares et al. 2015), among many other examples. The drawbacks of MALDI-MSI approaches are their destructive nature and their incompatibility with dynamic, *in vivo* studies as the sample needs to be frozen or fixed.

Different from proteins, specialized metabolites such as polyphenols cannot be genetically tagged to facilitate their visualization by microscopy. However, the various ways polyphenols interact with light can be used for their imaging within intact plant cells and tissues. For example, (poly)phenolic compounds with intrinsic fluorescence can be studied *in situ* by fluorescence-based microscopy, where multiple fluorescence fingerprints (multispectral emission, fluorescence lifetime) can be collected and analyzed. Both fluorescent and nonfluorescent metabolites can also be studied using light-scattering-based microscopy such as Raman microscopy (Min et al. 2011).

11.2 The Chemical Nature and Intrinsic Fluorescence Properties of Polyphenols

There are different definitions of polyphenols in the literature. In this chapter, we will follow the definition proposed by Stéphane Quideau and colleagues as compounds exclusively derived from the shikimate/phenylpropanoid and/or the acetate/malonate pathway, containing two or more phenolic units and deprived of nitrogen-based functions (Quideau et al. 2011). The phenylpropanoid pathway produces different hydroxycinnamic acids that differ from one another in their hydroxy and methoxy groups on the phenyl unit (*p*-coumaric acid, ferulic acid, sinapic acid, caffeic acid) and act as precursors for the synthesis of polyphenols (Quideau et al. 2011). Some examples of polyphenolic compounds are flavonoids (e.g. luteolin, apigenin), flavanones (e.g. hesperidin, naringenin), flavonols (e.g. kaempferol, quercetin), flavanols (e.g. catechin, epicatechin gallate), isoflavones (e.g. genistein, daidzein), anthocyanidins (e.g. malvidin, pelargonidin), anthocyanins, chalcones (e.g. butein), proanthocyanidins (or tannins), lignin, and lignans.

(Poly)phenols absorb light due to the presence of aromatic rings with conjugated systems of electrons. Some polyphenolic compounds absorb in the visible range of the spectrum and therefore they are colored. However, they all absorb UV light and typically exhibit intrinsic fluorescence. Whereas intrinsic fluorescence of metabolites is usually considered a hurdle for imaging fluorescent probes in plants, it can also offer many opportunities to analyze the localization and cellular trafficking of these compounds.

In the following sections, I will discuss some of the most useful approaches for noninvasive, label-free imaging of intact plant tissues and their applications to the analysis of polyphenols.

11.3 Microscopy-based Methods for Imaging Plant Phenolic Compounds

11.3.1 Multispectral Fluorescence Detection

Fluorescence is the emission of a photon of typically longer wavelength as a result of the excitation of a molecule by incident light (excitation light). The excited molecule remains in this excited state typically for nanoseconds before releasing a photon. As part of the energy of the incident photon is dissipated in the form of heat, the emitted photon is typical of lower energy (longer wavelength) than the excitation photon. The difference in wavelengths between the incident and the emitted photons is known as Stokes shift and it is the basis for the discrimination between the excitation and emission light in fluorescence microscopy (Lichtman and Conchello 2005).

The wavelength and intensity of the emitted light can provide two-dimensional (2D) and 3D datasets with information on the spatial and temporal distribution of intrinsically fluorescent plant compounds. UV light, which is needed to excite *p*-coumaric acid, ferulic acid, and various polyphenols, has limited penetrance in plant tissues and it is highly damaging in long-exposure observations. To overcome this limitation, it is possible to use multiphoton-based excitation combined with spectral analysis. In this case, two infrared photons of the same combined energy of a UV photon can successfully excite plant (poly)phenols, with increased penetration into tissues and reduced photodamage.

Most (poly)phenols can be excited in the 340–360 nm range for fluorescence imaging. Hydroxycinnamoyl acids and some flavonoids (quercetin, apigenin) have been reported to emit in the blue region (450–500 nm), vanillin (phenolic aldehyde from vanilla) in green (500–550 nm), and mangiferin (C-glucosylxanthone, a pigment found in wild coffee species) in yellow-orange (550–600 nm), whereas anthocyanin and anthocyanindins emit in the red region (600–630 nm) (Conejero et al. 2014; Talamond et al. 2015) (Figure 11.1). Partially overlapping emission spectra can be better separated by using linear unmixing algorithms when reference emission spectra of isolated phenolic compounds are available (Conejero et al. 2014).

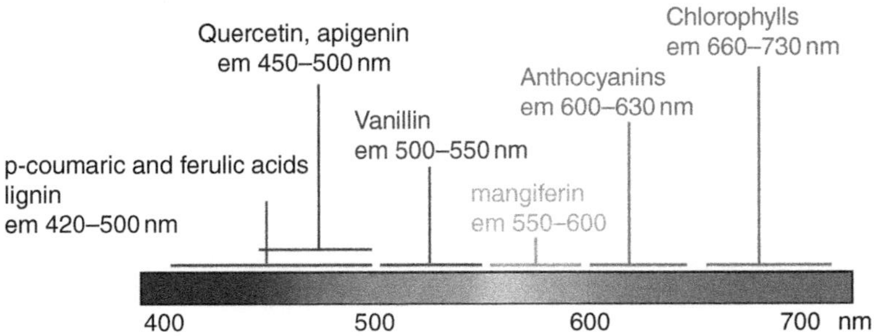

Figure 11.1 Emission wavelengths of some (poly)phenolic compounds and chlorophylls. (*See insert for color representation of the figure.*)

11.3.2 Fluorescence Lifetime Microscopy (FLIM)

The intrinsic fluorescence from specialized metabolites can be used to generate not only simple intensity-based images, but also multidimensional quantitative maps of their fluorescent lifetime using FLIM. When a fluorophore absorbs energy from light, an electron is excited and makes a transition to a higher-energy electronic state. This electron can then return to the ground state via two decay processes, radiative or nonradiative decay. Radiative decay is accompanied by photon emission, whereas nonradiative decay describes the de-excitation processes where the energy is dissipated by thermal processes or energy transfer (Chang et al. 2007). Both decay processes play a role in determining a fluorophore's fluorescence lifetime. Fluorescence lifetime reflects how long it takes the electron to go from the excited energy state to the ground state. This delay is in the order of picoseconds to nanoseconds (So and Dong 2001) and can be measured by fluorescence lifetime electronics. Temperature, pH, oxygen concentration, molecular associations, and ion concentration can all modulate the intrinsic lifetime of the fluorophore. FLIM generates images that reflect the differences in the fluorescence decay rate of a given fluorophore, informing about the fluorophore microenvironment or molecular binding state.

Most FLIM systems use time-correlated single photon counting (TCSPC) to determine the fluorescence lifetime (Becker 2012). This means that the instrument measures the time between fluorophore excitation by a pulsed laser and the detection of the emitted photon by the detector. This is repeated multiple times for each pixel in the image to estimate the fluorescence decay rate of the fluorophore. In addition, current FLIM systems can acquire fluorescence lifetime data for multiple emission wavelengths simultaneously (multispectral detectors).

Thus, a typical FLIM system consists of a pulsed laser source (either a diode laser for single photon excitation or an infrared laser for two-photon excitation), a single photon-sensitive detector (single or a multispectral detector), and a TCSPC module (Becker et al. 2007).

Although FLIM can be done by one-photon or multiphoton excitation, in the case of plant tissues, excitation with infrared light leads to less scattering and absorption by the sample and deeper penetration into the tissue with reduced photodamage.

11.3.3 Raman Microscopy

The Raman effect, named after the Indian physicist Sir Chandrasekhara Venkata Raman (Raman and Krishnan 1928), refers to energy changes in the light scattered by a molecule due to molecular vibrations (e.g., bond stretching, rotation, torsion) that can then inform on the chemical nature of such molecule. In other words, when a molecule interacts with a photon, the scattered light resulting from this interaction can have either the same energy (called Rayleigh scattering) or different energy (called Raman scattering) from the incident photon. If the energy of the scattered light is lower than that of the incident light, it is called Stokes scattering; if it is higher, it is called anti-Stokes scattering (Figure 11.2). This change in the energy of the Raman scattered light contains information about the chemical nature of the functional groups and bonds within the molecule (Syed and Smith 2017).

The Raman effect and fluorescence have a feature in common (Figure 11.2). As a result of the interaction between a photon and a molecule, a photon with a different energy from that of the incident light is produced. However, whereas the Raman effect is based on light scattering and can occur at any frequency of incident light, during fluorescence the incident light (usually within a range of wavelengths) is completely absorbed by the electrons in the molecule, transferring them to an excited state. After a resonance lifetime in the range of nanoseconds, the system goes back to a lower energy state, releasing a photon. Raman offers some advantages over fluorescence imaging since (i) it does not require fluorescent molecules; (ii) it less impacted by photobleaching (a common problem in fluorescence imaging); and (iii) Raman peaks are narrower than those resulting from fluorescence emission, enabling the detection of multiple molecular fingerprints (Syed and Smith 2017). This latter point is relevant considering that 90% of the Raman peaks from most biological molecules are in the spectral region of 500 to 1800 cm^{-1} (Syed and Smith 2017).

The Raman effect takes place spontaneously as photons are scattered by the material, and therefore this effect is commonly referred to as spontaneous Raman scattering.

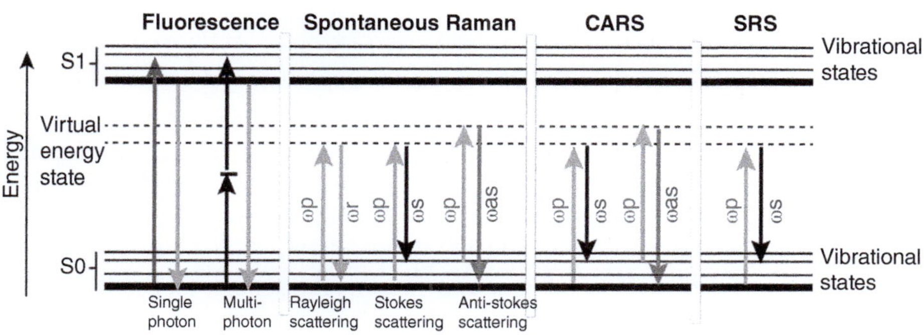

Figure 11.2 Jablonski diagrams showing energy transitions during fluorescence (single and multiphoton excitation) and Raman scattering (spontaneous Raman, coherent anti-Stokes Raman scattering or CARS, and stimulated Raman scattering or SRS). When excited by the incident photon, the scattering molecule is taken to an excited energy state known as a virtual energy state. The molecule decays back to its ground electronic state (S0), which contains multiple vibrational energy levels. If the molecule decays to the same vibrational level, there is no transfer of energy between the photon and the molecule (elastic scattering or Rayleigh scattering), and the wavelength of the scattered photon remains the same as the incident photon. If there is energy transfer between the incident photon and the scattering molecule, causing a change in the molecule vibrational energy, inelastic scattering occurs. If the photon transfers energy to the scattering molecule and the molecule gains vibrational energy, the wavelength of the scattered photon increases (Stokes shift) as it loses energy. If the energy is transferred from the molecule to the photon, the energy of the scattered photon increases and its wavelength decreases (anti-Stoke shift). (*See insert for color representation of the figure.*)

However, spontaneous Raman occurs at a very low frequency (one scattered photon every 10^6 to 10^8 incident photons), drastically limiting its applications for microscopy imaging. In addition, Raman-based imaging is affected by fluorescence background interference (Zhang et al. 2015). To overcome these limitations, approaches that enhance spontaneous Raman have been developed in recent years. One of the most commonly used approaches is coherent Raman scattering (CRS), which relies on either coherent anti-Stokes Raman scattering (CARS) or stimulated Raman scattering (SRS) (Figure 11.2).

In CARS, the incident and output photons exchange energy with each other while the molecule remains in the ground state after this interaction (Cheng and Xie 2015; Min et al. 2011). Most CARS-based imaging devices collect information from three laser fields: pump (ωp), Stokes (ωs), and probe (ωpr). In most cases, the pump and probe fields are generated from the same laser source. The pump–probe beam and the Stokes beam (ωS) interact with molecules in the specimen, generating anti-Stokes scattered photons. The difference in frequency between the pump and Stokes fields (ωp − ωs) informs about the specific vibrational mode of the sample and its chemical nature. The CARS signal is calculated based on ωpr + (ωp − ωs) (Syed and Smith 2017).

Whereas the spontaneous Raman signal is isotropic, the CARS signal is largely directional, facilitating its detection. In addition, because this technique is based on the detection of anti-Stokes signals of shorter wavelengths than the incident light, CARS imaging is not affected by the interference of fluorescence background. However, CARS does suffer from an inherent nonresonant background that makes it difficult to retrieve the signal. Several improvements to eliminate nonresonant background have been proposed and new ones continue to be developed (Zhang et al. 2015; Cheng and Xie 2004)

Unlike CARS, during SRS imaging molecules do exchange energy with incident laser fields after interaction (Min et al. 2011). In SRS, two laser beams at ωp (pump) and ωS (Stokes) focus on the sample. When the difference frequency, Δω = ωp − ωS (Raman shift), matches a particular molecular vibrational frequency, there is an amplification of the spontaneous Raman signal. Consequently, the intensity of the Stokes beam experiences a gain (stimulated Raman gain), and the intensity of the pump beam experiences a loss (stimulated Raman loss) (Freudiger et al. 2008). SRS is not affected by nonresonant background and can drastically increase the Raman signal over the interference caused by intrinsic fluorescence typical of plant tissues. In addition, unlike CARS, the SRS signal depends linearly on the concentration of the molecule and therefore it is suitable for obtaining concentration maps of multiple molecular species (Syed and Smith 2017).

Raman-based microscopes can obtain Raman spectral information in plant tissues and individual cells, informing about the spatial distributions and chemical nature of multiple cellular components simultaneously. Thus, a Raman image is composed of thousands of spectra in which each pixel carries its own chemical information.

The resolution in Raman imaging is limited by light diffraction, as in any other form of light-based microscopy. Raman spectroscopy uses laser wavelengths ranging from UV to near infrared. Raman scattering intensity is proportional to $1/\lambda^4$, where λ is the laser wavelength, and therefore shorter laser wavelengths give the best signal-to-noise ratio (Freudiger et al. 2008). However, excitation in the UV range can cause photodamage in biological samples, and therefore most Raman-based microscopes use green or near infrared lasers (most commonly 532 nm and 785 nm, respectively).

11.4 Polyphenols and Microscopy Imaging

11.4.1 Anthocyanins

Anthocyanins are glycosylated flavonoid-derived pigments consisting of an anthocyanidin backbone with conjugated sugar and acyl (both aliphatic and aromatic) groups. The most common anthocyanidins are pelargonidin, cyanidin, delphinidin, peonidin, malvidin, and petunidin (Brouillard 1982); as common features, they share the presence of two aromatic rings (A and B) and an oxygenated heterocyclic ring (C) (Figure 11.3a). Besides the structure of the anthocyanidin core, the type, quantity, and position of conjugated moieties contribute to anthocyanin variation. In fact, more than 600 anthocyanins have been described in nature (Smeriglio et al. 2016).

Anthocyanins accumulate inside vacuoles of flowers, fruits, and seeds, as well as vegetative tissues (Tanaka et al. 2008). In flowers, anthocyanin coloration aids in attracting pollinators, whereas in vegetative tissues anthocyanins play photoprotecting roles by absorbing excess visible and UV light (Tanaka et al. 2008).

Unlike other flavonoids, anthocyanins strongly absorb visible light (between 465 and 550 nm) and exhibit a large variation of colors, typically orange, red, purple, or blue (Glover and Martin 2012). Their color and hue are influenced by the vacuolar environment, mostly by pH. This is because, depending on the pH, the anthocyanidin core can adopt different chemical forms. At pH 1, the reddish flavylium ion with a positively charged oxygen atom in the C ring is predominant; at pH 2–4, the blue quinoidal forms are more abundant, whereas at pH 5–6 the colorless carbinol base and chalcone forms become the dominant species (Castañeda-Ovando et al. 2009). In addition, the color and stability of anthocyanins is also dependent on the presence of copigments (other organic compounds or metallic ions), and glycosylation and/or acylation of the anthocyanidin backbone.

Anthocyanins are synthesized in the cytoplasm and subsequently accumulate inside vacuoles (Harborne and Williams 2000; Grotewold 2006). At the acidic pH of the vacuole, anthocyanins are deeply pigmented and therefore easy to visualize. However, the imaging of anthocyanins in the neutral pH of the cytoplasm has been more challenging. Besides being colored, anthocyanins also show intrinsic fluorescence in the 600–650 nm emission range (red), even at neutral pH (Poustka et al. 2007; Chanoca et al. 2015), allowing for the detection of both vacuolar and cytoplasmic anthocyanins in intact plant cells through fluorescence microscopy (Chanoca et al. 2016, 2015).

In addition, FLIM has also been shown to be very useful for the detection and analysis of anthocyanins *in situ*. Fluorescence lifetime is an intrinsic property of a fluorophore independent of its concentration, sample thickness, or fluorescence intensity. However, the fluorescence decay rate is sensitive to the environmental conditions surrounding the fluorophore, its association to other molecules, molecular modifications, conformational changes, and pH (Chang et al. 2007). Consistently, although the wavelength of the emitted light from excited anthocyanin molecules in *Arabidopsis* does not seem to be affected by pH or the type of decorations of the cyanidin backbone, the fluorescence lifetime is (Chanoca et al. 2015).

The fluorescence decay of anthocyanins measured both *in vitro* and *in vivo* is best modeled as a double exponential function, with the majority of photons showing shorter

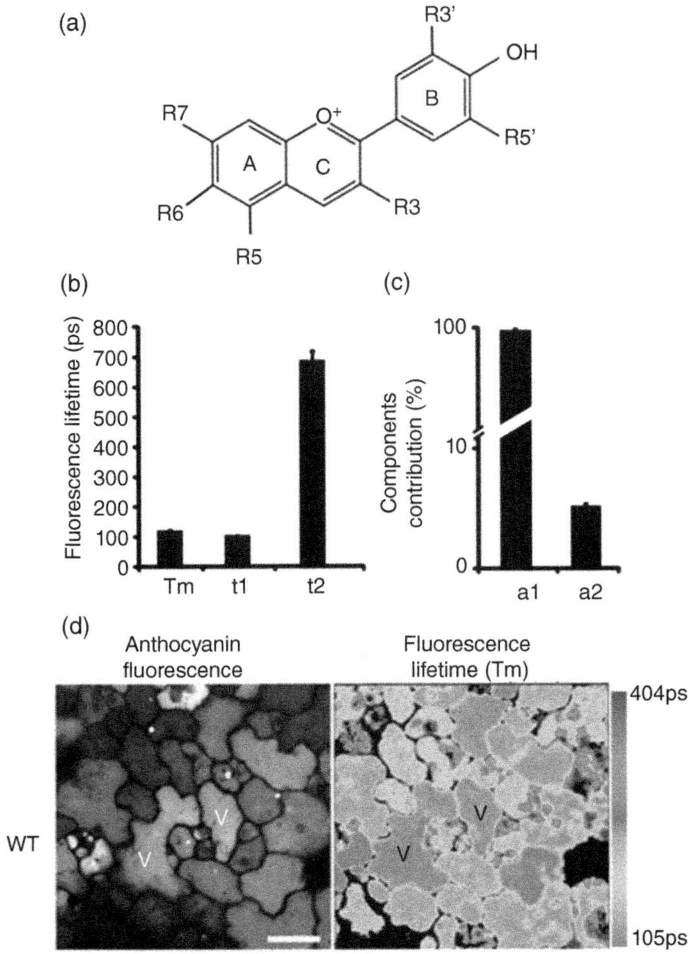

Figure 11.3 Anthocyanin imaging. (a) General structure of an anthocyanidin backbone. (b) Fluorescence lifetime of isolated wild-type anthocyanins. τm, weighted mean lifetime. $\tau 1$ and $\tau 2$ are the lifetimes in picoseconds of the two identified fluorescence lifetime components. (c) Relative contributions of the two identified components. (d) Steady-state fluorescence image (left panel) and pseudo-colored image according to anthocyanin fluorescence lifetime values (right panel) of cotyledon pavement cells. Color map shows τm, ranging from 105 ps to 404 ps. V, vacuole. Scale bar: 20 μm. *Source:* (a, b, c) Modified from Chanoca et al. 2015. © American Society of Plant Biologists. (d) Reproduced from Chanoca et al. 2015. © American Society of Plant Biologists. (*See insert for color representation of the figure.*)

fluorescence lifetimes and a minor component showing longer fluorescence lifetimes (Figure 11.3b and c) (Chanoca et al. 2016, 2015).

As FLIM can quantify the frequency of photons emitted at different time intervals from the excited fluorophore, it is a powerful quantitative tool to detect changes in the behavior of the same fluorophore under different conditions or of two closely related molecular species. For example, FLIM imaging allows us to differentiate the vacuolar pool of anthocyanins from the cytoplasmic pool (Chanoca et al. 2018, 2016). The shorter fluorescence lifetime values (mean lifetime = 110 ps) were mostly found in the central vacuole (pH ~4.9

to 6.0) whereas the longer fluorescence lifetime values (mean lifetime = 400 ps) were more abundant at the peripheral cytoplasm (pH ~7.1 to 7.8) of epidermal cells of *Arabidopsis* cotyledons accumulating anthocyanins (Figure 11.3d) (Chanoca et al. 2018, 2016). Consistently with these observations made in tissues, the fluorescence lifetime of extracted anthocyanins in *in vitro* conditions also tends to be shorter at lower pH, with the shortest fluorescence lifetime values at pH 4.5 (Chanoca et al. 2016).

In addition, decorations of the anthocyanidin backbone also seems to alter fluorescence decay. For example, in the *Arabidopsis 5gt* mutant, which is unable to glucosylate anthocyanidins at the 5-*O* position of the A ring, cyanidin-3-sambubiose exhibits a longer fluorescence lifetime than its related coumarylated derivatives (Chanoca et al. 2015).

For the chemical determination of anthocyanins in solution, previous studies have generally used the Raman spectral features between 1300 and 1650 cm^{-1} (Zaffino et al. 2015). However, in the context of tissues and intact cells, discrimination of specific signatures can be more difficult due to overlapping peaks in this region. For example, the 1545 cm^{-1} peak of anthocyanins overlaps with a carotenoid peak at 1525 cm^{-1}, whereas the anthocyanin peaks between 1250 and 1450 cm^{-1} overlap with chlorophyll and fatty acid signatures. The anthocyanin peaks at 539, 623, and 733 cm^{-1} have been used to follow changes in anthocyanin accumulation during various types of abiotic stresses in Coleus plants (*Plectranthus scutellarioides*) (Altangerel et al. 2017)

11.4.2 Lignin

Lignin is a complex polymer usually made of three monolignols; *p*-coumaryl, coniferyl, and sinapyl alcohols synthesize in the cytoplasm (Li and Chapple 2010). Once exported to the apoplast, peroxidases and laccases catalyze monolignol polymerization through free radical reactions within the cell wall, impregnating the cell wall matrix by forming covalent bonds to the surrounding noncellulosic carbohydrates (Voxeur et al. 2015). In addition, ferulic acid binds to lignin, polysaccharides, and structural proteins of cell walls, crosslinking these components. Cell types that undergo cell wall lignification include tracheary elements and other xylem cells, fibers, and endodermis. Lignification shows variations in different cells, including timing, associated enzymes and substrates, local deposition patterns, and monolignol composition (Barros et al. 2015), highlighting the importance of analyzing lignification in individual cell types and in the context of intact tissues (Meents et al. 2018).

Lignin absorbs light in the UV range (absorption maxima close to 350 nm) as well as in the visible range. However, lignin is degraded by short UV radiation; therefore, excitation in the blue range (450–490 nm) is preferred since chemical degradation of lignin is less pronounced with wavelengths longer than 434 nm (Kataoka et al. 2007). Lignin fluorescence emission is multimodal, indicating that there is more than one fluorescent structure within the complex lignin polymers. In fact, a study employing multispectral analysis and spectral unmixing of fluorescence signals from tracheids in normal and compression wood of Pinus radiata described five distinct emission peaks in the 420–600 nm region (Donaldson et al. 2010).

FLIM of lignin has also been used to analyze local heterogeneities in normal and compression pine wood (Donaldson and Radotic 2013). The fluorescence lifetime of pine wood sections fits a multiexponential decay with three components, with a tendency towards shorter lifetimes in the compression wood compared to the normal wood (mean lifetime of 473 ps in normal and 418 ps in compression wood) (Donaldson and Radotic 2013).

Although imaging of lignin intrinsic fluorescence is a noninvasive and very sensitive approach, it does not allow for chemical analysis of lignin synthesis and deposition. For this purpose, Raman-based imaging is more commonly used. Specific Raman vibrational signatures from plant cell walls can discriminate among lignin, cellulose, xylans, and pectin (Zhu et al. 2018; Zeng et al. 2017). Two dominant Raman signatures at 1600 cm^{-1} and 1660 cm^{-1} are attributed to aromatic ring stretching in the phenyl rings and to the ethenyl bond of coniferyl alcohol (monolignol), respectively (Prats-Mateu et al. 2018; Agarwal 1999; Gierlinger and Schwanninger 2006). Since the ethenyl bond is lost when coniferyl alcohol is incorporated into lignin (guaiacyl unit), the ratio of the Raman bands at 1660 and 1600 cm^{-1} has been used to estimate the polymerization degree of lignin (Morikawa et al. 2010). However, other studies have shown that this ratio can be altered due to photodamage caused by laser exposure, highlighting the importance of careful determination of data acquisition conditions (Prats-Mateu et al. 2018).

11.4.3 (Poly)phenolic Compounds in Cell Walls, Cuticles, and Suberin

Many phenolic compounds, predominantly derivatives of hydroxycinnamic acids, are found impregnating in plant cell walls. The fluorescence decay of ferulic acid bound to the cell walls of wheat leaves has been shown to exhibit four components ranging from 110 to 5600 ps, with a mean lifetime of 660 ps. The different lifetime components are likely to represent pools of ferulic acid molecules in different molecular environments (Meyer et al. 2003).

Hydroxycinnamic acids such as p-coumaric acid and ferulic acid have also been detected in the cuticle and suberin layers of plant cell walls. Cuticles are viscoelastic polymers that cover the surface of the plant; they are made of a cutin matrix infused with waxes, hydroxycinnamic acid-derivatives, and silica (in grasses), which increase stiffness but decreases extensibility; and cell wall polysaccharides, which are responsible for its elastic behavior (Khanal and Knoche 2017; Fich et al. 2016). In fact, the presence of caffeic acid-derivatives in the cuticle of mosses indicates that the impregnation of cuticles with phenolic compounds was an important acquisition for plant adaptation to terrestrial environments (Renault et al. 2017). Cuticles often show intrinsic fluorescence due to the presence of hydroxycinnamic acids. During development, the composition and properties of the cuticle change extensively in response to both developmental and environmental factors (Ingram and Nawrath 2017). Cutin is composed of a network of interlinked omega hydroxy acids derived from C18 and C16 fatty acids, forming a polyester polymer of indeterminate size (Fich et al. 2016). The presence and distribution of p-coumaric acid in tomato cuticles has been assessed by Raman microscopy (Prats Mateu et al. 2016)

Suberin impregnates the cell walls as continuous lamellas of specific cell types, such as root endodermis and exodermis. Similar to cutin, suberin is also a polyester of predominantly C16 and C18 fatty acids but it contains higher amounts of ester-bound aromatic components, predominantly ferulic acid (Andersen et al. 2015). The endodermal cells (innermost layer of the root cortex), which act as a diffusion barrier for water and minerals, first differentiate local impregnation of their primary cell wall (called Casparian strips) that seals the apoplastic space. Although Casparian strips were thought to be made of suberin, more recent studies in *Arabidopsis* showed that they are lignin impregnations connected to the local activity of specific peroxidases and the oxidases (Hosmani et al. 2013; Lee et al. 2013). Suberin is deposited later in endodermal cell walls as continuous inner lamellas (Doblas et al. 2017; Andersen et al. 2015). Interestingly, lignin and suberin do not seem to coexist in the same cell wall domains and the absence of early lignin deposition in Casparian strips is correlated to earlier and more extensive deposition of suberin (Hosmani et al. 2013), suggesting compensatory mechanisms among different strategies for cell wall impregnation. In grasses, there is a third stage of endodermal cell wall modification in which additional layers of cellulose heavily impregnated in lignin (tertiary cell wall) are deposited in the inner periclinal and the radial sides of the endodermal cells. Grass endodermal cell walls contain lignin, suberin, and silica (Zeier et al. 1999; Soukup et al. 2017). The deposition of suberin and lignin in the endodermis of maize roots has been studied by Raman microscopy (Man et al. 2018)

11.5 Future Challenges and Opportunities in Imaging Plant Metabolites

The distribution of plant metabolites in tissues has been traditionally analyzed by light microscopy using specific dyes, intrinsic fluorescence, and addition of fluorescent precursors such as labeled monolignols (Nadiminti et al. 2015; Buda et al. 2009). Noninvasive, nondestructive techniques based on chemical imaging, such as Raman microscopy, are also being applied to the analysis of plant metabolites (Zeng et al. 2016; Man et al. 2018). Although all these imaging approaches independently have provided useful means to analyze (poly)phenolic compounds' distribution and trafficking, future developments will include platforms able to integrate multiparametric imaging modalities. The simultaneous and integrated imaging of plant cells and tissues by multiple noninvasive, label-free approaches (e.g. multiple multispectral imaging, FLIM, Raman microscopy) will allow for coregistration of multiple fingerprints with high sensitivity, and spatial and temporal resolution to inform on the chemical nature, molecular environment, synthesis, trafficking, and accumulation sites, as well as binding state from multiple compounds in their native conditions.

11.6 Acknowledgments

Research on imaging of plant metabolites in the Otegui laboratory is supported by grant DOE (BER) DE-SC0019013.

References

Agarwal, U. (1999). An overview of Raman spectroscopy as applied to lignocellulosic materials. In: *Advances in Lignocellulosics Characterization* (ed. D. Argyropoulos), 201–225. Tappi Press.

Agati, G., Azzarello, E., Pollastri, S., and Tattini, M. (2012). Flavonoids as antioxidants in plants: Location and functional significance. *Plant Science* 196: 67–76.

Agati, G., Matteini, P., Goti, A., and Tattini, M. (2007). Chloroplast-located flavonoids can scavenge singlet oxygen. *New Phytologist* 174: 77–89.

Agati, G. and Tattini, M. (2010). Multiple functional roles of flavonoids in photoprotection. *New Phytologist* 186: 786–793.

Altangerel, N., Ariunbold, G.O., Gorman, C., et al. (2017). In vivo diagnostics of early abiotic plant stress response via Raman spectroscopy. *Proceedings of the National Academy of Sciences of the United States of America* 114: 3393–3396.

Andersen, T.G., Barberon, M., and Geldner, N. (2015). Suberization – the second life of an endodermal cell. *Current Opinion in Plant Biology* 28: 9–15.

Barros, J., Serk, H., Granlund, I., and Pesquet, E. (2015). The cell biology of lignification in higher plants. *Annals of Botany* 115: 1053–74.

Becker, W. (2012). Fluorescence lifetime imaging – techniques and applications. *Journal of Microscopy* 247: 119–136.

Becker, W., Bergmann, A., and Biskup, C. (2007). Multispectral fluorescence lifetime imaging by TCSPC. *Microscopy Research and Technique* 70: 403–409.

Brillouet, J.-M., Romieu, C., Schoefs, B., et al. (2013). The tannosome is an organelle forming condensed tannins in the chlorophyllous organs of Tracheophyta. *Annals of Botany* 112: 1003–1014.

Brillouet, J.-M., Verdeil, J.-L., Odoux, E., et al. (2014). Phenol homeostasis is ensured in vanilla fruit by storage under solid form in a new chloroplast-derived organelle, the phenyloplast. *Journal of Experimental Botany* 65: 2427–2435.

Brouillard, R. (1982). Chemical structure of anthocyanins. In: *Anthocyanins as food colors* (ed. P. Markakis). Academic Press.

Buda, G.J., Isaacson, T., Matas, A.J., et al. (2009). Three-dimensional imaging of plant cuticle architecture using confocal scanning laser microscopy. *The Plant Journal* 60: 378–385.

Castañeda-Ovando, A., Pacheco-Hernández, M.D. L., Páez-Hernández, M.E., et al. (2009). Chemical studies of anthocyanins: A review. *Food Chemistry* 113: 859–871.

Cha, S., Zhang, H., Ilarslan, H.I., et al. (2008). Direct profiling and imaging of plant metabolites in intact tissues by using colloidal graphite-assisted laser desorption ionization mass spectrometry. *The Plant Journal* 55: 348–360.

Chang, C.W., Sud, D., and Mycek, M.A. (2007). Fluorescence lifetime imaging microscopy. *Methods in Cell Biology* 81: 495–524.

Chanoca, A., Burkel, B., Eliceiri, K.W., et al. (2018). Imaging vacuolar anthocyanins with fluorescence lifetime microscopy (FLIM). *Methods in Molecular Biology* 1789:131–141.

Chanoca, A., Burkel, B., Kovinich, N., et al. (2016). Using fluorescence lifetime microscopy to study the subcellular localization of anthocyanins. *The Plant Journal* 88: 895–903.

Chanoca, A., Kovinich, N., Burkel, B., et al. (2015). Anthocyanin vacuolar inclusions form by a microautophagy mechanism. *The Plant Cell* 27: 2545–2559.

Cheng, J.-X. and Xie, X.S. (2004). Coherent anti-Stokes Raman scattering microscopy: Instrumentation, theory, and applications. *The Journal of Physical Chemistry B* 108: 827–840.

Cheng, J.-X. and Xie, X.S. (2015). Vibrational spectroscopic imaging of living systems: An emerging platform for biology and medicine. *Science*, 350: aaa8870.

Conejero, G., Noirot, M., Talamond, P., and Verdeil, J.L. (2014). Spectral analysis combined with advanced linear unmixing allows for histolocalization of phenolics in leaves of coffee trees. *Frontiers in Plant Science* 5: 39.

Dettmer, K., Aronov, P.A., and Hammock, B.D. (2007). Mass spectrometry-based metabolomics. *Mass Spectrometry Reviews* 26: 51–78.

Dixon, R.A. and Paiva, N.L. (1995). Stress-induced phenylpropanoid metabolism. *The Plant Cell* 7: 1085.

Doblas, V.G., Geldner, N., and Barberon, M. (2017). The endodermis, a tightly controlled barrier for nutrients. *Current Opinion in Plant Biology* 39: 136–143.

Donaldson, L., Radotić, K., Kalauzi, A., et al. (2010). Quantification of compression wood severity in tracheids of Pinus radiata D. Don using confocal fluorescence imaging and spectral deconvolution. *Journal of Structural Biology* 169: 106–115.

Donaldson, L.A. and Radotic, K. (2013). Fluorescence lifetime imaging of lignin autofluorescence in normal and compression wood. *Journal of Microscopy* 251: 178–87.

Fich, E.A., Segerson, N.A., and Rose, J.K. (2016). The plant polyester cutin: biosynthesis, structure, and biological Roles. *Annual Review of Plant Biology* 67: 207–233.

Freudiger, C.W., Min, W., Saar, B.G., et al. (2008). Label-free biomedical imaging with high sensitivity by stimulated raman scattering microscopy. *Science* 322: 1857–1861.

Gierlinger, N. and Schwanninger, M. (2006). Chemical imaging of poplar wood cell walls by confocal Raman microscopy. *Plant Physiology* 140: 1246–1254.

Glover, B.J. and Martin, C. (2012). Anthocyanins. *Current Biology* 22: R147–R150.

Grotewold, E. (2006). The genetics and biochemistry of floral pigments. *Annual Review of Plant Biology* 57: 761–780.

Gunnaiah, R., Kushalappa, A.C., Duggavathi, R., et al. (2012). Integrated metabolo-proteomic approach to decipher the mechanisms by which wheat QTL (Fhb1). contributes to resistance against *Fusarium graminearum*. *PLOS ONE* 7: e40695.

Harborne, J.B. and Williams, C.A. (2000). Advances in flavonoid research since 1992. *Phytochemistry* 55: 481–504.

Hill, C.B., Czauderna, T., Klapperstuck, M., et al. (2015). Metabolomics, standards, and metabolic modeling for synthetic biology in plants. *Frontiers in Bioengineering and Biotechnology* 3: 167.

Hosmani, P.S., Kamiya, T., Danku, J., Naseer, S., et al. (2013). Dirigent domain-containing protein is part of the machinery required for formation of the lignin-based Casparian strip in the root. *Proceedings of the National Academy of Sciences of the United States of America* 110: 14498–14503.

Hsieh, K. and Huang, A.H. C. (2007). Tapetosomes in Brassica tapetum accumulate endoplasmic reticulum-derived flavonoids and alkanes for delivery to the pollen surface. *The Plant Cell* 19: 582–596.

Ingram, G. and Nawrath, C. (2017). The roles of the cuticle in plant development: organ adhesions and beyond. *Journal of Experimental Botany* 68: 5307–5321.

Kataoka, Y., Kiguchi, M., Williams, R.S., and Evans Philip, D. 2007. Violet light causes photodegradation of wood beyond the zone affected by ultraviolet radiation. *Holzforschung* 61: 23–27.

Khanal, B.P. and Knoche, M. (2017). Mechanical properties of cuticles and their primary determinants. *Journal of Experimental Botany* 68: 5351–5367.

Kim, H.K., Choi, Y.H., and Verpoorte, R. (2011). NMR-based plant metabolomics: where do we stand, where do we go? *Trends in Biotechnology* 29: 267–275.

Korte, A.R., Yandeau-Nelson, M.D., Nikolau, B.J., and Lee, Y.J. (2015). Subcellular-level resolution MALDI-MS imaging of maize leaf metabolites by MALDI-linear ion trap-Orbitrap mass spectrometer. *Analytical and Bioanalytical Chemistry* 407: 2301–2309.

Lee, Y., Rubio, M.C., Alassimone, J., and Geldner, N. (2013). A mechanism for localized lignin deposition in the endodermis. *Cell* 153: 402–412.

Li, X. and Chapple, C. (2010). Understanding lignification: Challenges beyond monolignol biosynthesis. *Plant Physiology* 154: 449.

Lichtman, J.W. and Conchello, J.A. (2005). Fluorescence microscopy. *Nature Methods* 2: 910–919.

Man, Y., Zhao, Y., Ye, R., et al. (2018). In vivo cytological and chemical analysis of Casparian strips using stimulated Raman scattering microscopy. *Journal of Plant Physiol* 220: 136–144.

Meents, M.J., Watanabe, Y., and Samuels, A.L. (2018). The cell biology of secondary cell wall biosynthesis. *Annals of Botany* 121: 1107–1125.

Meyer, S., Cartelat, A., Moya, I., and Cerovic, Z.G. (2003). UV-induced blue-green and far-red fluorescence along wheat leaves: a potential signature of leaf ageing. *Journal of Experimental Botany* 54: 757–769.

Min, W., Freudiger, C.W., Lu, S., and Xie, X.S. (2011). Coherent nonlinear optical imaging: Beyond fluorescence microscopy. *Annual Review of Physical Chemistry* 62: 507–530.

Morikawa, Y., Yoshinaga, A., Kamitakahara, H., et al. (2010). Cellular distribution of coniferin in differentiating xylem of Chamaecyparis obtusa as revealed by Raman microscopy. *Holzforschung* 64: 61–67.

Nadiminti, P.P., Rookes, J.E., Boyd, B.J., and Cahill, D.M. (2015). Confocal laser scanning microscopy elucidation of the micromorphology of the leaf cuticle and analysis of its chemical composition. *Protoplasma* 252: 1475–1486.

Polster, J., Dithmar, H., Burgemeister, R., et al. (2006). Flavonoids in plant nuclei: detection by laser microdissection and pressure catapulting (LMPC), in vivo staining, and uv–visible spectroscopic titration. *Physiologia Plantarum* 128: 163–174.

Poustka, F., Irani, N.G., Feller, A., et al. (2007). A trafficking pathway for anthocyanins overlaps with the endoplasmic reticulum-to-vacuole protein-sorting route in *Arabidopsis* and contributes to the formation of vacuolar inclusions. *Plant Physiology*, 145.

Prats-Mateu, B., Bock, P., Schroffenegger, M., et al. (2018). Following laser induced changes of plant phenylpropanoids by Raman microscopy. *Scientific Reports* 8: 11804–11804.

Prats Mateu, B., Hauser, M.T., Heredia, A., and Gierlinger, N. (2016). Waterproofing in *Arabidopsis*: Following phenolics and lipids In situ by confocal Raman microscopy. *Frontiers in Chemistry* 4: 10

Qin, L., Zhang, Y., Liu, Y., et al. (2018). Recent advances in matrix-assisted laser desorption/ionisation mass spectrometry imaging (MALDI-MSI) for in situ analysis of endogenous molecules in plants. *Phytochemical Analysis* 29: 351–364.

Quideau, S., Deffieux, D., Douat-Casassus, C., and Pouységu, L. (2011). Plant Polyphenols: Chemical Properties, Biological Activities, and Synthesis. *Angewandte Chemie International Edition* 50: 586–621.

Raman, C.V. and Krishnan, K.S. (1928). A new type of secondary radiation. *Nature* 121: 501.

Renault, H., Alber, A., Horst, N.A., et al. (2017). A phenol-enriched cuticle is ancestral to lignin evolution in land plants. *Nature Communications* 8: 14713–14713.

Smeriglio, A., Barreca, D., Bellocco, E., and Trombetta, D. (2016). Chemistry, pharmacology and health benefits of anthocyanins. *Phytotherapy Research* 30: 1265–1286.

So, P.T.C. and Dong, C.Y. (2001). Fluorescence spectrophotometry. *Encyclopedia of Life Sciences*. Chichester: Wiley.

Soares, M.S., Da Silva, D.F., Forim, M.R., et al. (2015). Quantification and localization of hesperidin and rutin in *Citrus sinensis* grafted on *C. limonia* after *Xylella fastidiosa* infection by HPLC-UV and MALDI imaging mass spectrometry. *Phytochemistry* 115: 161–170.

Soukup, M., Martinka, M., Bosnic, D., et al. (2017). Formation of silica aggregates in sorghum root endodermis is predetermined by cell wall architecture and development. *Annals of Botany* 120: 739–753.

Sturtevant, D., Lee, Y.-J., and Chapman, K.D. (2016). Matrix assisted laser desorption/ionization-mass spectrometry imaging (MALDI-MSI) for direct visualization of plant metabolites in situ. *Current Opinion in Biotechnology* 37: 53–60.

Syed, A. and Smith, E.A. (2017). Raman imaging in cell membranes, lipid-rich organelles, and lipid bilayers. *Annual Review of Analytical Chemistry* 10: 271–291.

Talamond, P., Verdeil, J.-L., and Conéjéro, G. (2015). Secondary metabolite localization by autofluorescence in living plant cells. *Molecules* 20: 5024.

Tanaka, Y., Sasaki, N., and Ohmiya, A. (2008). Biosynthesis of plant pigments: anthocyanins, betalains and carotenoids. *The Plant Journal* 54: 733–749.

Voxeur, A., Wang, Y., and Sibout, R. (2015). Lignification: different mechanisms for a versatile polymer. *Current Opinion in Plant Biology* 23: 83–90.

Zaffino, C., Russo, B., and Bruni, S. (2015). Surface-enhanced Raman scattering (SERS) study of anthocyanidins. *Spectrochimica Acta A Molecular and Biomolecular Spectroscopy* 149: 41–47.

Zeier, J., Ruel, K., Ryser, U., and Schreiber, L. (1999). Chemical analysis and immunolocalisation of lignin and suberin in endodermal and hypodermal/rhizodermal cell walls of developing maize (*Zea mays* L.) primary roots. *Planta* 209: 1–12.

Zeng, Y., Himmel, M.E., and Ding, S.-Y. (2017). Visualizing chemical functionality in plant cell walls. *Biotechnology for Biofuels* 10: 263–263.

Zeng, Y., Yarbrough, J.M., Mittal, A., et al. (2016). In situ label-free imaging of hemicellulose in plant cell walls using stimulated Raman scattering microscopy. *Biotechnology for Biofuels* 9: 256.

Zhang, C., Zhang, D., and Cheng, J.-X. (2015). Coherent Raman scattering microscopy in biology and medicine. *Annual Review of Biomedical Engineering* 17: 415–445.

Zhao, J. and Dixon, R.A. (2010). The 'ins' and 'outs' of flavonoid transport. *Trends Plant Sci* 15: 72–80.

Zhao, J., Pang, Y., and Dixon, R.A. (2010). The mysteries of proanthocyanidin transport and polymerization. *Plant Physiology* 153: 437–443.

Zhu, N., Wu, D., and Chen, K. (2018). Label-free visualization of fruit lignification: Raman molecular imaging of loquat lignified cells. *Plant Methods* 14: 58.

Index

Entries in *italics* relate to figures and schemes. Entries in **bold** relate to tables.

Recent Advances in Polyphenol Research, Volume 7, First Edition. Edited by Jess Dreher Reed, Victor Armando Pereira de Freitas, and Stéphane Quideau.

q

q

r

s

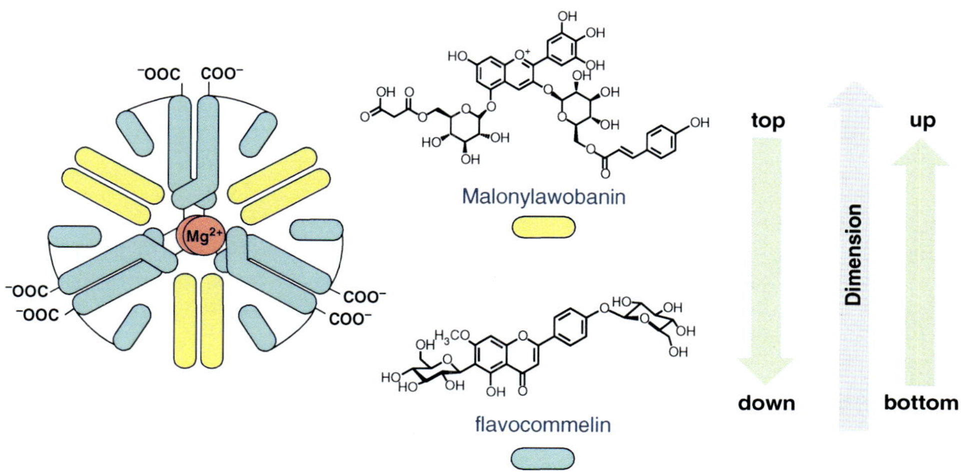

Figure 1.1 Sketch of the metalloanthocyanin responsible for the color in *Cummelina communis*. By courtesy of Prof. Kumi Yoshida. The building blocks self-associate to create the supramolecule in a bottom-up approach. *Source:* Courtesy of Prof. Kumi Yoshida.

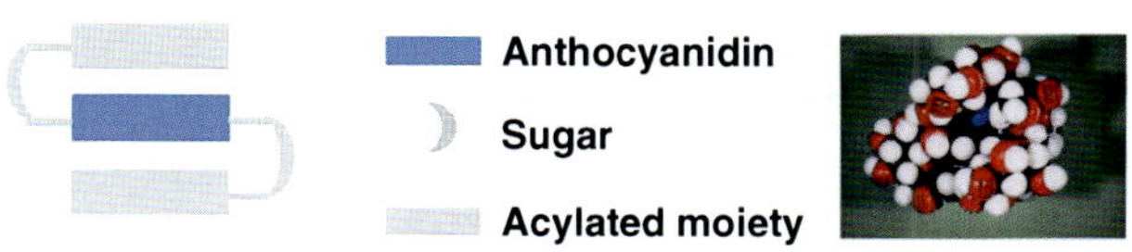

Scheme 1.7 Sketch representing the intramolecular copigmentation in polyacylated anthocyanins; CPK models of heavenly blue anthocyanin. *Source:* Mendoza et al. 2018.

Recent Advances in Polyphenol Research, Volume 7, First Edition. Edited by Jess Dreher Reed, Victor Armando Pereira de Freitas, and Stéphane Quideau.

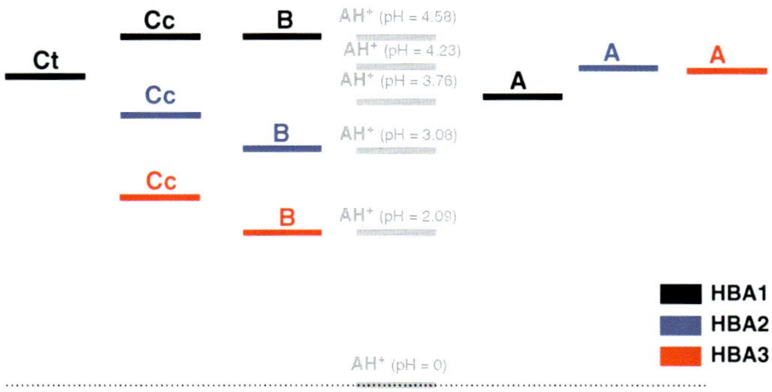

Scheme 1.8 Energy level diagrams of HBA**1** (black), HBA**2** (blue), and HBA**3** (red). In HBA**1** there is an inversion of the relative stability between the quinoidal base and hemiketal. The energy levels of **Ct** cannot be measured with the necessary accuracy due to some decomposition in HBA**2** and HBA**3** for longer reaction times. *Source:* Mendoza et al. 2018.

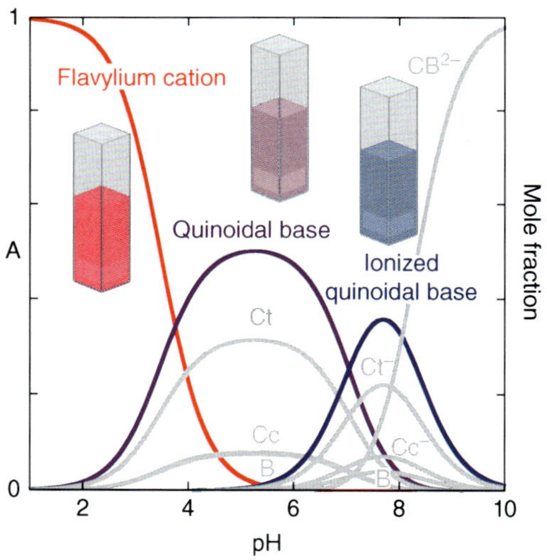

Figure 1.5 Mole fraction distribution of heavenly blue anthocyanin. *Source:* Mendoza et al. 2018.

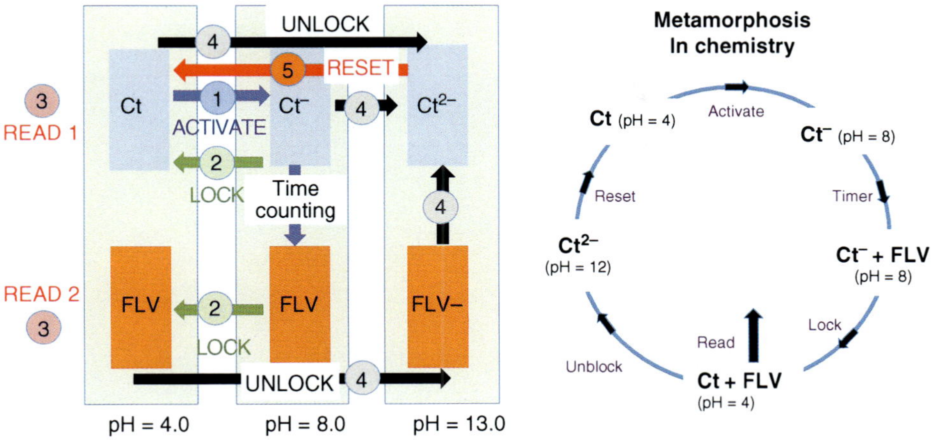

Scheme 1.12 Illustrating the concept of a timer with reset capacity through a metamorphosis cycle. *Source:* Slavcheva, et al. 2018. © 2018 Elsevier.

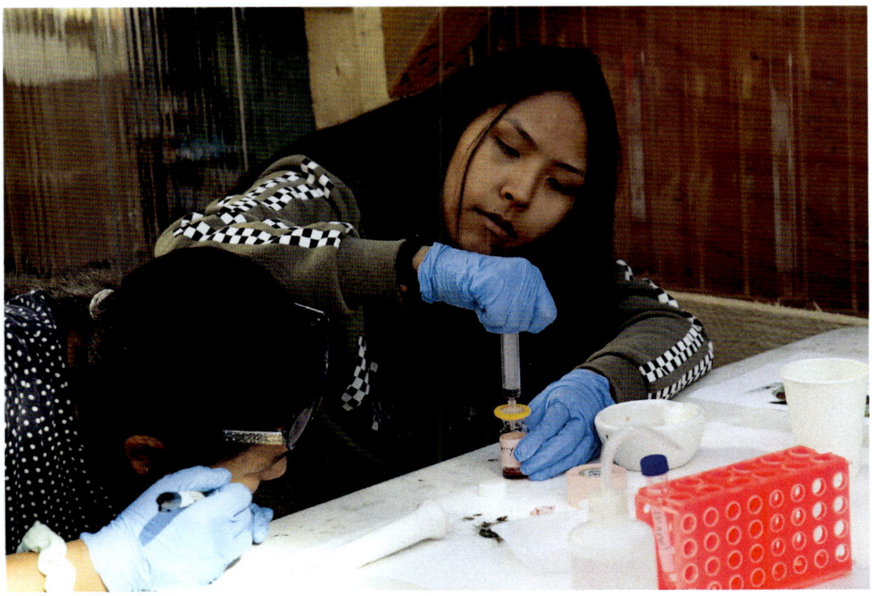

Figure 3.2 Alaska Native youth engage in a workshop featuring simple mobile biodiscovery assays during a 2019 fish camp on the Yukon River in Alaska.

Trimeric fucol | Trimeric phloroethol | Trimeric fucophloroethol

Figure 3.3 (a) *Fucus distichus* (bladder wrack), a traditionally used phlorotannin-rich seaweed harvested in Whittier, AK. (b) Structural units of phlorotannins from the Alaskan brown algae *F. distichus*.

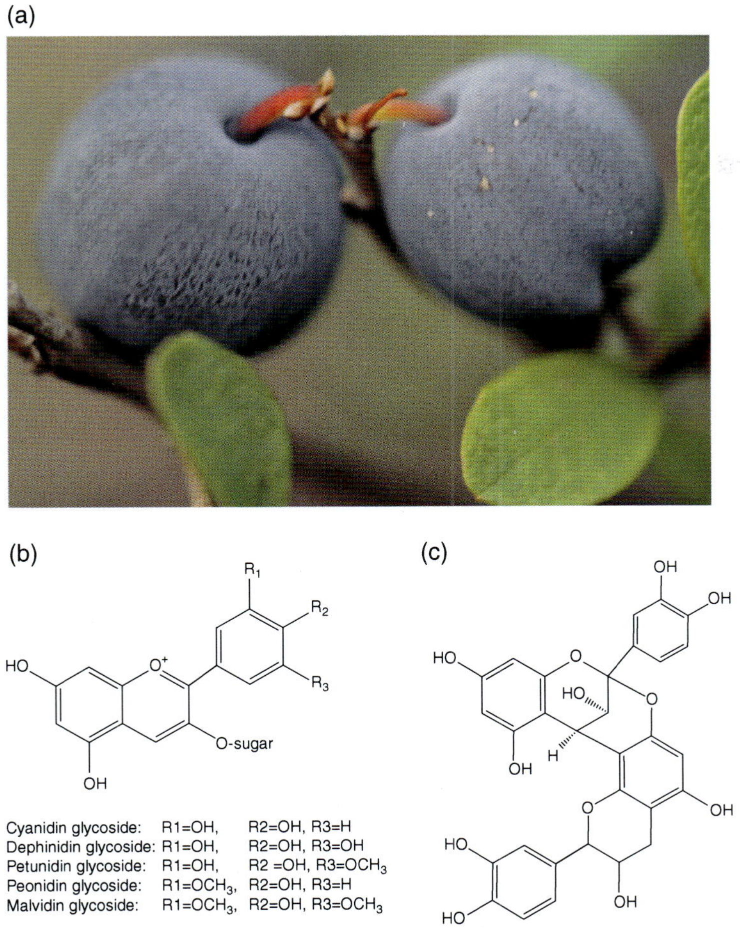

Figure 3.4 (a) Wild *Vaccinium uliginosum* (bog blueberry) growing on the Alaskan tundra. (b) Representative anthocyanins from Alaskan berries. (c) Proanthocyanidin A-type structure characteristic of many Alaskan berry genotypes.

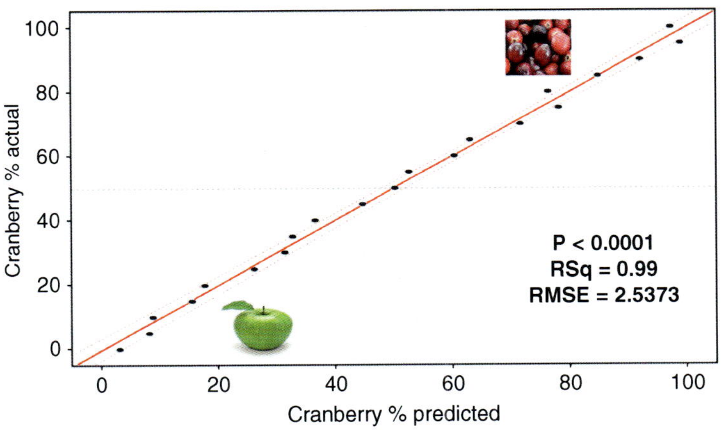

Figure 5.6 Deconvolution of MALDI-TOF MS of 21 different ratios of isolated cranberry and apple PAC shows predicted percentage cranberry PAC is within 3.9% of actual mixed ratios. *Source:* Christian G Krueger.

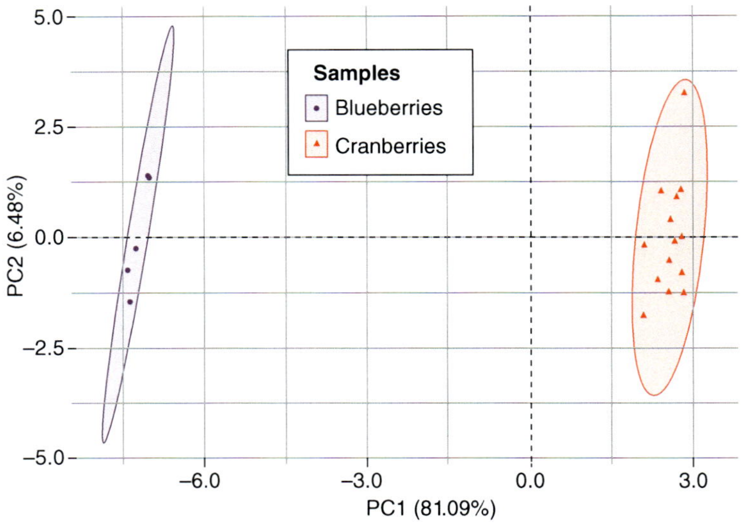

Figure 5.7 Principal component analysis score plot of proanthocyanidins from blueberries and cranberries based on the deconvoluted MALDI-TOF MS spectra. PC1 (x-axis) and PC2 (y-axis) accounts for 87.6% of the total variance. The colored shadows correspond to the 95% confidence ellipse, based on the distribution of A-type interflavan bonds in the PAC oligomers.

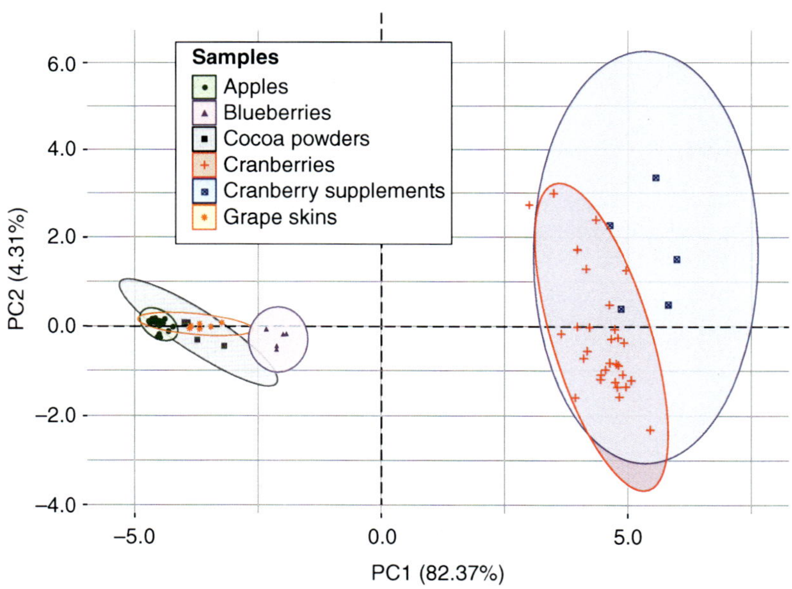

Figure 5.8 Principal component analysis score plot of proanthocyanidins from apples, blueberries, cocoa, powders, cranberries, cranberry supplements, and grape skins based on the deconvoluted MALDI-TOF MS spectra. PC1 (x-axis) and PC2 (y-axis) accounts for 86.7% of the total variance. The colored shadows correspond to the 95% confidence ellipse, based on the distribution of A-type interflavan bonds in the PAC oligomers.

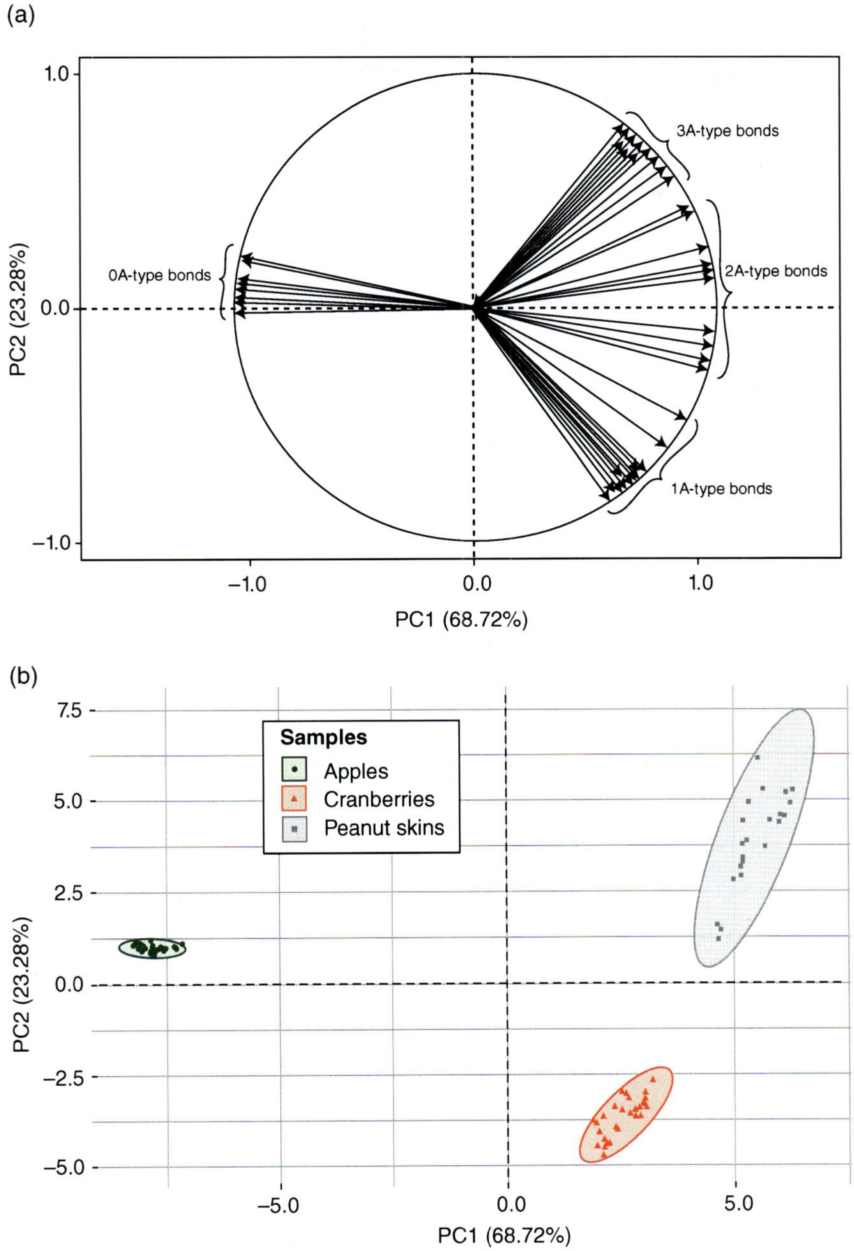

Figure 5.9 Principal component analysis of proanthocyanidins from apples, cranberries, and peanut skins based on the autoscaled MALDI-TOF MS spectral data. PC1 (x-axis) and PC2 (y-axis) accounts for 92.0% of the total variance. (a) PCA loading plot. The arrows indicate the location of all the PAC oligomers with the number of A-type interflavan bonds, as represented by the brackets, for oligomers from DP3 to DP8. (b) PCA score plot. The colored shadows correspond to the 95% confidence ellipse, based on the distribution of A-type interflavan bonds in the PAC oligomers.

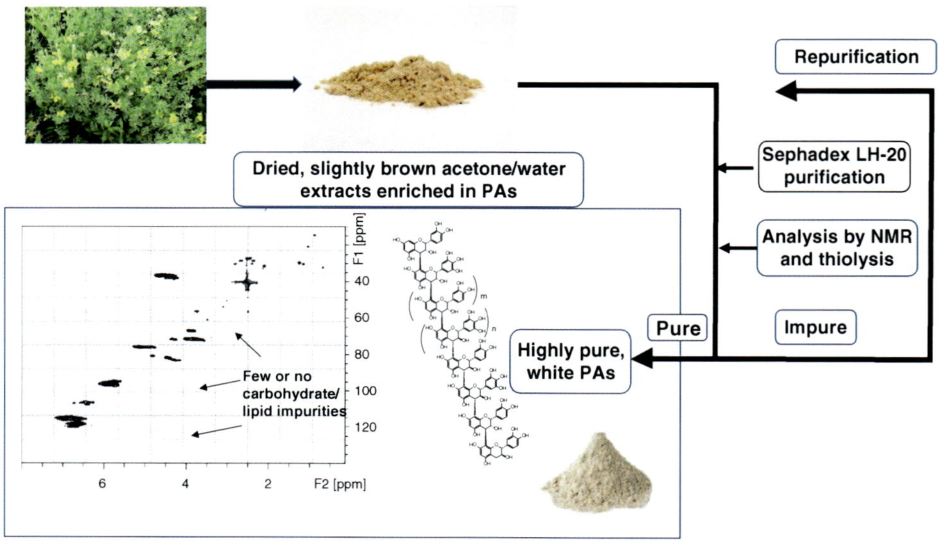

Figure 6.5 Illustration of a rapid approach to proanthocyanidin isolation and purification at the gram scale. *Source:* Adapted from Brown et al. 2017.

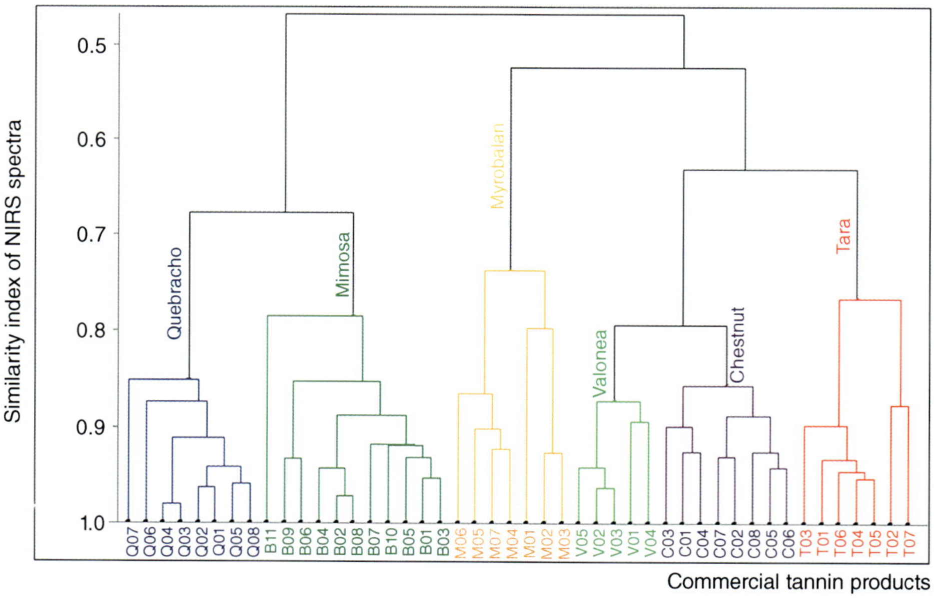

Figure 6.20 Clustering of near-infrared reflectance spectra of commercial tannin products illustrating distinct proanthocyanidin and hydrolyzable tannin groups and tannin-specific subgroups. *Source:* Grasel and Ferrao 2016.

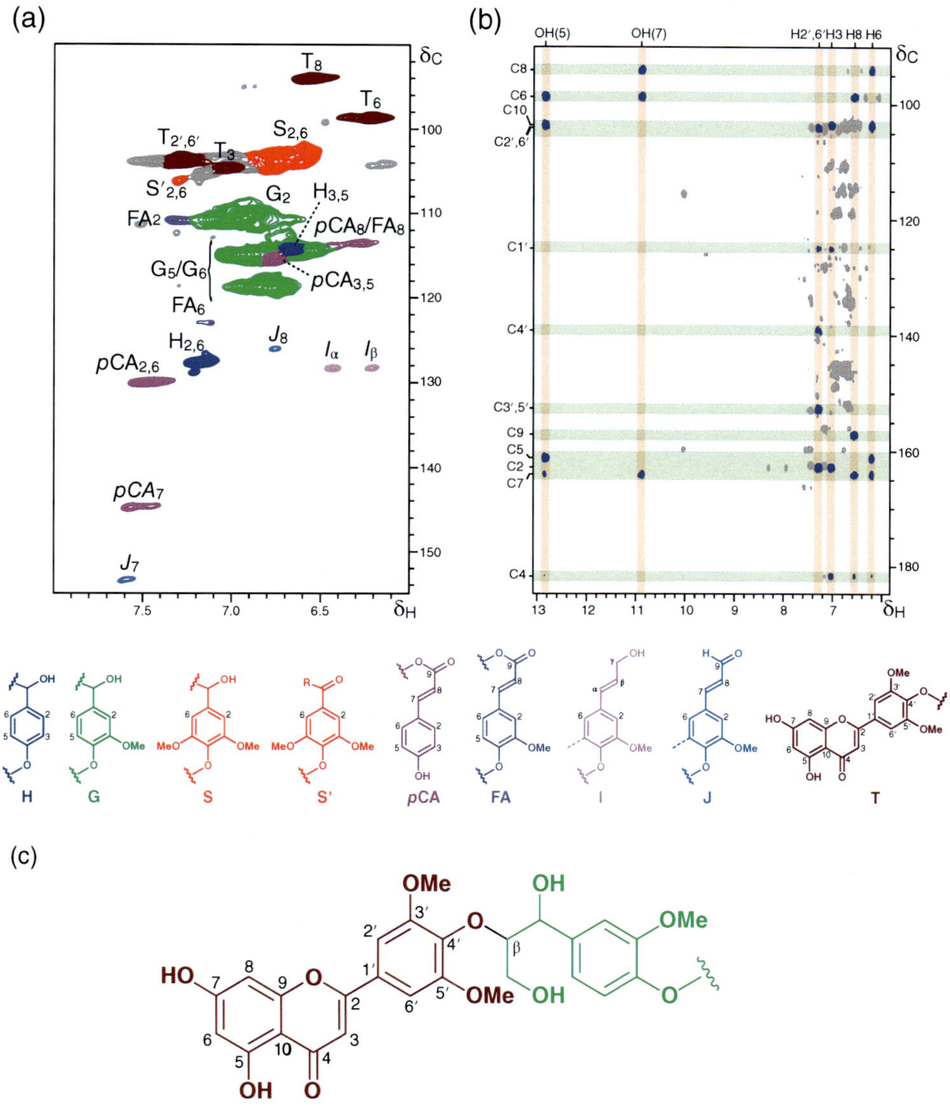

Figure 7.6 (a) Aromatic region of the 2D-HSQC-NMR spectrum of the lignin isolated from wheat straw showing the main correlation signals and their structural assignments to the color-coordinated structures below. (b) Partial HMBC spectrum (δ_C/δ_H 90–185/6.0–13.0) of wheat straw lignin that was crucial in establishing that the non-monolignol component of the lignin was the flavone tricin (5,7,4′-trihydroxy-3′,5′-dimethoxyflavone, **T**). (c) Mode of incorporation of tricin into the lignin structure through 4′–*O*–β ether linkages. *Source:* (a, b, c)Adapted from del Río et al. 2012.

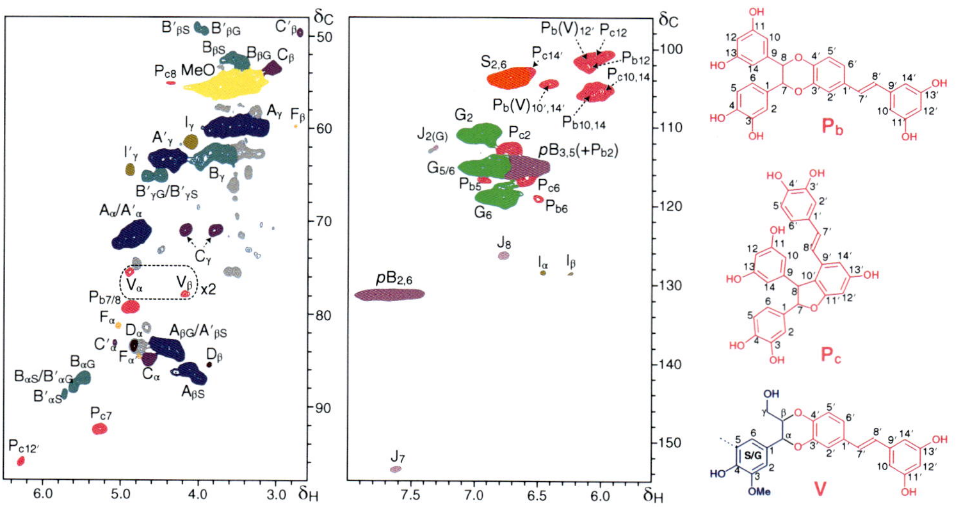

Figure 7.12 Side-chain (δ_C/δ_H 48-98/2.6-6.5) and aromatic (δ_C/δ_H 96-155/5.6-8.0 ppm) regions of the 2D-HSQC-NMR spectrum (in DMSO-d_6) of the lignin isolated from macaúba (*Acrocomia aculeata*) fruit endocarp showing the main signals from lignin and piceatannol (**P**) units. The structures of the main piceatannol homodimeric coupling (**P$_b$**, **P$_c$**) and piceatannol-lignin cross-coupling structures (**V**) are depicted. Signals from typical lignin structures are as follows: **A**: β–*O*–4′ alkyl-aryl ethers; **A′**: β-ether structures with *p*-hydroxybenzoates acylating the γ-OH; **B**: phenylcoumarans; **B′**: phenylcoumarans with *p*-hydroxybenzoates acylating the γ-OH; **C**: resinols; **C′**: tetrahydrofuran structures formed by β–β′-coupling of monolignols acylated at the γ-OH; **D**: dibenzodioxocins; **F**: spirodienones; **I**: cinnamyl alcohol end-groups; **I′**: cinnamyl alcohol end-groups acylated at the γ-OH; **J**: cinnamaldehyde end-groups; ***p*B**: *p*-hydroxybenzoates; **G**: guaiacyl units; **S**: syringyl units. Source: Adapted from del Río et al. 2017.

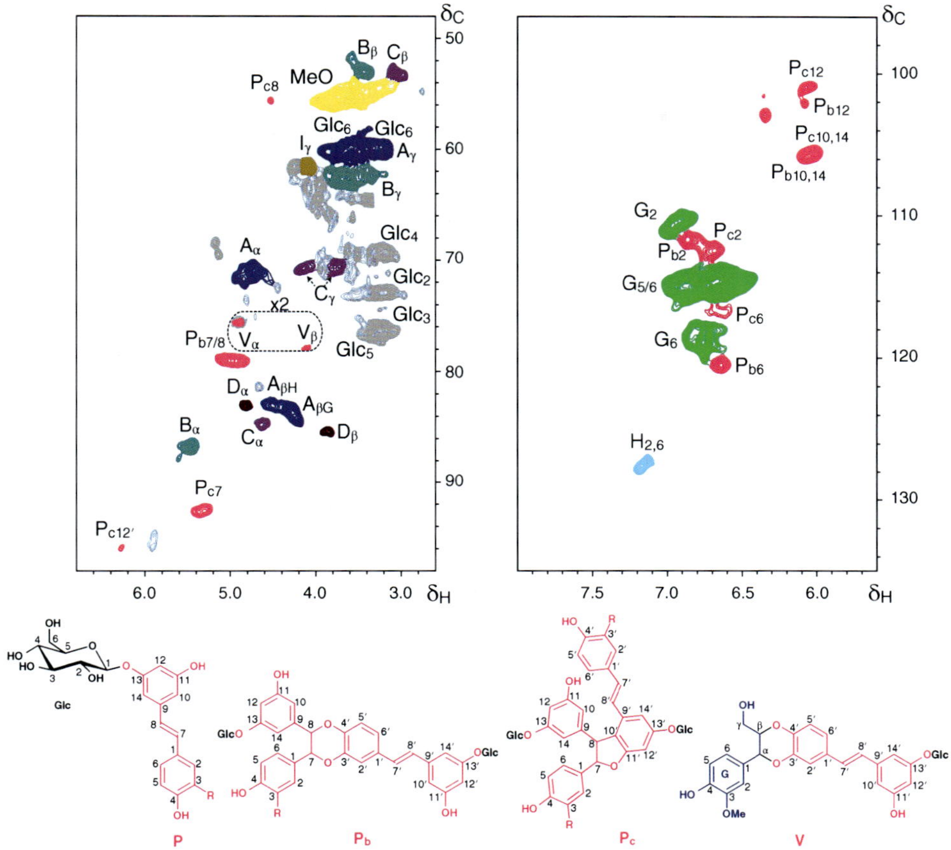

Figure 7.15 Aliphatic-oxygenated (δ_C/δ_H 48–98/2.6–6.8), and aromatic (δ_C/δ_H 96–135/5.6–8.0) regions of the 2D-HSQC-NMR spectrum (in DMSO-d_6) of the 'milled-wood' lignin preparation isolated from Norway spruce bark showing the main signals from lignin and hydroxystilbene glucoside units. The main signals from hydroxystilbene glucosides are: **P**: hydroxystilbene glucosides (isorhapontin, R= OCH_3; astringin, R = OH; piceid, R = H); **P_b**: 8–*O*–4′/3′–*O*–7 benzodioxane structures involving isorhapontin (R = OCH_3), astringin (R = OH) or piceid (R = H) units; **P_c**: 8–10′/11′–7 phenylcoumaran structures involving isorhapontin (R = OCH_3), astringin (R = OH) or piceid (R = H) units (please note that other phenylcoumaran structures arising from other linkages, such as 8–5′ and 8–12′, may also occur, but only the 8–10′ linkage has been authenticated due to the occurrence of signal **$P_{c12'}$**); **V**: β–*O*–4′/3′–*O*–α benzodioxane structure formed by cross-coupling of astringin and coniferyl alcohol. Note that the signals from glucose (**Glc**) units appear in the aliphatic-oxygenated region. Signals from typical lignin structures are as follows: **A**: β–*O*–4′ alkyl-aryl ethers; **B**: phenylcoumarans; **C**: resinols; **D**: dibenzodioxocins; **I**: cinnamyl alcohol end-groups; **H**: *p*-hydroxyphenyl units; **G**: guaiacyl units. *Source:* Adapted from Rencoret et al. 2019.

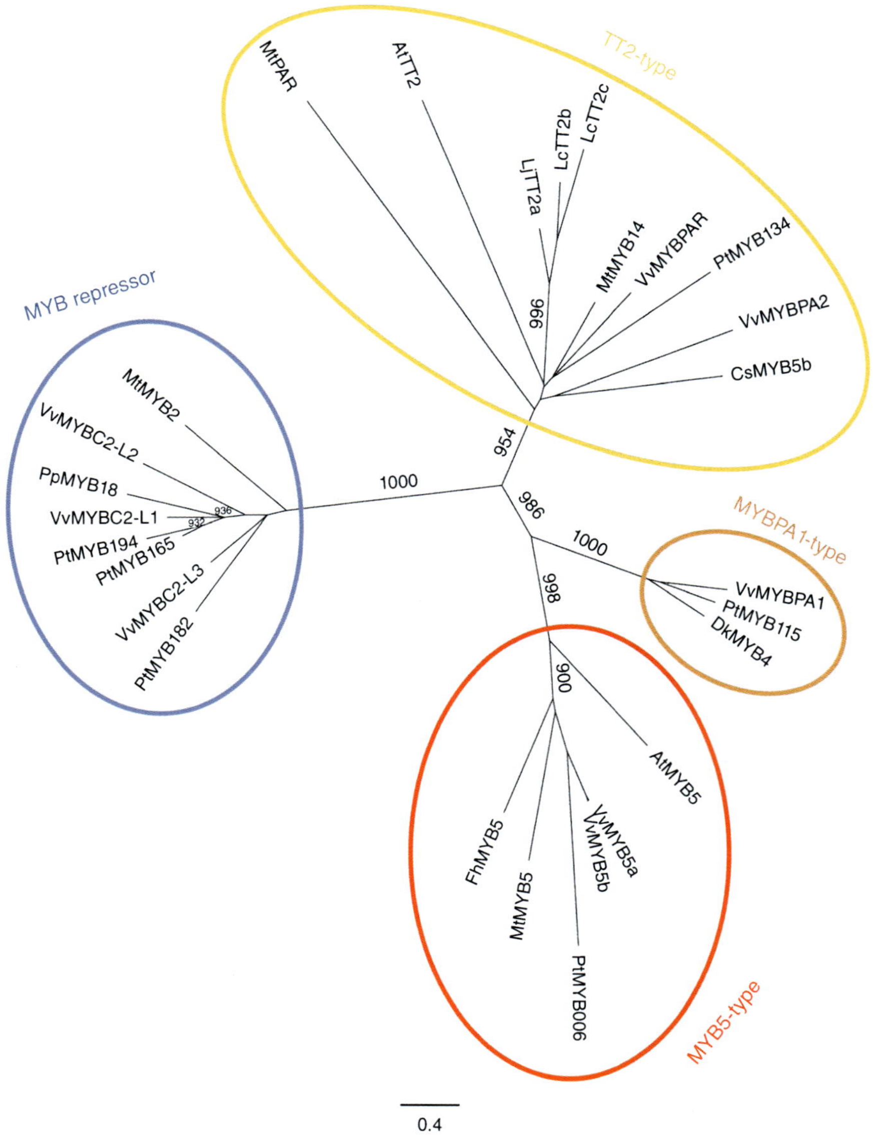

Figure 8.1 Phylogenetic analysis of characterized PA-regulating MYBs. Phylogenetic analysis of three major groups of PA MYB activators and one group of PA MYB repressors. The protein sequences were taken from NCBI (www.ncbi.nlm.nih.gov/). The multiple sequence alignment was performed using the Multiple Sequence Comparison by Log-Expectation (MUSCLE) method in Mega-X (version 10.1 www.megasoftware.net/). The best-fit model was determined by Prottest3 (Darriba et al. 2011) to be Jones-Taylor-Thornton with gamma distribution and invariant sites (JTT+I+G). The phylogenetic tree was generated using the maximum likelihood method in Mega-X. Scale bar represents mean substitutions per site. Protein sequences were obtained from NCBI (www.ncbi.nlm.nih.gov/). AtMYB5 (At3G13540, *Arabidopsis thaliana*); AtTT2 (NM_122946.3, *A. thaliana*); CsMYB5b (KY827397.1, *Camellia sinensis*); DkMYB4 (BAI49721.1, *Diospyros kaki*), FhMYB5 (MK168337.1, *Feesia hybrida*); LjTT2a (BAG12893.1, *Lotus japonicus*); LcTT2b (KY196475.1, *L. corniculatus*); LcTT2c (KY196476.1, *L. corniculatus*); MtMYB2 (EF468475.1, *Medicago truncatula*); MtMYB5 (XM_003601561.3, *M. truncatula*); MtMYB14 (JN225514.1, *M. truncatula*); MtPAR (HQ337434.1, *M. truncatula*); PpMYB18 (KT159234.1, *Prunus persica*); PtMYB006 (KF811025.1, *Populus tomentosa*); PtMYB115 (XP_002302644.1, *P. tremula* x *P. tremuloides*); PtMYB134 (ACR83705.1, *P. tremuloides*); PtMYB165 (XP_002315890.2, *P. tremula* x *P. tremuloides*); PtMYB182 (XP_002305872.1, *P. tremula* x *P. tremuloides*); PtMYB194 (XP_002311495.2, *P. tremula* x *P. tremuloides*); VvMYB5a (AAS68190.1, *Vitis vinifera*); VvMYB5b (AAX51291.1, *V. vinifera*); VvMYBC2-L1 (JJX050227.1, *V. vinifera*); VvMYBC2-L2 (NM_001281251.1, *V. vinifera*); VvMYBC2-L3 (KM046932.1, *V. vinifera*); VvMYBPA1 (CAJ90831.1, *V. vinifera*); VvMYBPA2 (ACK56131.1, *V. vinifera*); VvMYBPAR (AB911341.1, *V. vinifera*).

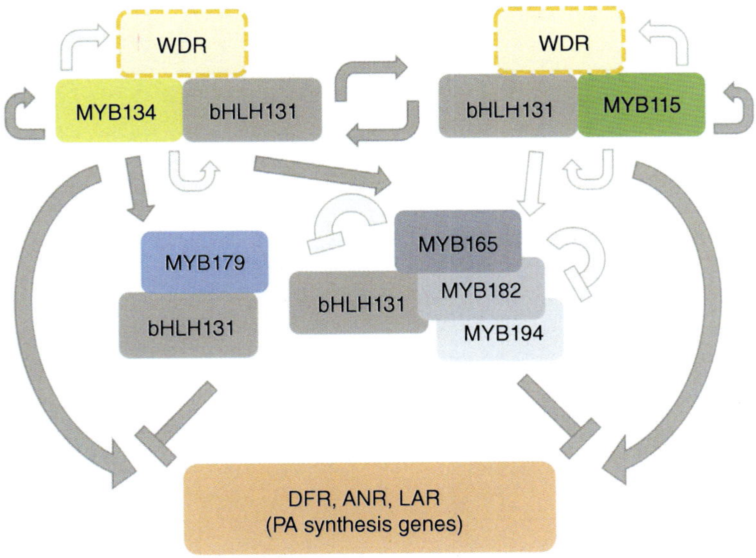

Figure 8.2 Model depicting the interaction of MYB activators, MYB repressors, and bHLH cofactors in poplar. PtMYB134- and PtMYB115-containing MBW complexes (top) activate genes encoding PA enzymes (bottom) as well as numerous other regulators. Solid arrows represent activation or repression demonstrated directly by promoter activation assays in transiently transformed poplar cells. Hollow arrows represent activation or repression inferred solely from transcriptomic analysis of MYB-overexpressing transgenic poplars. Microarray analysis suggests that PtMYB134 and PtMYB115 upregulate PtWDR, PtbHLH131, PtMYB165, PtMYB194, PtMYB182, and PtMYB179 expression (James et al. 2017). Transient promoter activation assays further confirm PtMYB165 and PtMYB194 induction by MYB134 (see Figure 8.3), and that MYB134 and MYB115-containing MBW complexes activate each other's and their own promoters (James et al. 2017). RNA-seq analysis of MYB165-overexpressor plants indicates that MYB182, MYB194, and bHLH131 expression is downregulated by MYB165. *Source:* Ma et al. 2018.

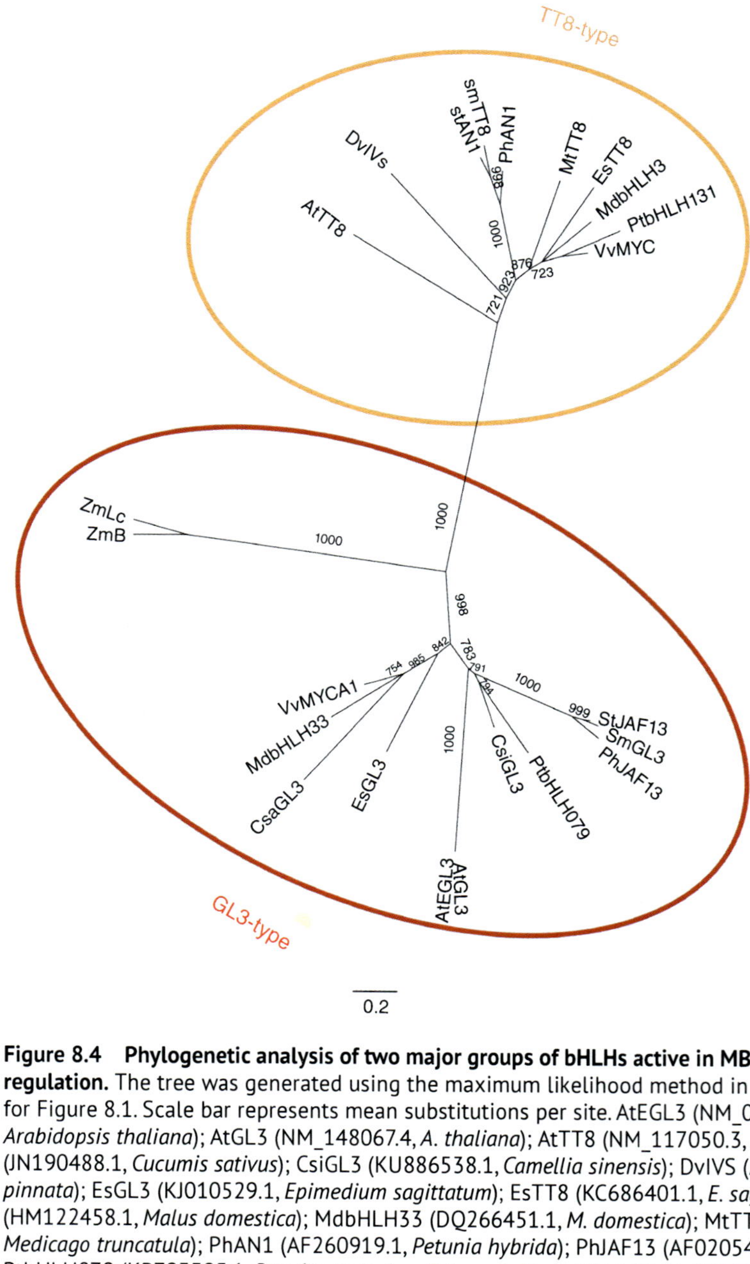

Figure 8.4 Phylogenetic analysis of two major groups of bHLHs active in MBW complexes for PA regulation. The tree was generated using the maximum likelihood method in Mega-X as outlined for Figure 8.1. Scale bar represents mean substitutions per site. AtEGL3 (NM_001198373.2, *Arabidopsis thaliana*); AtGL3 (NM_148067.4, *A. thaliana*); AtTT8 (NM_117050.3, *A. thaliana*); CsaGL3 (JN190488.1, *Cucumis sativus*); CsiGL3 (KU886538.1, *Camellia sinensis*); DvIVS (AB787560.1, *Dahlia pinnata*); EsGL3 (KJ010529.1, *Epimedium sagittatum*); EsTT8 (KC686401.1, *E. sagittatum*); MdbHLH3 (HM122458.1, *Malus domestica*); MdbHLH33 (DQ266451.1, *M. domestica*); MtTT8 (KM892777.1, *Medicago truncatula*); PhAN1 (AF260919.1, *Petunia hybrida*); PhJAF13 (AF020545.1, *P. hybrida*); PtbHLH079 (KP723395.1, *Populus tremula* x *P. tremuloides*); PtbHLH131 (KJ909757.1, *P. tremula* x *P. tremuloides*); SmGL3 (KP006502.1, *Solanum melongena*); SmTT8 (KT591486.1, *S. melongena*); StAN1 (NP_001275098.1, *S. tuberosum*); StJAF13 (KP317176.1, *S. tuberosum*); VvMYC (NM_001281253.1, *Vitis. vinifera*); VvMYCA1 (NM_001281025.1, *V. vinifera*); ZmB (X57276.1, *Zea mays*); ZmLc (DQ414252.1, *Z. mays*).

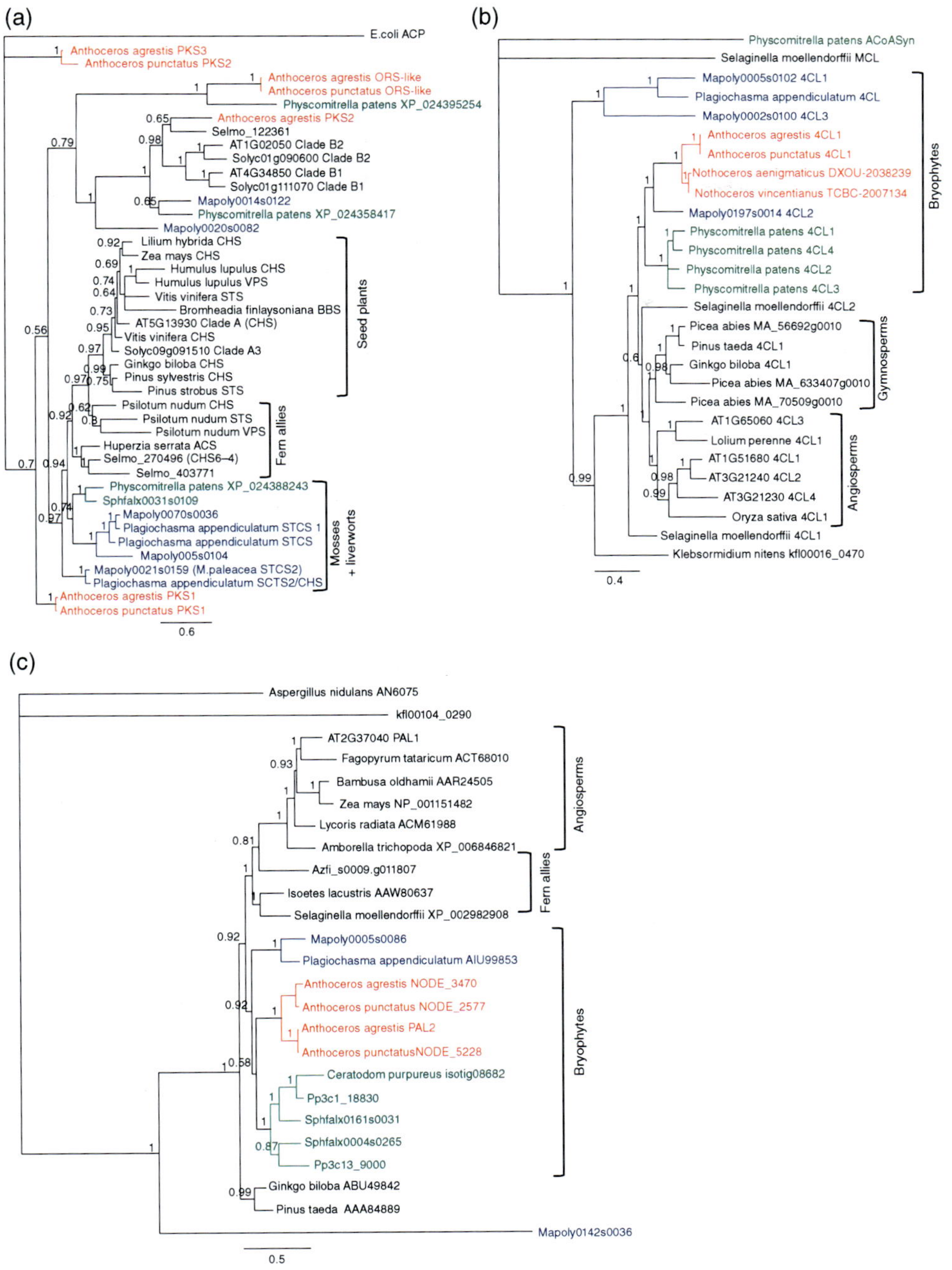

Figure 9.2 Phylogenetic analysis of sequences related to phenylpropanoid biosynthesis. Deduced amino acid sequences were aligned using ClustalW and trees formed using MrBayes within the Geneious (v 10.0.9) sequence analysis software package using default parameters, with the posterior probabilities of the branches shown. Trees are for **(a)** Polyketide Synthases, **(b)** 4-Coumarate-CoA ligase, **(c)** Phenylalanine ammonia-lyase, and **(d)** Chalcone isomerase. Sequence accession details not given directly on the tree or in the chapter text are as follows: *Amborella trichopoda* CHI (XP_006850622), CHIL (XP_011622657), FAPa1 (AmTr_v1.0_scaffold00103.21), FAPa2 (AmTr_v1.0_scaffold00024.257), FAPa3 (AmTr_v1.0_scaffold00061.192); *Anthoceros agrestis* PAL2 (SPO49995); *Arabidopsis thaliana* CHI (At5g05270), CHIL (At3g55120), FAPa1 (AT3G63170), FAPa2 (At2g26310), FAPa3 (XP_020245856), FAPb (At1g53520); *Aspergillus nidulans* AN6075 (XP_663679); *Bazzania trilobata* CHIL (WZYK 6680, from the data of Wickett et al. 2014), FAPb1 (WZYK 8096), FAPb2 (WZYK 7662); *Bromheadia finlaysoniana* BBS (CAA10514); *Cannabis sativa* CHIL (AFN42529); *Carica papaya* FAP1 (XP_021898058), FAPa3 (XP_021911943); *Ceratodon purpureus* CHIL (isotig15641), FAPa1 (isotig05846), FAPb1 (isotig15077), FAPb2 (isotig14426); *Ginkgo biloba* 4CL1 (AFJ73426), CHS (AY647263);

(d)

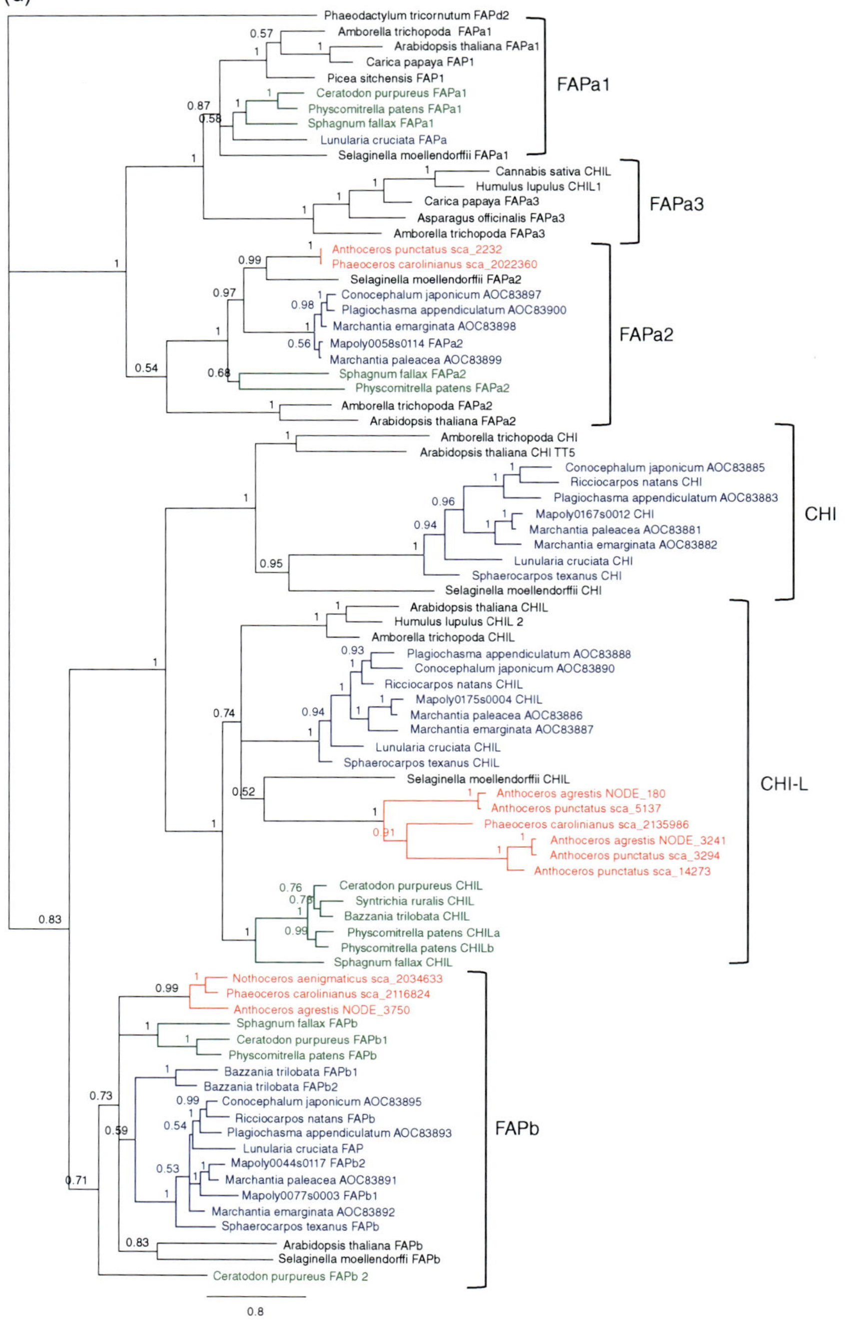

Figure 9.2 (Cont'd) *Huperzia serrata* ACS (ABI94386); *Humulus lupulus* CHIL1 (AVR53896), CHIL2 (AVR53897), CHS (BAB47195), VPS (BAB12102); *Lilium hybrida* CHS (AAD49355); *Lolium perenne* 4CL1 (AAF37732); *Oryza sativa* 4CL1 (CAA36850); *Phaeodactylum tricornutum* FAPd2 (XP_002181351); *Physcomitrella patens* CHILa (XP_001773128), CHILb (XP_001769093), FAPa1 (DC930902), FAPa2 (XP_001783129), FAPb (DC956762), 4CL1 (ABV60447), 4CL2 (Pp019G044600), 4CL3 (Pp022G056000), 4CL4 (Pp021G054800), ACoASyn (XP_024389050); *Picea sitchensis* FAP1 (ABK25337); *Pinus strobus* STS (CAA87013); *Pinus sylvestris* CHS (CAA43166); *Pinus taeda* 4CL1 (P41636); *Plagiochasma appendiculatum* 4CL (AJT43268), STCS (AUG98244), STCS1 (AHY39237), SCTS2 (AIV42295); *Psilotum nudum* CHS (BAA87922), STS (BAA87924), VPS (BAA87923); *Ricciocarpos natans* CHI (WJLO 9530), CHIL (WJLO 8822), FAPb (WJLO 406); *Selaginella moellendorffii* CHI (XP_002980623), CHIL (FE452267), FAPa1 (FE510754), FAPa2 (XP_002987458), FAPb (FE511008), 4CL1 (XP_002969881), 4CL2 (Smoe_177393), MCL (XP_024540059); *Sphaerocarpos texanus* CHI (HERT 7773), CHIL (HERT 829), FAPb (HERT 8792); *Sphagnum fallax* CHIL (Sphfalx0068s0053), FAPa1 (Sphfalx0013s0179), FAPa2 (Sphfalx0032s0188), FAPb (Sphfalx0071s0061); *Syntrichia ruralis* CHIL (CN201507); *Vitis vinifera* CHS (BAB84112), STS (ABD64685); *Zea mays* CHS (P24824). *Source:* (a, b, c, d) Kearse et al. 2012.

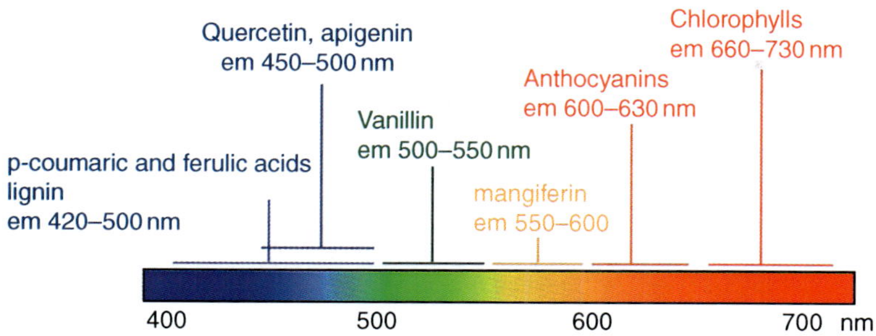

Figure 11.1 Emission wavelengths of some (poly)phenolic compounds and chlorophylls.

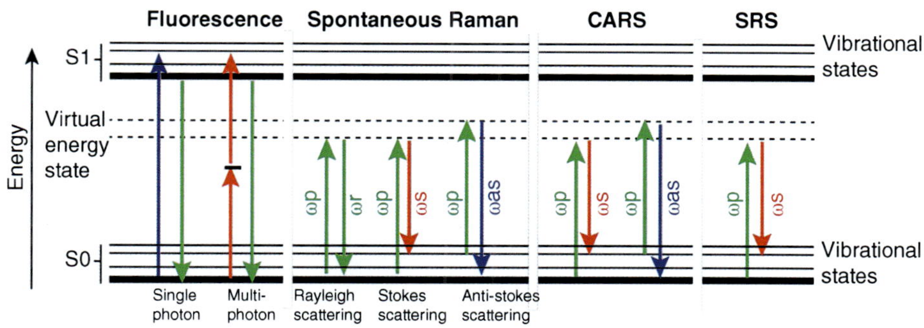

Figure 11.2 Jablonski diagrams showing energy transitions during fluorescence (single and multiphoton excitation) and Raman scattering (spontaneous Raman, coherent anti-Stokes Raman scattering or CARS, and stimulated Raman scattering or SRS). When excited by the incident photon, the scattering molecule is taken to an excited energy state known as a virtual energy state. The molecule decays back to its ground electronic state (S0), which contains multiple vibrational energy levels. If the molecule decays to the same vibrational level, there is no transfer of energy between the photon and the molecule (elastic scattering or Rayleigh scattering), and the wavelength of the scattered photon remains the same as the incident photon. If there is energy transfer between the incident photon and the scattering molecule, causing a change in the molecule vibrational energy, inelastic scattering occurs. If the photon transfers energy to the scattering molecule and the molecule gains vibrational energy, the wavelength of the scattered photon increases (Stokes shift) as it loses energy. If the energy is transferred from the molecule to the photon, the energy of the scattered photon increases and its wavelength decreases (anti-Stoke shift).

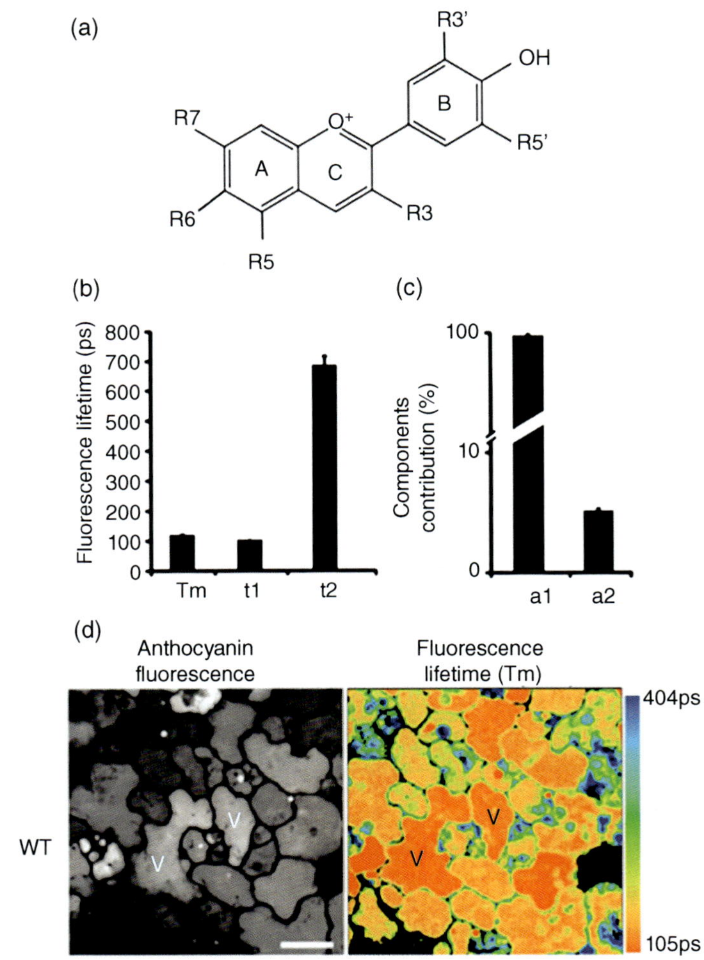

Figure 11.3 Anthocyanin imaging. (a) General structure of an anthocyanidin backbone. (b) Fluorescence lifetime of isolated wild-type anthocyanins. τm, weighted mean lifetime. τ1 and τ2 are the lifetimes in picoseconds of the two identified fluorescence lifetime components. (c) Relative contributions of the two identified components. (d) Steady-state fluorescence image (left panel) and pseudo-colored image according to anthocyanin fluorescence lifetime values (right panel) of cotyledon pavement cells. Color map shows m, ranging from 105 ps to 404 ps. V, vacuole. Scale bar: 20 m. *Source:* (a, b, c) Modified from Chanoca et al. 2015. © American Society of Plant Biologists. (d) Reproduced from Chanoca et al. 2015. © American Society of Plant Biologists.